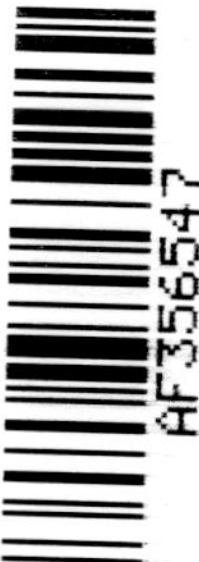
AF356547

Battery Systems and Energy Storage beyond 2020

Battery Systems and Energy Storage beyond 2020

Editors

Kai Peter Birke
Duygu Karabelli

MDPI • Basel • Beijing • Wuhan • Barcelona • Belgrade • Manchester • Tokyo • Cluj • Tianjin

Editors
Kai Peter Birke
University of Stuttgart
Germany
Fraunhofer Institute for Manufacturing
Engineering and Automation IPA
Germany

Duygu Karabelli
Fraunhofer Institute for Manufacturing
Engineering and Automation IPA
Germany

Editorial Office
MDPI
St. Alban-Anlage 66
4052 Basel, Switzerland

This is a reprint of articles from the Special Issue published online in the open access journal *Batteries* (ISSN 2313-0105) (available at: https://www.mdpi.com/journal/batteries/special_issues/Battery_Systems_Energy_Storage).

For citation purposes, cite each article independently as indicated on the article page online and as indicated below:

LastName, A.A.; LastName, B.B.; LastName, C.C. Article Title. *Journal Name* **Year**, *Volume Number*, Page Range.

ISBN 978-3-0365-3025-3 (Hbk)
ISBN 978-3-0365-3024-6 (PDF)

Contents

About the Editors

Kai Peter Birke is a physicist, materials scientist and a full Professor at the University of Stuttgart, Germany, researching the field of electrical energy storage systems, including new energy storage cell materials and technologies, advanced Li-ion batteries and Power-to-X. He obtained his Ph.D. in Materials Science (ion-conducting ceramics) from the University of Kiel, Germany, in 1998. In 1999, he joined the Fraunhofer-Institute for Silicon Technology, Itzehoe, Germany, to work on the development of proprietary Li-ion laminated hybrid-solid cells with novel functional ceramic separators. He also co-founded two spin-offs to put this technology into production. After being involved in the development and production of pouch-type laminated Li-ion cells (PoLiFlex) in leading positions at VARTA AG for five years (2000–2005), Prof. Birke joined Continental AG, Business Unit Hybrid Electric Vehicle, in Berlin in 2005, as a Senior Expert and Project Leader in Energy-Storage systems. He subsequently became a Senior Technical Expert in Battery Technology and a Team Leader in Cell Technology, and in 2010 he was appointed as the Head of Battery Modules and Electromechanics. In 2013, he joined the JV SK-Continental e-motion as Head of Advanced Development. He was one of the most relevant pioneers for this JV. In 2015, Prof. Birke became a full professor at the University of Stuttgart, Institute for Photovoltaics, ipv, as head of Electrical Energy Storage Systems. Since 2018 he additionally leads the Center for Battery Cell Manufacturing at the Fraunhofer Institute for Manufacturing Engineering and Automation, IPA, with an extended focus on battery cell production and digitalization.

Duygu Karabelli After obtaining her PhD in electrochemistry in 2011 at University of Grenoble (France), Dr. Karabelli started her postdoctoral research in the field of next-generation batteries at Franhofer ICT (Germany). In 2019, she joined the Center of Battery Cell Manufacturing at Fraunhofer IPA (Germany) as a project manager. Here, Dr. Karabelli supported and coordinated many national and international projects related to the manufacturing of both Li-ion battery and all-solid-state Li-ion battery technologies. Research interests: Li-ion batteries; solid-state batteries; polymer electrolytes; battery production; Industry 4.0.

Preface to "Battery Systems and Energy Storage beyond 2020"

Currently, the transition from using the combustion engine to electrified vehicles is a matter of time and drives the demand for compact, high-energy-density rechargeable lithium ion batteries as well as for large stationary batteries to buffer solar and wind energy. The future challenges, e.g., the decarbonization of the CO_2-intensive transportation sector, will push the need for such batteries even more.

The cost of lithium ion batteries has become competitive in the last few years, and lithium ion batteries are expected to dominate the battery market in the next decade. However, despite remarkable progress, there is still a strong need for improvements in the performance of lithium ion batteries. Further improvements are not only expected in the field of electrochemistry but can also be readily achieved by improved manufacturing methods, diagnostic algorithms, lifetime prediction methods, the implementation of artificial intelligence, and digital twins. Therefore, this Special Issue addresses the progress in battery and energy storage development by covering areas that have been less focused on, such as digitalization, advanced cell production, modeling, and prediction aspects in concordance with progress in new materials and pack design solutions.

Kai Peter Birke, Duygu Karabelli
Editors

Article

Lithium-Ion Battery Thermal Management Systems: A Survey and New CFD Results

Morena Falcone [1], **Eleonora Palka Bayard De Volo** [1], **Ali Hellany** [2], **Claudio Rossi** [3] and **Beatrice Pulvirenti** [1,*]

[1] Department of Industrial Engineering, Alma Mater Studiorum Università di Bologna, Viale Risorgimento 2, 40136 Bologna, Italy; morena.falcone@unibo.it (M.F.); eleonora.palkabayar2@unibo.it (E.P.B.D.V.)
[2] School of Engineering, Design and Built Environment, University of Western Sydney, Penrith, NSW 2751, Australia; A.Hellany@westernsydney.edu.au
[3] Department of Electric Energy Engineering, Alma Mater Studiorum Università di Bologna, Viale Risorgimento 2, 40136 Bologna, Italy; claudio.rossi@unibo.it
* Correspondence: beatrice.pulvirenti@unibo.it

Abstract: The environment has gained significant importance in recent years, and companies involved in several technology fields are moving in the direction of eco-friendly solutions. One of the most discussed topics in the automotive field is lithium-ion battery packs for electric vehicles and their battery thermal management systems (BTMSs). This work aims to show the most used lithium-ion battery pack cooling methods and technologies with best working temperature ranges together with the best performances. Different cooling methods are presented and discussed, with a focus on the comparison between air-cooling systems and liquid-cooling systems. In this context, a BTMS for cylindrical cells is presented, where the cells are arranged in staggered lines embedded in a solid structure and cooled through forced convection within channels. The thermal behavior of this BTMS is simulated by employing a computational fluid dynamics (CFD) approach. The effect of the geometry of the BTMS on the cell temperature distribution is obtained. It is shown that the use of materials with additives for the solid structure enhances the performance of the system, giving lower temperatures to the cells. The system is tested with air-cooling and water-cooling, showing that the best performances are obtained with water-cooling in terms of cell packing density and lowest cell temperatures.

Keywords: lithium-ion cells; battery thermal management systems; CFD simulations; liquid cooling

Citation: Falcone, M.; Palka Bayard De Volo, E.; Hellany, A.; Rossi, C.; Pulvirenti, B. Lithium-Ion Battery Thermal Management Systems: A Survey and New CFD Results. *Batteries* **2021**, *7*, 86. https://doi.org/10.3390/batteries7040086

Academic Editors: Seung-Wan Song, Kai Peter Birke and Duygu Karabelli

Received: 4 August 2021
Accepted: 8 December 2021
Published: 14 December 2021

1. Introduction

Driven by the need to lower the pollution at street level, the interest in electric vehicles is growing increasingly. Lithium-ion cells are the most commonly used batteries in electric vehicles (EVs) due to their high power density (255 Wh/kg for commercial products, and, in the short term, cells with 270 Wh/kg are expected on the market), low self-discharge rate, high recyclability, light weight and compact size, which is very helpful when space is lacking. They also have a longer life cycle with respect to the other rechargeable batteries [1–3]. On the other hand, Li-ion cells show highly non-isotropic thermal properties due to the different layers of the electrodes [4]. In addition, the automotive field necessity of having compact battery packs leads to modules with small space among the cells and between the strings. This yields high temperatures and non-uniform thermal distribution situations, which make the battery efficiency worse and can cause them to degenerate in dangerous and uncontrolled environments. For instance, thermal runaway is caused by successive processes that can lead to fire and combustion [5]. Today, 18,650 Li-ion cells (with diameter 18 mm and length 65 mm) are among the most used batteries in the automotive field because of their small size: as the cells contain a limited amount of energy, if a failure event occurs, the effect is much less than that expected from a larger cell [6]. These cells were introduced by Tesla more than ten years ago on the pioneer Roadster model, and now they are massively used in the 21,700 format on Tesla Model X and Model 3. Recent diffusion of cylindrical cells in realizing large packs are mainly driven by the exploitation of simple integration technology, allowing the realization of modules containing between 50 and 100 cells in parallel. Many efforts have been made to increase the battery efficiency. Research in this area started with investigations about why Li-ion batteries should have represented the best choice for EVs [7], then proceeded with a considerable number of simulations (for example, Huang et al. used 3D simulations to study the convection in the spacing between the cells and what their best arrangement is [8]). The final step was the validation of the simulations (for instance, experiments about the effects of different arrangements and shapes of the cells [9–11]). In the context of battery pack cooling management, it is necessary to take into consideration several methods and systems to control the temperature of a battery pack, maintaining it in a precise range. In this work, battery thermal management systems (BTMSs) are considered. Many approaches are involved in various aspects of BTMSs: air cooling, hydrogen cooling, helium cooling, ammonia cooling, liquid cooling, phase change materials [12,13], and hybrid cooling, all used to reduce energy waste and to achieve the battery best performances. Each method has its own pros and cons and is more suitable for a precise application rather than another. The best operating temperature range for Li-ion batteries is between 15 °C and 40 °C, as suggested by many recent studies [14]. If the temperature rises above 50 °C, the charging efficiency and the longevity of batteries decreases [15]. The purpose of a BTMS is to keep the cell temperature within the optimal range while maintaining the temperature's uniformity within the modules [16]; heating is also essential in those rare situations when the battery temperature is lower than the minimum acceptable temperature. At low temperatures, discharging of a charged Li-ion battery is easier than charging of a discharged one [17], and the capacity of Li-ion batteries can decrease up to 95% when the Energy Storage System (ESS) is operating at -10 °C [18]. Due to the exothermic nature of the chemical reactions inside the batteries, few studies have been done about heating [19]. However, there are some promising studies about internal self-heating strategies to overcome the poor performances at low temperatures [20] and the reduction of battery life [21]. While studying the heat dissipation performances of a battery pack, two main indexes are usually taken into account: the maximum temperature increase ($T_{ris,max}$) and the maximum temperature difference ($T_{dif,max}$). The maximum temperature rising is the maximum difference value between the battery pack temperature and the environmental temperature. The maximum temperature difference is the maximum difference value recorded inside the battery pack. If $T_{ris,max}$ is too high, it means that the environmental temperature is relatively low for the battery pack and that it is not possible to remove the heat generated by the battery pack through the cooling system. If $T_{dif,max}$ is too high, it means that there is no uniform temperature distribution inside the battery pack. An appropriate cooling system design is necessary to reduce both $T_{ris,max}$ and $T_{dif,max}$. It is desirable to have the maximum temperature increase be less than 10 °C and the maximum temperature difference be less than 5 °C [22]. Concerning air-cooling methods, ref. [23], Zhou et al. [16] show the advantages of an air distribution pipe thermal manage system, and Xu et al. [22] studied the optimization of the forced air cooling, testing the effects of different airflow duct modes; indeed, Xie et al. [24] do so by modifying factors such as the air inlet-angle and the outlet-angle. There are also studies that propose a multi-parameter control strategy for air-based BTMSs [25] to monitor the state of health (SOH) of the battery and energy consumption in function of the temperature fluctuations and the air mass-flow rates. Many experiments have been carried out on not only to optimize the whole BTMS, but also to analyze the thermal conditions inside the single battery modules improving the temperature uniformity. Several methods have been explored, with analytical and experimental tests, such as the installation of a baffle plate [26], different cell arrangement structures [10], the spacing optimization between the battery cells [27], and the installation of a secondary vent [28] in an air-cooled BTMS. Other examples are the addition of an inlet plenum [29], special axial-flow air-cooling systems [30], nanofluid-based cooling techniques with forced air-flow to remove the heat from the battery arrangement [31]. Among all the liquid cooling BTMSs, the cooling plates are those on which it has been written less considering to cool the cylindrical 18650 lithium-ion batteries often used in the automotive field. Several studies have been carried on about prismatic cells cooled by cooling plates, about how this system can be designed [32], improved, and optimized [33], focusing on which parameters more influence the optimum working point. Recently, some studies [34] have shown that it makes sense to compare cylindrical cells to much larger prismatic ones. This is first because the governing chemistry reactions inside both kinds of the batteries are the same. Secondly, it has to be assumed that the experiments are performed with strict temperature control, which means considering an extensive thermal mass test system with exceptional temperature control sensors. Cooling plates were shown to be a good solution in particular for those applications where high thermal conductivity and compact design are required. This is why they are a common choice in the automotive field. An advanced liquid cooling system can involve the use of heat pipes [35,36], leading to energy saving in electric vehicles, with better BTMS performances and increasing heat flux loads [37]. In particular, recent studies have explored heat pipes used in innovative systems such as inclined U-shaped flat micro heat pipes [38] and hybrid oscillating heat pipes (OHPs) with an ethanol aqueous solutions of carbon nanotubes as working fluid [39]. Other innovative studies have focused their attention on the thermal performance achievable using a two-phase refrigerant BTMS, comparing it with an ordinary liquid cooled one [40]. The thermal performances of liquid-based BTMSs coupled with heat-pump air-conditioning systems (HPACSs) were investigated by Tang et al. [41] to predict the cooling capability of the liquid system on a basis of a machine learning method. Hydrogen-based cooling systems were realized by Alzareer et al. [42] to maximize the BTMS cooling efficiency in hybrid fuel cell electric vehicles (HEVs) with prismatic battery packs. These systems are also able to increase the driving range of these vehicles in which the hydrogen is used both as coolant in aluminum cold plates and as fuel. Among the HEV BTMSs, Alipour et al. [43] studied the cooling capability of a helium-based cooling system (suitable for EVs as well) on pouch cells, comparing it with an ordinary air-based BTMS and optimizing its efficiency in function of several factors such as the inlet flow rate (the most influential one), the flow direction, and the inlet and outlet diameters. Ammonia is also considered as a refrigerant to cool the future HEV batteries: Al-Zareer et al. [44] create an electrochemical thermal model to study an innovative semi-immersive ammonia-based system for cylindrical cells. The system exploits the ammonia boiling process that involves the parts of the batteries immersed into the coolant, and then the natural convection process when the ammonia vapor cools the part of the cells out from the ammonia pool. Recently, hybrid BTMSs that combine natural air convection with liquid and thermoelectric cooling systems (TEC) have been introduced [45], together with innovative cooling methods such as the adoption of liquid immersive solutions by TESLA: these systems perform well both in hot and in cold environments [46]. Finally, a panoramic view of several emergency strategies needed to prevent the thermal runaway (TR) is given. The inception of TR is generally a consequence of an internal short circuit (ISC) inside a single cell that is caused by either mechanical or electrical abuse. It is important in that case to lower the damages or, in the worst cases, to extinguish fire [47–49]. This paper aims to give an overview of the different approaches for thermal management of lithium battery packs. In this context, a BTMS for cylindrical cells is presented, where the cells are arranged in staggered lines and embedded in a solid structure between wavy channels. A Computational Fluid Dynamics (CFD) approach is presented to simulate the thermal behavior of battery cells and to optimize the distance between the cells and the width of the channels with different materials used for the cells support and with different cooling fluids. The arrangement shown in this work is novel, both for the characteristics of the solid material in which the cells are embedded and for the coupling between thermal conduction and convective cooling methods adopted. Moreover, the CFD-based optimization approach to find the best performances of the BTMS is original and shows interesting results for a cells arrangement typical for automotive and that can be easily generalized. The temperature distribution on the cells is obtained with very high heat loads. The optimal coolant conditions that give the best thermal performances for the arrangement of the cells are obtained. It is shown that the use of materials with additives to increase the thermal conductivity enhances the heat overall heat transfer for this type of BTMS and gives lower temperatures of the cells. The best performances are obtained by using water instead of air, with small channel widths and with an optimal distance between cells in the direction parallel to the flow.

2. A Survey of the Battery Thermal Management Systems

In this section, some aspects concerning battery cooling with a focus on innovative systems are described. A comparison between some air-cooling technologies and liquid-cooling systems is introduced.

2.1. Air Cooling—Prismatic Cells

One of the easiest ways to control the battery pack temperature is by utilizing air-cooling systems. These can be realized with natural ventilation or with forced ventilation. Several simulations and experimental tests are available in the literature, which evidence the effect of different airflow duct modes [22] or important construction details such as the air-inlet angle, the air-outlet angle, and the width of the airflow channel between battery cells [16]. Many studies deal with the temperature distribution and streamlines obtained in different arrangements of the air-flow for battery pack made with densely packed prismatic cells.

Interesting results can be extrapolated from the research of Xu et al. [22]. In a parallel flow arrangement, the air flows parallel to the battery cells and is expelled by the fans from the air-outlets. The highest temperature areas are inside the battery near the air-outlet. With this configuration, the heat dissipation performance requirements are not satisfied with any of the environmental temperatures considered ($T_{env,1} = 20\ °C$, $T_{env,2} = 27\ °C$, $T_{env,3} = 40\ °C$).

In a cross-flow arrangement, the air flows perpendicularly to the battery cells and the shorter air-flow paths improve the heat dissipation. The temperature peaks are lower, showing that this configuration is better than a parallel-flow one in terms of heat dissipation. Mixed approaches, such as adding double-passage or U-passage channels at the bottom of the battery pack, can increase the heat transfer performances. Using the double-passage channel, the temperature values are lower with respect to the previous tests but not enough to satisfy the heat dissipation performance requirements for all the environmental temperatures considered (only for 27 °C and 40 °C). In the last case of a bottom U-passage, it clearly appears that the temperature field distribution is more uniform with respect to the double passage. Using the U-passage, the temperatures are lower with respect to all the previous cases, satisfying the heat dissipation performance requirements for all the environmental temperatures considered (the value of $T_{ris,max} = 10.10\ °C$ for $T_{env} = 20\ °C$ is sufficiently close to the specification). In Table 1, the values of $T_{ris,max}$ (A) and $T_{dif,max}$ (B) in the parallel flow test are compared with those recorded during the other tests in terms of temperature decreasing.

With the U-passage, the heat dissipation requirements are satisfied for SOC = 70% and SOC = 100%, but the results are better for the lowest SOC percentage: this means that in case of insufficient heat dissipation condition, it could be helpful to work with lower SOC values [22]. The heat dissipation requirements are satisfied for the charge and discharge rates of 0.6C, 0.8C and 1C with the best results for the lowest of these values: this means that, in the case of insufficient heat dissipation condition, it could be helpful to work with lesser charge and discharge rates [22]. The angle between air-inlet and air-outlet channels and the battery cells, as well as the air-flow channel width, are crucial for different aspects. Xie et al. [24] focus their attention on these aspects of prismatic lithium-ion cells arrangements. The environmental temperature for the experimental tests is set at 25 °C. During the tests, the air-inlet angle and the air-outlet angle are changed on the basis of the size of the inlet and outlet channels, respectively, while the layout of the air-flow channels is changed by the gap in the battery pack. Another important factor considered is the distance between the cells [24]. By taking into account all these parameters, the lowest values of $T_{ris,max}$ and $T_{dif,max}$ are obtained for the cases of the air-inlet angle of 2.5° and air-outlet angle of 2.5°, with evenly-spaced air flow channel between battery cells. The values of $T_{ris,max}$ and $T_{dif,max}$ are dropped by 12.82% and 29.72%, respectively, by the optimization approach [24].

Table 1. Temperature decreasing obtained moving from a parallel flow cooling method to a U-passage technology at different environmental temperatures [22].

	Environmental Temperature (°C)	20	27	40
Parallel flow	Maximum temperature rise (°C)	A_{20}	A_{27}	A_{40}
	Maximum temperature difference (°C)	B_{20}	B_{27}	B_{40}
Cross-flow	Maximum temperature rise (°C)	−2.25%	−1.98%	−1.50%
	Maximum temperature difference (°C)	−7.91%	−7.27%	−5.75%
Double passage	Maximum temperature rise (°C)	−36.80%	−37.04%	−37.55%
	Maximum temperature difference (°C)	−63.83%	−63.84%	63.87%
U-passage	Maximum temperature rise (°C)	−38.38%	−39.03%	−38.97 %
	Maximum temperature difference (°C)	−72.58%	−73.03%	−72.15%

Air Cooling—Cylindrical Cells

In this section, some air-cooling techniques tested on cylindrical-cell bricks are described. Zhou et al. [16] take into account 18,650 Li-ion batteries, monitoring the temperature of every single cell with k-type thermocouples in three different points: the top, the middle, and the bottom. The air distribution pipes have orifices arranged all along with the axial direction of the pipes themselves. Three kinds of pipes are tested: the first one presents three orifices with 1 mm diameter, the second one has four orifices with 15 mm diameter, and the third pipe has five orifices with 2 mm diameter. The air enters into the pipes from their inlets and flows out from the orifices. Considering a constant discharge rate process, it is shown that the temperature increases from the negative pole to the positive one, and this is explainable with a high battery cap internal resistance (high heating due to the Joule's effect). After this observation, four main factors are considered to optimize the air distribution pipe system: the number and the diameter of the orifices, the air inlet pressure, and the discharge rate. For all the diameters of the orifices, the temperature obtained in the middle of the cells is lower than that on the top and the bottom, and, in particular, it decreases as the orifice diameter increases. This is the effect of an enhanced heat transfer area between the battery surface and the airflow. Choosing pipes with bigger orifices leads to an increase in the forced cooling air inlet area. At the same time, the power consumption increases but it is not a significant effect. The effect of inlet pressure is considered. Higher inlet pressure values lead to higher inlet airflow rates and, consequentially, by increasing the inlet pressure, $T_{dif,max}$ decreases, while the power consumption increases [16]. As shown by the experiments of Xu et al. [22] on prismatic cells, the discharge rate affects the temperature. This is because as the heat dissipation rate remains the same, the heat generation rate changes. There is a degree of discharge (DOD) value for each discharge rate from which the temperature starts to drop, but, at the same time, the higher the discharge rate is, the higher the maximum temperature of the battery: the cooling rate declines with the increase of the discharge rate. The maximum temperature difference also increases with the discharge rate. The effect of baffle installation is important. Jiaqiang et al. [26] study the influence of the air-flow inlet and outlet positions and the benefits of a baffle in terms of heat dissipation. The tested module is made of 18,650 Li-ion batteries. The battery cells are 2 mm spaced from the others, 5 mm from the case bottom, and 20 mm from the case top. The inlet airflow speed is 2 m/s. Figure 1 shows the temperature distribution obtained within the brick under different forced ventilation conditions: lateral inlet and outlet, same side, lateral inlet and outlet, opposite side, and a baffle installation.

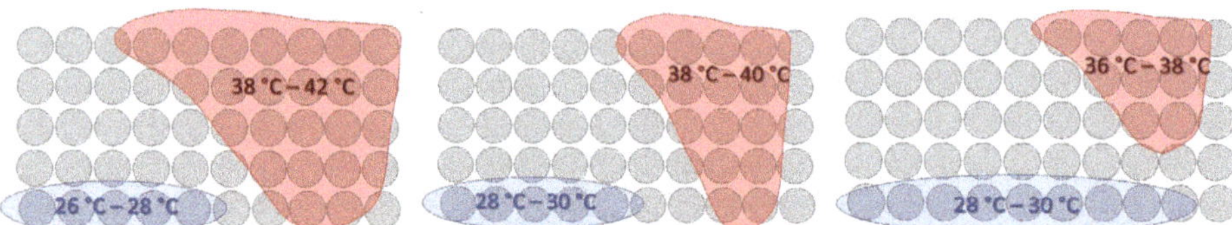

Figure 1. Temperature distribution obtained under different forced-convection conditions obtained in [26]. From left to right: lateral inlet and outlet, same side, lateral inlet and outlet, opposite side, and a baffle installation.

In the first condition, the cells near the air inlet have the lower temperatures, but the highest peaks are obtained on the right, where the inlet air hardly flows through the cells. $T_{ris,max}$ and $T_{dif,max}$ are still unacceptable. In the second condition, with the inlet and outlet air intakes in these positions, the temperature distribution remains the same, and the higher temperatures obtained before decrease but not enough to consider the values of $T_{ris,max}$ and $T_{dif,max}$ acceptable. In the third condition, it is possible to narrow and concentrate the airflow with the installation of a simple and cheaper baffle: the result is an improved airflow cooling ability. The temperature distribution is the same as the previous cases without the baffle with optimal working temperature range with $T_{ris,max} = 38.9\,°C$. Problems of temperature uniformity still have to be fixed to let the battery work in the optimum temperature range and, at the same time, to satisfy the requirement of $T_{dif,max} < 5\,°C$.

2.2. Liquid Cooling

The liquid cooling methods show promising performances thanks to the high thermal conductivity of refrigerant fluids. Indeed, the high weight (including the liquid coolant and the sealing elements to avoid the fluid leakage) and the high manufacturing costs for the fluid coolant circulation system represent some penalties [16], but when high cooling performances are required, air cooling is not sufficient. Here, two leading liquid cooling technologies are presented: cooling plates and heat pipes.

2.2.1. Cooling Plate Technology

A typical cooling plate is made of a metal plate on which flow paths are machined. The cooling liquid flows along the flow paths absorbing the cells' heat waste and dissipating it through the plate (by means of conduction and convection heat transfer). The factors that influence the efficiency of this technology are the shape and the size of the flow paths, the available contact surface between the coolant and the flow path walls, the type of coolant, the flow rate, and the material of the plates. Cooling plates can be divided into two categories: the ice plates, which show the best performances and are placed between the battery cells, and the cold plates, which have lower performances but are easier to install because they are placed as a floor under the cells. Ice plates are often preferred in BEVs for their high efficiency, but the tight spaces represent a hard challenge. Darcovich et al. [50] compare the two technologies showing what happens to the maximum cell temperature if we change the cell case material and the coolant. They also take into account two different drive cycles: the US06, an urban-like driving-cycle, and the HWY, for cars which drive along the highways. They also obtain the maximum cell temperature and the battery lifetime as a function of a range of values of SOH. They obtain better temperature uniformity with ice plates than with cold plates. The higher the value of the heat transfer coefficient is, the lower the maximum temperature of the battery is for any kind of cell case materials and driving cycles. This means a longer battery lifetime of about 2 years if we consider US06 driving-cycles, 1 year for HWY driving-cycles. Studies about new technologies suitable to improve cold plates' cooling systems have been done on new kinds of channels such as leaf-like channels [51]. The design project provides four collection channels arranged along diagonal lines where the decrease in the temperature gradient is required. After many simulations on the channels analyzing the influence of width and length ratio, as well as channel thickness, and also investigating the optimum inlet mass-flow rate, optimal values of heat dissipation and power consumption have been found. Leaf-like channels can be helpful in the study of fractal networks for cold plates [51].

2.2.2. Heat Pipes

Heat pipes (HP) are passive capillary-driven two-phases systems that represent another solution to manage the temperature of electric vehicles battery packs. The two-phase system means that the heat transfer occurs due to a phase change and, in this case, due to the liquid–vapor one. A heat-pipe based cooling system is made of three main parts: the evaporator (heat source), the adiabatic section (which links the first and the third parts and along which the heat transport happens), and the condenser (heat sink). Evaporation and condensation rule the thermodynamic cycle: the coolant within the pipes (usually made of copper) absorbs the heat of the battery cells, which causes the evaporation of the cooling liquid itself, then the fluid moves along the pipes, towards the condenser, with an efficient heat transfer thanks to the latent heat of vaporization. Once the vapor reaches the condenser, another phase change occurs, and it turns to liquid again and heat is dissipated. Simulations [35] have shown that BTM systems based on heat pipes provide energy savings in electric vehicles, keeping the maximum temperature of the battery under 50 °C, the temperature difference under 5 °C, and a good temperature distribution. The possibility to curve and bend the pipes makes them suitable in almost any battery design, realizing unique and efficient cooling systems. Figure 2 shows a scheme of a heat-pipe-based BTMS, equipped with a further U-pipe system (remote heat transfer heat pipe system, RHE-HP) which helps in transporting heat away.

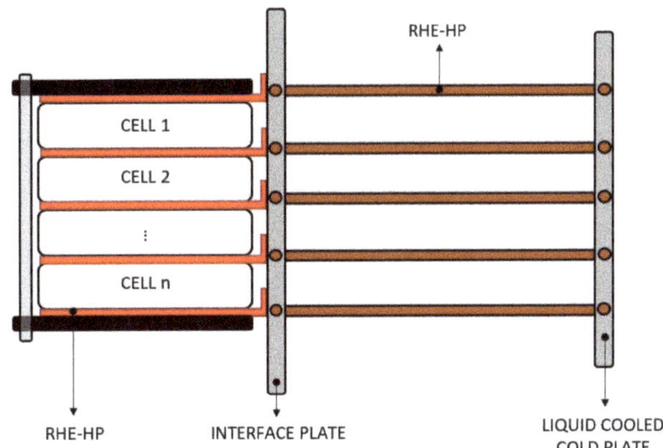

Figure 2. Schene of a heat pipe-based BTMS assembly equipped with a RHE-HP system, as studied in [52].

The interface plate connects the heat pipe cooling plates system (HPCP) and the RHE-HP one. Several types of research have been carried on studying the influence of many factors on the efficiency of HP systems. For their experiments, Putra et al. [37] use different coolants and modify the heat flux load studying the maximum temperature and the best system performance. The coolants chosen are distilled water, alcohol 96%, and acetone 95% with a filling ratio of 60%. Looking at the evaporator temperatures during the transient case, it is shown that they increase quickly followed by a drop. This is due to super-heating, which occurs when the boiling point is reached. As the heat flux load increases, the duration of the transient decreases. Another factor that influences the transient is the capacitance of the system, which is a function of the mass of the system itself. For all the evaporator and condenser temperatures, the shortest transients are obtained with acetone and alcohol as coolants. During the steady-state case, the same temperature distribution can be seen for distilled water and alcohol, except when the highest heat flux load is reached: in that case, alcohol and acetone have an equal temperature distribution. The evaporator temperature is always under 50 °C for all the heat flux loads and almost all the coolants. The lowest temperature difference between the evaporator and the condenser is obtained with acetone in all the experiments carried out. Another consideration is that as the heat flux load increases, at the beginning, it is possible to see the alcohol evaporator temperature close to the distilled water one and then closer to the acetone one. This is because a rise in heat flux load is followed by a pressure increase, which is the cause of higher temperatures. Depending on the coolant, the evaporator temperature will be different. It has been found that, for the highest heat flux load, the acetone condenser temperature is very different from the other coolant ones: this is due to super-heating in the evaporator and for the vapor infiltration with a consequent increasing of the liquid temperature and much more heat transported towards the condenser. However, the lowest temperature difference is obtained with acetone for the most moderate heat flux load. The best performance is reached using acetone as a coolant and with the highest heat flux load. Wang et al. [53] studied cylindrical cells' battery performances, taking into account the effect of several structural parameters such as the spacing, the thickness of the conductive elements, and the angle between the battery and the conductive elements. Simulations show that the height of the conductive element is the parameter that most influences the temperature distribution. Moreover, the angle is a second influencing factor, while the thickness of the conductive element and the spacing have a minimum effect.

2.2.3. Immersive Solutions

This method helps to make the performance of the heat pipes system more efficient. Adopted by Tesla, it puts about 25% of the cell lateral surface in direct contact with the cold plate, with the cooling liquid circulating inside. The liquid circulates in metal thin pipes embedded in an extruded compound with the function of the thermal interface layer and electric insulation between the cells and the cooling medium. The remaining internal volume among cells is filled by a thermal interface that is constituted by intumescent gel, which is not electrically conductive. This material coats the sidewall of the cell casing, which is not in contact with the cold plate, and the space at both cell sides between the terminal and the metal connection plate.

2.3. Innovative Cooling Strategies

Among the different cooling systems, the most common ones are the air-based and the liquid-based BTMSs. Innovative studies brought to new promising cooling systems (carbon-fiber-based phase change materials, hybrid cooling systems) which, however, are not yet as widespread due to high costs.

Phase Change Materials

Phase change materials (PCMs) are materials that change their physical state, from solid to liquid and vice versa, to store and release heat. A solid-phase PCM starts absorbing heat from the battery and, at a certain temperature (it depends on the specific PCM), the melting begins. As the PCM absorbs heat, the battery is maintained to a target temperature. When the PCM is cooled by the environmental changes, the phase changes to solid again, releasing energy in the form of latent heat. There are around 200 different PCMs that can target temperatures between −50 °C and 150 °C [54]. For electric vehicle BTMSs, it should be necessary to test PCMs in a smaller temperature range, but the starting costs are still high. This is one of the reasons why PCMs are not as common in BTMSs. On the other hand, PCMs are shown to be one of the most efficient coolants thanks to their high heat storage capacity. The operating costs are quite low, and the retention of properties in sequential cycles is a very appreciable characteristic [55]. There are several kinds of PCMs, different in their composition and for the temperatures at which they change their phase. The most used are water-based PCMs, salt hydrates, paraffin, or innovative and eco-friendly plant-based PCMs. There are also innovative methods to improve PCMs' characteristics such as loading them with composite materials and metal foams. Samimi et al. studied innovative PCM loaded with carbon fiber [56] to increase the PCMs' thermal diffusivity (which is usually low) and, consequently, improving the temperature uniformity in the battery cells. Carbon-based materials have been chosen as proper conductive fillers because of their excellent thermal and electrical properties, their chemical stability, and their lower density than metals. For the experiments, different cooling media (air, wax as unloaded PCM, carbon-fiber-loaded PCM) have been compared during the discharge of a lithium-ion battery, provide evidence for the results of the research. It is shown that the higher the percentage of carbon fiber loading is, the lower the resulting cell temperature is. Then, PCMs not only act like heat sinks but, if loaded with carbon fiber, also lead to a better temperature distribution in the cells thanks to their improved thermal conductivity.

2.4. Hybrid Cooling Systems

Some hybrid BTMSs combine natural air convection with liquid and thermoelectric cooling systems [45]. With a thermoelectric cooler (TEC), all the processes concern transformations from electricity to heat and vice versa. These systems can work like coolers as well as like heaters (useful for those technologies that have to work in the extremely low-temperature environment). Lyu et al. compare the temperature reached by a cylindrical lithium-ion cell if it is cooled with air, with liquid, and finally with a hybrid BTMS, which combines air and liquid cooling methods and a TEC system. Results show that during the test with air-cooling, the cell temperature increases with the voltage, and the higher is it, the faster the temperature rising. With the liquid-cooling, it is shown that the cell temperature and the water temperature rise almost together and with a more moderate trend. Lower cell temperatures can be obtained using the hybrid TEC system.

2.5. Fire-Prevention Strategies

The incredible spread of lithium-ion battery packs seen in recent years is easily understandable if we think about their specific capacity, energy density, and power density. All these characteristics are higher in Li-ion batteries rather than in the outdated batteries, but this new technology leads to some issues. One of the hardest challenges in the use of lithium-ion batteries is to find valid methods to prevent or face the so-called thermal runaway. It is a dangerous phenomenon during which the temperature rises and exothermic reactions are triggered, and these could lead to fires or explosions. There may be several causes that generate the thermal runaway: short circuits, battery overcharging, and design and manufacturing defects [49].

2.5.1. Methods to Prevent Thermal Runaway

There are several ways of acting against the thermal runaway and these can be divided into three categories, taking into account the effects on the process. The first category regards preventive measures, such as the addition of flame retardants. The second category involves fail-safe measures that stop or decrease the damage caused by thermal runaway, such as separators or cell venting techniques. The third category concerns actions for extinguishing fires once the thermal runaway has occurred [49]. There is another classification about the way to prevent thermal runaway or reduce its damages. Three main ways of acting corresponding to three levels of protection have been identified: cell-to-cell, module-to-module, and battery pack level.

- Cell-to-cell. This is the highest level of protection using engineered materials between every single cell. Obviously, if we consider space constraints, this represents a challenge but the advantages are relevant: in case of thermal runaway, the material surrounding the cells absorbs heat and minimize the propagation of thermal effects to the adjacent cells [47]. Thanks to their heat-absorbing nature, phase-change materials (PCMs) are often used in cell-to-cell protection. Moreover, if they change phase from liquid to gas, they also bring with them the cell gases out from the battery modules. The shape of the cells has to be taken into account because it influences the PCM that it is possible to use. For cylindrical cells, the PCM can be solid; it is not the same for pouch batteries that need liquid phase material as they expand and contract continuously.

- Module-to-module. This kind of protection follows the same philosophy of the cell-to-cell one, but it separates the modules one from the other. This is a lower level of protection with respect to the previous one but it is lighter and easily fittable on several kinds of batteries.
- Battery pack. This represents the lowest level of protection because it does not work against the heat propagation between the cells or between the modules, but it only gives more time to the car occupants to put themselves in safe.

2.5.2. Emergency Spray Cooling

There are some emergency situations during which rapid battery pack cooling is required. With a common BTMS, it is usually hard to deal with these events with the right timing. An efficient solution for these issues is adding a supplemental refrigerant spray cooling system [48], not only useful to let the temperature drop fast but also to suppress oxygen, one of the possible causes of the exothermic reaction. When the thermal sensors detect an incoming thermal runaway, the refrigerant is sprayed inside the battery box, and it gasifies due to the high temperature strengthening the heat convection. In this way, a rapid temperature decrease occurs and the generated refrigerant gasses push the oxygen out of the box. Vents positioned in the right places can also contribute to the oxygen flow out. Several tests have been carried on about different spray modes: continuous, fixed-interval intermittent, and non-fixed-interval intermittent mode. With the continuous mode, the result is the most efficient cooling method and the maximum value of temperature difference uniformity. On the other hand, oxygen suppression is not better than the one obtained with the intermittent modes. With intermittent modes, the higher is the frequency of the spray, the better is the efficiency of the cooling. Concerning the temperature difference uniformity, the non-fixed-interval intermittent mode is less influenced by the spray frequency than the fixed-interval intermittent one. Another advantage of the intermittent modes is that using them it is easier to maintain the low-oxygen status.

3. CFD Analysis of a Hybrid Solution

The battery arrangement studied in this section is a hybrid solution with staggered cylindrical cells embedded in solid support and cooled by a fluid that flows in wavy channels. This study is part of a deep investigation of the performances of battery packs conducted by our research group in the automotive field, with dedicated experimental campaigns in which the theoretical results shown in this work are deployed and validated in the laboratory. The novelty of this solution is the strict coupling of solid plastic matrix and channels for liquid cooling [57]. The geometry of a portion of the hybrid solution is shown in Figure 3.

In the figure, the gray part is the solid plastic support with the function to fix the cells, and the blue zones are the wavy channels where the coolant flows. This study aims to optimize the distance p between two cells and the width c of the channels. Twelve combinations of p and c have been chosen for the optimization, shown in Table 2.

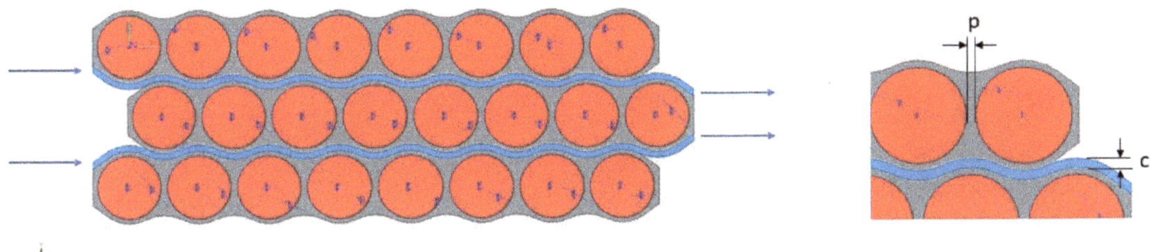

Figure 3. Section of a portion of the battery arrangement (**left**) and a zoomed-in image of the sketch (**right**).

Table 2. Parameters chosen for the optimization of the parameters p and c.

Geometry Number	p (mm)	c (mm)
1	1	1
2	1	2
3	1	3
4	2	1
5	2	2
6	2	3
7	3	1
8	3	2
9	3	3
10	1	5
11	2	5
12	3	5

A representative part of the battery pack has been chosen for the simulations, as shown in Figure 4, with two lines of cells cut in their symmetry plane, with the channel between the two lines.

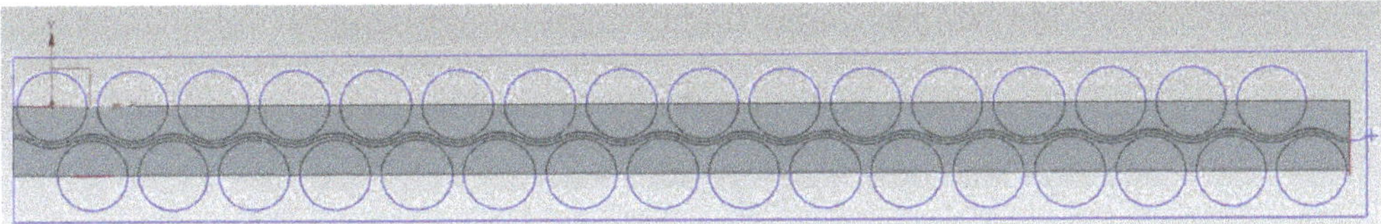

Figure 4. Fundamental unit analyzed in the simulations.

The real geometry of the battery is obtained by repeating periodically this part in the direction perpendicular to the flow. The number of cells in the direction parallel to the coolant flow was varied from 8 to 16 to show the linear behavior of the cell temperature as a function of the distance from the coolant inlet. In the present paper, the cells have a format 21,700, but the approach can be easily up-scaled to cells with a different format. The Li-ion cells are considered as made by homogeneous material with density 2680 kg/m^3, thermal conductivity 3.4 W/(m K), and heat capacity 1280 J/(kg K). Two materials are compared for the cell support, a common plastic with thermal conductivity $k_1 = 0.28$ W/(m K) and density 1380 kg/m^3 (here called "plastic 1"), and a conductive plastic with additives with thermal conductivity $k_2 = 1$ W/(m K) and density 1450 kg/m^3 (here called "plastic 2"). The two materials have specific heat 1100 J/(kg K). These values refer to plastic material that are used in our laboratory to realize the support for the cell in the battery pack. The CFD code STAR-CCM+ has been employed to solve the steady-state Reynolds-averaged

Navier–Stokes (RANS) equations with realizable k-ε turbulence model. Bidimensional meshes with polygonal elements have been created for each configuration, as shown in Figure 5.

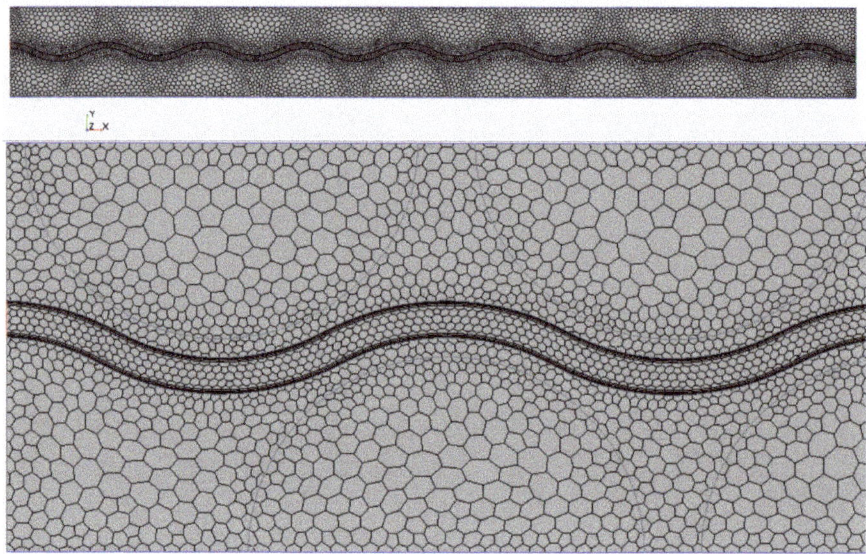

Figure 5. Mesh of a portion of the battery arrangement (**top**) and zoom of the mesh refinement and prism layer (**bottom**).

A base size of 0.8 mm and a surface grow rate of 1.3 have been set for each mesh. Once the target surface size is defined, a mesh refinement is necessary to concentrate spacial resolution in the fluid flow zone. The mesh in the fluid zone mesh has a finer resolution with a base size of 0.5 mm in order to get at least 10 elements along the channel height. In addition, a boundary layer has been modeled in the fluid zone with five layers with a total thickness of 33% of the base size, as shown in Figure 5. The final mesh has 14,939 cells. The studied cases are steady-state problems for incompressible flows, with conjugate heat transfer. The equations that describe this problem are the incompressible Navier–Stokes equations, together with the energy conduction for the solid region and convection equations for the fluid region. The continuity equation, together with the momentum balance equation, were solved for the liquid phase:

$$\frac{\partial u_i}{\partial t} + u_j \frac{\partial u_i}{\partial x_j} = -\frac{1}{\rho}\frac{\partial p}{\partial x_i} + \nu \left[\nabla^2 u_i + \frac{1}{3}\vec{\nabla}(\vec{\nabla}\cdot\vec{u}) \right] \tag{1}$$

The energy equation in temperature formulation has been solved for the flow region:

$$\frac{\partial T}{\partial t} + u_i \frac{\partial T}{\partial x_j} = \alpha \nabla^2 T + \frac{q_g}{\rho c} + \frac{\nu}{\rho c}\Phi \tag{2}$$

where $\alpha = \frac{k}{\rho c}$ is the fluid thermal diffusivity and $\Psi = \nu\phi$ is called the *Dissipation Function* and is defined as $\nu\phi = D_{ij}\tau_{ij}$. The governing equation for the solid is

$$k\frac{\partial^2 T}{\partial x_i^2} = 0 \tag{3}$$

where k is the fluid thermal conductivity and c is the specific heat. The temperature at the interface between the solid and liquid region is equal, while the heat flux entering one region at one side of the interface is equal to the flux leaving the other region on the other side of the interface. A heat source is defined at the center of each cell so that the temperature of the system rises according to the governing equations and the materials' properties. Contact interface conditions are applied at interfaces between solid and fluid regions. Additionally, a thermal contact resistance of 0.001921 W/m^{2}K is considered between the plastic support and the cells for mechanical backlash. In all the cases, the inlet fluid has a temperature of 20 °C. The cases of air-cooling and water-cooling have been considered in order to evidence the main strengths and challenges related to the two approaches. Different cooling-fluid velocity ranges have been considered in the two cases in order to have similar pressure drop ranges along the channels. A steady thermal power $Q = 4$ W has been assigned to each cell, because these values were measured in our laboratory for some extreme automotive regimes [57].

3.1. Air-Cooling

In the first case the cooling fluid is air. The inlet velocities considered in this case are $v = 13.9$ m/s, $v = 27.8$ m/s, and $v = 38.9$ m/s, which correspond to the Reynolds number is in the range (900–13,000). The characteristic length for the Reynolds number is the channel width c. The maximum temperature of the cells obtained for the case of plastic 1 with air-cooling are summarized in Figure 6.

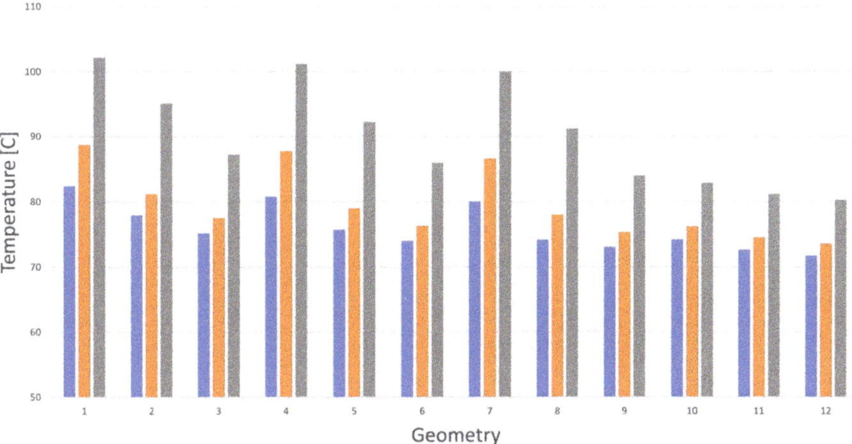

Figure 6. Maximum temperature of the cells obtained for the case of plastic 1 and air as a cooling fluid. Inlet velocities are $v = 13.9$ m/s (grey), $v = 27.8$ m/s (orange), and $v = 38.9$ m/s (blue).

The figure shows that the maximum temperature of the cells decreases as the air velocity increases. Moreover, the maximum temperature of the cells decreases as the channel width c increases, while its dependence on the axial distance between the cells is weak. The best geometry is n. 12, with $c = 5$ mm and $p = 3$ mm. These results can be compared with the maximum temperatures on the cells obtained for the case with plastic 2, as shown in Figure 7.

Figures 6 and 7 show similar trends, but the values of the maximum temperatures obtained with plastic 2 are about 10 °C lower than those obtained with plastic 1. The best geometry with air-cooling is again the n. 12 for low velocities ($v = 13.9$ m/s or $v = 27.8$ m/s), while for high velocities ($v = 38.9$ m/s) the geometry with $c = 3$ mm and $p = 2$ mm is the one which gives the lowest maximum temperature of the cells. The temperature distribution on the cells and in the channel, obtained for the geometry n. 12, is shown by Figure 8 for the case of plastic 1.

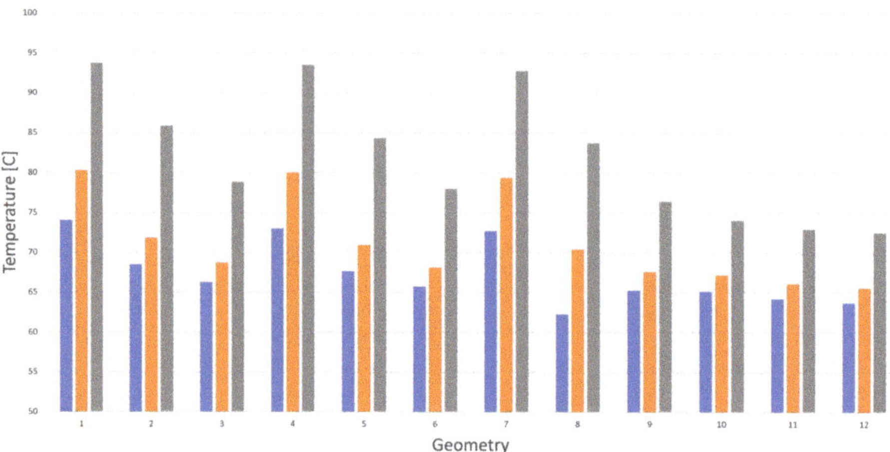

Figure 7. Maximum temperature of the cells obtained for the case of plastic 2 and air as a cooling fluid. Inlet velocities are $v = 13.9 \, \text{m/s}$ (grey), $v = 27.8 \, \text{m/s}$ (orange), and $v = 38.9 \, \text{m/s}$ (blue).

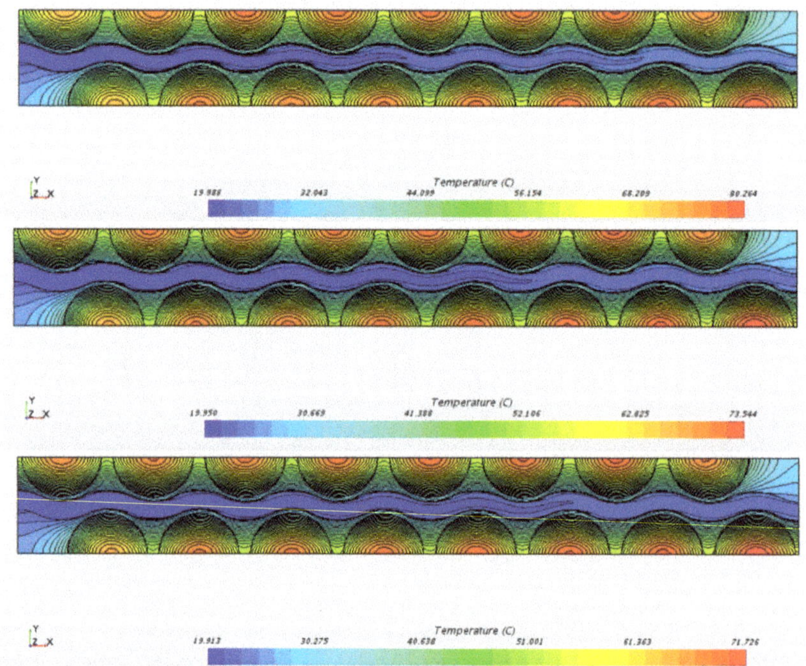

Figure 8. Temperature of the cells obtained for the case of geometry n. 12 and air as cooling fluid. Inlet velocities are $v = 13.9 \, \text{m/s}$ (**top**), $v = 27.8 \, \text{m/s}$ (**middle**), and $v = 38.9 \, \text{m/s}$ (**bottom**).

The figure shows that the maximum temperature difference between the first and the last cell is $9 \, °\text{C}$ for the case with $v = 13.9 \, \text{m/s}$, while it is $6 \, °\text{C}$ for the case with $v = 27.8 \, \text{m/s}$ and $5 \, °\text{C}$ for the case with $v = 38.9 \, \text{m/s}$. The number of cells along the direction of the flow has been varied from 8 to 16 to show that the cell temperature is a linear function of the cell number. Figure 9 shows the cell temperature as a function of the cell number obtained for geometry n. 12.

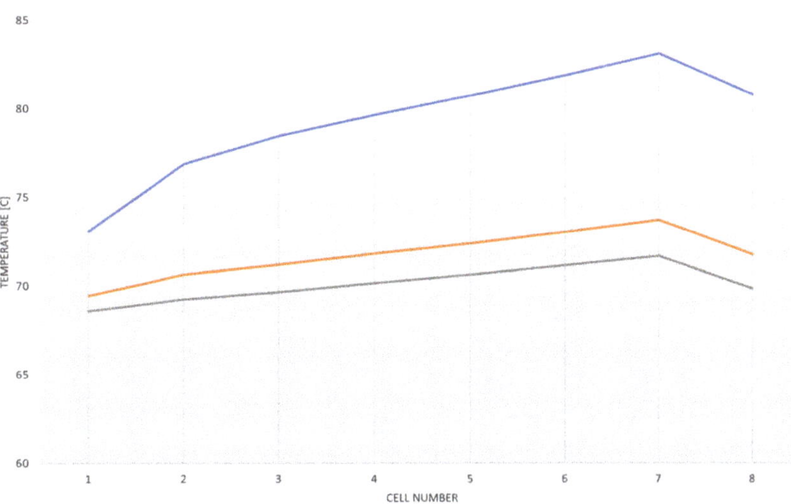

Figure 9. Temperature of the cells as a function of the cell number, obtained for the case of geometry n. 12, and air as cooling fluid. Inlet velocities are $v = 13.9$ m/s (blue line), $v = 27.8$ m/s (red line), and $v = 38.9$ m/s (grey line).

The slope of the central part of the curves shown in Figure 9 is a crucial parameter for the battery design, which has to be minimized to find the optimal parameters. For the case of air-cooling, the results of the simulations can be summarized as follows:

- The maximum temperature of the cells decreases weakly as the axial distance between the cells, and it strongly decreases with the channel width,
- The temperature of the cells linearly increases with the number of the cells in the direction of the flow, and the trend is a function of the air-flow velocity within the channels,
- By changing the solid material in which the cells are embedded, one obtains similar trends with lower absolute values of the temperature distribution as the thermal conductivity increases.

3.2. Water-Cooling

In the second case the cooling fluid is water. The inlet velocities considered in this case are $v = 0.1$ m/s and $v = 1$ m/s; i.e., the Reynolds number is in the range $(100 \div 6000)$. The characteristic length in the Reynolds number is the channel width c. The results for the case of plastic 1 with water cooling are summarized in Figure 10.

The figure shows that the maximum temperature of the cells increases as the channel width c increases and decreases as the axial distance p between the cells increases. The best geometry is n. 7, with $p = 3$ mm and $c = 1$ mm. These results can be compared with the results obtained with plastic 2 with the same geometry, as summarized in Figure 11.

The Figures 10 and 11 show similar trends but the values of the maximum temperature are about 6–7 °C lower with the plastic 2. The best geometry with water cooling is the n. 7, i.e., the one with the smallest channel. The temperature difference on the cells for the geometry n. 7 is shown by Figure 12 for the case of plastic 1.

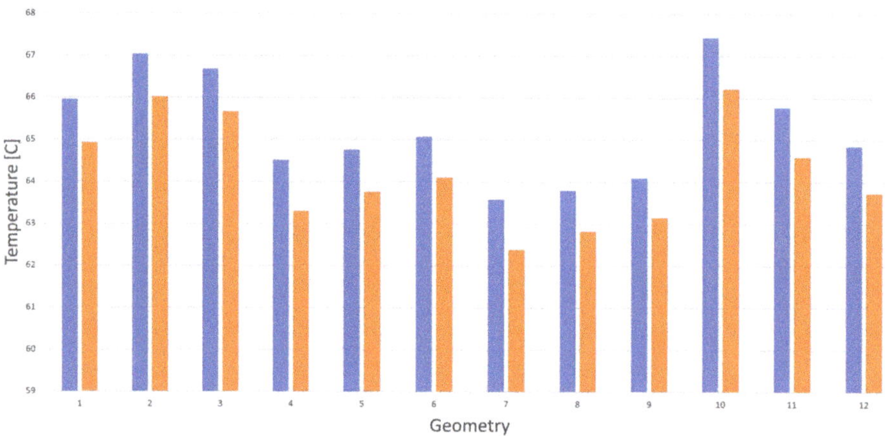

Figure 10. Maximum temperature of the cells obtained for the case of plastic 1 and water as a cooling fluid. Inlet velocities are $v = 0.1$ m/s (blue) and $v = 1$ m/s (orange).

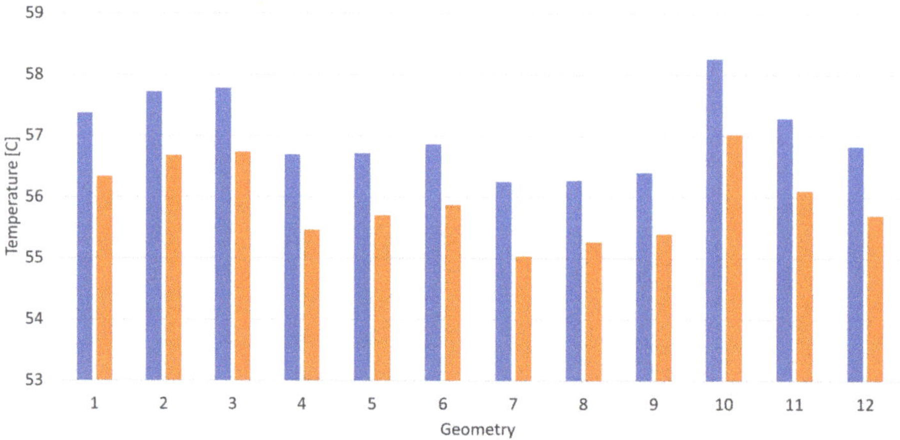

Figure 11. Maximum temperature of the cells obtained for the case of plastic 2 and water as a cooling fluid. Inlet velocities are $v = 0.1$ m/s (blue) and $v = 1$ m/s (orange).

The figure shows that the maximum temperature difference between the first and the last cell is 4 °C for the case with $v = 0.1$ m/s, and it is 2 °C for the case with $v = 1$ m/s. The number of cells along the direction of the flow has been varied from 8 to 16 to show that the cell temperature is a linear function of the cell number. Figure 13 shows the cell temperature as a function of the cell number obtained for a channel width $c = 1$ mm. The inlet velocity is $v = 0.1$ m/s.

The slope of the central part of the curves shown in Figure 13 is a crucial parameter for the battery design, which has to be minimized to find the optimal parameters. For the case of water-cooling, the results of the simulations can be summarized as follows:

- The maximum temperature of the cells decreases as the axial distance between the cells, and it increases with the channel width;
- The temperature of the cells linearly increases with the number of the cells in the direction of the flow, and the trend is a function of the water-flow velocity within the channels and of the axial distance between the cells;

- By changing the solid material in which the cells are embedded, one obtain similar trends with lower absolute values of the temperature distribution as the thermal conductivity increases.

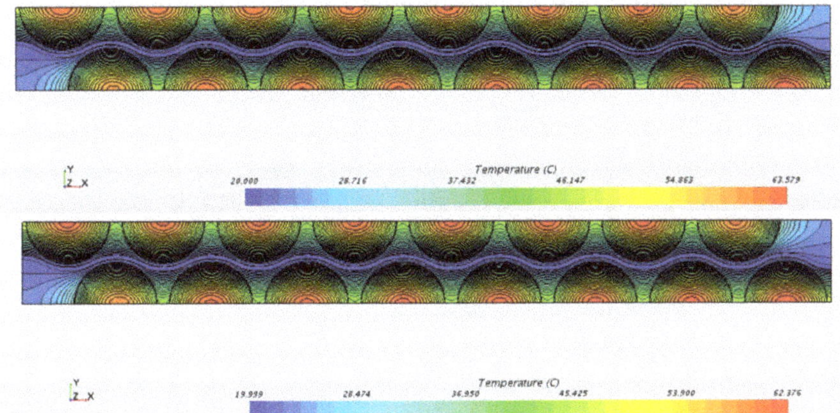

Figure 12. Temperature of the cells obtained for the case of geometry n. 7 and water as cooling fluid. Inlet velocities are $v = 0.1$ m/s (**top**) and $v = 1$ m/s (**bottom**).

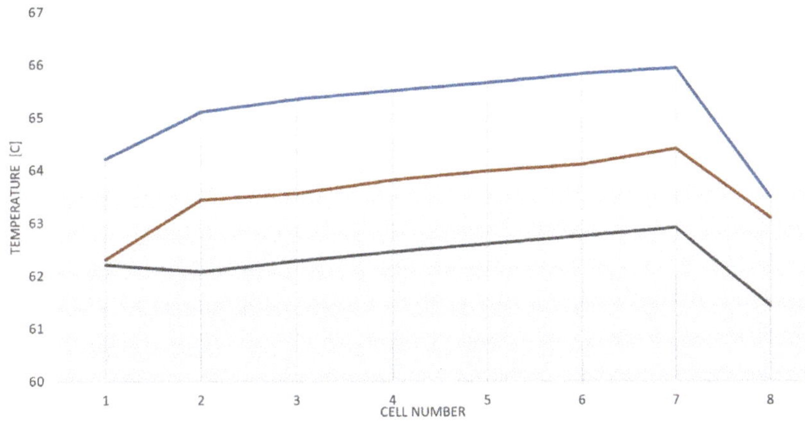

Figure 13. Temperature of the cells as a function of the cell number, obtained for the case of channel width $c = 1$ mm and water as cooling fluid. The distances between the cells are $p = 1$ mm (blue line), $p = 2$ mm (red line) and $p = 3$ mm (grey line).

4. Conclusions

A survey of the existent thermal management systems for lithium batteries has been presented, showing, in particular, some air-cooling and liquid-cooling approaches. The benefits resulting with the installation of a baffle plate and the importance of the design of the cell arrangements' structures have been shown. In this context, a hybrid system that combines heat conduction between cells in the longitudinal direction and forced convection in channels between the cells lines is presented. This approach is studied numerically using a CFD approach. It is shown that a combination between cooling fluid, solid material that connects the cells, and distances between the cells leads to the determination of the optimal thermal management arrangement. This means that the lowest temperature of the cells and the lowest temperature differences within the battery are obtained by this methodology. The best dimensions of the channels with air-cooling and water-cooling are discussed, showing that with air-cooling the best choice is to increase the width of the channel, while with water-cooling, the best choice is to reduce it. A linear temperature increase in the cells along the direction of the flow is found, and the dependence on the solid matrix thermal conductivity is found. These results show that a multi-parameter optimization approach gives the best arrangement in terms of number of cells and their packing density, as a function of the thermal characteristics of the solid matrix where the cells are embedded.

Author Contributions: Data curation, M.F., E.P.B.D.V., A.H. and B.P.; methodology, A.H. and B.P.; investigation, M.F., E.P.B.D.V.; validation, E.P.B.D.V.; formal analysis, A.H., C.R. and B.P.; writing—original draft, M.F. and B.P.; supervision, B.P.; funding acquisition, B.P. All authors have read and agreed to the published version of the manuscript.

Funding: This research was fund by Emilia-Romagna Region, under the PORFESR program, years 2018–2019, thanks to the LiBER project.

Institutional Review Board Statement: Not applicable.

Informed Consent Statement: Not applicable.

Data Availability Statement: Not applicable.

Acknowledgments: The authors would also like to thank Tommaso Brugo, Luca Frigerio and Andrea Bitto from the Department of Industrial Engineering, University of Bologna, for their support in the CAD design of the brick geometries analyzed.

Conflicts of Interest: The authors declare no conflict of interest.

References

1. Liu, H.; Wei, Z.; He, W.; Zhao, J. Thermal issues about Li-ion batteries and recent progress in battery thermal management systems: A review. *Energy Convers. Manag.* **2017**, *150*, 304–330. [CrossRef]
2. Kim, J.; Oh, J.; Lee, H. Review on battery thermal management system for electric vehicles. *Appl. Therm. Eng.* **2019**, *149*, 192–212. [CrossRef]
3. Lowe, M.; Tokuoka, S.; Trigg, T.; Gereffi, G. *Lithium-ion Batteries for Electric Vehicles: The U.S. Value Chain*; Center on Globalization, Governance and Competitiveness; Duke University: Durham, NC, USA. 2010.
4. Bolsinger, C.; Birke, K.P. Effect of different cooling configurations on thermal gradients inside cylindrical battery cells. *J. Energy Storage* **2019**, *21*, 222–230. [CrossRef]
5. An, Z.; Shah, K.; Jia, L.; Ma, Y. Modeling and analysis of thermal runaway in Li-ion cell. *Appl. Therm. Eng.* **2019**, *160*, 113960. [CrossRef]
6. Berdichevsky, G. Kelty, K.; Straubel, J.B.; Toomre, E. *The Tesla Roadster Battery System*; Testa Motors: Palo Alto, CA, USA, 2006.
7. Opitz, A.; Badami, P.; Shen, L.; Vignarooban, K.; Kannan, A.M. Can Li-Ion batteries be the panacea for automotive applications? *Renew. Sustain. Energy Rev.* **2017**, *68*, 685–692. [CrossRef]
8. Huang, Y.; Lu, Y.; Huang, R.; Chen, J.; Chen, F.; Liu, Z.; Yu, X.; Roskilly, A.P. Study on the thermal interaction and heat dissipation of cylindrical Lithium-Ion Battery cells. *Energy Procedia* **2017**, *142*, 4029–4036. [CrossRef]
9. Erb, D.; Kumar, S.; Carlson, E.; Ehrenberg, I.; Sarma, S. Analytical methods for determining the effects of lithium-ion cell size in aligned air-cooled battery packs. *J. Energy Storage* **2017**, *10*, 39–47. [CrossRef]
10. Wang, T.; Tseng, K.; Zhao, J.; Wei, Z. Thermal investigation of lithium-ion battery module with different cell arrangement structures and forced air-cooling strategies. *Appl. Energy* **2014**, *134*, 229–238. [CrossRef]
11. Lu, Z.; Yu, X.; Wei, L.; Qiu, Y.; Zhang, L.; Meng, X.; Jin, L. Parametric study of forced air cooling strategy for lithium-ion battery pack with staggered arrangement. *Appl. Therm. Eng.* **2018**, *136*, 28–40. [CrossRef]
12. Subramanian, M.; Hoang, A.T.; Kalidasan, B.; Nižetić, S.; Solomon, J.M.; Balasubramanian, D.C.S.; Metghalchi, H.; Nguyen, X.P. A technical review on composite phase change material based secondary assisted battery thermal management system for electric vehicles. *J. Clean. Prod.* **2021**, *322*, 129079. [CrossRef]
13. Sharma, D.K.; Prabhakar, A. A review on air cooled and air centric hybrid thermal management techniques for Li-ion battery packs in electric vehicles. *J. Energy Storage* **2021**, *41*, 102885. [CrossRef]
14. Wang, Q.; Jiang, B.; Li, B.; Yan, Y. A critical review of thermal management models and solutions of lithium-ion batteries for the development of pure electric vehicles. *Renew. Sustain. Energy Rev.* **2016**, *64*, 106–128. [CrossRef]
15. Sato, N. Thermal behavior analysis of lithium-ion batteries for electric and hybrid vehicles. *J. Power Sources* **2001**, *99*, 70–77. [CrossRef]

16. Zhou, H.; Zhou, F.; Xu, L.; Kong, J.; Yang, Q. Thermal performance of cylindrical Lithium-ion battery thermal management system based on air distribution pipe. *Int. J. Heat Mass Transf.* **2019**, *131*, 984–998. [CrossRef]

17. Zhang, S.S.; Xu, K.X.; Jow, T.R. The low temperature performance of Li-ion batteries. *J. Power Sources* **2003**, *225*, 137–140. [CrossRef]

18. Shabani, B.; Biju, M. Theoretical Modelling Methods for Thermal Management of Batteries. *Energies* **2015**, *8*, 10153–10177. [CrossRef]

19. Moghaddam, H.; Mazyar, S. *Designing Battery Thermal Management Systems (BTMS) for Cylindrical Lithium-Ion Battery Modules Using CFD*; KTH School of Industrial Engineering and Management: Stockholm, Sweden, 2019.

20. Wu, W.; Wang, S.; Wu, W.; Chen, K.; Hong, S.; Lai, Y. A critical review of battery thermal performance and liquid based battery thermal management. *Energy Convers. Manag.* **2019**, *182*, 262–281. [CrossRef]

21. Jiang, J.; Ruan, H.; Sun, B.; Wang, L.; Gao, W.; Zhang, W. A low-temperature internal heating strategy without lifetime reduction for large-size automotive lithium-ion battery pack. *Appl. Energy* **2018**, *230*, 257–266. [CrossRef]

22. Xu, X.; He, R. Research on the heat dissipation performance of battery pack based on forced air cooling. *J. Power Sources* **2013**, *240*, 33–41. [CrossRef]

23. Zhao, G.; Wang, X.; Negnevitsky, M.; Zhang, H. A review of air-cooling battery thermal management systems for electric and hybrid electric vehicles. *J. Power Sources* **2021**, *501*, 230001. [CrossRef]

24. Xie, J.; Ge, Z.; Zang, M.; Wang, S. Structural optimization of lithium-ion battery pack with forced air cooling system. *Appl. Therm. Eng.* **2017**, *126*, 583–593. [CrossRef]

25. Fan, Y.; Zhan, D.; Tan, X.; Lyu, P.; Rao, J. Optimization of cooling strategies for an electric vehicle in high-temperature environment. *Appl. Therm. Eng.* **2021**, *195*, 117088. [CrossRef]

26. E, J.; Yue, M.; Chen, J.; Zhu, H.; Deng, Y.; Zhu, Y.; Zhang, F.; Wen, M.; Zhang, B.; Kang, S. Effects of the different air cooling strategies on cooling performance of a lithium-ion battery module with baffle. *Appl. Therm. Eng.* **2018**, *144*, 231–241. [CrossRef]

27. Chen, K.; Wang, S.; Song, M.; Chen, L. Configuration optimization of battery pack in parallel air-cooled battery thermal management system using an optimization strategy. *Appl. Therm. Eng.* **2017**, *123*, 177–186. [CrossRef]

28. Hong, S.; Zhang, X.; Chen, K.; Wang, S. Design of flow configuration for parallel air-cooled battery thermal management system with secondary vent. *Int. J. Heat Mass Transf.* **2018**, *116*, 1204–1212. [CrossRef]

29. Shahid, S.; Agelin-Chaab, M. Experimental and numerical studies on air cooling and temperature uniformity in a battery pack. *Int. J. Energy Res.* **2018**, *42*, 2246–2262. [CrossRef]

30. Yang, T.; Yang, N.; Zhang, X.; Li, G. Investigation of the thermal performance of axial-flow air cooling for the lithium-ion battery pack. *Int. J. Therm. Sci.* **2016**, *108*, 132–144. [CrossRef]

31. Sefidan, A.M.; Sojoudi, S.; Saha, S.C. Nanofluid-based cooling of cylindrical lithium-ion battery packs employing forced air flow. *Int. J. Therm. Sci.* **2017**, *177*, 44–58. [CrossRef]

32. Jarrett, A.; Kim, I.Y. Design optimization of electric vehicle battery cooling plates for thermal performance. *J. Power Sources* **2011**, *196*, 10359–10368. [CrossRef]

33. Jarrett, A.; Kim, I.Y. Influence of operating conditions on the optimum design of electric vehicle battery cooling plates. *J. Power Sources* **2014**, *245*, 644–655. [CrossRef]

34. Darcovich, K.; MacNeil, D.D.; Recoskie, S.; Cadic, Q.; Ilinca, F.; Kenney, B. Coupled Numerical Approach for Automotive Battery Pack Lifetime Estimates with Thermal Management. *J. Electrochem. Energy Convers. Storage* **2018**, *15*, 021004. [CrossRef]

35. Rao, Z.; Wang, S.; Wu, M.; Lin, Z.; Li, F. Experimental investigation on thermal management of electric vehicle battery with heat pipe. *Energy Convers. Manag.* **2013**, *65*, 92–97. [CrossRef]

36. Mbulu, H.; Laoonual, Y.; Wongwises, S. Experimental study on the thermal performance of a battery thermal management system using heat pipes. *Case Stud. Therm. Eng.* **2021**, *26*, 101029. [CrossRef]

37. Putra, N.; Ariantara, B.; Pamungkas, R.A. Experimental investigation on performance of lithium-ion battery thermal management system using flat plate loop heat pipe for electric vehicle application. *Appl. Therm. Eng.* **2016**, *99*, 784–789. [CrossRef]

38. Liang, L.; Zhao, Y.; Diao, Y.; Ren, R.; Jing, H. Inclined U-shaped flat microheat pipe array configuration for cooling and heating lithium-ion battery modules in electric vehicles. *Energy* **2021**, *235*, 121433. [CrossRef]

39. Zhou, Z.; Lv, Y.; Qu, J.; Sun, Q.; Grachev, D. Performance evaluation of hybrid oscillating heat pipe with carbon nanotube nanofluids for electric vehicle battery cooling. *Appl. Therm. Eng.* **2021**, *196*, 117300. [CrossRef]

40. Hong, S.H.; Jang, D.S.; Park, S.; Yun, S.; Kim, Y. Thermal performance of direct two-phase refrigerant cooling for lithium-ion batteries in electric vehicles. *Appl. Therm. Eng.* **2020**, *173*, 115213. [CrossRef]

41. Tang, X.; Guo, Q.; Li, M.; Wei, C.; Pan, Z.; Wang, Y. Performance analysis on liquid-cooled battery thermal management for electric vehicles based on machine learning. *J. Power Sources* **2021**, *494*, 229727. [CrossRef]

42. Al-Zareer, M.; Dincer, I.; Rosen, M.A. Performance assessment of a new hydrogen cooled prismatic battery pack arrangement for hydrogen hybrid electric vehicles. *Energy Convers. Manag.* **2018**, *173*, 303–319. [CrossRef]

43. Alipour, M.; Hassanpouryouzband, A.; Kizilel, R. Investigation of the Applicability of Helium-Based Cooling System for Li-Ion Batteries. *Electrochem* **2021**, *2*, 135–148. [CrossRef]

44. Al-Zareer, M.; Dincer, I.; Rosen, M.A. Electrochemical modeling and performance evaluation of a new ammonia-based battery thermal management system for electric and hybrid electric vehicles. *Electrochim. Acta* **2017**, *247*, 171–182. [CrossRef]

45. Lyu, Y.; Siddique, A.R.M.; Majid, S.H.; Biglarbegian, M.; Gadsden, S.A.; Mahmud, S. Electric vehicle battery thermal management system with thermoelectric cooling. *Energy Rep.* **2019**, *5*, 822–827. [CrossRef]

46. Lyu, Y.; Siddique, A.R.M.; Gadsden, S.A.; Mahmud, S. Experimental investigation of thermoelectric cooling for a new battery pack design in a copper holder. *Results Eng.* **2021**, *10*, 100214. [CrossRef]
47. Liebscher, A. Preventing Thermal Runaway in Electric Vehicle Batteries. 2018. Available online: www.machinedesign.com/materials/article/21837402/preventing-thermal-runaway-in-electric-vehicle-batteries (accessed on 1 December 2021).
48. Gao, Q.; Liu, Y.; Wang, G.; Deng, F.; Zhu, J. An experimental investigation of refrigerant emergency spray on cooling and oxygen suppression for overheating power battery. *J. Power Sources* **2019**, *415*, 33–43. [CrossRef]
49. Kong, L.; Li, C.; Jiang, J.; Pecht, M.G. Li-Ion Battery Fire Hazards and Safety Strategies. *Energies* **2018**, *11*, 2191. [CrossRef]
50. Darcovich, K.; MacNeil, D.; Recoskie, S.; Cadic, Q.; Ilinca, F. Comparison of cooling plate configurations for automotive battery pack thermal management. *Appl. Therm. Eng.* **2019**, *155*, 185–195. [CrossRef]
51. Deng, T.; Ran, Y.; Zhang, G.; Yin, Y. Novel leaf-like channels for cooling rectangular lithium ion batteries. *Appl. Therm. Eng.* **2019**, *150*, 1186–1196. [CrossRef]
52. Smith, J.; Singh, R.; Hinterberger, M.; Mochizuki, M. Battery thermal management system for electric vehicle using heat pipes. *Int. J. Therm. Sci.* **2018**, *134*, 517–529. [CrossRef]
53. Wang, J.; Gan, Y.; Liang, J.; Tan, M.; Li, Y. Sensitivity analysis of factors influencing a heat pipe-based thermal management system for a battery module with cylindrical cells. *Appl. Therm. Eng.* **2019**, *151*, 475–485. [CrossRef]
54. Sutterlin, W.R. A Phase Change Materials Comparison: Vegetable-Based vs. Paraffin-Based PCMs. 2015. Available online: www.chemicalonline.com/doc/a-phase-change-materials-comparison-vegetable-based-vs-paraffin-based-pcms-0001 (accessed on 1 December 2021).
55. Babapoor, A.; Azizi, M.; Karimi, G. Thermal management of a Li-ion battery using carbon fiber-PCM composites. *Appl. Therm. Eng.* **2015**, *85*, 281–290. [CrossRef]
56. Samimi, F.; Babapoor, A.; Azizi, M.; Karimi, G. Thermal management analysis of a Li-ion battery cell using phase change material loaded with carbon fibers. *Energy* **2016**, *96*, 355–371. [CrossRef]
57. Rossi, C.; Pontara, D.; Falcomer, C.; Bertoldi, M. Simplified parameters estimation for the dual polarization model of lithium-ion cell. In *Lecture Notes in Electrical Engineering, Proceedings of the 13th International Conference of the IMACS TC1 Committee, Electrimacs 2019, Salerno, Italy, 21–23 May 2019*; Springer: Berlin/Heidelberg, Germany, 2020; Volume 697, pp. 129–144.

Article

Impedance Based Temperature Estimation of Lithium Ion Cells Using Artificial Neural Networks

Marco Ströbel * , Julia Pross-Brakhage, Mike Kopp and Kai Peter Birke

Electrical Energy Storage Systems, Institute for Photovoltaics, University of Stuttgart, Pfaffenwaldring 47, 70569 Stuttgart, Germany; Julia.Pross-Brakhage@ipv.uni-stuttgart.de (J.P.-B.); Mike.Kopp.2@ipv.uni-stuttgart.de (M.K.); Peter.Birke@ipv.uni-stuttgart.de (K.P.B.)
* Correspondence: Marco.Stroebel@ipv.uni-stuttgart.de

Abstract: Tracking the cell temperature is critical for battery safety and cell durability. It is not feasible to equip every cell with a temperature sensor in large battery systems such as those in electric vehicles. Apart from this, temperature sensors are usually mounted on the cell surface and do not detect the core temperature, which can mean detecting an offset due to the temperature gradient. Many sensorless methods require great computational effort for solving partial differential equations or require error-prone parameterization. This paper presents a sensorless temperature estimation method for lithium ion cells using data from electrochemical impedance spectroscopy in combination with artificial neural networks (ANNs). By training an ANN with data of 28 cells and estimating the cell temperatures of eight more cells of the same cell type, the neural network (a simple feed forward ANN with only one hidden layer) was able to achieve an estimation accuracy of $\Delta T = 1$ K ($10\,^{\circ}\text{C} < T < 60\,^{\circ}\text{C}$) with low computational effort. The temperature estimations were investigated for different cell types at various states of charge (SoCs) with different superimposed direct currents. Our method is easy to use and can be completely automated, since there is no significant offset in monitoring temperature. In addition, the prospect of using the above mentioned approach to estimate additional battery states such as SoC and state of health (SoH) is discussed.

Keywords: lithium-ion batteries; temperature estimation; sensorless temperature measurement; artificial intelligence; artificial neural network

Citation: Ströbel, M.; Pross-Brakhage, J.; Kopp, M.; Birke, K.P. Impedance Based Temperature Estimation of Lithium Ion Cells Using Artificial Neural Networks. *Batteries* **2021**, *7*, 85. https://doi.org/10.3390/batteries7040085

Academic Editor: Catia Arbizzani

Received: 31 July 2021
Accepted: 8 December 2021
Published: 12 December 2021

Publisher's Note: MDPI stays neutral with regard to jurisdictional claims in published maps and institutional affiliations.

1. Introduction

The performance of lithium ion batteries (LIBs) is strongly dependent on the cell temperature, particularly with regard to battery aging and safety issues. With low temperatures there is a risk of lithium plating due to reduced reaction kinetics, which results in decreased lithium availability. However, operating LIBs at a high temperature can cause a rise in undesirable side reactions that cause rapid degradation, including capacity and power loss [1]. Furthermore, there is a risk of material decomposition which can trigger a so-called thermal runaway and may lead to self-ignition and even an explosion [2]. High temperature issues are caused by cell-internal heat generation, and low temperature operation is generated due to environmental temperatures.

There are various temperature indication methods in existence. Raijmakers et al. provided a broad overview on various temperature indication methods for LIBs [3]. The most common approach uses a conventional temperature sensor, such as a thermocouple or thermistor, placed on the housing of the cell. However, the core temperature varies widely from the surface temperature in cases of heavy loading, and the temperature rise can only be detected with a time-shift or requires extensive thermal models [4–6]. Moreover, the accuracy varies with thermal contact and the position of the temperature sensor. In addition, in most cases not every cell is equipped with a sensor, and therefore the pack design needs to be considered to detect feasible hot spots [7]. Thus, a thermal runaway can only be detected stochastically [2]. By placing temperature sensors internally, these

21

problems can be avoided. This might lead to an increase in costs, more complexity for manufacturers and possible negative effects in terms of battery life [1,8].

The impedance based methods have gained substantial interest because of their characteristic of measuring the average internal temperature without using internal or external hardware [9]. Therefore, the method is also known as sensorless temperature measurement. Figure 1 clarifies the temperature's strong dependency on the battery's characteristic impedance response. Additionally, with increased temperature, a significant reduction in impedance response can be observed. Such measurements were conducted by the authors. EIS measurements were performed on Samsung INR18650-15L cells at different temperatures. Details about the experimental process are described in Section 2.1.

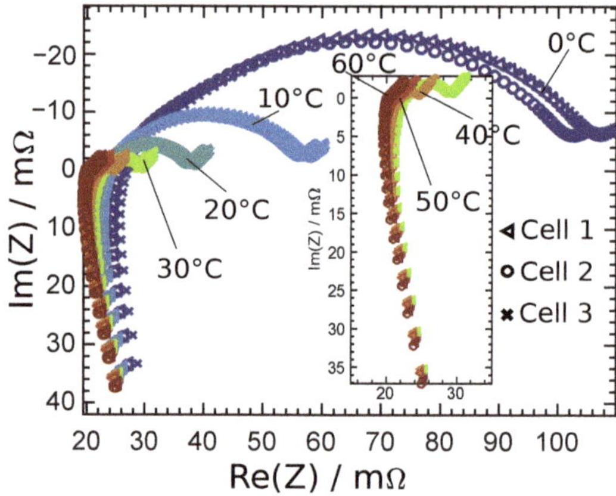

Figure 1. Nyquist plot for 3 Samsung INR18650-15L1 cells at different temperatures from 0 to 60 °C.

The impedance can also be detected in other battery states, such as the state of health with respect to nominal capacity (C/C_N, SoHC) and the state of charge (SoC); therefore, those other states can be crucial input variables for battery management systems [3,10]. In Figure 2 the progression of the impedance for different cycles and three different cells are presented, and it points out the continuous increase in impedance with over the life-cycle. Details about the experimental process are described in Section 2.1. However, it is important to distinguish the various influence factors from each other and find the optimal basis for predictions [11–14].

Srinivasan et al. were the first to find the relation between impedance at a specific frequency and the temperature—more precisely, the phase shift at a frequency that is associated with the solid electrolyte interface [12]. Contrarily, Schmidt et al. made use of the real part of the impedance measurement at higher frequencies, because the time constant has a significantly lower level of correlation with the SoC, which enables improved temperature estimation at unknown SoCs [11]. Similarly, Richardson et al. analyzed the influence of internal thermal gradients on the impedance. It can be shown that this technique estimates the volume-average temperature and is therefore able to detect internal hotspots without any time delays [8].

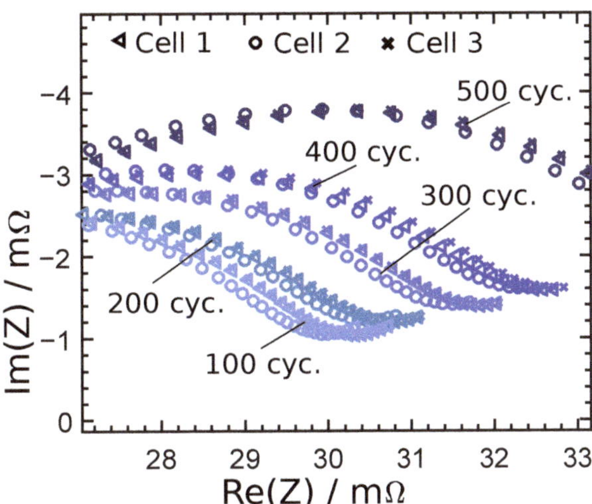

Figure 2. Nyquist plot for 3 Samsung INR18650-15L1 cells after 100, 200, 300, 400, and 500 cycles.

Most publications in this field either show a correlation between a measurable value (e.g., impedance) and the state of interest (e.g., SoC or temperature), or present the state estimation results for a single cell. The state estimation of a number of cells of the same cell type is more complicated due to the existence of variations among individual cells. In this case it is crucial to select the required input parameters. Several studies have raised concerns regarding the question of finding the optimum input variable for determining the temperature. Beelen et al. compared some of these approaches and performed a sensitivity analysis to optimize the prediction accuracy [15]. In the case of an unknown SoC, the achieved average bias was ±0.4 K with an average standard deviation of ±0.7 K. The study highlights the importance of selecting the appropriate input parameters for temperature determination.

This study investigates an approach using ANNs, which are promising for handling multidimensional feature problems. The procedure can be automated and can be easily transferred to other cell chemistries. There are only a few studies available which present data based temperature estimation methods using artificial neural networks. Feng et al. combined the advantages of physical models with artificial neural networks to enhance the performance for estimating the SoC and temperature [16]. Hasan et al. estimated cell temperature based on a nonlinear, autoregressive exogenous artificial neural network and time series data, namely, current and ambient temperature for a battery container [17]. However, there are a number of studies that present ANN based methods for estimating the SoC and SoH [18–21]. Khumprom et al. confirmed ANNs' ability to approximate a nonlinear system by comparing a deep ANN against other machine learning algorithms for SoH prediction; the former could either match or outweigh the other algorithms' performances [22]. Furthermore, some researchers have used impedance data in combination with ANNs. Messing et al. used impedance data for equivalent circuit parameterization and input data for the ANN [23]. However most approaches use methods which require a lot of computational effort due to the need for solving partial differential equations and the fitting of physical models or time series.

For this study, we chose an approach using impedance data from directly measurable indicators (voltage, current, time) as input data and linking impedance based temperature estimation with ANNs. The implied advantage is that error-prone parameterization may be dispensed with. Since it is necessary to monitor the voltage of a lithium ion cell constantly, it should be possible to perform a four-point measurement on each cell by using an AC current source in the battery system to create the EIS spectra. Our main focus lies in demonstrating the technical feasibility of this concept. In contrast to many other publications, the state estimation was not performed for a single cell but for a number of cells. We show that it is possible to train an ANN with data from a number of cells to estimate the temperatures of other cells of the same cell type. An advantage of the EIS-ANN method is that once the ANN is trained, the temperature estimation is completed within milliseconds since there is no need to solve partial differential equations. In addition, a perspective is given on the possibility of utilizing this method of impedance based state estimation using ANNs to estimate the SoC and the SoHC.

In Section 2, we describe the data acquisition and the architecture of the ANN. In Section 3 we present the results of the temperature, SoC and SoHC estimations and discuss the limitations of the ANN method.

2. Materials and Methods

Lithium ion cells were set to well defined states, where SoHC, SoC and temperature parameters were varied. For each state, an electrochemical impedance measurement (EIS) was performed; every EIS spectrum is related to a defined state. To simulate dynamic working conditions, the EIS measurements were performed as soon as the SoC was adjusted, without relaxation. Since it was not possible to perform EIS measurements for every possible combination of SoHC, SoC and temperature, different series of measurements were performed; each series mainly focused on one state, e.g., temperature. These series of measurements are described in the following subsections. The EIS datasets were separated into a training dataset and a test dataset. The training dataset was used to train the ANN. In the first step, the ANN was trained by using the EIS spectra as input and the related data of the cell states as target values. After the training process, the ANN was evaluated with the test data set. The ANN needed to estimate the related state by itself, and at last the estimated states were compared with the measured states to evaluate the estimation quality.

2.1. Electrochemical Impedance Spectroscopy

EIS involves a non-destructive technique for characterizing electrochemical systems by applying a sinusoidal excitation and measuring the corresponding response, as shown in Figure 3.

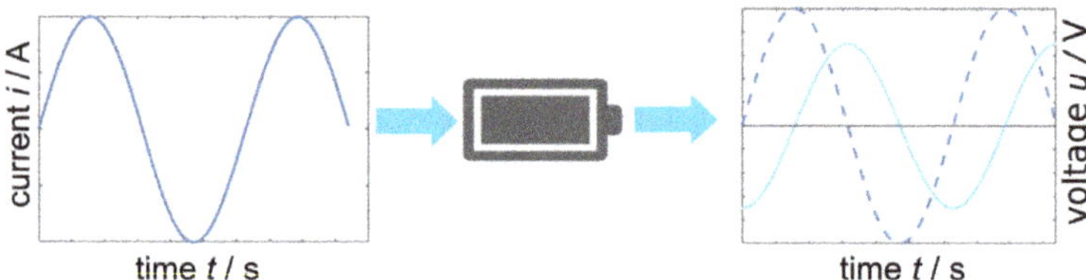

Figure 3. Schematic figure of a sinusoidal current excitation and the voltage response of the system with the phase shift.

The impedance was calculated using the complex voltage, complex current and the phase shift between those values:

$$\underline{Z} = \frac{\underline{u}}{\underline{i}} = R + jX \, ; \tag{1}$$

and from the complex impedance equation, the real part R and the imaginary part jX were calculated. Applying AC currents at different frequencies (usually between 10 kHz and 10 mHz) creates an electrochemical impedance spectrum. The results of an EIS measurement are usually plotted in a Nyquist plot, as shown in Figure 4.

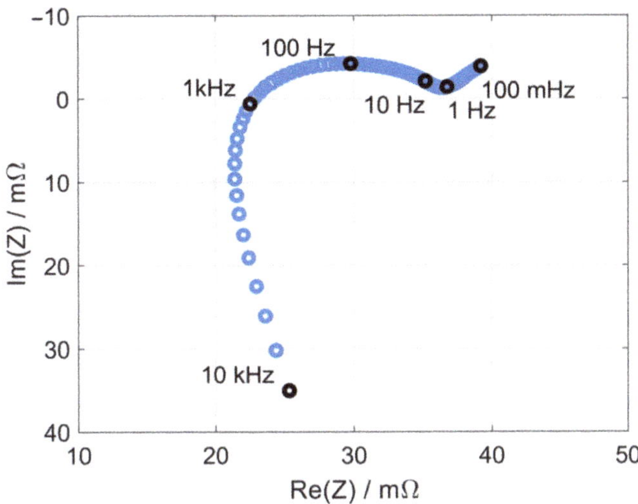

Figure 4. Nyquist plot of a single impedance spectrum for a lithium ion cell.

In order to test the performance in different loading conditions and guarantee charge conservation, the galvanostatic mode was chosen. All EIS measurements were performed using a Gamry reference 3000 AE potentiostat multiplexed on a Basytec CTS. Each internal process can be allocated a characteristic time constant. Therefore, the operating frequency range was varied between 1 Hz and 10 kHz with 15 frequencies per decade (61 frequencies in total). Moreover, the frequency band is limited to the change of charge while running a possibly superimposed current. As a compromise, the AC current was set to a C-rate of 1/10 C for all measurements, in order to guarantee linearity while achieving a good noise-to-signal ratio. The temperature was controlled by a Memmert ICP 110 thermal chamber ($\Delta T = \pm 0.1$ K). The SoC adjustment was performed by charging/discharging with constant current/constant voltage (CC/CV) (current limit: C/30). The selected voltage corresponds to the open circuit potential. For measurements with superimposed DC current during the EIS measurements, the SoC was set to be 5% higher than the SoC of interest by CC/CV. From there, the cell was discharged by CC with the same value of current, which was used for the superimposed EIS measurement. After reaching the target SoC, the EIS measurement was performed without relaxation, to simulate dynamic working conditions.

To make sure that the presented results can be generalized, cells of different types, such as cylindrical high power and prismatic high energy cells, were investigated. To evaluate the impact of the cell to cell variance, at least nine cells per cell type were measured using the same load. The investigated cell types are shown in Table 1.

Table 1. Cells investigated for state estimation.

Cell	Capacity [mAh]	Design
Samsung INR18650-15L1	1500	cylindric
Sony US18650VTC6	3000	cylindric
Panasonic NCA 103450	2350	prismatic

2.1.1. Temperature Estimation

Investigations were performed on the Samsung INR18650-15L1 1500 mAh lithium ion cells. Therefore, 36 cells at different SoHCs (SoHC = state of health regarding actual capacity to nominal capacity C/C_N) were used. The cells were aged by cycling (at $T = 40°C$, constant current charge at 2C, discharge at 3C) using a PEC ACT0550. More detailed information about the SoHC of all 36 cells is shown in Table 2.

Table 2. State of health with respect to nominal capacity (C/C_N). Thirty-six cells were investigated. For each SoHC range, the data of 2 cells were defined as test data; the data of the other cells were used as the training dataset for the neural network (ANN). This guaranteed that each SoHC range would be taken into account. The labels Hx refer to the investigated cells.

SoHC C/C_N	Number of Cells Used for	
	ANN Training	ANN Test, (Cell Label)
98–102%	12	2, (H18, H38)
93–97%	5	2, (H31, H39)
88–92%	4	2, (H1S, H3S)
83–87%	7	2, (H41, H42)

Figure 5 shows the measurement procedure. EIS measurements were performed at 14 temperatures (T in °C: 10, 15, 20, 25, 30, 35, 40, 45, 50, 52, 54, 56, 58, 60), with 5 different SoCs for each temperature (SoC: 90%, 70%, 50%, 30%, 10%).

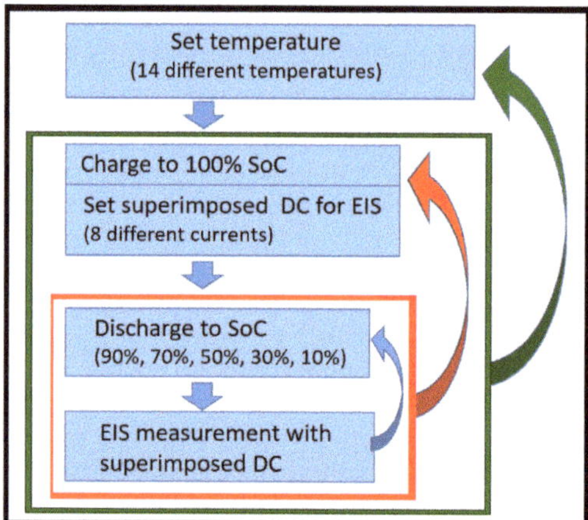

Figure 5. Procedure of data acquisition.

For each temperature and SoC state, EIS measurements were performed with 8 different superimposed DC-currents (C-rate: 0C, −1/4C, −1/2C, −3/4C, −1C, −3/2C, −7/8C, +1C). For the training and testing process, a multi-dimensional dataset was created by performing more than 20,000 EIS-measurements at different states. A relaxation time of 1 hour was used after changing the temperature to ensure that the cells were at the same temperature as the thermal chamber. To simulate a real application, there was no relaxation time between charging/discharging and EIS measurements.

A shorter series of measurements for the temperature estimation were created on the Panasonic NCA 103450 (9 cells SoHC = 100%, C_N = 2350 mAh, prismatic high energy) and on the Sony US18650VTC6 (9 cells SoHC = 100%, C_N = 3000 mAh, cylindrical high energy). The temperature was varied from 10 to 50 °C in 5 K steps. For each temperature, the SoC was varied from 10% to 90% in steps of 20%. For each temperature setting and SoC state, EIS measurements were performed with 4 different superimposed DC-currents (C-rate: 0C, −1/4C, −1/2C, −1C).

2.1.2. State of Charge Estimation

The SoC estimation was performed for all cell types shown in Table 1. The SoHC of every cell was nearly 100%. EIS measurements were performed at 4 different temperatures

(20, 25, 30, and 35 °C). To take the charging/discharging history (hysteresis) of the cells into account, the cells were discharged in steps from 95% to 5% (in total 36 SoCs: 95%, 92%, 90%, 87%, 85%, 82%, . . . , 5%) and then went through the same SoCs for charging (again 36 SoCs). During the EIS measurement, there was no DC applied, but only an AC of C/10.

2.1.3. State of Health Estimation

The SoHC estimation with respect to total capacity (SoHC = C/C_N) was performed only for the Samsung INR18650-15L1 cells. To generate the training data, three lithium ion cells were aged by cycling (at $T = 40$ °C, constant current: charge at 2C, discharge at 3C, 1400 cycles) using a Basytec CTS. After every hundredth cycle, the capacity was determined (charge constant current C/10 constant voltage C/20, discharge constant current C/10) and EIS measurement was performed (at 25 °C, at 70% SoC, 0C DC-current). The data of the investigated cells are described in Section 2.1.1 (at 25 °C, at 70% SoC, 0C DC-current).

2.2. Artificial Neural Network Architecture

The ANN used in this study is based on the MATLAB NN toolbox. The architecture consists of a feed forward architecture with a single output depending on the state of interest (temperature, SoHC, SoC), as shown in Figure 6 .

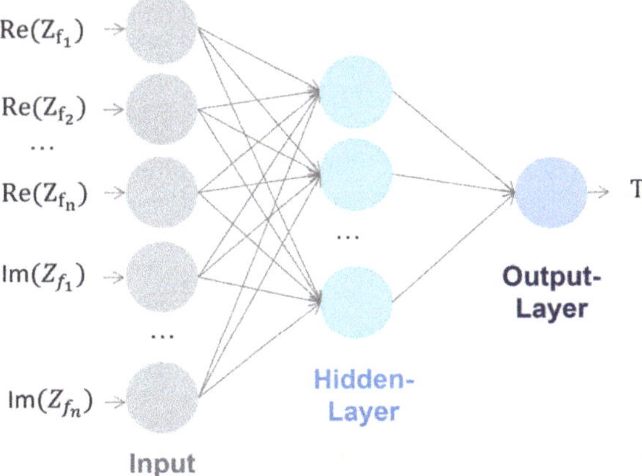

Figure 6. Schematic figure of the used artificial neural network. The real and/or imaginary part of the impedance measurement is used as the input

The used ANN is subcategorized into three main layers: input layer, hidden layer and output layer. Each neuron is connected with a neuron at the following layer in the forward direction. In the context of this work, the number of hidden layers was limited to a single layer. When the number of hidden layers used is more than one, the network is a deep neural network. Within the layer, the number of neurons was varied. In order to determine the optimal number of neurons, a grid search approach was used, where the number of neurons was continually increased until the prediction accuracy was no longer improving without restricting the generalization ability. The number of neurons within the hidden layer is presented in the discussion for each investigated case separately. In the hidden layers, a hyperbolic tangent sigmoid transfer function is used, which is given by:

$$tansig(n) = \frac{2}{(1 + exp(-2n)) - 1}$$

As an input parameter, the real or imaginary portion of the EIS measurements with respect to the frequency domain was used. The training dataset was selected in such a way that the generalization of the battery cell variance can be verified and validated. The bias values and weights were updated according to two different optimization strategies: Bayesian regularization backpropagation optimization (BRBP) and Levenberg–Marquardt backpropagation optimization (LMBP). The prediction accuracy was assessed using the root mean square error and the coefficient of determination. The code was developed based on ANN toolbox.

3. Results

The following discussion shows the results of the state estimations of the temperature, SoC and SoHC predictions. Note that the test dataset was separated from the training dataset before training the neural network in order to ensure that the test data were completely unknown to the ANN.

3.1. Temperature Estimation

Figure 7 presents more than 4000 temperature estimations by the neural network for the EIS data of eight Samsung INR18650 15L1 cells at five different SoCs, eight superimposed DC currents during the EIS measurements and different SoHCs. The SoHC of each cell is shown in Table 2. For each investigated SoHC, two cells were selected for testing. By including various DC currents, SoHCs and SoCs, the realistic performance of a cell was simulated. The overall root mean square error (RMSE) was less than 1 K. The maximum RMSE for a single temperature estimation was about 5 K.

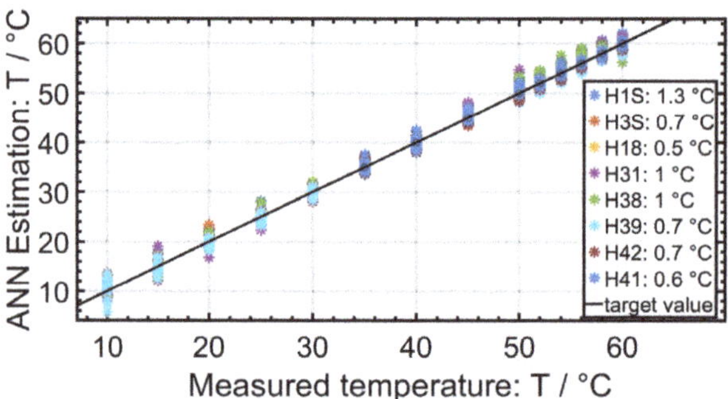

Figure 7. More than 4000 temperature estimations by a neural network using EIS data as input are presented. The estimated temperature is plotted over the measured temperature. The black line is a guide to the eye for the target values. The overall RMSE for 8 Samsung INR18650-15L1 cells was less than 1 K. The legend shows the RMSE for each cell.

Only the real part of the impedance was used as an input parameter, since the influence of the temperature was much greater on the real part than on the imaginary part, as shown in Figure 1. A combination of real and imaginary parts and only using imaginary part as input were investigated. However, as expected, the temperature estimation became more inaccurate when using the imaginary part. For the presented results, the actual SoHCs, the SoCs and the applied DC current during the EIS measurement were used as input parameters. The ANN was able to estimate the cell temperature without additional input parameters. However, the time required to train the ANN increased significantly, and the overall RMSE increased to 1.5 K. During the measurements, there were no temperature sensors attached to the cells. The temperature was taken directly from the thermal chamber. The self heating effect in Samsung 18650-15L cells by applying DC current was not taken into account for the state estimation. When using temperature sensors within the cells, center measurements showed that the temperature difference between the center and the surface was less than 2 K. For our proof of concept study, this temperature discrepancy was assumed to be negligible, since every measurement was performed in the same way. The results show that in general the ANN was able to estimate the temperature from a corresponding EIS spectrum.

Figure 8 shows the evolution of the ANN during training. After 230 epochs, the RMSE reached its lowest value and was stabilized. The RMSE evolution was used to find a suitable configuration of the ANN. For the presented results, only one hidden layer consisting of 11 neurons within the hidden layer was used in a Bayesian regularization-backpropagation neural network. Using more than 11 neurons in the hidden layer had a tendency to overfit.

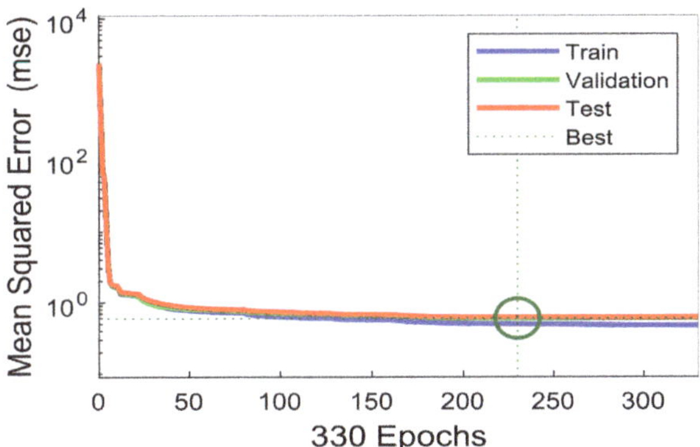

Figure 8. RMSE evolution during training epochs of the temperature estimation of the Samsung INR18650-15L.

Additional investigations were performed on prismatic Panasonic NCA 103450 high energy cells and cylindrical Sony US18650VTC6 high energy cells to show that the presented method is independent of cell geometry and cell type (high energy/high power).

Figure 9a presents the results for the Panasonic NCA 103450 cells with an overall RMSE of 0.7 K, and Figure 9b shows the results for the Sony VTC6 cells with an overall RMSE of 0.5 K. Each figure shows 360 temperature estimations at different SoCs and different applied superimposed DC currents. For both cells, a Bayesian regularization-backpropagation neural network with one hidden layer consisting of five neurons was used. As input parameters, both the real and the imaginary parts of the impedance were fed to the ANN; the information about SoHC, applied DC current and SoC were restrained.

The SoHC of every Sony and Panasonic cell was about 100%. The temperature estimations for different SoHCs were performed only for the Samsung cells.

Nevertheless, we showed that the temperature estimation by an ANN using EIS data can be realized for different cell types. However, it is necessary to create an individual ANN for each cell type with individual hyperparameters. To finally generalize this method, more cells with different aging profiles and higher DC currents need to be investigated. Due to the self heating effect of the cells through high currents, temperature sensors should be installed within the cells to measure the exact temperature in real time.

The superimposed DC current affects the EIS spectrum, especially at low temperatures. Therefore, the presented ANN method is a powerful tool, especially at high temperatures where the influence of the DC current is reduced. This makes it perfectly suitable for real-life applications. In comparison to other sensorless temperature estimation methods, the main advantages are that there is no need for storing time series data, and that the computational effort is reduced, since there is no need to solve complicated equations, such as partial differential equations. It is suitable for different cell types, and it also takes a superimposed DC current and actual SoHC into account.

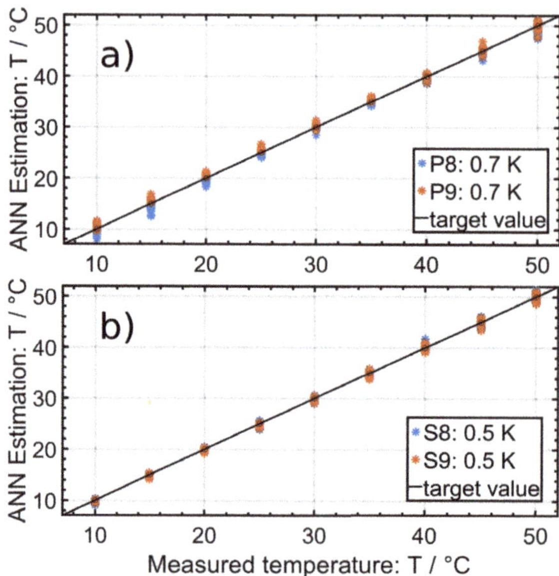

Figure 9. Temperature estimations for (**a**) Panasonic NCA 103450 with an RMSE of 0.7 K, and (**b**) Sony US18650VTC6 with an RMSE of 0.5 K. In both figures, 360 temperature estimations are shown. The estimated temperature is plotted over the measured temperature. The black line is a guide to the eye of the target values. The legend shows the RMSE for each cell.

The use of EIS data greatly improved the estimation accuracy. Various internal processes in the lithium ion cell show different temperature dependencies. These processes can be allocated to various frequency domains measured via EIS [24]. For some electrochemical processes in lithium ion cells, the influence of the temperature predominates over the influence of the cell to cell variance. The estimation accuracy can be further improved by adding impedance data at various frequency domains in addition to the cell resistance.

3.2. State of Charge Estimation

The state estimation methodology using EIS data and ANN was investigated for the estimation of the SoC. Figure 6 shows the results for the Panasonic NCA 103450 cells. Since the influence of the SoC on the impedance varies, it was necessary to split the SoC into area 1 from 5% to 45%, as shown in Figure 10a, for which the RMSE was 1.9%; and area 2 from 48% to 95%, as shown in Figure 10b, for which the RMSE was 2.2%. In both cases, a Bayesian regularization-backpropagation neural network with one hidden layer consisting of four neurons was used. For the lower SoCs, only the imaginary part was used as input parameter, and at higher SoCs the real and the imaginary parts were used. Using only one network for the whole SoC area increased the RMSE significantly.

The RMSE shows higher deviations in the mid SoC range (30–70%), where the EIS data mostly overlap. This makes it harder for the ANN to distinguish SoCs.

For the Sony VTC6 cells it was also necessary to split the SoC in two areas. For the lower SoCs from 5% to 35% the RMSE was 3.1%, as shown in Figure 11a. Figure 11b presents the results from 40% to 95% with an overall RMSE of 2%. In both cases a Levenberg–Marquardt backpropagation artificial neural network with one hidden layer and five neurons was used. Only the real part of the impedance was used as an input parameter.

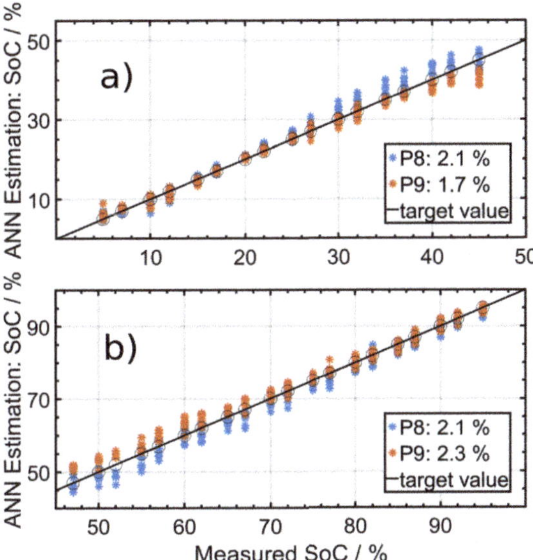

Figure 10. SoC estimation for Panasonic NCA 103450 cells with an RMSE of about 2%. The SoC was split into (**a**) 5% to 45% and (**b**) 47% to 95% and estimated by individual neural networks. Overall, almost 600 SoC estimations are shown. The estimated SoC is plotted over the measured SoC. The black line is a guide to the eye of the target values. The legend shows the RMSE for each cell.

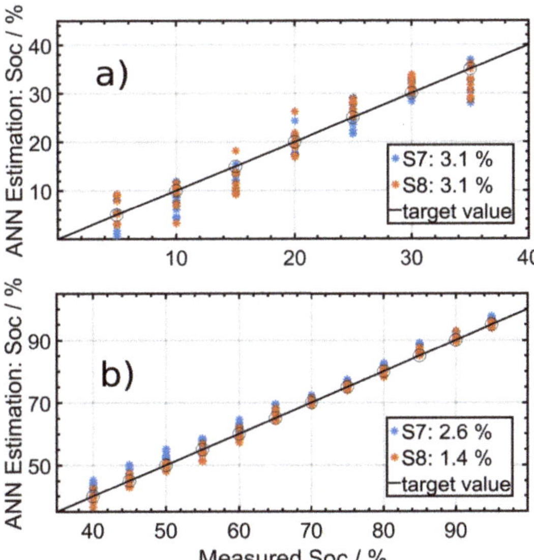

Figure 11. SoC estimation for Sony US18650VTC6 cells with an RMSE of about 3%. The SoC was split into (**a**) 5% to 35% and (**b**) 40% to 95% and estimated by individual neural networks. Overall, more than 300 SoC estimations are shown. The estimated SoC is plotted over the measured SoC. The black line is a guide to the eye of the target values. The legend shows the RMSE for each cell.

The RMSE was increased for the estimation of lower SoCs because the impedance varies only slightly with low SoC.

The SoC estimation for the Samsung 15L cells is not shown, since the best overall RMSE was about 8%, with a maximum estimation discrepancy up to 20%, which is not sufficient for any application. This was caused by the influence of the SoC on the impedance spectra being in the same order of magnitude as the influence of the cell to cell variance. The presented results for the SoC estimation clearly show that it is necessary to create an individual ANN for each cell type. If there is a clear dependency between the SoC and the EIS spectra and only a little cell to cell variance, the signal to noise ratio is big enough to allow the state estimation, as shown for the Panasonic NCA 103450 cells.

In general, SoC estimation is much more complex than temperature estimation. The reason is the behavior of the EIS spectra depends on the state. As shown in Figure 1, an increase in temperature causes a decrease in impedance. The SoC dependency of the EIS spectra varies among different cell types. For some cell types, an increase in SoC causes a decrease in the value of impedance at a low SoC or an increase in value at a high SoC. For such a case, several ANNs for different SoC ranges are required.

The advantages of this method compared to other methods are, again, the little computational effort, since no partial differential equations to be solved and no error-prone parameterization needs to be performed. Furthermore, there is no need for storing and handling time series data. The utilizable capacity of a cell depends strongly on the cell temperature. Therefore, the SoC varies with the temperature. The SoC was calculated by the ANN using an EIS spectrum that is characteristic of the measured state.

3.3. State of Health Estimation

Another interesting application is the estimation of the SoHC with respect to the capacity (SoHC). Therefore, three cells were aged and characterized (EIS and capacity) and used to train the ANN. The cells which were aged and used for the investigation of the temperature estimation were characterized afterwards and used to test the SoHC estimation. Figure 12 shows the estimated cell SoHC and the measured SoHC. The RMSE for every SoHC estimation was below 2%.

A Bayesian regularization-backpropagation neural network with one hidden layer consisting of four neurons was used. Both the real and the imaginary part of the impedance were used as input parameters. We were unable to estimate the SoHC at the very beginning of life because the shape and the value of the EIS spectrum varies strongly for the first cycles. The impedance spectrum first decreases and than increases. After a number of cycles, stabilization of impedance growth was achieved, as shown in Figure 2. Only the data of cells which performed 100 or more cycles were used. Since the cells that were used to collect the training data were characterized only for 1400 cycles, data from cells with more cycles could not be used for testing, because the ANN is not able to extrapolate. The aging profiles of the cells that were used to collect training data and the cells for the test data were partially comparable. After aging, the testing cells experienced further aging during measurements at different temperatures and currents. Nevertheless, to generalize this system it will be necessary to investigate the presented state estimation method with data from different aging profiles. The advantage of the SoHC estimation method compared to other algorithms is the ability to estimate the total utilizable capacity from an EIS spectrum within a few seconds. Unlike the current pulse method where only one data point is determined, the impedance spectrum offers many data points belonging to a single state. That increases the estimation accuracy and reduces other influences, such as that of contact resistances.

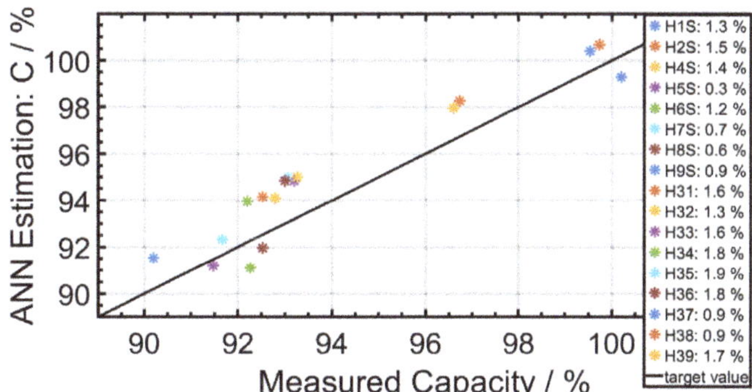

Figure 12. SoHC estimates for Samsung INR18650-15L1 cells by a neural network using EIS data as input are presented. Here the SoHC is defined as the actual capacity related to the nominal capacity. The estimated SoHC is plotted over the measured SoHC. The black line is a guide to the eye of the target values. The overall RMSE was less than 2%. The legend shows the RMSE for each cell.

3.4. Further Discussion

We have developed a sensorless ANN-method based on EIS data to estimate the temperature of a lithium ion cell. The temperature estimation was performed with different SoCs, SoHCs and discharge currents. Furthermore, we have given a perspective on the possibility of using the presented method to estimate the SoC and the SoHC. One of the biggest advantages of the presented method is that a well-known ANN with a very simple architecture which requires little computational effort is able to estimate the cell temperature successfully. This fact makes it even more interesting and practical for industrial applications. The ANN was trained within minutes for each system. The time needed to train the ANN depends on the availability of supplementary input data, such as SOC, SOH or applied DC current. Once the ANN is trained, the calculations for the state estimation by the ANN are performed within milliseconds. The EIS measurement from 1 Hz to 10 kHz was performed in less than 1 min. However, we predict that the measurement time could be reduced to milliseconds by selecting suitable frequencies. The calculations for the state estimation by the ANN were performed in less than a second.

Further, we showed that the presented model is independent of the cell geometry by investigating cylindrical and prismatic cells. The focus of this study lay in the investigation of 18,650 lithium ion cells. To generalize the presented method, we will investigate larger cells with higher capacity in future work. Furthermore, we will use temperature sensors in the cell core for data acquisition and investigate connected cells in battery modules. Due to its data-driven nature, we suggest that it is possible to adapt the model to every other cell chemistry, as long as there is a strong correlation between the EIS spectrum and the investigated state. We were able to achieve reasonable results with LFP cells. As our data-driven model is easily applicable to other systems, it is attractive for practical applications, since cell manufacturers usually do not reveal exact cell chemistry.

4. Conclusions

In this work we presented a sensorless method for predicting the temperatures of lithium ion cells that uses ANNs which take electrochemical impedance spectra as input data. Investigation were performed on Samsung INR18650-15L1, Sony US18650VTC6 and Panasonic NCA 103450 cells. To simulate real applications, the SoC was varied; a superimposed DC current during the EIS measurement was applied; and for every cell type at least nine cells were investigated to include the cell to cell variance. In addition, Samsung 15L cells with different SoHCs were investigated. The RMSE for all temperature estimations were around 1 K, which makes the presented method attractive for practical applications. SoC estimation was also investigated likewise. For the Sony VTC6 and the Panasonic NCA 103450 cells, the RMSE was about 3%. Different cell temperatures during the EIS measurement were taken into account. The SoC estimation for the Samsung 15L cells was not successful, with an RMSE of above 8%. In this case the influence of the SoC on the impedance spectrum was of the same order of magnitude as the influence of the cell to cell variance. Therefore, for the SoC estimation, it is necessary to investigate each cell system individually for its applicability. At last, the ANN was applied to estimate the SoHCs of Samsung 15L cells. Since the estimation errors for all cells were below 2%, this seams to be a feasible use of the method as well. However further investigations on cells with different aging profiles are necessary to give a definitive evaluation of the suitability. Its advantages compared to other temperature estimation methods are that there is no need to fit a battery model to the data, and no differential equation needs to be solved. Furthermore, the ANN needs only a single EIS spectrum to estimate the cell temperature. There is no need for handling time series data. The presented prediction method seems to be a promising way to estimate the inner cell temperature with high accuracy in a short time period, as little effort regarding measurements and calculations is required. Further work is suggested to investigate the ability of the neural network to estimating the temperatures of cells within electrical circuits. Furthermore, a reduction in the number of input parameters will be investigated to improve this method by reducing and simplifying the computational effort and the measurement time.

Author Contributions: Conceptualization, M.S., J.P.-B., M.K. and K.P.B.; methodology, M.S., J.P.-B., M.K. and K.P.B.; investigation, M.S., J.P.-B.; writing—original draft preparation, M.S. and J.P.-B.; writing—review and editing, M.K. and K.P.B.; visualization, M.S.; supervision, K.P.B.; project administration, M.S.; funding acquisition, K.P.B. All authors have read and agreed to the published version of the manuscript.

Funding: This research was funded by Robert Bosch GmbH due to the Bosch Promotionskolleg.

Institutional Review Board Statement: Not applicable.

Informed Consent Statement: Not applicable.

Data Availability Statement: Not applicable.

Conflicts of Interest: The authors declare no conflict of interest. The funders had no role in the design of the study; in the collection, analyses, or interpretation of data; in the writing of the manuscript, or in the decision to publish the results.

References

1. Ma, S.; Jiang, M.; Tao, P.; Song, C.; Wu, J.; Wang, J.; Deng, T.; Shang, W. Temperature effect and thermal impact in lithium-ion batteries: A review. *Prog. Nat. Sci. Mater. Int.* **2018**, *28*, 653–666. [CrossRef]
2. Wang, Q.; Ping, P.; Zhao, X.; Chu, G.; Sun, J.; Chen, C. Thermal runaway caused fire and explosion of lithium ion battery. *J. Power Sources* **2012**, *208*, 210–224. [CrossRef]
3. Raijmakers, L.; Danilov, D.L.; Eichel, R.A.; Notten, P. A review on various temperature-indication methods for Li-ion batteries. *Appl. Energy* **2019**, *240*, 918–945. [CrossRef]
4. Richardson, R.R.; Howey, D.A. Sensorless Battery Internal Temperature Estimation using a Kalman Filter with Impedance Measurement. *IEEE Trans. Sustain. Energy* **2015**, *6*, 1190–1199. [CrossRef]
5. Surya, S.; Marcis, V.; Williamson, S. Core Temperature Estimation for a Lithium ion 18650 Cell. *Energies* **2021**, *14*, 87. [CrossRef]
6. Liu, K.; Li, K.; Peng, Q.; Zhang, C. A brief review on key technologies in the battery management system of electric vehicles. *Front. Mech. Eng.* **2019**, *14*, 47–64. [CrossRef]
7. Lelie, M.; Braun, T.; Knips, M.; Nordmann, H.; Ringbeck, F.; Zappen, H.; Sauer, D. Battery Management System Hardware Concepts: An Overview. *Appl. Sci.* **2018**, *8*, 534. [CrossRef]
8. Richardson, R.R.; Ireland, P.T.; Howey, D.A. Battery internal temperature estimation by combined impedance and surface temperature measurement. *J. Power Sources* **2014**, *265*, 254–261. [CrossRef]
9. Wang, L.; Lu, D.; Song, M.; Zhao, X.; Li, G. Instantaneous estimation of internal temperature in lithium-ion battery by impedance measurement. *Int. J. Energy Res.* **2020**, *44*, 3082–3097. [CrossRef]
10. Waag, W.; Käbitz, S.; Sauer, D.U. Experimental investigation of the lithium-ion battery impedance characteristic at various conditions and aging states and its influence on the application. *Appl. Energy* **2013**, *102*, 885–897. [CrossRef]
11. Schmidt, J.P.; Arnold, S.; Loges, A.; Werner, D.; Wetzel, T.; Ivers-Tiffée, E. Measurement of the internal cell temperature via impedance: Evaluation and application of a new method. *J. Power Sources* **2013**, *243*, 110–117. [CrossRef]

12. Srinivasan, R.; Carkhuff, B.G.; Butler, M.H.; Baisden, A.C. Instantaneous measurement of the internal temperature in lithium-ion rechargeable cells. *Electrochim. Acta* **2011**, *56*, 6198–6204. [CrossRef]
13. Spinner, N.S.; Love, C.T.; Rose-Pehrsson, S.L.; Tuttle, S.G. Expanding the Operational Limits of the Single-Point Impedance Diagnostic for Internal Temperature Monitoring of Lithium-ion Batteries. *Electrochim. Acta* **2015**, *174*, 488–493, [CrossRef]
14. Raijmakers, L.; Danilov, D.L.; Van Lammeren, J.; Lammers, M.; Notten, P. Sensorless battery temperature measurements based on electrochemical impedance spectroscopy. *J. Power Sources* **2014**, *247*, 539–544. [CrossRef]
15. Beelen, H.; Raijmakers, L.; Donkers, M.; Notten, P.; Bergveld, H.J. A comparison and accuracy analysis of impedance-based temperature estimation methods for Li-ion batteries. *Appl. Energy* **2016**, *175*, 128–140. [CrossRef]
16. Feng, F.; Teng, S.; Liu, K.; Xie, J.; Xie, Y.; Liu, B.; Li, K. Co-estimation of lithium-ion battery state of charge and state of temperature based on a hybrid electrochemical-thermal-neural-network model. *J. Power Sources* **2020**, *455*, 227935. [CrossRef]
17. Hasan, M.M.; Ali Pourmousavi, S.; Jahanbani Ardakani, A.; Saha, T.K. A data-driven approach to estimate battery cell temperature using a nonlinear autoregressive exogenous neural network model. *J. Energy Storage* **2020**, *32*, 101879. [CrossRef]
18. Eddahech, A.; Briat, O.; Bertrand, N.; Delétage, J.Y.; Vinassa, J.M. Behavior and state-of-health monitoring of Li-ion batteries using impedance spectroscopy and recurrent neural networks. *Int. J. Electr. Power Energy Syst.* **2012**, *42*, 487–494. [CrossRef]
19. Yang, D.; Wang, Y.; Pan, R.; Chen, R.; Chen, Z. A Neural Network Based State-of-Health Estimation of Lithium-ion Battery in Electric Vehicles. *Energy Procedia* **2017**, *105*, 2059–2064. [CrossRef]
20. You, G.W.; Park, S.; Oh, D. Diagnosis of Electric Vehicle Batteries Using Recurrent Neural Networks. *IEEE Trans. Ind. Electron.* **2017**, *64*, 4885–4893. [CrossRef]
21. Tong, S.; Lacap, J.H.; Park, J.W. Battery state of charge estimation using a load-classifying neural network. *J. Energy Storage* **2016**, *7*, 236–243. [CrossRef]
22. Khumprom, P.; Yodo, N. A Data-Driven Predictive Prognostic Model for Lithium-Ion Batteries based on a Deep Learning Algorithm. *Energies* **2019**, *12*, 660. [CrossRef]
23. Messing, M.; Shoa, T.; Ahmed, R.; Habibi, S. Battery SoC Estimation from EIS using Neural Nets. In Proceedings of the 2020 IEEE Transportation Electrification Conference & Expo (ITEC), IEEE, Chicago, IL, USA, 23–26 June 2020; pp. 588–593.
24. Jossen, A. Fundamentals of battery dynamics. *J. Power Sources* **2006**, *154*, 530–538. [CrossRef]

Article

Implementation of Battery Digital Twin: Approach, Functionalities and Benefits

Soumya Singh [1,*]**, Max Weeber** [1] **and Kai Peter Birke** [1,2]

[1] Fraunhofer Institute for Manufacturing Engineering and Automation IPA, Nobelstr. 12, 70569 Stuttgart, Germany; max.weeber@ipa.fraunhofer.de (M.W.); Peter.Birke@ipv.uni-stuttgart.de (K.P.B.)

[2] Institute for Photovoltaics, Electrical Energy Storage Systems, University of Stuttgart, Pfaffenwaldring 47, 70569 Stuttgart, Germany

* Correspondence: soumya.singh@ipa.fraunhofer.de

Abstract: The concept of Digital Twin (DT) is widely explored in literature for different application fields because it promises to reduce design time, enable design and operation optimization, improve after-sales services and reduce overall expenses. While the perceived benefits strongly encourage the use of DT, in the battery industry a consistent implementation approach and quantitative assessment of adapting a battery DT is missing. This paper is a part of an ongoing study that investigates the DT functionalities and quantifies the DT-attributes across the life cycles phases of a battery system. The critical question is whether battery DT is a practical and realistic solution to meeting the growing challenges of the battery industry, such as degradation evaluation, usage optimization, manufacturing inconsistencies or second-life application possibility. Within the scope of this paper, a consistent approach of DT implementation for battery cells is presented, and the main functions of the approach are tested on a Doyle-Fuller-Newman model. In essence, a battery DT can offer improved representation, performance estimation, and behavioral predictions based on real-world data along with the integration of battery life cycle attributes. Hence, this paper identifies the efforts for implementing a battery DT and provides the quantification attribute for future academic or industrial research.

Keywords: digital twin; battery model; battery management system; Doyle-Fuller-Newman model; equivalent circuit model; parameter estimation

Citation: Singh, S.; Weeber, M.; Birke, K.P. Implementation of Battery Digital Twin: Approach, Functionalities and Benefits. *Batteries* **2021**, *7*, 78. https://doi.org/10.3390/batteries7040078

Academic Editors: Carlos Ziebert and Seung-Wan Song

Received: 30 July 2021
Accepted: 15 November 2021
Published: 16 November 2021

Publisher's Note: MDPI stays neutral with regard to jurisdictional claims in published maps and institutional affiliations.

1. Introduction

Digital Twin (DT) is a virtual dynamic model of a system, process, or service, with real-world data interactions that facilitate improved system analysis and comprehensive representation [1]. NASA defined DT as an integrated multi-physics, multi-scale simulation of a system that uses the best available physical models, sensor data, and historical data to mirror the life of its physical twin [2]. The applications of DT are: (1) to simulate the behavior of the physical twin before its usage, where even without the benefit of continuous sensor updates, the DT can study the effects of various parameters, determine the various anomalies and validate the degradation mitigation strategies; (2) to simulate the system behavior during operation, through the continuous update of actual load, temperature, and other environmental factors, as input to the models, enabling continuous predictions for the physical twin; (3) to perform diagnostics in the event of a fault or damage; (4) to serve as a platform where the effects of parameter modifications, not considered during the design phase, can be studied. In practice, DT implementation involves allocating real-world data to the virtual model or platform. Boschert et al. [3] identified that simulation would become the primary tool for decision support once a DT is fully integrated. Likewise, Kunath et al. [4] summarized the three main functions of a DT: Prediction—execution of studies ahead of the system run; Safety—monitoring and control of the system state in terms of a continuous prediction during the system run; Diagnosis—analysis of unpredicted disturbances during the system run.

By definition, a **model** is a simplified abstraction of the structure or processes that define a real system. In that sense, models do not aim to replicate the original system in intensive detail [5]. The idea of moving a digital model closer to the real system is in fact a basic rationale for building computer models. Some models are extreme simplifications of the real system, while some are much closer to the real system, through multiscale simulations and interdisciplinary collaborations. The difference between a DT and a simulation model has been discussed previously [1,6], but it is essential to address the features that qualify a model as a DT: (1) model of the product—physical or data-driven; (2) evolving set of real-world data about/related to the product; (3) method of adjusting the model in accordance with the data. According to [7], the DT evaluation framework consists of four metrics, autonomy, intelligence, learning, and fidelity.

Some articles [8,9] describe seemingly similar concepts to DTs, such as digital shadow, virtual model, product avatar and digital thread, but they do not necessarily indicate the complete concept of DT, but rather fragments of overlapping functions. DT implementation is often mapped to IoT (Internet of Things) devices and CPS (Cyber-Physical Systems) [10,11], due to its high dependence on a compatible mode of data acquisition. A DT can have single or multiple stakeholders and may make use of 3D simulations, IoT devices, 4G and 5G networks, blockchain, edge computing, cloud computing, and artificial intelligence. Depending on the complexity, a DT may have access to past and present operational data along with predictive capabilities.

A multitude of literature [12,13] has been published, defining and characterizing the concept of DT in a variety of domains. However, missing from the literature is a consistent view on what the DT is and how the concept is helping to meet the challenges of the many use-cases to which it is being associated as a solution approach [14]. Some articles [15–17] suggest the methodologies, frameworks, and interpretation of DT for specific use cases. While this may help understand the system boundaries, it also leads to inconsistent ideas of the requirements of DT implementation and thus generating the risk of diluting the concept. The potential costs, infrastructure challenges, clarity of return on investments for product/process DT are not transparent. Without substantial effort to describe and quantify the DT benefits, it is challenging even to suggest that the concept itself may be the most appropriate solution to the challenges faced by each particular industry [14].

DT is generally perceived with the core idea that it is a model that can replicate the behavior of an existing system through the acquisition of real system data. This highlights the importance and completeness of the phrase "replicate the behavior". To which extent is the replication satisfactory? DTs at first may appear to be a replica but replicating every behavioral aspect might not necessarily be realistic. For example, in battery DTs, it is unnecessary to digitally replicate each of the molecular, fluid, and structural behavior of each cell component. DTs need not attempt to mirror everything about the original system [5]. Hence, the reality of an exhaustive high-fidelity DT, which replicates every aspect of the physical system and maximizes services while minimizing expense and technical difficulty of implementation, is ambiguous. The level of model fidelity, cost, and effort for implementation is subject to limitations which will vary depending on the use case and its applications. Furthermore, the idea of DTs across the lifecycle is not yet fully understood. The number of DTs required across the entire lifecycle, or the transitions and interactions between the software components of the DT are yet to be explored.

Therefore, the stakeholders in industries or academia looking to invest in or develop a DT face the following challenges:

- Limited use cases and implementation results available to learn from others;
- No clear guidance on how much to budget;
- Difficult to know where to start to get value quickly;
- Initiatives that are misleadingly branded as "Digital Twin";
- Limited know-how.

The hypothesis that DT would be a driver for product lifecycle management and smart manufacturing in the future needs to be tested. This paper is a part of an ongoing study that investigates the DT functionalities and quantifies the DT attributes across the life cycle phases of a battery system. Essentially, the pressing question is whether battery DT is a practical and realistic solution to meeting the growing challenges of the battery industry, such as degradation evaluation, usage optimization, manufacturing inconsistencies, second-life application possibility, etc. Within the scope of this paper, the concept of DT will be explicitly explored for battery cells and the functionalities it can offer during the operation and end-of-life (EoL) phases. Research on battery DT has already gained much popularity, and with this paper, we aim to provide a consistent approach to DT implementation for battery cells. In doing so, we identify the efforts for implementing a battery DT. The contributions of this paper are as follows: (1) literature-analysis of the potential functionalities of a battery DT during operation and EoL; (2) battery DT implementation approach; (3) KPIs (Key Performance Indicators) to quantify the value-add of battery DTs; (4) testing the main functions of the implementation approach on a DFN model.

This paper is structured as follows: Section 2 reviews past literature followed by a discussion about battery DT functionalities during operation and EoL. Section 3 outlines the approach of battery DT implementation with subsections elaborating each step. Section 4 demonstrates the results of the application of the described approach in Python. Lastly, Sections 5 and 6 is a discussion of the contributions of the paper and its future scope.

2. Battery DT Functionalities during Operation and End-of-Life

The number of literature specific to DTs has increased drastically in the last decade [8,18]. The topic is explored in various domains, such as product optimization, production planning/control, layout planning, maintenance, or product lifecycle. However, a microscopic look into the implementation of product DT, i.e., battery DTs has become prevalent since the past decade with increased utilization of IoT devices, CPS, and cloud-based services. Sometimes battery model implementation learns from real-time operational data to evaluate the battery states but is not necessarily defined as a battery DT [19]. So, is it crucial to even understand if a model is, in fact, a DT? Generally, at the initial stage of product development, it is not an essential requirement. However, at the business level i.e., in order to draw profits and innovation in the existing business model, a DT can facilitate effective R&D.

In the context of battery systems, it is uncertain whether battery DT insinuates a single cell DT, module-level DT, or pack-level DT. Currently, DTs are the result of custom technical solutions that are difficult to scale [20]. The scalability of battery DTs depends on the extrapolation of cell behavior through physical, physics-based, or data-driven battery models. For this purpose, the term battery DT referred to in this paper implies the DT of a battery cell. However, module and pack-level battery DTs are worth pursuing in the later stages.

The literature review in this paper takes only those published literature into account, which consists of practical implementations of battery DTs with defined DT-functionalities and implementation methods. The focus is on reviewing the latest literature on DTs concerning applications in the battery industry, published in the past 5 years (starting from 2016 to 2021). With Google Scholar as the research literature database and the following logical expression of keywords: ("digital twin") AND ("battery") AND ("lithium-ion"), updates until July 2021, the total number of resulting articles were 392 (39 google scholar pages), among which several articles had to be excluded due to the following reasons:

- Some articles only mentioned battery DTs as a possible application
- Some of them did not explain the architecture to support battery DTs
- Others were only theoretical articles.

However, in combination with advanced search for the keywords ("battery digital twin") or ("digital twin" AND "battery") in the title, the literature was filtered down to

9 articles. As a result, Table 1 elaborates the available literature, their reported method of DT implementation, and the corresponding DT functionality.

Table 1. Literature review of DT-implementations relevant for battery system.

Reference	Implementation Method [1]	DT Functionality
[21]	HI and LSTM algorithm	Estimation of battery's actual discharge capacity
[22]	Cloud BMS with AEHF-based SOC estimation algorithm and PSO-based SOH estimation algorithm	Estimation of SOC, SOH, capacity fade, power fade
[23]	On-board diagnosis to cloud environment; ECM model parameter fitting, curve fitting and SOC-OCV curve	SOC, capacity, internal resistance, SOH-R, SOH-C
[24]	Visual software in LabVIEW; ECM with SVM and filter algorithms	DT platform for spacecraft lithium-ion battery pack degradation assessment; SOC estimation
[25]	Cloud connected BMS; electric-thermal model and empirical ageing model	Cell voltage and temperature
[26]	ECM and EFK algorithm	SOC estimation
[27]	Review paper on battery DT	Battery DT framework and its cyber-physical elements
[28]	Offline—Regression model using sparse-Proper Generalized Decomposition (s-PGD); Online—Dynamic Mode Decomposition technique	Cell voltage, anode/cathode bulk SOC, anode/cathode surface SOC
[29]	Linking reduced order model with ECM in Ansys Twin Builder	Real-time temperature of the battery pack at different locations; What-if scenarios for root cause analysis

[1] HI—Health Indicator; LSTM—Long Short-Term Memory; BMS—Battery Management System; AEHF—adaptive extended H-infinity filter; PSO—Particle Swarm Optimization; SOC—State of Charge; SOH—State of Health; OCV—Open Circuit Voltage; ECM—Equivalent Circuit Model; EKF—Extended Kalman Filter; SVM—Support Vector Machines.

While some of the references identified in Table 1 are pioneering attempts to map real-time battery data to the battery models through cloud services, the others have yet again used simulated driving cycle data to validate the state estimation algorithms. This procedure of using drive cycle simulations such as WLTP (Worldwide Harmonized Light Duty Vehicles Test Procedure) or UDDS (Urban Dynamometer Driving Schedule) for validation of battery state estimation algorithms is a state of the art state estimation approach [30]. Nevertheless, this is not ideal for designing battery DT because simulated driving cycles can only validate the algorithm accuracy of static models while battery DTs are dynamic. Additionally, a common observation among all the DT functionalities in Table 1 is that they all have approximately the same output, i.e., SOC, SOH, internal resistance, or capacity. Evaluation of the battery state through these output variables has already been an established requirement from electrical models, electric-thermal models, electrochemical models, and until recently, data-driven (or NN) battery models. In this context, it is worth noting the conundrums that this raises regarding the argument about the need for a battery DT only for performing state estimations. Other than the fact that if implemented correctly, a battery DT should deliver those output variables to the users, developers or testers in real-time, there are no significant utilities that only a battery DT can implement. So, what additional value-add does a battery DT contribute? In an attempt to answer this, the following 2 aspects are now elaborated:

- Battery DT influence on life cycle phases;
- Current BMS functionalities.

To achieve competitiveness with the internal combustion engine, the key requirement for battery development is high driving ranges, low-charging times, and low battery pack cost. The performance indicators of batteries during **usage** are characterized by cost, specific energy (Wh kg^{-1}), energy density (Wh L^{-1}), specific power (W kg^{-1}), and power density (W L^{-1}), and charging time (=fast charging ability) [31]. While a DT cannot take the responsibility of improving the energy density or specific power of a battery, it can significantly aid the design optimization process and EoL assessment process. Sensing the battery data and uploading that to a storage server gives the opportunity to easily access the battery data and create learning models, which directly guide the **product design, and optimization process** [27]. The battery data storage platform stores the design and usage history, which supports **behavioral integration in consequent life cycle phases** and simplifies the **prediction of the remaining useful life** (RUL) during operation and also at EoL for second life assessment [32].

The **Battery Management System (BMS)** is the central element for protecting, monitoring, and controlling the battery-powered system by ensuring safety, efficiency, and reliability [33]. BMS measurements are performed for cell voltages, pack current, pack voltage, and pack temperature and it usually uses these measurements to estimate SOC, SOH, DOD (Depth of Discharge) [34]. Battery DT requires the onboard-BMS to work together with the battery data storage platform.

The potential functionalities of a battery DT in combination with an onboard-BMS are identified in the literature. Identifying the stress factors from the time-series measurement data and calculating its effect on the model parameters facilitates **evaluating battery aging indicators** during operation [25]. Besides, the model update integrated with the charging data enables a battery DT to maximize the optimization objective and select the best parameters for an **optimal charging strategy** such as multi-stage constant current charging, pulse charging, multi-stage constant heat charging and AC charging [35]. Similarly, **thermal management** based on battery DT relies on prediction of aging effect of temperature distribution across the battery pack using thermal models. Detection and traceability of sensor faults, electrical faults, and thermal runaway in a battery DT can allow integration of **fault diagnosis** procedure of the BMS with the battery DT functionalities [32].

In order to identify the functionalities and potentially the value add of a battery DT, the above discussion is encapsulated in Figure 1. The **black circle** lists the functionalities of a BMS, taken from the datasheets of two commercial BMSs found in [36,37]. The extended **blue block** lists the battery DT outputs taken from Table 1. These are the applied battery DT functionalities. Lastly, the **green block** lists the potential DT functionalities identified from the literature (as highlighted above). Thus, Figure 1 compares the **existing BMS functionalities with the applied battery DT functionalities and the potential battery DT functionalities.**

The BMS functionalities taken from the referenced datasheets are to monitor the current and temperature sensors. It uses programmed settings to control the current flow into and out of the battery pack by broadcasting the charge and discharge current limits, cell balancing, and monitoring each cell tap to ensure that cell voltages are not too high or too low. Using the programmed values in the battery pack profile, the BMS calculates the pack and individual cell's internal resistance (SOH) and OCV. Current sensor data is used to calculate the battery pack's SOC (via coulomb counting CC).

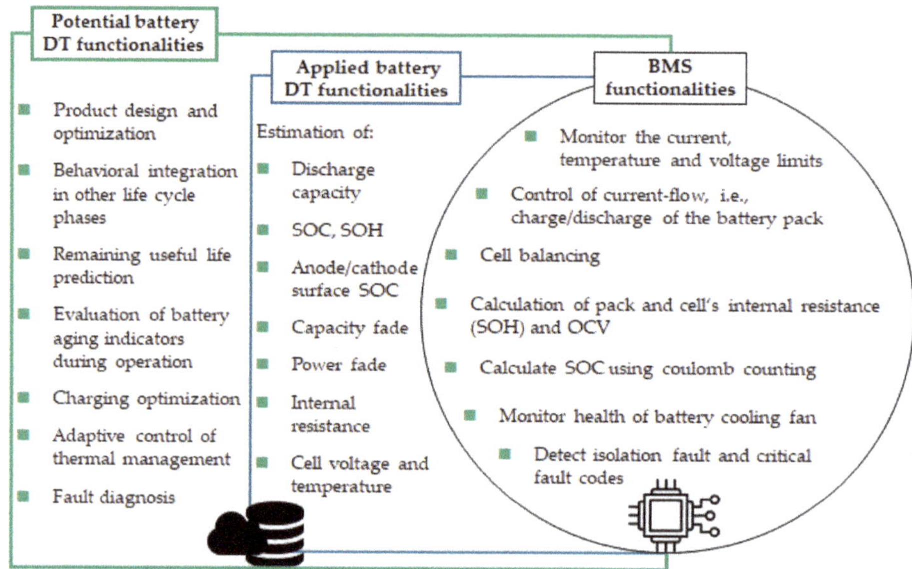

Figure 1. Comparison of BMS functionalities with the applied battery DT and the potential battery DT functionalities.

The SOC estimation method in the studied BMSs is CC. Literature shows alternative methods with higher estimation accuracy, such as adaptive EKF, impedance method, fuzzy logics, SVM, hybrid method (EKF combined with ANN) [38–40]. Here a DT can complement the current BMS functionalities by applying estimation algorithms with higher accuracy. Moreover, the main functionality of the BMS is to ensure that the battery stays with its specified limits. It takes immediate measurements to analyze the voltage and temperature of the cell to estimate the SOC. It does not consider the degradation effect of the charge/discharge cycles on the battery from an electrochemical perspective. The immediate battery user may not be interested in understanding the degradation effects of the battery, such as loss of lithium, diffusivity of electrolyte or SEI resistivity at the anode. However, for deeper knowledge and future innovations by the battery designers, this would serve as a stepping-stone towards ensuring that the battery lasts until its maximum possible capacity and optimal performance. Hence, a DT can complement the functionalities of a BMS by taking the load for large computation requirements. BMS diagnostics over a long period can be enhanced and even simplified by using a DT.

To sum up, the added value that battery DTs can offer is improved representation, performance estimation, behavioral predictions, optimization strategies, and integration of battery life cycle attributes to the remaining DT functions.

3. Approach

In this section, an approach for implementing a battery DT is introduced. The purpose is to define a functional procedure to move from battery model to battery DT systematically. Figure 2 illustrates the 5-step approach, and each step is then further elaborated. By piecing together the existing methodologies of battery modeling, model parameter estimation, battery state prediction, the efforts needed for implementing a battery DT will be investigated.

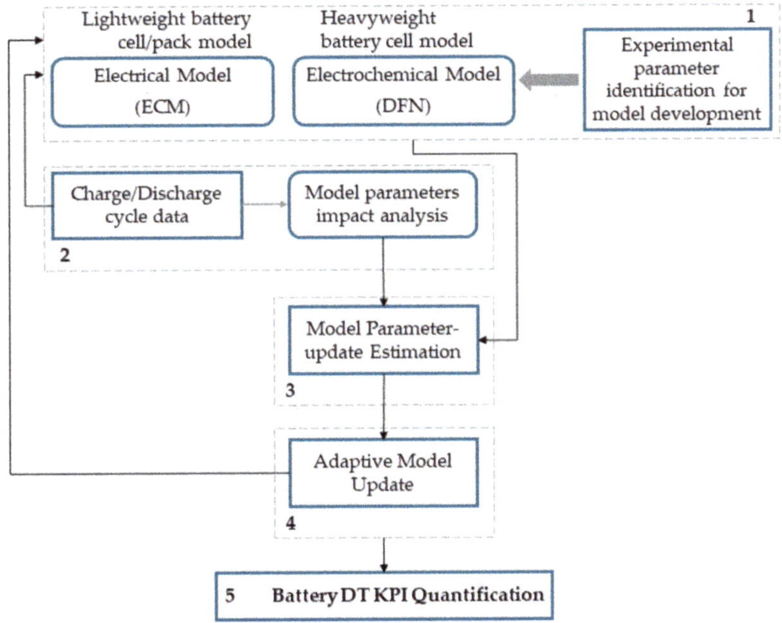

Figure 2. Approach for battery DT implementation.

3.1. Step 1: Lightweight or Heavyweight Battery Model Development

The first step to implement a battery DT is inevitably the development of a reliable battery model. Battery models have become an essential tool in battery-powered applications, which are safety and performance-critical. Depending on how the model inputs and outputs are related, battery models can be classified as empirical, semi-empirical, physical and, data-driven [41,42], while the different types of battery models are:

- Electrical model (ECM);
- Electrochemical model (P2D);
- Thermal model;
- Mechanical model;
- Interdisciplinary combined model.

Fast and accurate identification of the BMS model parameters is a vision for battery developers and engineers. With the outlook of computational expense, the battery model types can be differentiated as either lightweight battery models or heavyweight battery models. Based on [43], the factors which differentiate between the two are as follows:

- Battery dynamics represented by the model
- Number of parameters
- Computation time
- Accuracy
- Ease of understanding and complexity for implementation.

We limit discussion to only electrochemical and electrical models for this paper. As a lightweight model, ECM of a battery is relatively easy to scale to the module or pack level and is widely used in BMS algorithms. They are derived from the empirical measurements of external characteristics of the cell [44]. However, by observing only the external behavior of the battery, the internal electrochemical dynamics cannot be entirely represented, and it is challenging to provide insights into electrochemical or life-reduction phenomena occurring inside the battery. Additionally, ECMs are developed based on data obtained from specific operating conditions of the target batteries, and their accuracy abruptly decreases while performing calculations on other operating conditions or if the battery is replaced [45].

As heavyweight models, the electrochemical models mainly fall into two categories, single particle model or pseudo two-dimensional (P2D) model (also known as the Doyle-Fuller-Newman—DFN model). DFN model for lithium-ion batteries use the combination of the porous electrode theory and the concentrated solution theory [46]. Compared to ECM, which has less physically relevant parameters, a DFN model contains a large number of parameters with a physical meaning. It calculates electrical, chemical, and electrochemical phenomena occurring inside the battery to predict its performance and lifespan Hence, DFN models provide the opportunity for a deep understanding of lithium-ion batteries' aging mechanisms, accurately predicting battery performance by considering the material characteristics and the electrode design.

Experimental Parameter Identification Techniques

Accurate fitting of the battery model with experimental data is not the focus of this paper, so the authors rely on the existing techniques. Instead, the focus is parameterizing the model once the battery is in the usage phase and eventually at EoL (discussed in Section 3.3). Experimental parameter identification is naturally the first step to model development.

ECM or DFN model development requires the human and hardware effort to set up the experiments and perform the necessary tests followed by the subsequent calculation of the battery parameters and its state [47]. The state of the art method of optimizing the parameter set consists of: (1) initially solving the model for a given set of model parameters; and (2) finding the parameter set that minimizes the sum of squared error between the simulated voltage response of the cell and the experimentally observed voltage for a specific drive cycle [48]. It is worth noting that, many techniques have been proposed to identify the necessary parameters for electrical models, but a lower number of identification techniques are available in the literature for electrochemical and aging models.

The term **parameter** refers to the characteristic of the battery, including chemical (solid-phase conductivity, diffusion coefficients, etc.) and electric quantities (internal resistance, capacitance, etc.), while the term **state** refers to the variables which illustrate the behavior of the battery such as SOC and SOH.

In the conventional battery model parameter identification methods, experimental data is used to reference the model parameters, which are then brought closer to the experimental results using the following methods: Kalman filter (KF) method, the gradient method, and the gradient-free method. KF approach is usually applied in the parameter estimation of ECMs due to its recursive computation process, while gradient and gradient-free methods are often employed for a DFN model. Evolutionary computation-based identification methods such as particle swarm optimization (PSO) and genetic algorithm (GA) are gradient-free methods, immune to local minimum traps and are usually used to solve the cell's governing equations faster.

The objective or fitness function for parameter identification via GA, PSO, DE, and KF algorithms is defined by Equation (1) [49–52]:

$$L^2 = \frac{1}{N} min \sum_{i=1}^{N} \left[V_{exp}(t_i) - V_{sim}(\theta, t_i) \right]^2 \tag{1}$$

where, V_{exp} and V_{sim} are the experimental and simulated cell output voltage with the same input current, N is the total number of input current data samples, and i is the time index, L is a representation of the RMS error.

However, the estimations done using these methods may deviate from the actual values due to the fact that the degradation physics caused by SEI (solid electrolyte interphase) layer or lithium plating is not included. Moreover, the variations of the concentration-dependent parameters are usually ignored or assumed as the constant values, which makes the battery state estimation deviate further from reality.

3.2. Step 2: Impact Analysis of Real-World Charge/Discharge Cycles on Battery Model Parameter

Voltage, current, and temperature are the real-world charge/discharge data types collected from the battery. Data in the time-series comes directly from the BMS. Direct connection with the BMS, plug-and-play devices, or cloud servers are some of the data interaction methods. Although the data interaction method is not standardized, cloud services have been integrated with onboard-BMS in the past (Table 1) for seamless data exchange. Initial investment and time for implementation have made real-work data interactions and control of battery systems challenging.

The simulated experimental inputs used for model parameterization (as described in Section 3.1) in contrast with real-world charge/discharge cycles do not have the same effect on battery model parameters. While driving procedures such as WLTP or USSD are used to validate the accuracy of model parameter identification, after prolonged use of the battery, the battery model parameters do not always mimic the actual state of the battery. Hence, with continued usages, the state estimation results of the battery model go farther and farther away from the actual battery state. The effect of prolonged real-world charge/discharge cycle on the battery behavior and the changes that it causes in the inherent characteristics are experimented and researched by academics [53]. However, these effects are not transparent or readily available.

For the impact analysis, first, the governing equations of both DFN model and ECM model are obtained (Tables A1 and A2). Based on these, the list of parameters of both models is summarized in Table 2. For a battery DT, it is essential to identify the parameters that undergo drastic change due to a long period of usage. Hence, step 2 of the approach is to understand the impact of charge/discharge cycles on battery model parameters. The argumentations provided in this section are based on Uddin et al.'s work [48], which applies non-destructive experimental techniques to quantify the detailed degradation associated with different aging stress factors. Model parameter estimation was done using a non-linear fitting algorithm i.e., minimizing the square of the error between simulated and measured voltage. The authors assert that since model parameters are connected to intrinsic properties of the battery, the evolution of these parameters will highlight physical changes within the battery. Thus, by tracking the evolution of model parameters, it will be possible to deduce the mechanisms by which the battery has degraded over time.

Factors that are known to cause degradation effects on Li-ion batteries are: calendar age (t_{age}), cycle number (N), temperature, SOC, DOD, cycle bandwidth (ΔSOC), charge voltage, C-rate, cycle frequency. Expected parameter changes reviewed in the reference paper and further published literature is depicted in Table 2. Knowing the factors that affect the battery life during cycling is essential to design a DT model that evolves along with the degradation of the actual battery. Hence, the model parameters which are directly and most largely affected by the cycling of the battery system are identified.

For reduced-order modeling of DFN model, no particular consensus exists on which parameter needs to be estimated. Some of the parameters are considered while keeping the others at a nominal value. A systematic approach for selecting which parameters can be reliably estimated is presented in [54] in the form of parameter sensitivity analysis. The parameter sensitivity analysis determines how sensitive is the output of the model with respect to variation in values of parameters. Rather than looking into the sensitivity of the model output with variations in values of the parameters, in the following Table 2, we look into the sensitivity of model parameters with respect to cycling data. Table 2 mainly elaborates the effect of high cycle number and charging current (C-rate) on the parameters of the DFN model and the ECM—based on the referenced literature. High cycle number is a primary factor for cyclic aging, and high C-rate directly impacts the cell temperature, which in turn influences battery degradation. Hence, only these two stress factors are considered and not all the mentioned battery degradation factors because the experimental and quantitative sensitivity analysis for each model parameter will be too large to handle in this paper. With the outlook for battery DTs, understanding the sensitivity of model parameters to the environmental conditions and usage practices is a requisite. This implies experimental or simulative confirmation of effect of prolonged cycling on model parameters.

Table 2. Impact analysis of model parameters vs. battery degradation factors based on [48,54,55]. Legend for the table: x—impact identified in the reference article, but the impact level is not specified; **C**—Constants; **Text**—impact level identified by the reference article; **Blank (-)**—No direct information found on the sensitivity or dependency for this parameter.

Parameters	Symbol (Unit) [1]	High Cycle Number	High C-Rate
DFN			
Thickness	L_p, L_n, L_s (µm)	x [55]	Moderate [55]
Surface area	A_p, A_n, A_s, (m^2)	x [48]	Moderate [55]
Particle radius	$R_p{}^+, R_p{}^-$ (µm)	x [54]	x [54]
Active/Inactive material volume fraction	$\varepsilon_s{}^p, \varepsilon_s{}^n$	x [55]	Moderate [55]
Electrolyte phase volume fraction	$\varepsilon_e{}^p; \varepsilon_e{}^n$	-	-
Maximum Li$^+$ concentration	$c_s{}^{p,n,se}$ (mol cm^{-3})	x [48]	Moderate [55]
Average electrolyte concentration	c_e (mol cm^{-3})	-	x [48]
Stoichiometry of n, p at 0% and 100% SOC	$x_{p,n}{}^{0,100}$	-	-
Diffusion coefficient in solid and liquid phase	$D_s{}^p, D_s{}^n, D_e$ (m^2 s^{-1})	x [54]	-
Solid phase conductivity	$\sigma_s{}^p, \sigma_s{}^n$ (µm)	x [48]	x [48]
Li transference number	$t_+{}^0$	Not sensitive [54]	Not sensitive [54]
Resistivity of film layers (including SEI)	R_f (Ω)	Not significant [54]	x [48]
Negative electrode potential, U− coefficients	-	x [48]	-
Positive electrode potential, U+, coefficients	-	x [48]	-
Open circuit potential	V	x [48]	-
Overpotential	η	Not significant [55]	-
Reaction flux at the solid particle surface	j (mol cm^{-1} s^{-1})	-	-
Exchange (electrolyte and solid) current density	i_e (A cm^{-2})	-	-
Electrolyte activity coefficient	$\pm f$	C	C
Bruggeman porosity exponent	p	C	C
Anodic/Cathodic charge transfer coefficient	α_a, α_c	C	C
Intercalation/deintercalation reaction-rate coefficient	$k_{n,p}$ (A cm$^{2.5}$ mol$^{-1.5}$)	C	C
Universal gas constant	R	C	C
Absolute temperature	T	C	C
Faraday's constant	F	C	C
ECM			
Internal ohmic resistance	R_O (Ω)	Sensitive [56]	Sensitive [56]
OCV	V_{OCV} (V)	Sensitive [56]	Sensitive [56]
Polarization Resistances	$R_1, R_2 \ldots$ (Ω)	Sensitive [56]	Sensitive [56]
Polarization Capacitances	$C_1, C_2 \ldots$ (F)	Sensitive [56]	Sensitive [56]
Coulomb efficiency	η	Almost constant [57]	Sensitive [57]
Hysteresis voltage, hysteresis decaying factor	H (V), k	Not significant [58]	Impact of overvoltage [59]

[1] n = Negative electrode; p = Positive electrode; se = Separator; s = Solid phase; e = liquid phase.

A natural question here is that after continued usage of the battery, shouldn't all the parameters change. Not all the coefficients are majorly altered, and the constants do not change. Also, based on the usage practice, only a particular subset of parameters may undergo significant changes, such as the performance parameters or the structural parameters.

Knowledge of the impact of cyclic aging on model parameters helps in model reduction, intermediate battery state estimation at different stages of battery lifecycle, and control algorithms for the BMS. Additionally, it helps the battery designers evaluate how the cell degrades for various applications in different working conditions because the model is up-to-date with the real system in battery DTs. Thus, supporting product optimization for designers and product state estimation during battery usage and end of life.

3.3. Step 3: Model Parameter-Update Estimation

The third step of the approach should not be mistaken as the parameter identification described in step 1. The first parameter identification of a battery requires an experimental setup, i.e., testing the cell in a battery cycler and temperature chamber. However, for identifying the parameters of partially (in operation) or completely aged cells (at EoL), the option of removing the cell from its application and testing it does not necessarily exist. Instead, the sensor measurements from the BMS and knowledge of how the battery has been used (charge/discharge cycles and environmental data) are the basis for the parametrization of the battery DT.

The procedures used to estimate the model parameters and states primarily limit the model usability. From the existing parameter identification techniques for ECM and DFN models, what role do they play when identifying the parameters of a battery DT that continuously needs to evolve as the battery is aging? Ultimately, the parameter estimation procedure that can track model parameters evolution as the cell ages is ideal for battery DT.

The battery DT parameter-update estimation procedure cannot entirely rely on the existing identification methods (Equation (1)) because, with the battery-DT-workflow, a cell in operation cannot be dismounted in order to perform experiments. Although the first step of accurate model development involves estimating parameters using experimental datasets and validation datasets, in practice, many parameters are constantly changing. For battery DT the parameter-update estimation needs to be repeated during operation after a certain number of cycles (N) or time (t). The input data from the BMS is not necessarily retrieved continuously in real-time. N and t will differ based on the battery application (EV, grid storage) and its usage practices. We leave the evaluation of optimal values of N and t for future studies, but an apparent range of N as evaluated from [60,61] is 500–1000 cycles, after which a significant change in electrochemical model parameters is observed. Parameter update for DTs during battery operation and at EoL includes but is not limited to the following methods:

- Calculate the model parameters at the end of N cycles, and repeat the update process iteratively. Identify the reduced set of parameters (such as in Table 1) directly influenced by the number of cycles and operating conditions. The initial conditions (from the governing equations) of the model are certainly updated. Thus, new parameters set and initial conditions are available to the model for its next simulation (N cycles). For DFN, the mathematical estimations of parameters mainly involves revaluating the governing equations which employs Fick's law of diffusion, charge and mass conservation, concentrated solution theory and Butler-Volmer electrochemical kinetic expression.

- Calculate the rate of degradation physics caused by lithium plating and SEI growth through the reaction equations and rate expressions [62]. Lithium-plating passive film layers formed by consuming of cyclable Li-ions is influenced by the charge transfer mechanism. The rate of SEI formation reaction is affected by mass transport within the anode and by surface kinetics. Effects of degradation physics are integrated in the model after every N cycles.

- Utilize the fast minimization algorithms such as Gauss-Newton method, prediction error minimization by estimating the parameter-update through synthetic experimental data [63]. Synthetic experimental data can be obtained using simulated battery output with computer-generated randomness. However, this method has an unjustified validation scheme because the input would also be simulated; hence this approach is mainly beneficial for initial testing purposes of the battery DT.
- Apply data-driven parameter identification methods estimation which employs the terminal voltage and load current for parameter update (partially applied in [64]). A comprehensive literature survey of the data-driven parameter identification methods is not conducted. Therefore, this paper does not attempt to review the data-driven parameter identification methods thoroughly. Instead, we choose to review if data-driven approaches can support the parameter-update step. There is no doubt that a large amount of training data (collected at the beginning of life) is a requirement for data-driven parameter-update during usage. Nonetheless, the cost and computation time of the data-driven algorithms [65,66] for application in battery DT need to be compared.

Looking at the state of art **parameter** estimation algorithms, the gaps between the currently used battery models and the proposed battery DT are as follows; (1) Availability of cycling data to the battery model; (2) Model parameter-update method that does not entirely rely on experimental inputs, but instead on the charge/discharge characteristics and environmental data. Nevertheless, the **state** estimation algorithms would inherently remain the same in both battery model and DT.

3.4. Step 4: Adaptive Model Update

Step 4 of the approach addresses the execution efforts applied on the software segment for implementing a battery DT. The commonly used software and programming languages of battery modeling are MATLAB, COMSOL or Python. Irrespective of the chosen platform, the flowchart of the implementation steps is shown in Figure 3. The green arrows denote the workflow of battery DT while the blue arrows denote that of a battery mode.

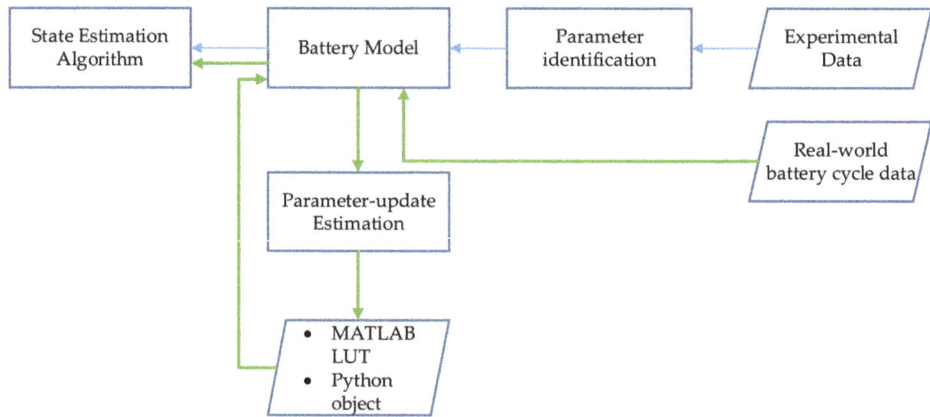

Figure 3. Software segment for implementation steps.

The conventional steps of building a battery model using experimental inputs for its parameter identification followed by the state estimation is the preliminary implementation step for battery DT. Real-world battery data (i.e., terminal voltage, load current and temperature) is then integrated with the model, which means they are added as time series input by importing the battery data (in .csv or other compatible formats). The parameter-update estimation step can be computationally handled in a dedicated code or in the differential solver tools in MATLAB or Python. A standard method in MATLAB for accessing the parameters is through lookup tables (LUT). For model parameter-update in MATLAB, the parameter (in the required format) is uploaded to the workspace and directly imported to the LUT. Alternatively, in Python, re-assigning or updating an input parameter is relatively easy by direct access and assignment of the parameter in its respective object. Finally, the battery DT is subjected to the state estimation algorithm.

3.5. Step 5: Battery DT KPI Quantification

Lastly, but most importantly, it is necessary to identify the benefits of the battery DT to draw light on its significance for the battery industry. Referring to the problem identified in the Section 1 of the paper that without substantial effort to describe and quantify benefits, it is challenging to suggest that the DT concept itself may be the most appropriate solution to the challenges faced by each particular industry. To support a holistic battery DT implementation in the future, both qualitative and quantitative KPIs are identified and elaborated below:

1. Investment

   - Effect on optimization cost due to battery DT functionalities.
   - Cost to establish data acquisition from BMS to the battery model. Here, we assume the preexisting cost of sensors installed on the BMS and the cells.
   - Cost of data storage method, i.e., cloud server, memory drive, etc.
   - Computational cost of simulating the algorithms of the battery DT.

2. Time

   - Time needed for the state estimation algorithms, optimization algorithms and other battery DT functionalities
   - Time to retrieve battery data from its application and assign it to the DT
   - Speed of battery DT alignment with actual battery, i.e., total time for executing the parameter-update step.

3. Accuracy

   - Accuracy of parameter identification.
   - Accuracy of parameter-update estimation parameter identification.
   - Accuracy of state estimation

4. Functionalities

   - DT functionalities that support the battery designers (battery design optimization)
   - DT functionalities that support the battery users
   - DT functionalities that support the battery EoL handler (RUL assessment)

Note: Functionalities is a qualitative KPI. We determine the services that a DT is capable of providing its users through the battery DT functionalities. The accuracy of those services across the battery lifecycle is a KPI for evaluating the benefits of using a battery DT.

4. Results

Here, we highlight the usage of approach in Section 3. The number and variety of existing open-source battery modeling software packages (MATLAB, COMSOL, DUALFOIL, fastDFN, to name a few) have made it convenient for academics and industrial battery designers to begin with partially developed and functional models rather than building the battery model from scratch.

Utilization of experimental results for estimating the battery model parameters is necessary during initial design of the model. However, for estimating the battery state that is already in use for N cycles, experimental validation of terminal voltage followed by a minimization optimization algorithm is impossible. Here, the alternative is that either the BMS provides the voltage values of each cell through the voltage sensor or the parameter-update is identified through the aging physics of the battery.

PyBaMM (Python Battery Mathematical Modelling) is an open-source Python package that can solve continuum models. It can be coupled with other software packages and is capable of solving standard electrochemical battery models. Documentations for using PyBaMM are available [67,68] and the results show that PyBaMM can facilitate the model parameter-update estimation.

The data structure of battery models in PyBaMM is a collection of symbolic expression trees (variable, parameter, addition, multiplication, and gradient). These enable mathematical representation of the model components. The addition of a new battery model in PyBaMM requires initialization of attributes such as: (1) variable boundary conditions; (2) governing equations; (3) initial conditions; (4) output variables of the model; (5) other optional attributes (geometry, computation solver, parameter values, termination events, and battery region).

PyBaMM provides parameter sets based on experimental data provided in literature [69,70] which are used as the reference parameter values in this paper. Types of parameter for all the components (cells, electrolytes, negative electrode, positive electrode, SEI, separators) includes, macroscale geometry [m] or microstructure, current collector conductivities [S m^{-1}], current collector density [kg m^{-3}], current collector specific heat capacity [J$\cdot$ kg^{-1} K^{-1}], current collector thermal conductivity [W m^{-1} K^{-1}], nominal cell capacity [A.h], current function [A], electrode properties, and interfacial reactions.

According to Figure 3, the model is first fed with the charge/discharge cycle data through an excel file. In the following execution, a high acceleration aggressive driving schedule—US06 is used to discharge the DFN model. Code segments (2), (3) and (4) are the cycle integration steps.

$$\text{Experiment1} = \text{pybamm.Experiment} ([\text{"Charge at 1 A until 4.1 V", "Hold at 4.1 V until 50 mA"}]) \tag{2}$$

$$\text{Experiment2} = \text{pybamm.Experiment}([[(\text{"Discharge at 1C for 0.5 h", "Discharge at C/20 for 0.5 h"})] \times 2 + \\ [(\text{"Charge at 0.5 C for 45 min",})] \tag{3}$$

$$\text{Experiment3} = \text{pybamm.Experiment}([[(\text{"Discharge at 2C until 3.3 V", "Rest for 0.5 h",} \\ \text{"Charge at 1 A until 4.1 V", "Hold at 4.1 V until 50 mA"})]]) \tag{4}$$

The Python program was written based on the provided examples in the documentation, hence the results can be reproduced. PyBaMM operates with respect to a defined set of input parameters (~97) and output variables (~400). The PyBaMM program steps highlight the ease of following the implementation steps of Figure 3. It can be considered the means to develop battery DTs, mainly due to fast access to the model parameters in Python. Further analysis and research with different input data and parameter identification are definitely needed in order to implement a holistic battery DT. The correctness of the findings in these results are dependent on the precision of the PDE and ODE solvers in PyBaMM.

In Figure 4, the default DFN model is assigned with the two types of charging preconditions—(2) and (3). The battery model is fully charged using 2, followed by the final state of the experimental solution being mapped to the initial conditions of the next discharge cycle. Similarly, using (3), the battery model is completely discharged followed by slow charge at 0.5 C for 45 min. Likewise, the final state of the experimental solution is mapped as the initial condition for the consequent discharge cycle. The surface concentration, electrode potentials, and terminal voltage values are compared for the three scenarios, i.e., default battery initial state, fully charged, and slow charged. The comparison implies that the effect of charging under varying conditions can be mirrored to the consequent discharge cycle. This helps in running the state estimation procedures after certain discharge cycles with relevant initial conditions.

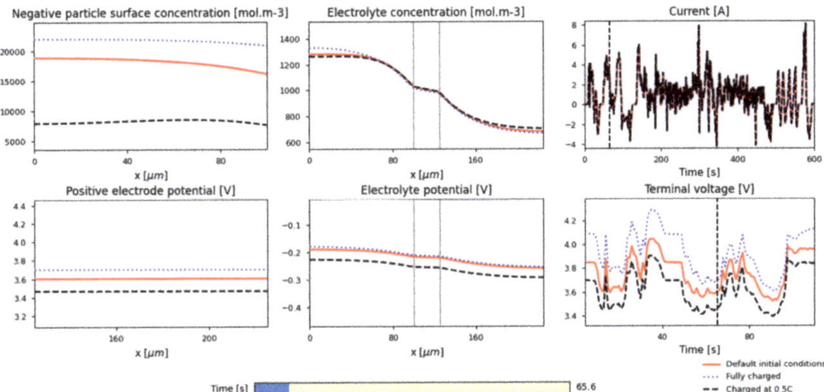

Figure 4. DFN model discharge results with different preconditions. Fully charged—code statement (2). Complete discharge and slow charge—code statement (3).

Code segment (4) is a variation of (3), where the model is discharged at 2 C until 3.3 V, followed by 30 min rest and full charge until 4.1 V. Figures 5 and 6 show that a discharge/charge cycle with respect to (4) is followed by the US06 drive cycle. Similar to the processing of Figure 4, the final state of the model at the end of the charge cycle (Experiment 3) is used to update the initial conditions at the beginning of the discharge cycle. Figure 5 compares the model behavior with direct discharge and discharge after the Experiment 3. The plot of terminal voltage reflects that an accurate terminal voltage estimation is possible when the model parameters are appropriately updated with respect to the operation of the actual battery. Figure 6 partially repeats a section of Figure 5. It mainly shows the DFN model output variables for charge and discharge cycle based on code segment (4) and the US06 drive cycle.

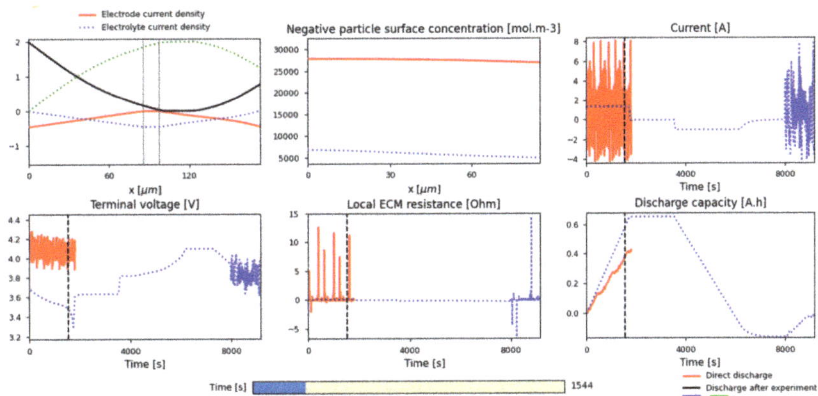

Figure 5. Comparison between output parameters of two discharge characteristics.

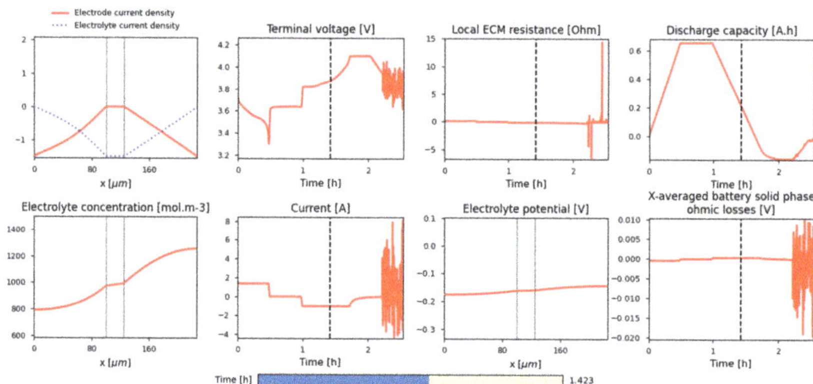

Figure 6. DFN model output variables for discharge based on code statement—(4).

Some of the variables are two-dimensional, with respect to time and the distance from the anode current collector (x). Thus, the graphs change as time is modified. From Figure 5, the output variables for both the scenarios (i.e., only discharge and charge followed by discharge) are retrieved in MATLAB files in 1D or 2D arrays. The output variable set shown in the following code segment can be used to calculate the degradation mechanisms of the electrode, separator, and electrolyte.

```
solution.save_data("Output_AfterDischarging.mat",
["Time [h]", "Current [A]", "Terminal voltage [V]","x [m]",
"Electrolyte concentration [mol.m-3]",
"Electrolyte potential [V]",
"Negative particle surface concentration [mol.m-3]",
"Negative electrode potential [V]",
"Positive particle surface concentration [mol.m-3]",
"Positive electrode potential [V]",
"Negative electrode surface area to volume ratio [m-1]",
"Positive electrode surface area to volume ratio [m-1]",
"Positive electrode active material volume fraction",
"Negative electrode active material volume fraction"], to format="matlab")
```

Hence, the first parameter-update method of 3.3 is realizable in PyBaMM. This example was based on only a small range of driving schedules, while the input charge/discharge cycles are much longer in reality. There also arises an alternative of a straightforward implementation through continuous concatenation of the driving cycles. While that will work for small durations, a very high computation power will be needed to cycle the model through the complete usage phase using a single concatenated file. The model parameter-update during usage keeps the battery DT up-to-date with the actual cell after every N cycle.

Knowledge of which driving cycle parameters affects the model parameters might not always be computationally practical because evaluating the exact values of each parameter is not practically possible. Model reduction is, therefore, an important component, especially of electrochemical battery models. However, understanding potential dependencies will directly impact the formulation of the estimation algorithms' objective function and cost functions.

The output files generated from the results contribute to parameter retrieval and comparison, it can further be integrated with existing battery state estimation techniques. The extensive parameter set that the DFN model provides complements the requirements of the high fidelity prediction algorithms that aim for high accuracy, irrespective of the computation costs. Additionally, the output parameter set can be used for generating training data for data-driven predictions in real-time applications.

Due to the unavailability of operational data, the KPIs identified in Step 5 of the approach are not be quantified for this implementation. However, comprehensive testing and KPI quantification will be investigated in future research.

5. Discussion

The following points focus on the unaddressed issues that remain unresolved but are considerably broad in scope to be handled in one paper:

- Level of fidelity expected from a battery DT—A model that captures the electrical, thermal, electrochemical, mechanical and aging aspects of a battery is deemed a high fidelity model. The reality and practicality of such a model are not clear. The cost and time needed for an exhaustive high-fidelity battery DT are high, and the estimate of accuracy improvement is also missing;
- Number of DTs across the battery lifecycle—The idea of a DT across the lifecycle of a product is not entirely understood. This is due to the uncertainty of the number of DTs needed in such cases. Either there is one DT with a large capacity, or there are many small-sized DTs coupled together. For battery DT, the coupling of process and product DT is a possible use case during manufacturing;
- Scaling the battery DT to module and pack-level DT—Achieving battery DTs at scale will require a reduction in technical barriers for their adoption. This implies that for a pack-level battery DT, the number of sensors and the amount of data retrieved will drastically increase. Hence, the data acquisition and storage needs to be seamless;
- Accuracy of behavioral prediction using battery DT—For commercial utilization of battery DTs, it is necessary to compare and quantify the accuracy of existing BMS predictions vs. the prediction of battery DT. Quantification and comparison of percentage error in DT estimations should be the primary focus in future works.

6. Conclusions

An approach for battery DT implementation was presented. The 5-step approach allows the readers to recognize the difference between a battery model and battery DT implementation. The first challenge recognized for battery DT implementation is making the battery operational data available to the model. Cloud services have been integrated with onboard-BMSs in the past, but this is not common. In the coming years, the data integration method for battery DTs has to be standardized, even though it might entail initial investment and implementation time. The second challenge recognized for battery DT implementation is that the method of model parameter-update during usage is not well established and still needs further research. The results of the paper indicate that, if tracked, the DFN model parameters will keep changing after a certain period of usage and cycling. The other methods proposed for parameter-update in the paper will be investigated in future work.

This paper provides a consistent view of a battery DT and the added value, i.e., its functionalities that it offers during battery operation and EoL. The benefits of battery DTs are: improved representation, performance estimation, behavioral predictions, optimization strategies, and integration of battery life cycle attributes to the remaining DT functions. Based on the results provided in this paper, a battery DT can widen the scope of current BMS functionalities by evaluating the degradation effect that the driving cycle has on the battery from an electrochemical or electrical perspective.

The quantitative uncertainty of the potential costs, infrastructure challenges, and return on investments for battery DTs still exist. However, the KPIs identified in this paper will play a significant role in quantifying the battery DT attributes. Promising future scope exists in evaluating the KPIs for DT across the life cycle phases of a battery. As part of the ongoing research to evaluate the feasibility of battery DTs, its functionalities and quantification of its attributes across the lifecycle will be explored in future works.

Author Contributions: Conceptualization, S.S.; approach, S.S.; implementation, S.S.; writing—original draft preparation, S.S.; writing—review and editing, M.W., K.P.B.; visualization, S.S.; project administration, M.W.; funding acquisition, M.W. and K.P.B.; supervision, M.W. and K.P.B. All authors have read and agreed to the published version of the manuscript.

Funding: This research was funded by the Federal Ministry for Economic Affairs and Energy (BMWi), funding code 19I21014C. The authors wish to thank the Federal Ministry for Economic Affairs and Energy, Germany for funding this work as part of the accompanying research of the project "iBMS".

Institutional Review Board Statement: Not applicable.

Informed Consent Statement: Not applicable.

Data Availability Statement: The data presented in this study are openly available in GitHub at [69].

Conflicts of Interest: The authors declare no conflict of interest.

Appendix A

Table A1. Governing equations of DFN model along with the phenomenon that describes the diffusion, charge/mass conservation and flux density in DFN model [46,71].

DFN Model	Derived through	Governing Equations [1]
Solid phase mass transport equation—Li$^+$ concentration in electrodes and separator	Fick's law of diffusion	$\frac{\partial c_s(x,r,t)}{\partial t} = \frac{D_s}{r^2} \frac{\partial}{\partial r}\left(r^2 \frac{\partial c_s(x,r,t)}{\partial r}\right)$
Liquid phase mass transport equation—Li$^+$ concentration in electrolyte	Conservation of Li+ ions (Conservation of mass)	$\varepsilon_e \frac{\partial c_e(x,t)}{\partial t} = \frac{\partial}{\partial x}\left(D_e^{eff} \frac{\partial c_e(x,t)}{\partial x}\right) + \left(1 - t_+^0\right) A_s j(x,t)$
Solid phase charge transport equation—Potential in electrode	Ohm's law (Conservation of charge)	$\frac{\partial}{\partial x}\left(\sigma^{eff} \frac{\partial \phi_s(x,t)}{\partial x}\right) - A_s F j(x,t) = 0$
Liquid phase charge transport equation—Potential in electrolyte	Ohm's law and Kirchhoff's law (Concentrated solution theory, conservation of charge)	$\frac{\partial}{\partial x}\left(\kappa^{eff} \frac{\partial \phi_e(x,t)}{\partial x}\right) + \frac{\partial}{\partial x}\left(\kappa_D^{eff} \frac{\partial \ln(c_e(x,t))}{\partial x}\right) + A_s F j(x,t) = 0$
Flux density between solid and liquid phase	Butler-Volmer Equation	$j = k_0 c_e^{1-\alpha} (c_{s,max} - c_{s,e})^{1-\alpha} c_{s,e}^{\alpha} \left(\exp\left(\frac{(1-\alpha)F}{RT}\eta\right) - \exp\left(-\frac{\alpha F}{RT}\eta\right)\right)$

[1] Concentration of lithium in the solid phase $c_s(x, r, t)$ and electrolyte $c_e(x,t)$. Electric potential in the solid phase $\phi_s(x,t)$ and electrolyte $\phi_e(x,t)$. Flux density between solid phase and electrolyte $j(x,t)$. Remaining symbols are described in Table 1.

Table A2. Equations of ECM model in the time domain for pulse current [72].

ECM Model Equations	Variables
$p_i = f_i(SOC, SOH, T, I)$ $p_1 = \{V_{OCV}, R_1, C_1, R_S\}$ $V_t = V_{OCV} - V_1 - V_{Rs};\ V_{Rs} = I * R_S$ $SOC = SOC_0 - \int_0^t \frac{I_b}{C} dt$	i is the i-th parameter of the model. R_1 and C_1 are the polarization resistance and capacitance and R_S is the ohmic resistance. V_{OCV} is the open circuit voltage where V_{Rs} refers to the voltage reduction from R_s and V_t is the terminal voltage SOC calculation using CC, where C is capacity, I_b is current, and SOC_0 is the initial SOC

References

1. Singh, S.; Weeber, M.; Birke, K.-P. Advancing digital twin implementation: A toolbox for modelling and simulation. *Procedia CIRP* **2021**, *99*, 567–572. [CrossRef]
2. Shafto, M.; Conroy, M.; Doyle, R.; Glaessgen, E.; Kemp, C.; LeMoigne, J.; Wang, L. Modeling, simulation, information technology & processing roadmap. *Natl. Aeronaut. Space Adm.* **2012**, *32*, 1–38.
3. Boschert, S.; Rosen, R. Digital twin—The simulation aspect. In *Mechatronic Futures*; Springer: Cham, Switzerland, 2016; pp. 59–74.
4. Kunath, M.; Winkler, H. Integrating the Digital Twin of the manufacturing system into a decision support system for improving the order management process. *Procedia CIRP* **2018**, *72*, 225–231. [CrossRef]
5. Batty, M. Digital twins. *Environ. Plan. B Urban. Anal. City Sci.* **2018**, *45*, 817–820. [CrossRef]
6. Wright, L.; Davidson, S. How to tell the difference between a model and a digital twin. *Adv. Model. Simul. Eng. Sci.* **2020**, *7*, 13. [CrossRef]
7. Arup Research. Digital Twin—Towards a Meaningful Framework. Available online: https://research.arup.com/publications/digital-twin-towards-a-meaningful-framework/ (accessed on 26 July 2021).

8.   Singh, M.; Fuenmayor, E.; Hinchy, E.P.; Qiao, Y.; Murray, N.; Devine, D. Digital Twin: Origin to Future. *ASI* **2021**, *4*, 36. [CrossRef]

9.   Saracco, R. Digital Twins: Bridging Physical Space and Cyberspace. *Computer* **2019**, *52*, 58–64. [CrossRef]

10.   Al-Ali, A.R.; Gupta, R.; Zaman Batool, T.; Landolsi, T.; Aloul, F.; Al Nabulsi, A. Digital twin conceptual model within the context of internet of things. *Future Internet* **2020**, *12*, 163. [CrossRef]

11.   Lin, W.D.; Low, M.Y.H. Concept design of a system architecture for a manufacturing cyber-physical Digital Twin system. In Proceedings of the 2020 IEEE International Conference on Industrial Engineering and Engineering Management (IEEM), Singapore, 14–17 December 2020; IEEE: Piscataway, NJ, USA, 2020; pp. 1320–1324, ISBN 978-1-5386-7220-4.

12.   Wagg, D.J.; Worden, K.; Barthorpe, R.J.; Gardner, P. Digital Twins: State-of-the-Art and Future Directions for Modeling and Simulation in Engineering Dynamics Applications. *ASCE-ASME J. Risk. Uncert. Eng. Syst. Part. B Mech. Eng.* **2020**, *6*, 101001. [CrossRef]

13.   Lim, K.Y.H.; Zheng, P.; Chen, C.-H. A state-of-the-art survey of Digital Twin: Techniques, engineering product lifecycle management and business innovation perspectives. *J. Intell. Manuf.* **2020**, *31*, 1313–1337. [CrossRef]

14.   Jones, D.; Snider, C.; Nassehi, A.; Yon, J.; Hicks, B. Characterising the Digital Twin: A systematic literature review. *CIRP J. Manuf. Sci. Technol.* **2020**, *29*, 36–52. [CrossRef]

15.   Merkle, L.; Segura, A.S.; Grummel, J.T.; Lienkamp, M. Architecture of a digital twin for enabling digital services for battery systems. In Proceedings of the 2019 IEEE International Conference on Industrial Cyber Physical Systems (ICPS), Taipei, Taiwan, 6–9 May 2019; IEEE: Piscataway, NJ, USA, 2019; pp. 155–160.

16.   Schluse, M.; Atorf, L.; Rossmann, J. Experimentable digital twins for model-based systems engineering and simulation-based development. In Proceedings of the 2017 Annual IEEE International Systems Conference (SysCon), Montreal, QC, Canada, 24–27 April 2017; IEEE: Piscataway, NJ, USA, 2017. ISBN 9781509046232.

17.   Josifovska, K.; Yigitbas, E.; Engels, G. Reference Framework for Digital Twins within cyber-physical systems. In Proceedings of the 2019 IEEE/ACM 5th International Workshop on Software Engineering for Smart Cyber-Physical Systems (SEsCPS), Montreal, QC, Canada, 28 May 2019; IEEE: Piscataway, NJ, USA, 2019; pp. 25–31, ISBN 978-1-7281-2282-3.

18.   Kritzinger, W.; Karner, M.; Traar, G.; Henjes, J.; Sihn, W. Digital Twin in manufacturing: A categorical literature review and classification. *IFAC-Pap.* **2018**, *51*, 1016–1022. [CrossRef]

19.   Sitterly, M.; Le Wang, Y.; Yin, G.G.; Wang, C. Enhanced identification of battery models for real-time battery management. *IEEE Trans. Sustain. Energy* **2011**, *2*, 300–308. [CrossRef]

20.   Niederer, S.A.; Sacks, M.S.; Girolami, M.; Willcox, K. Scaling digital twins from the artisanal to the industrial. *Nat. Comput. Sci.* **2021**, *1*, 313–320. [CrossRef]

21.   Qu, X.; Song, Y.; Liu, D.; Cui, X.; Peng, Y. Lithium-ion battery performance degradation evaluation in dynamic operating conditions based on a digital twin model. *Microelectron. Reliab.* **2020**, *114*, 113857. [CrossRef]

22.   Li, W.; Rentemeister, M.; Badeda, J.; Jöst, D.; Schulte, D.; Sauer, D.U. Digital twin for battery systems: Cloud battery management system with online state-of-charge and state-of-health estimation. *J. Energy Storage* **2020**, *30*, 101557. [CrossRef]

23.   Merkle, L.; Pöthig, M.; Schmid, F. Estimate e-Golf Battery State Using Diagnostic Data and a Digital Twin. *Batteries* **2021**, *7*, 15. [CrossRef]

24.   Peng, Y.; Zhang, X.; Song, Y.; Liu, D. A low cost flexible digital twin platform for spacecraft lithium-ion battery pack degradation assessment. In Proceedings of the 2019 IEEE International Instrumentation and Measurement Technology Conference (I2MTC), Auckland, New Zealand, 20–23 May 2019; IEEE: Piscataway, NJ, USA, 2019; pp. 1–6, ISBN 978-1-5386-3460-8.

25.   Baumann, M.; Rohr, S.; Lienkamp, M. Cloud-connected battery management for decision making on second-life of electric vehicle batteries. In Proceedings of the 2018 Thirteenth International Conference on Ecological Vehicles and Renewable Energies (EVER), Monte-Carlo, Monaco, 10–12 April 2018; IEEE: Piscataway, NJ, USA, 2018; pp. 1–6, ISBN 978-1-5386-5966-3.

26.   Ramachandran, R.; Ganeshaperumal, D.; Subathra, B. Parameter estimation of battery pack in EV using extended kalman filters. In Proceedings of the 2019 IEEE International Conference on Clean Energy and Energy Efficient Electronics Circuit for Sustainable Development (INCCES), Krishnankoil, India, 18–20 December 2019; IEEE: Piscataway, NJ, USA, 2019; pp. 1–5, ISBN 978-1-7281-4407-8.

27.   Wu, B.; Widanage, W.D.; Yang, S.; Liu, X. Battery digital twins: Perspectives on the fusion of models, data and artificial intelligence for smart battery management systems. *Energy AI* **2020**, *1*, 100016. [CrossRef]

28.   Sancarlos, A.; Cameron, M.; Abel, A.; Cueto, E.; Duval, J.-L.; Chinesta, F. From ROM of electrochemistry to ai-based battery digital and hybrid twin. *Arch. Comput. Methods Eng.* **2021**, *28*, 979–1015. [CrossRef]

29.   Soleymani, A.; Maltz, W. Real time prediction of Li-Ion battery pack temperatures in EV vehicles. In Proceedings of the ASME 2020 International Technical Conference and Exhibition on Packaging and Integration of Electronic and Photonic Microsystems, Virtual, 27–29 October 2020; American Society of Mechanical Engineers: New York, NY, USA, 2020; p. 10272020, ISBN 978-0-7918-8404-1.

30.   Ahmed, R.; Gazzarri, J.; Onori, S.; Habibi, S.; Jackey, R.; Rzemien, K.; Tjong, J.; LeSage, J. Model-Based Parameter Identification of Healthy and Aged Li-ion Batteries for Electric Vehicle Applications. *SAE Int. J. Alt. Power.* **2015**, *4*, 233–247. [CrossRef]

31.   Dühnen, S.; Betz, J.; Kolek, M.; Schmuch, R.; Winter, M.; Placke, T. Toward green battery cells: Perspective on materials and technologies. *Small Methods* **2020**, *4*, 2000039. [CrossRef]

32.   Wang, W.; Wang, J.; Tian, J.; Lu, J.; Xiong, R. Application of Digital Twin in Smart Battery Management Systems. *Chin. J. Mech. Eng.* **2021**, *34*, 1–19. [CrossRef]

33. Balasingam, B.; Ahmed, M.; Pattipati, K. Battery Management Systems—Challenges and Some Solutions. *Energies* **2020**, *13*, 2825. [CrossRef]

34. Gabbar, H.A.; Othman, A.M.; Abdussami, M.R. Review of Battery Management Systems (BMS) Development and Industrial Standards. *Technologies* **2021**, *9*, 28. [CrossRef]

35. Lin, Q.; Wang, J.; Xiong, R.; Shen, W.; He, H. Towards a smarter battery management system: A critical review on optimal charging methods of lithium ion batteries. *Energy* **2019**, *183*, 220–234. [CrossRef]

36. Orion BMS 2. Orion Li-Ion Battery Management System. Available online: https://www.orionbms.com/products/orion-bms-standard/ (accessed on 27 July 2021).

37. LION Smart. LION Smart—Battery Management System (BMS). Available online: https://lionsmart.com/en/battery-management-system/ (accessed on 27 July 2021).

38. van Dao, Q.; Dinh, M.-C.; Kim, C.S.; Park, M.; Doh, C.-H.; Bae, J.H.; Lee, M.-K.; Liu, J.; Bai, Z. Design of an effective State of Charge estimation method for a lithium-ion battery pack using extended kalman filter and artificial neural network. *Energies* **2021**, *14*, 2634. [CrossRef]

39. Danko, M.; Adamec, J.; Taraba, M.; Drgona, P. Overview of batteries State of Charge estimation methods. *Transp. Res. Procedia* **2019**, *40*, 186–192. [CrossRef]

40. Rivera-Barrera, J.; Muñoz-Galeano, N.; Sarmiento-Maldonado, H. SoC Estimation for Lithium-ion Batteries: Review and Future Challenges. *Electronics* **2017**, *6*, 102. [CrossRef]

41. Wang, Y.; Tian, J.; Sun, Z.; Wang, L.; Xu, R.; Li, M.; Chen, Z. A comprehensive review of battery modeling and state estimation approaches for advanced battery management systems. *Renew. Sustain. Energy Rev.* **2020**, *131*, 110015. [CrossRef]

42. Meng, J.; Luo, G.; Ricco, M.; Swierczynski, M.; Stroe, D.-I.; Teodorescu, R. Overview of lithium-ion battery modeling methods for State-of-Charge estimation in electrical vehicles. *Appl. Sci.* **2018**, *8*, 659. [CrossRef]

43. Schellenberg, S.; Berndt, R.; Eckhoff, D.; German, R. A Computationally inexpensive battery model for the microscopic simulation of electric vehicles. In Proceedings of the 2014 IEEE 80th Vehicular Technology Conference (VTC Fall), Vancouver, BC, Canada, 14–17 September 2014; IEEE: Piscataway, NJ, USA, 2014; pp. 1–6, ISBN 978-1-4799-4449-1.

44. Li, Y.; Vilathgamuwa, D.M.; Farrell, T.W.; Choi, S.S.; Tran, N.T. An equivalent circuit model of li-ion battery based on electro-chemical principles used in grid-connected energy storage applications. In Proceedings of the 2017 IEEE 3rd International Future Energy Electronics Conference and ECCE Asia (IFEEC 2017—ECCE Asia), Kaohsiung, Taiwan, 3–7 June 2017; IEEE: Piscataway, NJ, USA, 2017; pp. 959–964, ISBN 978-1-5090-5157-1.

45. Kim, H.-K.; Lee, K.-J. Scale-Up of Physics-Based Models for Predicting Degradation of Large Lithium Ion Batteries. *Sustainability* **2020**, *12*, 8544. [CrossRef]

46. Jokar, A.; Rajabloo, B.; Désilets, M.; Lacroix, M. Review of simplified Pseudo-two-Dimensional models of lithium-ion batteries. *J. Power Sources* **2016**, *327*, 44–55. [CrossRef]

47. Barcellona, S.; Piegari, L. Lithium Ion Battery Models and Parameter Identification Techniques. *Energies* **2017**, *10*, 2007. [CrossRef]

48. Uddin, K.; Perera, S.; Widanage, W.; Somerville, L.; Marco, J. Characterising Lithium-Ion Battery Degradation through the Identification and Tracking of Electrochemical Battery Model Parameters. *Batteries* **2016**, *2*, 13. [CrossRef]

49. Forman, J.C.; Moura, S.J.; Stein, J.L.; Fathy, H.K. Genetic parameter identification of the Doyle-Fuller-Newman model from experimental cycling of a LFP battery. In Proceedings of the 2011 American Control Conference, San Francisco, CA, USA, 29 June–1 July 2011; IEEE: Piscataway, NJ, USA, 2011; pp. 362–369, ISBN 978-1-4577-0081-1.

50. Rahman, M.A.; Anwar, S.; Izadian, A. Electrochemical model parameter identification of a lithium-ion battery using particle swarm optimization method. *J. Power Sources* **2016**, *307*, 86–97. [CrossRef]

51. Yang, G. Battery parameterisation based on differential evolution via a boundary evolution strategy. *J. Power Sources* **2014**, *245*, 583–593. [CrossRef]

52. Yang, K.; Tang, Y.; Zhang, Z. Parameter Identification and State-of-Charge Estimation for Lithium-Ion Batteries Using Separated Time Scales and Extended Kalman Filter. *Energies* **2021**, *14*, 1054. [CrossRef]

53. Harlow, J.E.; Ma, X.; Li, J.; Logan, E.; Liu, Y.; Zhang, N.; Ma, L.; Glazier, S.L.; Cormier, M.M.E.; Genovese, M.; et al. A Wide Range of Testing Results on an Excellent Lithium-Ion Cell Chemistry to be used as Benchmarks for New Battery Technologies. *J. Electrochem. Soc.* **2019**, *166*, A3031–A3044. [CrossRef]

54. Jin, N.; Danilov, D.L.; van den Hof, P.M.; Donkers, M. Parameter estimation of an electrochemistry-based lithium-ion battery model using a two-step procedure and a parameter sensitivity analysis. *Int. J. Energy Res.* **2018**, *42*, 2417–2430. [CrossRef]

55. Li, W.; Cao, D.; Jöst, D.; Ringbeck, F.; Kuipers, M.; Frie, F.; Sauer, D.U. Parameter sensitivity analysis of electrochemical model-based battery management systems for lithium-ion batteries. *Appl. Energy* **2020**, *269*, 115104. [CrossRef]

56. Miniguano, H.; Barrado, A.; Lazaro, A.; Zumel, P.; Fernandez, C. General Parameter Identification Procedure and Comparative Study of Li-Ion Battery Models. *IEEE Trans. Veh. Technol.* **2020**, *69*, 235–245. [CrossRef]

57. Madani, S.; Schaltz, E.; Knudsen Kær, S. Effect of Current Rate and Prior Cycling on the Coulombic Efficiency of a Lithium-Ion Battery. *Batteries* **2019**, *5*, 57. [CrossRef]

58. Tran, M.-K.; Mevawala, A.; Panchal, S.; Raahemifar, K.; Fowler, M.; Fraser, R. Effect of integrating the hysteresis component to the equivalent circuit model of Lithium-ion battery for dynamic and non-dynamic applications. *J. Energy Storage* **2020**, *32*, 101785. [CrossRef]

59. Ovejas, V.J.; Cuadras, A. Effects of cycling on lithium-ion battery hysteresis and overvoltage. *Sci. Rep.* **2019**, *9*, 14875. [CrossRef]

60. Severson, K.A.; Attia, P.M.; Jin, N.; Perkins, N.; Jiang, B.; Yang, Z.; Chen, M.H.; Aykol, M.; Herring, P.K.; Fraggedakis, D.; et al. Data-driven prediction of battery cycle life before capacity degradation. *Nat. Energy* **2019**, *4*, 383–391. [CrossRef]
61. Stroe, A.-I.; Knap, V.; Stroe, D.-I. Comparison of lithium-ion battery performance at beginning-of-life and end-of-life. *Microelectron. Reliab.* **2018**, *88–90*, 1251–1255. [CrossRef]
62. Atalay, S.; Sheikh, M.; Mariani, A.; Merla, Y.; Bower, E.; Widanage, W.D. Theory of battery ageing in a lithium-ion battery: Capacity fade, nonlinear ageing and lifetime prediction. *J. Power Sources* **2020**, *478*, 229026. [CrossRef]
63. Boovaragavan, V.; Harinipriya, S.; Subramanian, V.R. Towards real-time (milliseconds) parameter estimation of lithium-ion batteries using reformulated physics-based models. *J. Power Sources* **2008**, *183*, 361–365. [CrossRef]
64. Ali, M.; Kamran, M.A.; Kumar, P.S.; Himanshu; Nengroo, S.H.; Khan, M.A.; Hussain, A.; Kim, H.-J. An Online Data-Driven Model Identification and Adaptive State of Charge Estimation Approach for Lithium-ion-Batteries Using the Lagrange Multiplier Method. *Energies* **2018**, *11*, 2940. [CrossRef]
65. Hou, Y.; Zhang, Z.; Liu, P.; Song, C.; Wang, Z. Research on a novel data-driven aging estimation method for battery systems in real-world electric vehicles. *Adv. Mech. Eng.* **2021**, *13*, 168781402110277. [CrossRef]
66. Hashemi, S.R. An Intelligent Battery Managment System for Electric and Hybrid Electric Aircraft. Ph.D. Thesis, University of Akron, Akron, OH, USA, 2021.
67. Sulzer, V.; Marquis, S.G.; Timms, R.; Robinson, M.; Chapman, S.J. Python Battery Mathematical Modelling (PyBaMM). *J. Open Res. Softw.* **2020**, *9*, 14. [CrossRef]
68. Welcome to PyBaMM's Documentation!—PyBaMM 0.4.0 Documentation. Available online: https://pybamm.readthedocs.io/en/latest/ (accessed on 28 July 2021).
69. GitHub. PyBaMM/pybamm/input/parameters/lithium_ion/cells at developpybamm-team/PyBaMM. Available online: https://github.com/pybamm-team/PyBaMM/tree/develop/pybamm/input/parameters/lithium_ion/cells (accessed on 27 July 2021).
70. Chen, C.-H.; Brosa Planella, F.; O'Regan, K.; Gastol, D.; Widanage, W.D.; Kendrick, E. Development of Experimental Techniques for Parameterization of Multi-scale Lithium-ion Battery Models. *J. Electrochem. Soc.* **2020**, *167*, 80534. [CrossRef]
71. Luder, D.; Caliandro, P.; Vezzini, A. Enhanced physics-based models for state estimation of Li-Ion batteries. In Proceedings of the Comsol Conference 2020 Europe, Virtual, 14–15 October 2020.
72. Madani, S.; Schaltz, E.; Knudsen Kær, S. An Electrical Equivalent Circuit Model of a Lithium Titanate Oxide Battery. *Batteries* **2019**, *5*, 31. [CrossRef]

batteries

Article

A Method for Monitoring State-of-Charge of Lithium-Ion Cells Using Multi-Sine Signal Excitation

Jonghyeon Kim * and Julia Kowal

Department of Energy and Automation Technology, Electrical Energy Storage Technology, Technical University of Berlin, Einsteinufer 11, 10587 Berlin, Germany; julia.kowal@tu-berlin.de
* Correspondence: jonghyeon.kim@eet.tu-berlin.de

Abstract: In this paper, a method for monitoring SoC of a lithium-ion battery cell through continuous impedance measurement during cell operation is introduced. A multi-sine signal is applied to the cell operating current, and the cell SoH and SoC can be simultaneously monitored via impedance at each frequency. Unlike existing studies in which cell impedance measurement is performed ex situ through EIS equipment, cell state estimation is performed in situ. The measured impedance takes into account cell temperature and cell SoH, enabling accurate SoC estimation. The measurement system configured for the experiment and considerations for the selection of measurement parameters are described, and the accuracy of cell SoC estimation is presented.

Keywords: lithium-ion battery; battery management system; state monitoring; state-of-charge

Citation: Kim, J.; Kowal, J. A Method for Monitoring State-of-Charge of Lithium-Ion Cells Using Multi-Sine Signal Excitation. *Batteries* **2021**, *7*, 76. https://doi.org/10.3390/batteries 7040076

Academic Editors: Duygu Kaus and Kai Peter Birke

Received: 30 July 2021
Accepted: 3 November 2021
Published: 9 November 2021

Publisher's Note: MDPI stays neutral with regard to jurisdictional claims in published maps and institutional affiliations.

1. Introduction

A Li-ion battery is a type of secondary battery in which lithium ions move from a negative electrode to a positive electrode through an electrolyte during a discharge process. Lithium is the lightest solid element and has the lowest standard reduction potential. Therefore, when applied as an electrode material, the battery cell can obtain an electromotive force higher than 3 V, and a high energy density per weight and a high energy density per volume can be obtained. Li-ion batteries have less memory effect and less self-discharge as well. In addition to these advantages, due to their reliable performance, long life cycle, and advantages such as reduced pollution, Li-ion batteries are one of the most promising power sources for portable electronics, electric vehicles, renewable energy storage devices, et cetera [1,2]. However, high energy density batteries with low thermal stability electrode materials may have low safety performance [3–5]. In particular, battery failure caused by extreme conditions such as excessive external force, high temperature, low temperature, overcharge, and over-discharge becomes a serious problem [6–8].

Overcharging the Li-ion battery can be one of the most important safety issues. Overcharging occurs when a charging current is forced in after the battery reaches its upper cutoff voltage or state of charge (SoC) limit. It is usually caused by a malfunction of the battery charger or by an inaccurate estimate of the condition of the battery in the battery management system (BMS). When the Li-ion battery is overcharged, in addition to the increase in internal resistance, decomposition of the binder and electrolyte, formation of insoluble products, blocked electrode pores, and gas generation may occur sequentially [9]. Especially, severe expansion of the battery may occur due to gas accumulation [10], and when the internal pressure exceeds the limit, structural deformation, rupture, and an internal short circuit may occur [11,12]. Lithium metal and moisture in the air may react after the battery ruptures, and even flammable gases may ignite [8,13]. Heat generated by side reactions and internal short circuits can accelerate battery failure mechanisms through natural positive feedback, leading to thermal runaway, eventually causing battery ignition or explosion [8,14,15]. In the case of over-discharge, a significant loss of active lithium and positive electrode material occurs [16], and a decrease in battery capacity occurs [17,18].

Battery over-discharging also results in a change in the solid electrolyte interphase (SEI) on the anode surface and an increase in impedance [18]. Gas is also generated from the decomposition of the SEI, which can cause battery swelling [19].

Accurate SoC estimation has the following advantages:

- Battery cells can be used longer by preventing overcharging and over-discharging, which can cause permanent damage.
- It enables more aggressive cell operation. If the reliability of the SoC estimation is low, it must be operated conservatively in order to avoid overcharging and over-discharging the cell and to make the operation as safe as possible.
- As it shows reliable estimation results for any usage profile of the cell, it improves the reliability of the use of the battery cell in the application system.
- As the battery pack design does not have to be overengineered, it enables the production of smaller and lighter battery packs. It eventually lowers the price of the battery pack. In addition, the reliable battery system reduces battery maintenance costs.

Nonetheless, there is still no way to directly measure the SoC of a battery cell. Therefore, it must be estimated or inferred indirectly from the measured current, voltage, and temperature.

Li-ion cell SoC estimation methods can be listed as follows [20–22]:

- SoC estimation method based on open circuit voltage (OCV)

In order to represent the thermodynamic state of the energy of the cell electrode, the OCV according to the chemical composition of the electrode is used. OCV is the voltage of the electrode when it has been stable for a sufficient time without flowing current. Cell voltage is related to temperature and electrode particle surface concentration, whereas cell SoC is related to particle average concentration. In other words, OCV is measured when the electrodes of an electrochemical cell reach equilibrium and there is no voltage deviation depending on the position inside the electrode, which reflects the Gibbs free energy at thermodynamic equilibrium. The OCV represents a strong dependence on SoC in most batteries. Nevertheless, it is impractical for real-time or continuous state estimation since a relaxation time of several hours is generally required to reach electrochemical equilibrium. This is especially problematic for battery applications where resting time can never exist. Furthermore, Li-ion batteries have a flatter OCV compared to lead-acid batteries, making SoC estimation difficult. In addition, cell temperature and cell state of health (SoH) can also lead to SoC estimation errors based on OCV.

- SoC estimation method based on ampere counting

In the SoC estimation method based on ampere counting, the accumulated amount of charging and discharging current relative to the discharging capacity is defined as ΔSoC and calculated with the set initial SoC. Since only cumulative current information is used for SoC estimation, it has the advantage of requiring relatively low performance for the hardware and software of the BMS. However, this method also has weaknesses in estimating cell SoC. It becomes a problem if the SoC of the battery cell is entirely dependent on the initial SoC. Since only ΔSoC is calculated, it is unavoidable to misestimate the SoC if the initial SOC setting is incorrect. In addition, it is impossible to know exactly the total capacity and coulombic efficiency of a cell, which should be approximated, and both of these approximations contribute to the error in cell SoC estimation. Self-discharge currents and leakage currents from electronic circuits for measuring cell performance increase errors as well. Furthermore, errors in voltage, current, and temperature measurements contribute to increasing the estimation error. These errors are more integrated and intensified as battery cell operating time increases. As a result, the uncertainty in the SOC estimate is exacerbated by the accumulated measurement errors. Therefore, this method can show reliable estimation results for a short period of time only if the initial conditions are well known, unless there is a feedback mechanism for error correction.

- SoC estimation method based on heuristic data

The heuristic based SoC estimation method is a method based on experimental data. Statistical rules or patterns found from data obtained through various cell charge and discharge experiments are used. These methods include fuzzy logic, neural networks, and support vector machines (SVM). Reliable estimation results can be obtained when the learning technique is implemented with a large amount of experimental data under different conditions. However, it takes a lot of time for the necessary experimental data to be properly secured.

- SoC estimation method based on adaptive control

An adaptive control based SoC estimation method such as a Kalman filter (KF) [23] or a sliding mode observer [24,25] compares the actual SoC measurement result with the estimated value and gradually reduces the difference according to the feedback principle. Although this method has a high estimation performance, it is relatively complicated to implement and thus has a disadvantage of high cost.

- SoC estimation method based on equivalent circuit model

The electrochemical impedance spectroscopy (EIS) method is a well-established technique for determining the dynamic behavior of electrochemical systems [26,27]. It is used to characterize battery impedance behavior over a wide frequency range [28,29]. Using EIS, the measured spectrum in an electrochemical system can be interpreted as an impedance spectrum of a lumped element model consisting of resistors, inductors, and capacitors. It can be used to implement dynamic simulation models [30]. The SoC [21,22,31] and SoH [28,32,33] of the battery cell can be estimated through the equivalent circuit model obtained through EIS. Moreover, the commonly used definition for battery end of life (EoL) is the predefined battery impedance increase at nominal conditions [34–36]. Nevertheless, EIS equipment is generally used in laboratories for general propose impedance measurements [3], which are not suitable for battery monitoring purposes. For EIS measurements, battery cells need to be detached from the operating load, and it is usually time consuming for impedance measurements in a wide frequency range. In addition, EIS equipment can be an excessive investment for battery monitoring systems, especially heavy and bulky for portable devices. Moreover, without algorithms for cell state estimation, EIS measurements alone cannot estimate cell SoH and SoC.

Unlike the method using EIS, the proposed method can be used to estimate the SoC of the cell during operation. There is a growing interest in the use of cell impedance to monitor the condition of batteries. The papers by Qahouq [37] and Waag et al. [38] deal with the impedance measurement of the operating cell but do not cover cell SoC and SoH estimation using the measured impedance. In the paper by Do et al. [39], the measured impedance is not adjusted according to the temperature change, and the computational complexity increases by using the extended KF. In addition, this paper does not deal with the method of estimating the cell SoH and SoC from the measured impedance. In the papers by Huang et al. [40] and Howey et al. [41], the cell SoC is estimated by measuring the cell impedance, but the effect of temperature on the cell impedance is not considered. The paper by Fleischer et al. [33] shows a good result of estimating the state of a cell in operation. However, a physical model and an equivalent circuit model are used, and so-called mutation algorithms are used. Since nonlinear differential equations and matrix operations are required, the complexity of the operation is increased. On the other hand, the proposed method is relatively simple but accurate. Using the cell temperature and impedance at two frequencies, SoH and SoC of the cells in operation can be estimated.

Proposed Cell State Estimation Method

Battery cell SoC can be estimated by continuously measured impedance during discharge. To estimate the cell state, a multi-sine signal with a small amplitude is applied to the cell operating current, and the cell impedance is measured through the amplitude of the voltage response. The multi-sine signal is the sum of two different frequencies, and each frequency is used to estimate the cell SoH and SoC.

If the current i expressed by Equation (1) flows through the battery cell, the cell voltage e can be expressed by Equation (2).

$$i = I_{dc} + \Delta I_f \cdot \sin(2\pi ft) \tag{1}$$

$$e = E_{dc} + \Delta E_f \cdot \sin(2\pi ft + \phi_f) \tag{2}$$

where I_{dc} is direct current (DC) bias, ΔI_f is the amplitude of the excited test frequency f, E_{dc} is the offset voltage, ΔE_f is the amplitude of the output voltage, and ϕ_f is the phase difference.

Dividing the voltage by the current as Equation (3) produces a complex impedance Z_f.

$$Z_f = \frac{\Delta E_f}{\Delta I_f} \cdot e^{j\phi_f} = |Z_f| \cdot e^{j\phi_f} = Z_f' + j \cdot Z_f'' \tag{3}$$

The electrochemical impedance of batteries depends on frequency and characterized by its modulus $|Z_f|$ and phase angle $e^{j\phi}$. Another expression is given as the real and imaginary parts of the complex impedance.

EIS measurements usually use a single-sine signal in which individual frequencies are measured sequentially, which is also known as stepped sine or frequency sweep. Therefore, single-sine EIS has the disadvantage that it takes a long time to acquire impedance in a wide frequency range. This disadvantage can be overcome by measuring several frequencies simultaneously. The method of measuring multiple frequencies at the same time is called multi-sine EIS. Multi-sine signals have already been used for impedance spectroscopy and transfer function measurements in biomedical applications [42,43], material characterization [44], and other fields such as battery measurements [26,28]. Nonetheless, multi-sine EIS is not a common method for estimating the in situ state of a battery cell. In general, the multi-sine EIS, like the single-sine EIS, requires the cell to be separated from the application circuit for impedance measurements. In the proposed battery SoC monitoring method, the sum signal of two test frequencies is excited to the cell operating current and its response voltage is measured. A Fourier transform is used on the sampled cell voltage to obtain the amplitude at each test frequency. The impedance at each test frequency is obtained by substituting the amplitude of each response voltage into Equation (3).

2. Experiment

2.1. Measurement System

The measurement system is configured to measure cell impedance by applying a multi-sine signal to the operating current. Table 1 shows the specifications of the Li-ion battery cell used and Figure 1 shows a block diagram and a picture of the measurement system.

Table 1. Specifications of the Li-ion cell.

Item	Description
Anode	Based on intercalation graphite
Cathode	Based on lithiated metal oxide [1]
Product name	Samsung ICR 18650-26F
Battery system	Li-ion
Nominal voltage	3.7 V
Rated capacity	2.6 Ah

[1] Consists of cobalt, nickel, and manganese.

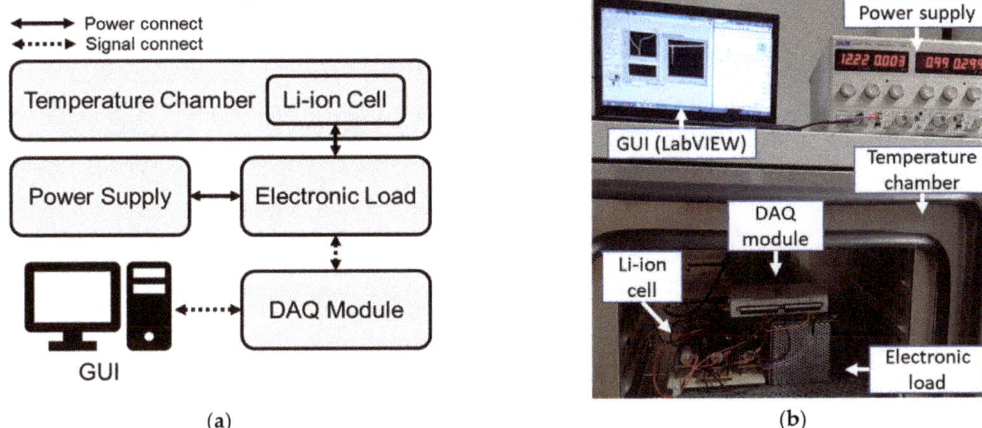

Figure 1. (**a**) A block diagram of the measurement system; (**b**) a photo of the measurement system.

The test battery cell is placed in the temperature chamber of the Binder GmbH set to 25 °C. NXP's silicon temperature sensor KTY 81-110 is attached to the cell to measure the actual cell surface temperature. The voltage signal output port of USB-6212, which is a data acquisition (DAQ) module of the Natural Instruments, is connected to an electronic load to apply test frequencies to the cell operating current. This DAQ module also acquires cell voltage, current, and temperature data. Experiments are controlled by adjusting measurement parameters through a graphical user interface (GUI) created using LabVIEW from National Instruments. Measured and calculated data is displayed on the monitor and can also be written to the hard disk. The analog inputs of the DAQ module have 16-bit analogue to digital converter (ADC) resolution, a maximum sample rate of 400 kS/s, and an input range of ±10 V. The analog output has 16-bit digital to analog converter (DAC) resolution, an output range of ±10 V, and a maximum update rate of 250 kS/s. The operation of the electronic load is described in the author's previous paper [45].

2.2. Definition of Key Terms

There is no standard definition that defines SoC and SoH for battery cells and may be defined differently depending on the battery application system. Therefore, it is necessary to clarify the terms defined in this paper before discussing cell state estimation.

- Definition of C-rate

In general, C-rate is used to indicate the rate of charge and discharge of a battery cell. The relationship shown in Equation (4) is given between the cell nominal capacity C_N and the operating current i.

$$h = C_N / i \tag{4}$$

where h is the time (hour) taken to completely discharge the battery cell, i is the current at which the cell operates, and C_N is the nominal capacity of the battery (Ah). Here, the reciprocal of the h value is defined as the C-rate.

- Definition of SoC

In this paper, the cell SoC is set through the following three steps:

(1) Full charge of the battery cell. The constant-current (CC) charging procedure at 1C and the constant-voltage (CV) charging procedure at 4.2 V are followed in sequence, where 4.2 V is the upper cut-off voltage specified by the cell manufacturer. This first step is complete when the charging current drops below C/10. Subsequently, each cell is given 90 min of relaxation time.

(2) Discharge the cell at 1C until it reaches a lower cut-off voltage. From this step, the actual cell capacity C_{cell} is obtained.

(3) Set the cell to the target SoC. To fully charge the cell, Stage 1 is performed again, and then the charge of the cell is consumed until the target SoC is reached. Since different cell relaxation times can introduce errors, the same relaxation times must be observed [46,47]. Each cell is given 90 min of relaxation time after the target SoC is set and before it is used in the experiment.

The SoC is then defined as Equation (5).

$$SoC = C_{residual}/C_{cell} \times 100~(\%) \tag{5}$$

The SoC of a cell is the ratio of the residual capacity ($C_{residual}$) to the total available capacity when the battery cell is fully charged (C_{cell}). The C_{cell} can be obtained from the second step of the procedure for setting up the cell SoC mentioned above. Depth of discharge (DoD) has the exact opposite definition of SoC, i.e., a cell of 80% SoC has the same meaning as a cell of 20% DoD.

- Definition of SoH

Battery cell SoH is defined as the ratio of the cell actual total available capacity C_{cell} to the cell initial total available capacity C_{init} and expressed by Equation (6).

$$SoH = C_{cell}/C_{init} \times 100~(\%) \tag{6}$$

In this paper, cell impedances from 100% to 80% SoH of cells are measured and compared. The EoL of a battery is reached when the energy content or power capacity is no longer sufficient for the application. This can vary depending on battery application; thus, there is no universal definition of how many cell SoHs have reached EoL. Nonetheless, publications assume that EoL is reached when the battery cell SoH is less than 80% [48–50].

2.3. Measurement Parameters

2.3.1. Selection of Test Amplitude

Depending on the battery cell application, the amplitude of the test frequency should be selected taking into account the trade-off between measurement accuracy and investment costs for battery management. In order for the appropriate test amplitude to be selected for impedance measurements, it must be selected to be small enough not to violate the linear criteria of the electrochemical battery system but large enough to obtain a suitable signal-to-noise ratio for robust measurements. The amplitude of the test signal must be small enough to satisfy the pseudo-linearity of the cell response voltage. This is crucial because the current–voltage curve of a Li-ion cell shows a nonlinear relationship that follows the Butler–Volmer kinetics. An excitation amplitude of 5 to 10 mV is generally recommended [51], e.g., if it is used for CC measurement (galvanostatic), the excitation amplitude of the current must be set so that the maximum absolute value of the voltage amplitude does not exceed 10 mV. Especially when a multi-signal signal is used, the maximum amplitude of the applied signal should be considered. Because the amplitudes at each frequency increase the total amplitude of the test signal, pseudo-linearity can be violated. For more test frequencies to be used, the amplitude of each frequency must be lowered, i.e., if a multi-sine signal of N frequencies is used, the output voltage amplitude at each frequency must be reduced to a maximum $10/N \cdot$mV. Amplitudes that are too small can be a problem as well. This is due to the finite resolution of the hardware that acquires and generates the test signal. Moreover, impedance measurements are not possible when the output voltage amplitude at individual frequencies becomes less than the system noise.

2.3.2. Selection of Test Frequencies

In a Li-ion battery cell, charged particles containing lithium ions are always involved in the electrode process. It is generally believed that the time constants of the movement or reaction of charged particles in these processes are different. Thus, impedance at different frequencies is related to different processes [52]. The influence of mass transfer, whose time constant is slower, is dominant at impedance at lower frequencies. Mass transfer (including diffusion) becomes faster as the temperature is higher and as the concentration of Li-ions increases (as DoD is lower). Faster transfer speed can be interpreted as lower impedance. On the other hand, the lower the cell temperature or the higher the DoD, the slower the mass transport [53], which appears to be a higher impedance in the lower frequency. This means that impedance at lower frequency can be more advantageous as it is used for cell SoC estimation. Studies on battery cell impedance consistently show that impedance at lower frequencies is more dependent on changes in cell SoC. However, since the frequency is the reciprocal of the period, the lower the frequency, the longer the period required to measure each continuous impedance. For example, if 1 mHz is used for the measurement, it takes about 17 min to collect only one period of signal. Therefore, the test frequency should be selected considering the time interval required for state estimation depending on the battery application. There is one more problem: the longer the measurement of one period, the greater the change in the internal state of the cell such as SoC and cell internal temperature. This also increases the error in impedance measurements. In this paper, 1 Hz where one period of signal can be collected per second is used as an example for SoC estimation.

There are also some considerations for higher frequencies to be used. In this case, the achievable sampling rate in the hardware can be a constraint. In theory, satisfying the Nyquist-Shannon criteria is sufficient for the signal to be restored, but, generally, oversampling is required due to noise and non-ideal properties in analog filters. Another problem is that the cell impedance is generally lower at higher frequencies, which can lead to the same problems when the test amplitude is too low. Most cell degradation results in an increase in internal resistance [54–57], and the 1 kHz impedance of the cell used in this paper represents the ohmic resistance of the cell in which the imaginary part of the impedance becomes 0. In paper [46], the accuracy of SoH estimation with impedance at different frequencies is compared, which shows that impedance at higher frequencies is more advantageous for use in SoH estimation, which is less affected by cell SoC and temperature. In this paper, a frequency of 1 kHz is used for SoH estimation as an example.

3. Measurement Results

The measurement system and parameters described in Section 2 are used for cell impedance measurements. The overall process of cell state estimation in this paper is shown in Figure 2 as a flow chart.

Table 2 shows the experimental conditions. Cell impedance is measured every second while discharging from 0% to 100% DoD for each cell with 100% to 80% SoH. A DC bias of 1C and an excitation amplitude of 130 mA are applied, i.e., an amplitude of 65 mA is given at each frequency of 1 Hz and 1 kHz. This amplitude is chosen for the cell output voltage amplitude to be less than 10 mV. The multi-sine test signal, the sum signal of the two frequencies, applies to the cell DC bias, and the impedance at each frequency is measured simultaneously.

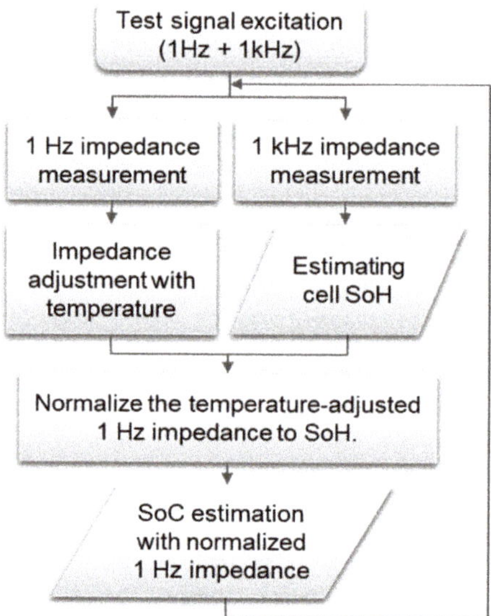

Figure 2. Flow chart showing the cell state estimation process.

Table 2. Experimental conditions for measuring cell impedance during discharge.

Parameter	Description
Chamber temperature	25 °C
Depth of discharge	From 0 to 100%
DC bias	2600 mA (1C)
f_1, f_2	1 Hz, 1 kHz
ΔI_1, ΔI_2	65 mA (each)

Figure 3 shows continuous 1 Hz and 1 kHz impedance during discharging of cells with different SoHs.

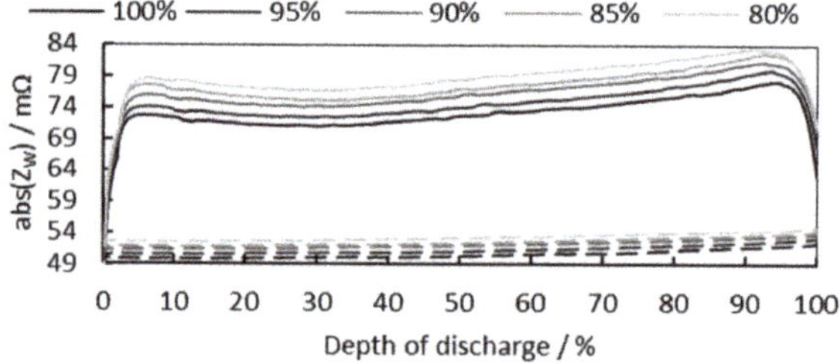

Figure 3. 1 Hz and 1 kHz cell impedance during discharge of cells with different SoHs. The solid lines represent the 1 Hz impedance, the dotted lines represent the 1 kHz impedance, and the different brightness of the line represents the different cell SoH.

As shown in Figure 3, the lower the cell SoH, the higher the cell impedance, and while 1 kHz impedance is relatively constant for DoD changes, 1 Hz impedance is more affected by cell DoD. While the cell is completely discharged, the increased rate of the highest value to the lowest value of 1 kHz impedance is 3.74%. At the same time, for 1 Hz impedance, the increased rate of the highest value to the lowest value between DoD 5% and 95% is 10.16%. The 1 kHz impedance shows a smaller increased rate, and it indicates less susceptibility to changes in cell DoD. Therefore, the 1 kHz impedance is more suitable to be used for

SoH estimation of cells for which the cell SoC is unknown. Contrastively, 1 Hz impedance shows larger increase rate, which means that it is more suitable for cell SoC estimation.

For the calculation of the increase rate, 1 Hz impedance between DoD 5% and 95% in Figure 3 are used (if the full range of DoD 0% to 100% is considered, the difference in impedance deviation at 1 Hz and 1 kHz will be more prominent.). The reason is that at the beginning and end of cell discharge, the 1 Hz impedance is not measured correctly and cannot be used for cell SoC estimation. The voltage response of an electrochemical cell is not only affected by the cell impedance. The cell discharge process causes an overpotential, which is the voltage loss described by polarization. Polarization refers to a phenomenon in which the electrode potential becomes excessive or insufficient in an equilibrium state. During the reaction process in a battery cell, the rate of charge transfer process occurring in each cell component is not the same. If this rate is relatively slow for a particular process, it becomes the rate-limiting process for the entire reaction of the cell. As the cell discharges, current flows between both terminals, causing the voltage to be measured below its equilibrium potential. In this case, the difference between the voltage at both terminals and the equilibrium voltage is called overpotential, indicating the degree of polarization. Figure 4 shows the three polarization regions appearing in the typical discharge curve of a Li-ion battery cell. In the cell discharge process, polarization can be classified into three categories: activation polarization, concentration polarization, and ohmic polarization [58].

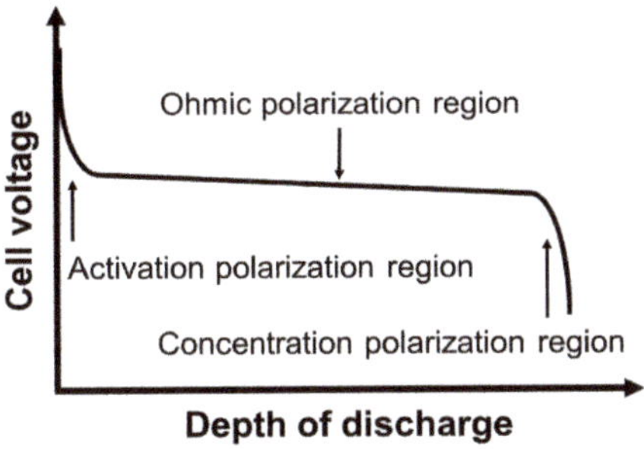

Figure 4. Typical discharge curve of a battery cell, showing three different regions of polarization.

Ohmic polarization, also referred to as ohmic loss, is caused by the current flowing through the internal resistance of the battery cell. The greater the cell internal impedance, the greater the operating voltage drop as the ohmic losses increase proportional to the current density. Therefore, in this region, it has a slope related to the cell impedance and shows the most linearity when charging and discharging the cell. Due to this linearity, impedance can be measured accurately; hence, it is the most suitable region for cell state estimation through impedance measurement. Meanwhile, the cell output voltage in the activation and concentration polarization regions has nonlinearity. Activation polarization is due to various delay factors inherent in the dynamics of electrochemical reactions, such as the work function that ions must overcome at the junction between the electrode and the electrolyte. It has a dominant effect at the beginning of cell discharge. Concentration polarization takes into account the resistance that ions face in the process of mass transfer (e.g., diffusion) as they move across the electrolyte and from one electrode to another. This polarization has a dominant effect at high cell DoD. In conclusion, the cell voltage drops significantly non-linearly as the cell discharges in the region of activation and concentration polarization. Because this nonlinearity is based on slow response, especially in battery cells, the lower the frequency, the more difficult it is to obtain an accurate cell impedance. As shown in Figure 3, the effect of activation and concentration polarization at a frequency of 1 Hz is more pronounced than at 1 kHz, because slow transport processes are well represented at lower frequencies. On the other hand, slow transport reactions cannot be detected at higher frequencies, but the effects of fast kinetics can be revealed.

3.1. Cell SoH Estimation

The measured 1 kHz impedance in Figure 3 is used to estimate the cell SoH. Figure 5 shows the average of the 1 kHz impedance at each SoH and the fitted line.

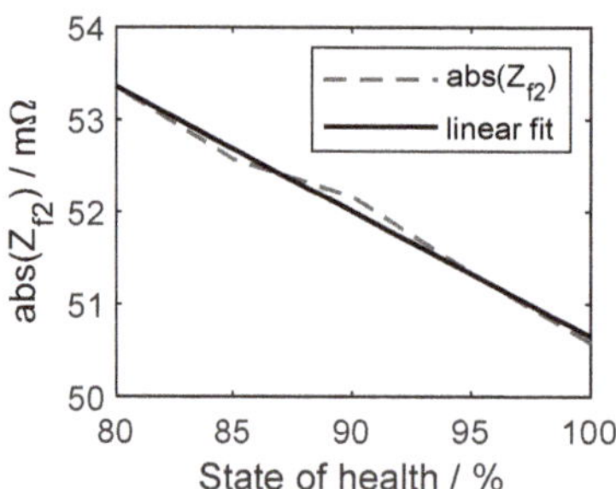

Figure 5. Averages of 1 kHz impedance at each cell SoH (dotted line) and linear fit (solid line).

The fitted line in Figure 5 is obtained through linear regression of these 1 kHz impedances measured at each SoH, and it is expressed as Equation (7).

$$Z_{f2} = -0.1355 \cdot \text{SoH} + 64.2 \tag{7}$$

The measured 1 kHz impedance can be used to estimate the cell SoH using Equation (7). Here, root mean squared error (RMSE) is 0.21 mΩ and R^2 is 0.99. R^2 represents the goodness of fit of the estimated model. The closer this value is to 1, the better the estimate matches the target value.

3.2. Cell SoC Estimation

The cell 1 Hz impedance is greatly influenced by not only the cell SoC but also the cell SoH and cell temperature. Therefore, to be used for cell SoC estimation, the 1 Hz impedance is adjusted as cell temperature in Section 3.2.1 and normalized as cell SoH in Section 3.2.2.

3.2.1. Consideration of Cell Temperature

The change in temperature of a battery cell mainly depends on the C-rate, discharge time, relaxation time, and ambient temperature [59]. Figure 6 shows the cell temperature that rises when the cell is discharged at 1C current from DoD 0% to 100% in a temperature chamber set to 25 °C (same experimental conditions as shown in Table 2).

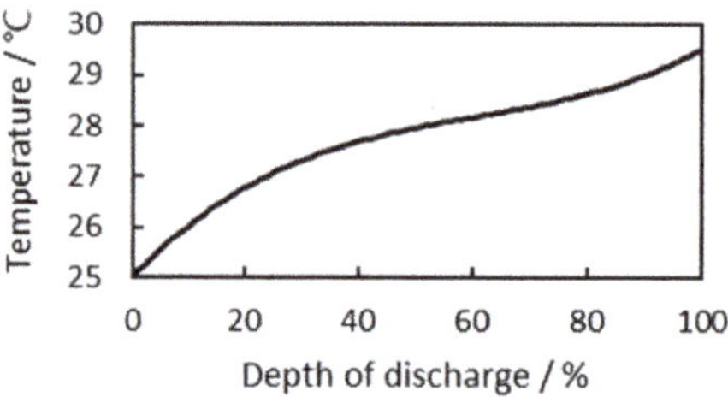

Figure 6. The temperature change of the battery cell during discharging at 1C current.

To measure the cell impedance at each constant temperature, cell impedances are measured through an EIS instrument (IM6ex from Zahner-Elektrik GmbH & CoKG) in a temperature chamber set to 20, 25, 30, and 35 °C. The measurement conditions are shown in Table 3.

Table 3. Parameters for EIS measurement (Galvanostatic).

Parameter	Description
Cell temperature	20, 25, 30, 35 °C
State of health	80%
State of charge	50%
DC bias	0 mA
Frequency range	0.2 to 2 kHz
AC amplitude	100 mA

The measurement results are shown in Figure 7 as a Nyquist plot and a Bode plot.

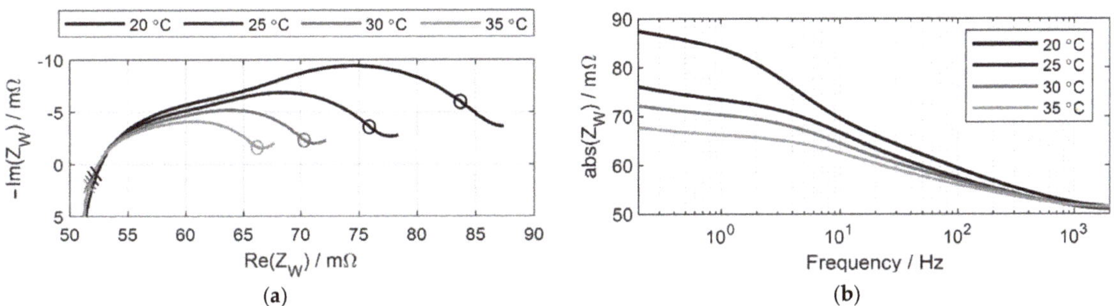

Figure 7. Cell impedance at each frequency at 20, 25, 30, and 35 °C. (**a**) Nyquist plot (O: 1 Hz, X: 1 kHz); (**b**) Bode plot.

As shown in Figure 7, the higher the cell temperature, the lower the cell impedance, and the 1 Hz impedance is more affected by temperature than the 1 kHz impedance. Figure 8 shows the 1 Hz impedance measured at each temperature and a curve fitted with a quadratic equation.

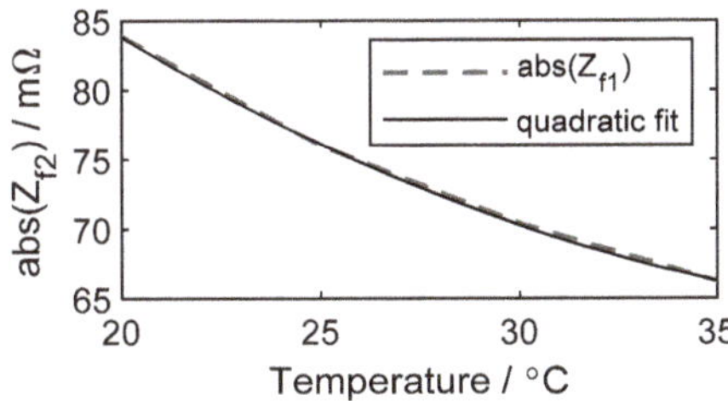

Figure 8. 1 Hz impedance measured at 20, 25, 30, and 35 °C (dotted line) and fitted curve (solid line).

Equation (8) is the quadratic equation fitted in Figure 8, where R^2 is 1.00 and RMSE is 0.19 mΩ.

$$Z(T)_{\text{fitted}} = -0.03717 \cdot T^2 - 3.217 \cdot T + 133.3 \tag{8}$$

where T is the cell temperature, and $Z(T)_{\text{fitted}}$ is the impedance obtained through the equation at the cell temperature T.

The measured 1 Hz impedance at various cell temperatures is uniformly adjusted to the cell impedance at 25 °C through Equation (9).

$$Z(T)_{\text{adjusted}} = Z(T)_{\text{measured}} - \left(Z(T)_{\text{fitted}} - Z(25)_{\text{fitted}} \right) \tag{9}$$

where $Z(T)_{\text{adjusted}}$ is the adjusted impedance to 25 °C and $Z(T)_{\text{measured}}$ is the measured impedance at temperature T.

The result of Equation (9) applied to the 1 Hz impedance in Figure 3 is shown in Figure 9.

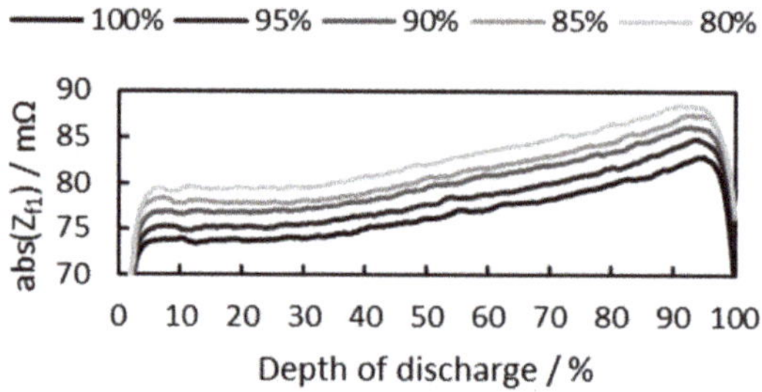

Figure 9. 1 Hz impedance adjusted to 25 °C.

In Figure 9, the 1 Hz impedance increases with increasing cell DoD. The higher the cell DoD, the higher the cell temperature (Figure 6), and as the cell temperature increases, the cell impedance decreases (Figure 8). Figure 9 shows the impedance when the impedance of Figure 3 is adjusted to a lower temperature of 25 °C. Therefore, the temperature-adjusted 1 Hz impedance in Figure 9 becomes higher than the 1 Hz impedance in Figure 3.

3.2.2. Consideration of Cell SoH

As mentioned above, cell 1 Hz impedance is affected by cell SoH as well as cell temperature. The cell SoH can be estimated via the 1 kHz impedance even if the cell SoC is unknown (Section 3.1). The cell SoH obtained in Equation (7) is used for normalizing the 1 Hz impedance. The measured cell 1 Hz impedance ($Z_{measured}$) can be normalized to Z_{norm} by Equation (10).

$$Z_{norm} = (Z_{measured} - Z_{min})/(Z_{max} - Z_{min}) \times 100 \tag{10}$$

In Equation (10), Z_{max} and Z_{min} represent the maximum and minimum values of the impedance between DoD 10% and 90% and can be obtained by Equations (11) and (12), respectively. In Equation (11), R^2 is 1.00 and RMSE is 0.12 mΩ, and in Equation (12), R^2 is 1.00 and RMSE is 0.26 mΩ.

$$Z_{max} = -0.00593 \cdot SoH^2 + 0.79922\,SoH + 63.09 \tag{11}$$

$$Z_{min} = -0.2969 \cdot SoH + 103.1 \tag{12}$$

The result of applying Equation (10) to the 1 Hz impedance in Figure 9 is shown in Figure 10.

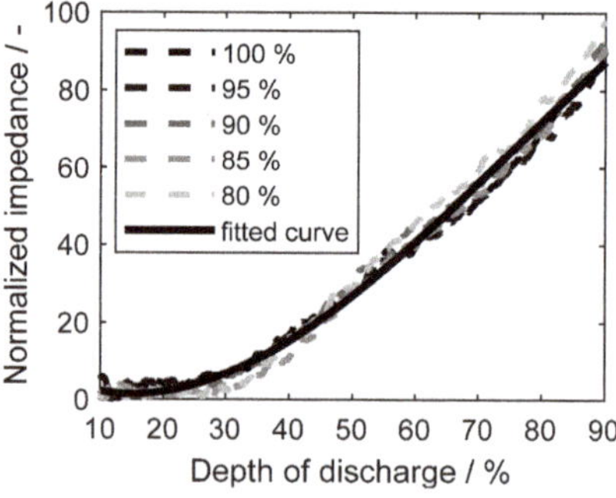

Figure 10. Normalized 1 Hz impedance vs. DoD.

The average of the normalized impedance at each SoH is fitted with a cubic equation and plotted as a curve in Figure 10. This fitted curve is expressed by Equation (13) and has an R^2 of 1.00 and an RMSE of 1.51.

$$Z_{norm} = -0.00014 \cdot SoC^3 + 0.032 \cdot SoC^2 - 0.85 \cdot SoC + 7.58 \tag{13}$$

Table 4 shows each R^2 and RMSE when Equation (13) is applied to cells with different SoHs.

Table 4. Accuracy of SoC estimation in cells with different SoHs.

SoH	R^2	RMSE
100%	0.99	2.42
95%	0.99	2.66
90%	0.99	2.02
85%	0.99	2.51
80%	0.98	3.85
Average	0.99	2.69

Because of the cell nonlinear response mentioned above, SoC estimation using 1 Hz impedance is made here between DoD 10% and 90%.

3.2.3. Battery Cell SoC Estimation at Different Initial SoCs

Section 3.2.2 shows the results of SoC estimation when a fully charged cell is fully discharged. However, battery cells are not always operated in a fully charged state. Here, battery cell SoC estimation at different cell initial SoCs is emulated. A total of 10 min of cell discharge and 60 min of cell relaxation are repeated until the cell is completely discharged. The cell impedance is measured every second while the cell is discharging, and each cell state after the relaxations represents the cell state at different initial SoCs. The experimental conditions are shown in Table 5.

Table 5. Experimental conditions for SoC estimation of cells that start operating at different SoCs.

Parameter	Description
State of health	80%
Chamber temperature	25 °C
Discharge time	10 min (each)
Relaxation time	60 min (each)
Depth of discharge	From 0 to 100%
DC bias	2600 mA (1C)
f_1, f_2	1, 1000 Hz
ΔI_1, ΔI_2	65 mA (each)

Figure 11a shows the cell voltage and cell temperature for the cell DoD, and Figure 11b shows the 1 Hz and 1 kHz impedances.

As shown in Figure 11a, the cell voltage drops while the cell is discharging and increases during each relaxation. At the same time, the cell temperature increases while the cell is discharged and decreases during the relaxation time. Figure 11b shows that cell 1 Hz impedance is more affected by DoD than 1 kHz impedance. The cell 1 kHz impedance is relatively constant for DoD changes and can be used for cell SoH estimation as shown in Section 3.1, while the 1 Hz impedance is highly influenced by DoD, and it is used for SoC estimation.

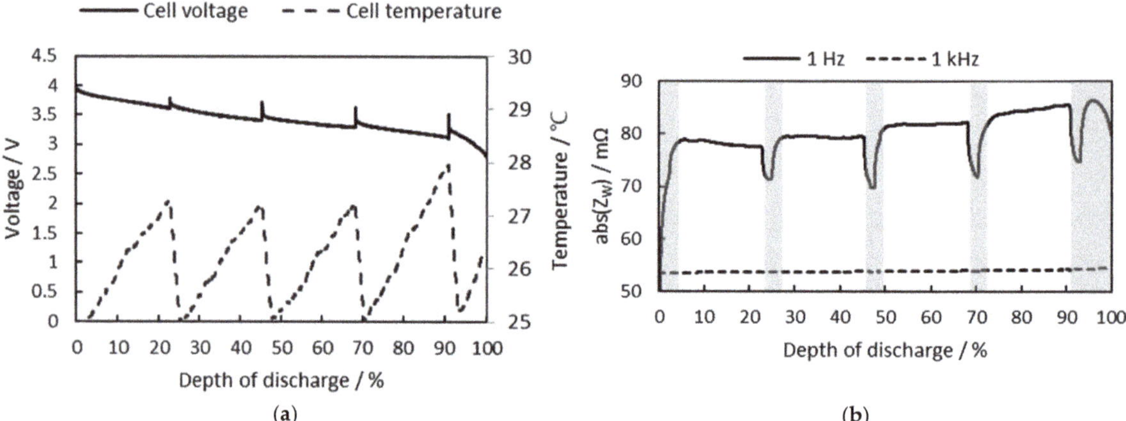

Figure 11. (**a**) Cell voltage (solid line) and cell temperature (dotted line) vs. DoD; (**b**) 1 Hz impedance (solid line) and 1 kHz impedance (dotted line) vs. DoD.

In Figure 11b, the DoD ranges where the 1 Hz impedance cannot be measured correctly are shown in gray. As explained earlier in Chapter 3, the cell 1 Hz impedance is not measured correctly at the beginning of each discharge, especially due to the nonlinearity of the cell discharge curve due to activation polarization shown in Figure 4. Each of these gray ranges in Figure 11b corresponds to 265 s. In addition, the range from DoD 95% in Figure 11b is also included in these gray ranges, as this range corresponds to the concentration polarization region in Figure 4. However, it should be noted that in most cell operations where cells are rarely fully discharged, a range that cell impedance cannot be measured correctly only appear once in the beginning of cell discharge. In Figure 11b, multiple gray ranges are shown because the impedance at the beginning of the discharge in different initial SoCs is shown in a single figure.

The 1 Hz impedance of Figure 11b adjusted to the cell temperature is shown in Figure 12a. Equation (9) is used to take into account the effects of cell temperature. The temperature-adjusted 1 Hz impedance is then normalized to the cell SoH, and the result is shown as a dotted line in Figure 12b. Equation (10) is used to normalize the temperature-adjusted impedance to the cell SoH.

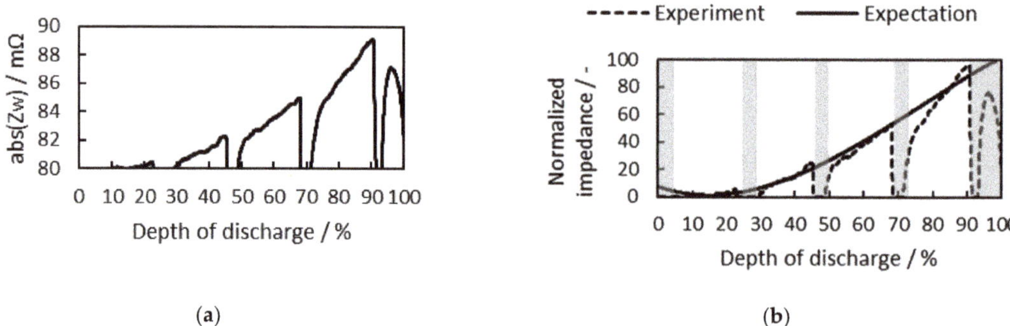

Figure 12. (**a**) 1 Hz impedance adjusted to cell temperature vs. DoD; (**b**) temperature-adjusted 1 Hz impedance normalized to cell SoH (dotted line) and expected value (solid line) vs. DoD.

The cell SoC can be estimated by applying Equation (13) to Figure 12b, where R^2 is 0.99 and RMSE is 3.07 mΩ. Cell impedance in grayed-out ranges is excluded from SoC estimation. These ranges include a range of 265 s after the start of the discharge and a range of cell DoD exceeding 95%. The purpose of this experiment is to show that SoC estimation is possible even when discharges are initiated from different cell SoCs, so it can lead to misunderstanding that there are too many grayed-out ranges where SoC estimation is impossible. Again, in most cases when battery cells are used, this grayed-out range appears only once a battery cell begins to discharge and rarely twice when the cell is completely discharged.

4. Conclusions and Discussion

This paper introduces a SoC monitoring method of Li-ion battery cells using impedance measurement. For accurate SoC estimation, the estimated cell SoH is considered with the measured temperature. Unlike traditional EIS measurement methods, the proposed method does not require impedance at wide frequencies, saving measurement time and simplifying measurement saves hardware costs. Especially, a multi-sine signal is applied to measure the cell impedance at two frequencies simultaneously. The cell impedance at 1 kHz for SoH estimation and the cell impedance at 1 Hz for SoC estimation are used. As a result, the proposed cell SoC monitoring method enables simultaneous estimation of unknown cell SoH and SoC. This is verified through an experiment in several different initial SoCs in a cell as well.

One problem to be pointed out is that at the start of battery cell discharge (ca. 265 s at 1C) and in the high DoD range (over ca. 95%), accurate cell impedance measurement is not possible due to the large nonlinearity in the cell voltage response; hence, the SoC cannot be correctly estimated. Nonetheless, the proposed SoC estimation method using impedance can be used together with the SoC estimation method using cell voltage without any additional hardware or measurement. This is because the proposed method already measures the cell voltage response to obtain cell impedance. At the same time, this SoC estimation method using cell impedance can compensate for the weaknesses of the existing SoC estimation methods using cell voltage. The voltage of the Li-ion battery cell drops significantly at the beginning and end of the discharge. This enables SoC estimation as a simple method through cell voltage measurement. However, the decrease in voltage in the middle region of the cell discharge curve, the so-called "flat plateau", is not significantly noticeable, and this is one of the factors that makes it difficult to estimate SoC simply with the measured cell voltage. In particular, this flat plateau is notorious for estimating the SoC of lithium iron phosphate (LFP) battery cells. Contrariwise, this flat plateau makes impedance measurements more accurate. This is because accurate impedance measurement is possible when the target system is linear. That is, when the proposed method is used, SoC and SoH estimation is possible on this flat plateau and even more accurate.

In this paper, only the experimental results of lithium nickel manganese cobalt oxide (NMC) type cells are shown. The proposed method is based on EIS, which has already been validated in other types of battery cells through numerous literatures. Thus, it can be expected to be applied to other types of Li-ion cells. Nevertheless, it is worth comparing the results of applying this method to other types of cells in future studies.

This paper only deals with SoC estimation during battery cell discharge. Accurate SoC estimation during cell charging is also worth studying. However, there is the problem that this method is only possible during CC charging. This is because the offset current changes frequently during CV charging, making it impossible to measure the impedance correctly. If SoC estimation is possible only during CC charging, the range of SoC that can be estimated is limited. The authors evaluate that monitoring SoC under this limited condition does not have a definite advantage; hence, this is not covered in this paper.

Lastly, the experiments in this paper consider the estimation of the state of a cell discharged at room temperature. Therefore, the experiment is conducted between 20 and 35 °C. However, since the cell state estimation method in this paper has the advantage of considering the effect of cell temperature on impedance, experiments will be conducted in a wider temperature range for wider application of this method in the future.

Author Contributions: Conceptualization, methodology, validation, software, hardware implementation, writing—original draft preparation J.K. (Jonghyeon Kim); writing—review and editing, supervision, counseling J.K. (Julia Kowal); All authors have read and agreed to the published version of the manuscript.

Funding: This research was funded by DAAD (German Academic Exchange Service), Research Grants—Doctoral Programmes in Germany.

Institutional Review Board Statement: Not applicable.

Informed Consent Statement: Not applicable.

Data Availability Statement: Not applicable.

Conflicts of Interest: The authors declare no conflict of interest. The funder had no role in the design of the study; in the collection, analyses, or interpretation of data; in the writing of the manuscript; or in the decision to publish the results.

References

1. Tarascon, J.-M.; Armand, M. Issues and challenges facing rechargeable lithium batteries. In *Materials for Sustainable Energy*; Nature Publishing Group: Berlin, Germany, 2011; pp. 171–179. [CrossRef]
2. Goodenough, J.B.; Kim, Y. Challenges for rechargeable batteries. *J. Power Source* **2011**, *196*, 6688–6694. [CrossRef]
3. Lu, L.; Han, X.; Li, J.; Hua, J.; Ouyang, M. A review on the key issues for lithium-ion battery management in electric vehicles. *J. Power Source* **2013**, *226*, 272–288. [CrossRef]
4. Goodenough, J.B.; Kim, Y. Challenges for rechargeable Li batteries. *Chem. Mater.* **2010**, *22*, 587–603. [CrossRef]
5. Armand, M.; Tarascon, J.-M. Building better batteries. *Nature* **2008**, *451*, 652–657. [CrossRef] [PubMed]
6. Biensan, P.; Simon, B.; Peres, J.P.; de Guibert, A.; Broussely, M.; Bodet, J.M.; Perton, F. On safety of lithium-ion cells. *J. Power Source* **1999**, *81*, 906–912. [CrossRef]
7. Doughty, D.H.; Roth, E.P. A general discussion of Li ion battery safety. *Electrochem. Soc. Interface* **2012**, *21*, 37.
8. Belov, D.; Yang, M. Failure mechanism of Li-ion battery at overcharge conditions. *J. Solid State Electrochem.* **2008**, *12*, 885–894. [CrossRef]
9. Yuan, Q.; Zhao, F.; Wang, W.; Zhao, Y.; Liang, Z.; Yan, D. Overcharge failure investigation of lithium-ion batteries. *Electrochim. Acta* **2015**, *178*, 682–688. [CrossRef]
10. Tobishima, S.; Yamaki, J. Aconsideration of lithium cell safety. *J. Power Source* **1999**, *81*, 882–886. [CrossRef]
11. Leising, R.A.; Palazzo, M.J.; Takeuchi, E.S.; Takeuchi, K.J. Abuse testing of lithium-ion batteries: Characterization of the overcharge reaction of LiCoO$_2$/graphite cells. *J. Electrochem. Soc.* **2001**, *148*, A838. [CrossRef]
12. Finegan, D.P.; Scheel, M.; Robinson, J.B.; Tjaden, B.; di Michiel, M.; Hinds, G.; Brett, D.J.L.; Shearing, P.R. Investigating lithium-ion battery materials during overcharge-induced thermal runaway: An operando and multi-scale X-ray CT study. *Phys. Chem. Chem. Phys.* **2016**, *18*, 30912–30919. [CrossRef] [PubMed]
13. Sun, J.; Li, J.; Zhou, T.; Yang, K.; Wei, S.; Tang, N.; Dang, N.; Li, H.; Qiu, X.; Chen, L. Toxicity, a serious concern of thermal runaway from commercial Li-ion battery. *Nano Energy* **2016**, *27*, 313–319. [CrossRef]
14. Larsson, F.; Mellander, B.-E. Abuse by external heating, overcharge and short circuiting of commercial lithium-ion battery cells. *J. Electrochem. Soc.* **2014**, *161*, A1611. [CrossRef]
15. Wang, Q.; Ping, P.; Zhao, X.; Chu, G.; Sun, J.; Chen, C. Thermal runaway caused fire and explosion of lithium ion battery. *J. Power Source* **2012**, *208*, 210–224. [CrossRef]
16. Zheng, Y.; Qian, K.; Luo, D.; Li, Y.; Lu, Q.; Li, B.; He, Y.-B.; Wang, X.; Li, J.; Kang, F. Influence of over-discharge on the lifetime and performance of LiFePO$_4$/graphite batteries. *RSC Adv.* **2016**, *6*, 30474–30483. [CrossRef]
17. Zhang, L.; Ma, Y.; Cheng, X.; Du, C.; Guan, T.; Cui, Y.; Sun, S.; Zuo, P.; Gao, Y.; Yin, G. Capacity fading mechanism during long-term cycling of over-discharged LiCoO$_2$/mesocarbon microbeads battery. *J. Power Source* **2015**, *293*, 1006–1015. [CrossRef]
18. Maleki, H.; Howard, J.N. Effects of overdischarge on performance and thermal stability of a Li-ion cell. *J. Power Source* **2006**, *160*, 1395–1402. [CrossRef]
19. Li, H.-F.; Gao, J.-K.; Zhang, S.-L. Effect of overdischarge on swelling and recharge performance of lithium ion cells. *Chin. J. Chem.* **2008**, *26*, 1585–1588. [CrossRef]
20. Piller, S.; Perrin, M.; Jossen, A. Methods for state-of-charge determination and their applications. *J. Power Source* **2001**, *96*, 113–120. [CrossRef]
21. Pop, V.; Bergveld, H.J.; Danilov, D.; Regtien, P.P.L.; Notten, P.H.L. State-of-the-art of battery state-of-charge determination. In *Battery Management Systems: Accurate State-of-Charge Indication for Battery-Powered Applications*; Springer Science & Business Media: Berlin/Heidelberg, Germany, 2008; pp. 11–45. [CrossRef]
22. Rodrigues, S.; Munichandraiah, N.; Shukla, A.K. A review of state-of-charge indication of batteries by means of ac impedance measurements. *J. Power Source* **2000**, *87*, 12–20. [CrossRef]

23. Plett, G.L. Extended Kalman filtering for battery management systems of LiPB-based HEV battery packs. *J. Power Source* **2004**, *134*, 277–292. [CrossRef]

24. Kim, I.-S. The novel state of charge estimation method for lithium battery using sliding mode observer. *J. Power Source* **2006**, *163*, 584–590. [CrossRef]

25. Kim, I.-S. Nonlinear State of Charge Estimator for Hybrid Electric Vehicle Battery. *IEEE Trans. Power Electron.* **2008**, *23*, 2027–2034. [CrossRef]

26. Barsoukov, E.; Macdonald, J.R. *Impedance Spectroscopy: Theory, Experiment, and Applications*, 3rd ed.; Barsoukov, E., Macdonald, J.R., Eds.; John Wiley & Sons: Hoboken, NJ, USA, 2017; ISBN 978-1-119-07408-3.

27. Orazem, M.E.; Tribollet, B. *Electrochemical Impedance Spectroscopy*; John Wiley & Sons: Hoboken, NJ, USA, 2008; pp. 383–389.

28. Tröltzsch, U.; Kanoun, O.; Tränkler, H.-R. Characterizing aging effects of lithium ion batteries by impedance spectroscopy. *Electrochim. Acta* **2006**, *51*, 1664–1672. [CrossRef]

29. Buller, S.; Thele, M.; Karden, E.; de Doncker, R.W. Impedance-based non-linear dynamic battery modeling for automotive applications. *J. Power Source* **2003**, *113*, 422–430. [CrossRef]

30. Buller, S. *Impedance-Based Simulation Models for Energy Storage Devices in Advanced Automotive Power Systems*; Shaker Verlag GmbH: Düren, Germany, 2003; ISBN 3832212256.

31. Huet, F. A review of impedance measurements for determination of the state-of-charge or state-of-health of secondary batteries. *J. Power Source* **1998**, *70*, 59–69. [CrossRef]

32. Fleischer, C.; Waag, W.; Heyn, H.M.; Sauer, D.U. On-line adaptive battery impedance parameter and state estimation considering physical principles in reduced order equivalent circuit battery models part 2. Parameter and state estimation. *J. Power Source* **2014**, *262*, 457–482. [CrossRef]

33. Vetter, J.; Novák, P.; Wagner, M.R.; Veit, C.; Möller, K.-C.; Besenhard, J.O.; Winter, M.; Wohlfahrt-Mehrens, M.; Vogler, C.; Hammouche, A. Ageing mechanisms in lithium-ion batteries. *J. Power Source* **2005**, *147*, 269–281. [CrossRef]

34. Abraham, D.P.; Knuth, J.L.; Dees, D.W.; Bloom, I.; Christophersen, J.P. Performance degradation of high-power lithium-ion cells—Electrochemistry of harvested electrodes. *J. Power Source* **2007**, *170*, 465–475. [CrossRef]

35. Thomas, E.V.; Bloom, I.; Christophersen, J.P.; Battaglia, V.S. Statistical methodology for predicting the life of lithium-ion cells via accelerated degradation testing. *J. Power Source* **2008**, *184*, 312–317. [CrossRef]

36. Belt, J.R.; Ho, C.D.; Miller, T.J.; Habib, M.A.; Duong, T.Q. The effect of temperature on capacity and power in cycled lithium ion batteries. *J. Power Source* **2005**, *142*, 354–360. [CrossRef]

37. Qahouq, J.A.A. Online battery impedance spectrum measurement method. In Proceedings of the 2016 IEEE Applied Power Electronics Conference and Exposition (APEC), Long Beach, CA, USA, 20–24 March 2016; pp. 3611–3615, ISBN 1467395501.

38. Waag, W.; Fleischer, C.; Sauer, D.U. On-line estimation of lithium-ion battery impedance parameters using a novel varied-parameters approach. *J. Power Source* **2013**, *237*, 260–269. [CrossRef]

39. Do, D.V.; Forgez, C.; Benkara, K.E.K.; Friedrich, G. Impedance observer for a Li-ion battery using Kalman filter. *IEEE Trans. Veh. Technol.* **2009**, *58*, 3930–3937.

40. Huang, W.; Qahouq, J.A.A. An online battery impedance measurement method using DC–DC power converter control. *IEEE Trans. Ind. Electron.* **2014**, *61*, 5987–5995. [CrossRef]

41. Howey, D.A.; Mitcheson, P.D.; Yufit, V.; Offer, G.J.; Brandon, N.P. Online measurement of battery impedance using motor controller excitation. *IEEE Trans. Veh. Technol.* **2013**, *63*, 2557–2566. [CrossRef]

42. Ojarand, J.; Min, M.; Annus, P. Crest factor optimization of the multisine waveform for bioimpedance spectroscopy. *Physiol. Meas.* **2014**, *35*, 1019. [CrossRef]

43. Sanchez, B.; Vandersteen, G.; Bragos, R.; Schoukens, J. Optimal multisine excitation design for broadband electrical impedance spectroscopy. *Meas. Sci. Technol.* **2011**, *22*, 115601. [CrossRef]

44. Breugelmans, T.; Tourwé, E.; van Ingelgem, Y.; Wielant, J.; Hauffman, T.; Hausbrand, R.; Pintelon, R.; Hubin, A. Odd random phase multisine EIS as a detection method for the onset of corrosion of coated steel. *Electrochem. Commun.* **2010**, *12*, 2–5. [CrossRef]

45. Kim, J.; Krüger, L.; Kowal, J. On-line state-of-health estimation of Lithium-ion battery cells using frequency excitation. *J. Energy Storage* **2020**, *32*, 101841. [CrossRef]

46. Barai, A.; Gael, H.C.; Guo, Y.; McGordon, A.; Jennings, P. A study on the impact of lithium-ion cell relaxation on electrochemical impedance spectroscopy. *J. Power Source* **2015**, *280*, 74–80. [CrossRef]

47. Frank, M.K.; Noel, A.; Simon, V.E.; Jossen, A. Long-term equalization effects in Li-ion batteries due to local state of charge inhomogeneities and their impact on impedance measurements. *Electrochim. Acta* **2015**, *185*, 107–116. [CrossRef]

48. Wood, E.; Alexander, M.; Thomas, H.B. Investigation of battery end-of-life conditions for plug-in hybrid electric vehicles. *J. Power Source* **2011**, *196*, 5147–5154. [CrossRef]

49. Faria, R.; Marques, P.; Garcia, R.; Moura, P.; Freire, F.; Delgado, J.; de Aníbal, T. Primary and secondary use of electric mobility batteries from a life cycle perspective. *J. Power Source* **2014**, *262*, 169–177. [CrossRef]

50. Han, X.; Ouyang, M.; Lu, L.; Li, J. A comparative study of commercial lithium ion battery cycle life in electric vehicle: Capacity loss estimation. *J. Power Source* **2014**, *268*, 658–669. [CrossRef]

51. Stoynov, Z.B.; Vladjkova, D.E. Measurement Methods | Electrochemical: Impedance spectroscopy. In *Encyclopedia of Electrochemical Power Source*; Elsevier: Amsterdam, The Netherlands, 2009; pp. 632–642. [CrossRef]

52. Wang, X.; Wei, X.; Zhu, J.; Dai, H.; Zheng, Y.; Xu, X.; Chen, Q. A review of modeling, acquisition, and application of lithium-ion battery impedance for onboard battery management. *eTransportation* **2021**, *7*, 100093. [CrossRef]
53. Jiang, K.; Liu, X.; Lou, G.; Wen, Z.; Liu, L. Parameter sensitivity analysis and cathode structure optimization of a non-aqueous Li–O$_2$ battery model. *J. Power Source* **2020**, *451*, 227821. [CrossRef]
54. Liu, L.; Zhu, M. Modeling of SEI Layer Growth and Electrochemical impedance spectroscopy response using a thermal-electrochemical model of li-ion batteries. *ECS Trans.* **2014**, *61*, 43–61. [CrossRef]
55. Liu, L.; Guan, P. Phase-field modeling of solid electrolyte interphase (SEI) evolution: Considering cracking and dissolution during battery cycling. *ECS Trans.* **2019**, *89*, 101–111. [CrossRef]
56. Stiaszny, B.; Ziegler, J.C.; Krauß, E.E.; Zhang, M.; Schmidt, J.P.; Ivers-Tiffée, E. Electrochemical characterization and post-mortem analysis of aged LiMn$_2$O$_4$–NMC/graphite lithium ion batteries part II: Calendar aging. *J. Power Source* **2014**, *258*, 61–75. [CrossRef]
57. Kim, J.-H.; Woo, S.C.; Park, M.-S.; Kim, K.J.; Yim, T.; Kim, J.-S.; Kim, Y.-J. Capacity fading mechanism of LiFePO4-based lithium secondary batteries for stationary energy storage. *J. Power Source* **2013**, *229*, 190–197. [CrossRef]
58. Bagotsky, V.S. *Fundamentals of Electrochemistry*; John Wiley & Sons: Hoboken, NJ, USA, 2005; ISBN 9780471741985.
59. Liu, S.; Liu, X.; Dou, R.; Zhou, W.; Wen, Z.; Liu, L. Experimental and simulation study on thermal characteristics of 18,650 lithium–iron–phosphate battery with and without spot–welding tabs. *Appl. Therm. Eng.* **2020**, *166*, 114648. [CrossRef]

batteries

Article

Optimization of Disassembly Strategies for Electric Vehicle Batteries

Sabri Baazouzi [1,*], Felix Paul Rist [1], Max Weeber [1] and Kai Peter Birke [1,2]

[1] Fraunhofer Institute for Manufacturing Engineering and Automation IPA, Nobel Str. 12, 70569 Stuttgart, Germany; felix@rist.net (F.P.R.); max.weeber@ipa.fraunhofer.de (M.W.); kai.peter.birke@ipa.fraunhofer.de (K.P.B.)

[2] Chair for Electrical Energy Storage Systems, Institute for Photovoltaics, University of Stuttgart, Pfaffenwaldring 47, 70569 Stuttgart, Germany

* Correspondence: sabri.baazouzi@ipa.fraunhofer.de

Abstract: Various studies show that electrification, integrated into a circular economy, is crucial to reach sustainable mobility solutions. In this context, the circular use of electric vehicle batteries (EVBs) is particularly relevant because of the resource intensity during manufacturing. After reaching the end-of-life phase, EVBs can be subjected to various circular economy strategies, all of which require the previous disassembly. Today, disassembly is carried out manually and represents a bottleneck process. At the same time, extremely high return volumes have been forecast for the next few years, and manual disassembly is associated with safety risks. That is why automated disassembly is identified as being a key enabler of highly efficient circularity. However, several challenges need to be addressed to ensure secure, economic, and ecological disassembly processes. One of these is ensuring that optimal disassembly strategies are determined, considering the uncertainties during disassembly. This paper introduces our design for an adaptive disassembly planner with an integrated disassembly strategy optimizer. Furthermore, we present our optimization method for obtaining optimal disassembly strategies as a combination of three decisions: (1) the optimal disassembly sequence, (2) the optimal disassembly depth, and (3) the optimal circular economy strategy at the component level. Finally, we apply the proposed method to derive optimal disassembly strategies for one selected battery system for two condition scenarios. The results show that the optimization of disassembly strategies must also be used as a tool in the design phase of battery systems to boost the disassembly automation and thus contribute to achieving profitable circular economy solutions for EVBs.

Keywords: electric vehicle battery; disassembly; disassembly planner design; disassembly strategy optimization

Citation: Baazouzi, S.; Rist, F.P.; Weeber, M.; Birke, K.P. Optimization of Disassembly Strategies for Electric Vehicle Batteries. *Batteries* **2021**, *7*, 74. https://doi.org/10.3390/batteries7040074

Academic Editor: Catia Arbizzani

Received: 21 July 2021
Accepted: 2 November 2021
Published: 7 November 2021

1. Introduction

Electrification of the transport sector is mandatory to achieve the Paris climate targets, as it currently accounts for around 24% of all global CO_2 emissions [1]. However, battery electric vehicles (BEVs) currently cause almost twice the greenhouse gas emissions in the manufacturing phase compared to equivalent combustion vehicles, mainly due to the resource-intensive production of the battery [2]. Nevertheless, electric vehicle batteries (EVBs) show a significantly better environmental performance if, first, renewable energy is used to charge the battery during the use phase and, second, if electrification occurs within a circular economy. In this context, the battery plays the most important role, as it is the most expensive component in BEVs and contains valuable materials and components suitable for reuse A battery pack generally consists of several modules made up of battery cells. Currently, Li-ion cells are the most common. They can be found in three shapes (cylindrical, pouch, and prismatic) with different cell chemistries (NMC, LCO, LMO, LFP, and NCA). However, the general structure of a Li-ion battery cell is independent of the cell format and the used chemistries. Its main components are anodes, cathodes, a separator, an electrolyte, and housing. In batteries for the automotive sector, all cell formats and a wide range of chemistries are used. Thereby, materials' purchase costs are the primary cost driver.

The circular use of components and materials offers big economic opportunities and has great potential to secure the supply of strategic raw materials for cell manufacturers [3]. The work of Sato and Nakata [4] showed that by 2035, high quantities of critical materials for the production of new Li-ion batteries in Japan will be obtained from the recycling of batteries at the end-of-life (EoL) stage (34% of lithium (Li), 50% of cobalt (Co), 28% of nickel (Ni), and 52% of manganese (M)). However, according to Kotak et al. [5], alternative circular economy strategies such as reuse and remanufacturing would extend the use phase of batteries and thus avoid the resource-intensive production of new batteries. Furthermore, they allow postponing recycling, which will result in improving the recycling efficiency due to the fact that the recycling processes are continuously being developed. In this context, disassembly plays a key role in the implementation of all alternative circular economy strategies at the EoL phase [6]. In addition, by using advanced disassembly technologies and strategies, material recyclers can significantly reduce the mix of materials to be handled in resource-intensive downstream material recycling processes.

Nevertheless, today, disassembly represents a bottleneck process that has to be performed faster [7]. Currently, EVB disassembly is done manually [8], which leads to high costs and poses safety risks to human workers. For these reasons, industrial and highly automated disassembly is mandatory in the future [6]. Automated disassembly is required to handle future quantities of returning battery systems in an economically viable and secure manner. Based on the review of several literature sources, Tan et al. [9] divided the battery disassembly process at the module-level into four steps. It starts with removing the battery casing, followed by the extraction of the battery management system (BMS), power electronics, and the thermal management system. After that, wires, cables, and connectors are removed. Finally, the modules are obtained after disassembling the securing holders. The modules can be further disassembled to obtain the battery cells. Thereby, the five main components have to be removed from the modules. These are cell contacting, cell fixation, housing, thermal management, and the BMS [10]. Gerlitz et al. [10] classified the challenges for automated disassembly at the module level into product-related and process-related challenges. Thereby, the main challenge posed by the product is the design variety. The main process-related challenges are the non-detachable joints and the hazards related to Li-ion batteries.

Figure 1 shows the different players in the life cycle of EVBs and their role in implementing a circular economy. Thereby, disassemblers specify the material flow at the EoL phase. Remanufacturers are very important for implementing high-value circularity solutions at the different system levels. Recyclers are obligatory to close the loop. Here, it is worth mentioning that there are different opinions in the ongoing research projects on who will carry out disassembly, remanufacturing, and recycling. It is assumed that these operations will either be performed by the same stakeholder, such as recyclers, or that new actors will be established due to the expected enormous return volumes and the diversified skills necessary to establish economic and high-quality circularity of EVBs. In this work, we adopt the second opinion. Disassemblers play a decisive role, whatever the recovery option at the EoL phase is. However, EVB disassemblers have to deal with several challenges in the future, such as:

- The increasing return volumes, the uncertainties in timing, quality, and quantity leading to uncertainties concerning the economic viability of future circular economy strategies,
- The wide range of battery system designs and cell chemistries and the short innovation cycles for new cell chemistries causing potential technological obsolescence and rapid decay of economic value of battery technologies currently prevalent in the field.

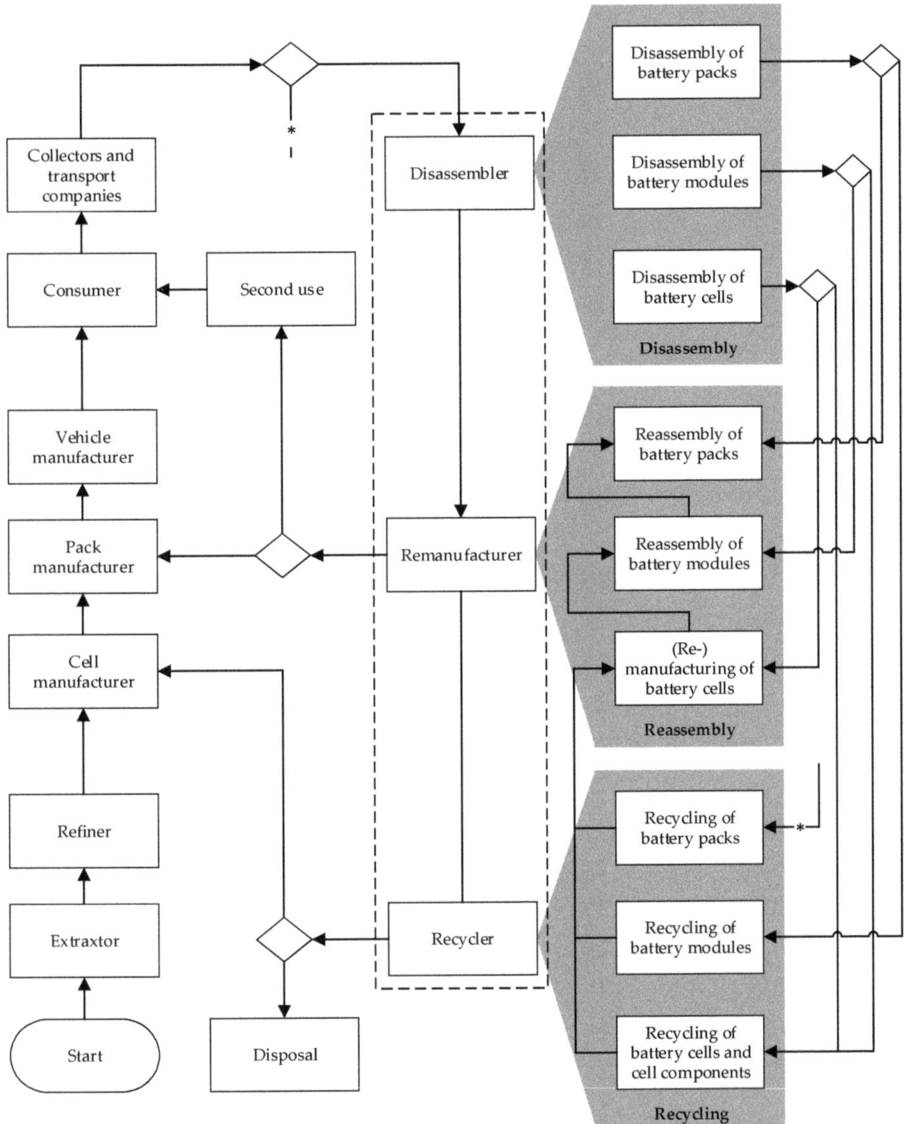

Figure 1. Participants along the life cycle of battery systems and their role in establishing a circular economy.

At the EoL phase, remanufacturers and recyclers are also crucial to extend the life of battery components or to recycle the battery parts if recycling is the only recovery option due to advanced aging or if recycling is the most suitable recovery option. The big challenge here is to find out the optimal route for an EVB at the EoL phase, as there are multiple alternative circular economy strategies and diverse recycling paths. Furthermore, this decision depends on the market demand. Therefore, a multicriteria decision platform is extremely essential [1].

Once the optimal EoL strategy has been determined, how should the battery be disassembled to implement the selected route economically? This question leads to an optimization problem that must be solved for individual batteries to significantly increase the economic efficiency of disassembly as the most expensive processing step in the current state [11]. Ke et al. [12] performed disassembly tests on the same battery type with the same skilled workers. They observed that the workers could disassemble the battery at least 11.5% faster when they had an optimized disassembly sequence.

Disassembly cannot be seen as the reverse of assembly because, first, disassembly is subject to many uncertainties and, second, there are different ways to perform disassembly. Here, different disassembly modes can be distinguished using several criteria. such as the disassembly depth (Complete/incomplete), the disassembly techniques (Destructive/non-destructive), the number of used manipulators (Sequential/parallel), and the automation level (Manual/automated). Thereby, the disassembly of complex products cannot be performed when it is only experience-based. Disassembly planning solutions that are adaptive and use optimization algorithms are necessary to determine optimal disassembly strategies.

This paper aims to contribute to designing adaptive disassembly planners for battery systems by combining the autonomous disassembly planner presented by Choux et al. [13] with a disassembly strategy optimizer, which will be implemented and tested using an Audi A3 Sportback e-tron hybrid battery pack. The battery, instructions about its disassembly, and several essential data for the disassembly planning, such as the disassembly times and revenues at component level after applying a specific circular economy strategy, have been described in detail in [14]. In this paper, the optimal disassembly strategy maximizes the optimal economic profit. It consists of the following decisions: (1) the optimal disassembly sequence, (2) the optimal disassembly depth, and (3) the optimal circular economy strategy for each component (reuse, remanufacturing, repurposing, and recycling). The proposed disassembly planner can significantly contribute to implementing high-value circularity levels at the EoL phase of EVBs in automated disassembly solutions in the future.

The following sections are organized as follows: Section 2 describes the main components of an automated disassembly solution. Thereby, the disassembly planner with an integrated disassembly strategy optimizer represents a core building block. Section 3 describes our methodology by presenting our design for an adaptive disassembly planner and a disassembly strategy optimizer. Finally, we present and discuss our use case results in Section 4.

2. Building Blocks of an Automated Disassembly Station

An automated disassembly station for EVBs can be reduced to two building blocks: (1) a mechanical system that directly interacts with the EoL products, and (2) a disassembly planner that adaptively calculates and updates the disassembly strategies (see Figure 2). The subcomponents are described in the following sections to show the big picture of our work. Thereby, publications in the context of battery disassembly are assigned to the respective subcomponent.

2.1. Mechanical System

2.1.1. Manipulators

Manipulators are responsible for moving several components during the disassembly process. These components can either be part of the battery or the disassembly station, such as tools and sensors. Robot arms are typical manipulators. The disassembly process can be carried out using a single manipulator resulting in a sequential disassembly, in which the parts are removed one by one. However, multiple manipulators have a great potential to reduce the disassembly time if parallel disassembly activities are possible. This is known as corporative disassembly. The disassembly sequence planning is more complicated if more than one manipulator is used [15].

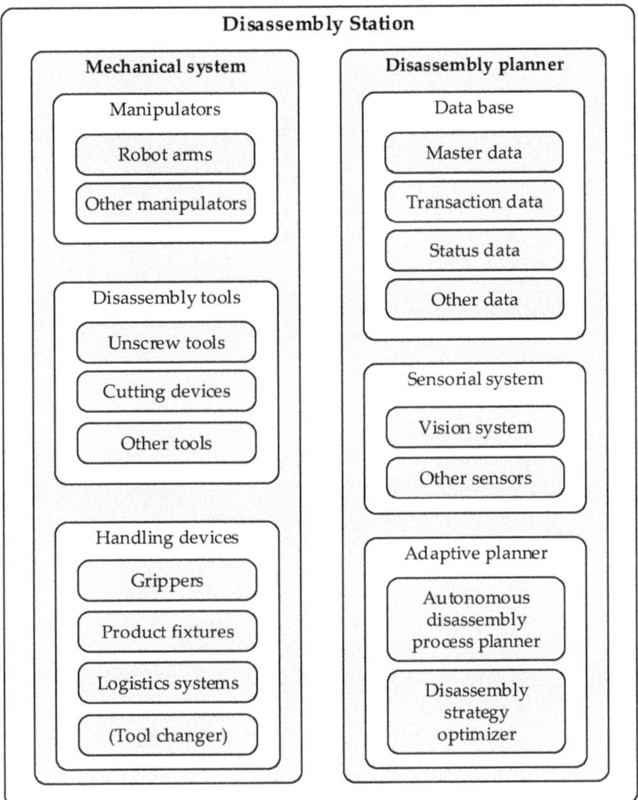

Figure 2. Building blocks of disassembly stations.

2.1.2. Disassembly Tools

EVBs are complex products whose disassembly is associated with multiple difficulties. Screw connections are frequently found in batteries. This allows the application of non-destructive disassembly. However, different screw types are often used, which are not accessible from the same direction [16]. This means that disassembly involves frequent tool and direction changes, which have to be considered while planning the disassembly process. Furthermore, many non-detachable connections are used in EVBs, such as welded joints. This is especially the case at the module level when connecting the cells [17], where welding processes have become established because they increase the electrical performance and improve the joints' thermal properties and long-term stability [18]. However, these joints are difficult to disassemble, especially when alternative circular economy strategies are targeted.

Moreover, various other joining techniques are used in the battery, such as adhesive bonding and riveting. In addition, there are several connector systems and flexible components that have to be disconnected or cut. That is why an automated disassembly of EVBs can only be achieved with a wide range of disassembly tools. In the literature, various tools have been presented that can be used in automating individual steps of battery disassembly. Tan et al. [9] presented a pneumatically actuated separation tool suitable for removing covers and stack holders and an unscrewing device with integrated torque sensors. Kay et al. [19] proposed a cost-effective cutting instrument based on a high-speed rotary cut-off wheel, which can be integrated into a disassembly station for battery modules. However, according to their evaluation, laser cutting is more suitable to achieve lower heat generation and vibration amplitudes and higher cutting accuracy. Li et al. [20] proposed an automated disassembly system for Z-folded pouch cells consisting of three modules for the pouch trimming, housing removal, and electrode sorting. Thereby, they presented a pouch trimming module consisting of a trimming blade set, a trimming base set, and a conveyor roller set. In this context, the main challenge in designing disassembly tools will be to develop universal tools that are, first, suitable for different battery variants and, second, capable of performing more than one disassembly task to reduce the number of tool changes during disassembly.

2.1.3. Handling Devices

They can be divided into three categories: grippers, product fixtures, and logistics systems [21]. The tool changer can also be seen as a handling device for the disassembly tools.

- Grippers: These are placed at the end of the manipulator and are used to handle the disassembled parts. They have to deal with objects with different geometries, volumes, weights, and surfaces, as well as with uncertainties concerning these properties, such as modifications in the surface due to usage and aging or contaminations during the disassembly. In the context of battery disassembly, handling flexible parts, such as cables, is a challenging task [16]. In addition, the extraction of the cells is associated with difficulties. This is due to the different cell types (pouch, cylindrical and prismatic cells) and the various design features, such as differences in volume, arrester position, and cell housing. Schmitt et al. [22] presented a flexible gripper with integrated voltage and internal resistance metering for automated handling of pouch cells. The gripper can grasp different cell geometries if the arresters are placed on the same side. Kay et al. [19] proposed a two-finger gripper with integrated force regulation to not damage the battery cells. From these publications, some essential features and requirements for battery cell gripper systems can be derived. These are the ability of condition assessment and monitoring, the handling of different cell formats and sizes, and controlling the gripping force.
- Product fixtures: These are necessary elements of an automated disassembly solution. They are needed to fix the product to be disassembled in a specific position and orientation so that the required forces are transmitted and to increase the process accuracy. Product fixtures can be designed to be stationary or movable. Thereby, the movement can take place with limited degrees of freedom, such as by means of a rotatable or translatory fixture table. They can also be seen as manipulators when they are mounted on a robot arm. In this case, they offer higher degrees of freedom in the positioning and orientation of the products to be disassembled. However, in the context of battery disassembly, this could be associated with technical difficulties. The weight of many EVBs can reach several hundreds of kilograms, which could pose a challenge for their mobile handling by means of robot arms during the disassembly process. Other challenges include the flexibility to clamp as many battery variants as possible. In addition, it must be ensured that the design features of the fixture devices do not prevent the detection of the product parts [21]. Detailed designs of fixture systems for EVBs cannot be found in the literature. This can be explained by the fact that battery disassembly is performed manually in the current state of the art.
- Logistics systems: These are in charge of transporting the products to be disassembled to the disassembly stations and transporting the disassembled parts and subassemblies in and from the disassembly net. Herrmann et al. [23] defined three disassembly scenarios for EVBs dependent on a product analysis and return quantities. In the present scenario, the batteries are disassembled in a single disassembly place. Thereby, the material transportation is carried out by forklifts, which also play a role in the near future and remote future scenarios. They transport the batteries to the disassembly net. The main difference to the present scenario is that the material transport inside the disassembly net is performed by roll conveyors.

2.2. Disassembly Planner

2.2.1. Database

The appropriate disassembly strategy for an individual EVB is the result of an optimization problem under consideration of a wide range of data, which must be available to the disassembly factories in order to increase the efficiency of a circular economy for battery systems. Therefore, structuring and managing these data in a database is of fundamental importance. Thereby, the need-based availability of some information is essential to protect the competitive advantages of battery manufacturers. Relevant data can be divided into process-related data and product-related data. Process-related data are, for example, disassembly times and costs and needed disassembly tools. The product-related data can be further classified into master data, transaction data, status data, and market data [24]. Master data comprise general information about the battery, such as the cell format and chemistry, the number of cells and modules, and information about other battery components, such as the battery management system. For disassembly, the precedence constraints, the joining techniques, the position of parts, and information about their accessibility are particularly important. Transaction data include information about the history of the battery. Status data provide information about the condition of the different components of the battery, such as the state of health of the modules and the cells. Market data are also essential to find out the optimal disassembly strategy. In particular, the potential revenues from selling components after applying a specific EoL strategy, such as remanufacturing, play a role.

2.2.2. Sensorial System

Before disassembly, the information from the database can be expanded with additional data using other sources, such as battery measurements or employees' experience. Nevertheless, not all the information may be available before starting the process. This is due to the variety of uncertainties during automated disassembly. Therefore, the sensorial system in an automated disassembly solution is mandatory to plan the process and to adapt it at the operational level [21]. Thereby, a vision system is needed to detect the components and their positions and monitor the progress of the disassembly process. In addition, other sensors are required, such as torque and force sensors, which can be used for both process control and monitoring.

2.2.3. Adaptive Planner

The adaptive planner is responsible for planning and optimizing the disassembly strategy. This process is known in the literature as disassembly sequence planning (DSP), and was described in detail in [25]. It finds application in both the planning and operating of disassembly lines, in addition to being used in the product development phase to ensure the guidelines of design for circularity (DfC). DSP starts with selecting the disassembly mode, followed by the modelling step, consisting of the two phases pre-processing and model building. Finally, the disassembly sequence can be optimized while considering a predefined objective function. Many publications have addressed only deriving disassembly sequences. Here, complete disassembly has been considered more frequently than incomplete disassembly [25]. However, the optimal circular economy strategies are supposed to be predefined in the literature and, therefore, are not seen as part of the disassembly planning [26]. An adaptive disassembly planner in an automated disassembly solution consists of an autonomous disassembly process planner that ensures the disassembly execution even when the required data are incomplete and a disassembly strategy optimizer to support decision-making by ensuring that data gaps are eliminated. In this way, different decisions can be made with the help of the optimizer, such as the optimal number of tool and direction changes as well as optimal circular economy strategies at the component level. In the context of battery disassembly, Ke et al. [12] proposed a disassembly sequence planning approach for EVBs based on a genetic algorithm using a frame-subgroup structure. The presented method has better convergence properties compared to other genetic algorithm implementations. However, it requires the existence of a frame (one component) that has connection and precedence relationships with all other parts (subgroups). In addition, only disassembly sequences for a complete disassembly are considered. Incomplete disassembly and decisions about the optimal circular economy strategies at the component level were not taken into account. Choux et al. [13] proposed an autonomous disassembly task planner, which can generate disassembly sequences autonomously. Thereby, no information about the battery to be disassembled is required. However, the completeness of the automatically detected precedence relationships in [14] was not guaranteed. Thus, possible disassembly sequences that may represent an improved disassembly strategy cannot be considered by the adaptive planner. In addition, the proposed task planner cannot make decisions about the disassembly depth and cannot provide information about the circular economy strategies for the different components, while it has been shown in [14] that they are decisive factors for the disassembly planning for EVBs. These disadvantages can be overcome by extending the autonomous disassembly process planner by a disassembly strategy optimizer. This represents the main research focus of this paper.

3. Methodology

3.1. Disassembly Planning Using an Adaptive Planner

The disassembly task planner presented by Choux et al. [13] consists of four steps: (1) image capturing by the installed vision system, (2) detection of the different components of the battery, (3) autonomous decision making about the possible disassembly sequences, and (4) position and path calculation to remove the components by the mechanical system of the disassembly station. Considering the identified issues in the previous section, we present our design for an adaptive disassembly planner (see Figure 3). Thereby, we extend the steps presented by Choux et al. with a disassembly strategy optimizer. However, disassembly strategy optimization requires a lot of data about the product, the disassembly process, and the market. For this reason, the disassembly strategy optimizer can be ignored if these data are not available.

Nevertheless, these gaps can be minimized in a disassembly factory after gaining some experience, e.g., by performing several disassembly experiments with autonomous decision-making or by taking advantage of the expertise of disassembly experts gained from past similar batteries. Once these data are available, an optimal disassembly strategy can be determined. In this way, the autonomous disassembly decision can be avoided. The component i to be disassembled is thus specified by the optimizer. After the calculation of the positions, disassembly operations can be performed.

If the disassembly proceeds normally, the disassembly strategy only needs to be calculated once. In case of failure, due to the many uncertainties during the disassembly process, or when complications occur that complicate the execution of disassembly with the existing strategy or pose any safety risks, alternative actions must be initiated. Here, two cases can be distinguished. The first case occurs when there is no optimized disassembly strategy due to incomplete data. In this case, the disassembly task planner must make an additional autonomous decision, such as changing disassembly tools or adjusting the disassembly sequence. The second case occurs if the disassembly strategy optimizer is active. The status data will then be updated if any components were damaged during the disassembly action. This is an important step, since it significantly impacts the optimized disassembly planning, as the appropriate circular economy strategy depends on the component condition. After that, a new disassembly strategy can be calculated to proceed with the disassembly by removing component i. If the disassembly is not yet complete, new images are captured to detect component i + 1 of the computed disassembly strategy. Thereby, it is necessary to check if the previous disassembly step of component i was performed correctly. If any connections were damaged or if the component i was destructively removed, that would require an update of the disassembly strategy. In this case, new precedence constraints may arise that need to be calculated. Moreover, the status data must be updated to calculate the next steps of the disassembly subsequently. In the next section, the design of the disassembly strategy optimizer implemented in this work is addressed.

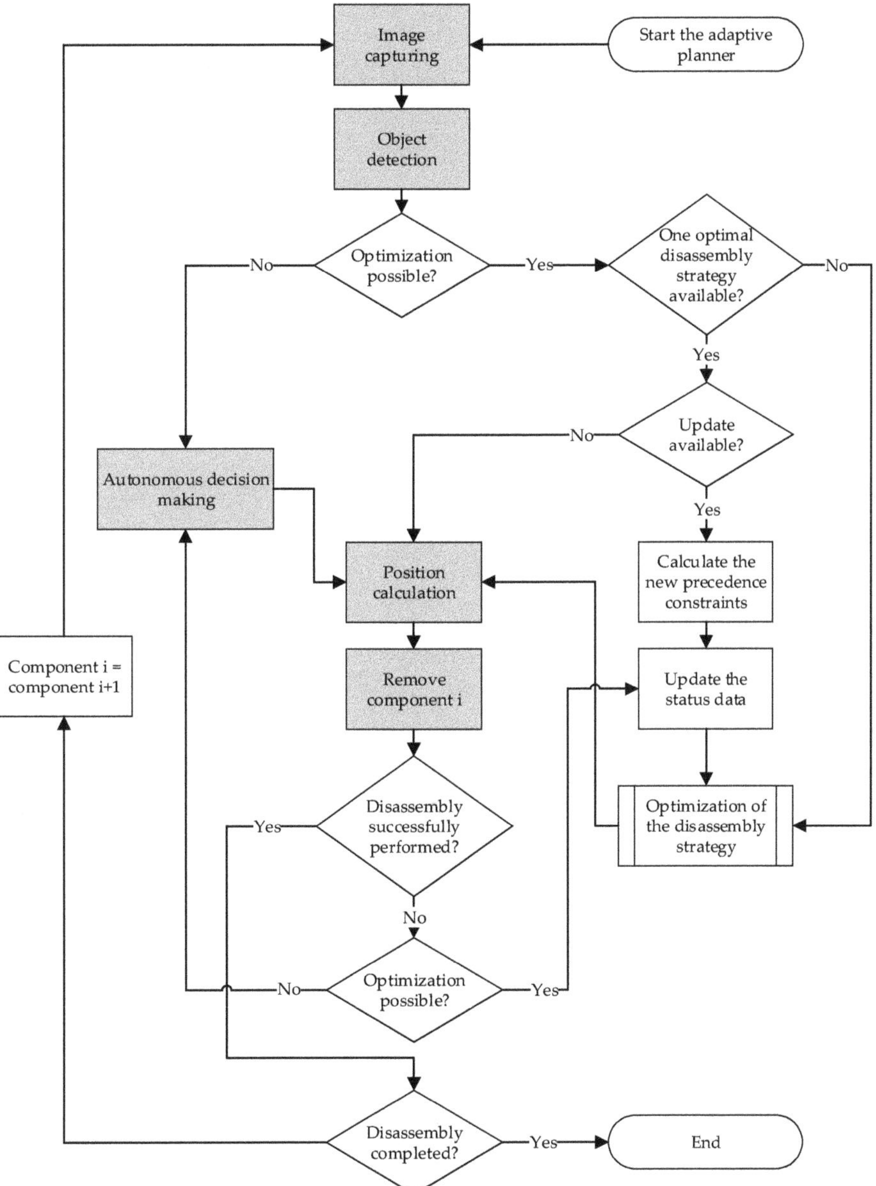

Figure 3. Concept of the adaptive planner.

3.2. Disassembly Optimizer

The first step in the disassembly strategy optimizer is selecting the most suitable optimization method depending on the available data and the objectives of the disassembly (see Figure 4). The technique chosen will then be applied to calculate an optimal disassembly strategy. Disassembly sequence planning (DSP) is a non-deterministic polynomial (NP) problem [25]. Here, the solution space is huge, especially for large products such as EVBs. Furthermore, the solution space becomes larger when further decisions have to be made when planning the disassembly strategy, such as the disassembly depth and the circular economy strategy at the component level. That is why primarily nature-inspired heuristic optimization methods are used in the literature to solve the DSP problem, such as genetic algorithms, particle swarm optimization, ant colony optimization, scatter search, and artificial bee colony optimization [25]. In this work, we will focus on the use of a modified genetic algorithm, since genetic algorithms are the most widely used optimization method for finding optimal disassembly strategies [25]. Furthermore, they offer multiple advantages compared to other metaheuristic methods, such as a wide application range, strong expansibility, and high robustness, since they usually do not fall into local optimal solutions [12]. In addition, genetic algorithms are attractive because, first, they quickly and cost-effectively produce high accuracy solutions, even when the solution space is huge, and second, they are easy to understand and implement since simple mathematics is involved [27].

Figure 4 shows the structure of the disassembly strategy optimizer designed in this paper. Here, the steps of the implemented genetic algorithm are shown in detail. It starts with generating the initial population of potential disassembly strategies coded in chromosomes. The chromosome structure will be described in the following subsection. Subsequently, the individuals of the first population are evaluated using an objective function, which can consist of different sub-objectives. In this context, Alfaro-Algaba and Ramirez [14] proposed a combined objective function composed of economic and environmental sub-objectives to maximize the economic profit while minimizing the environmental impact during the disassembly process of EVBs. A lot of data at the component level are needed, such as the disassembly costs, the costs to recondition disassembled components in order to implement a selected circular economy strategy, and environmental data.

Next, the selection step takes place to find out the fittest chromosomes to build a mating pool. The subsequent step is the mutation phase. Here, it should be ensured that all chromosome sections have the opportunity to mutate in order to increase the chances of discovering new solutions with higher performance.

The steps of the genetic algorithm are then performed until a termination condition is satisfied, such as a predefined number of generations or the fulfillment of specified convergence criteria.

In the following subsections, we describe our methodology for the different steps of the implemented genetic algorithm to optimize disassembly strategies for EVBs in terms of the disassembly sequence, disassembly depth, and circular economy strategies at the component level.

3.2.1. Generating the Initial Population

The initial population consists of feasible disassembly strategies coded into chromosomes. Thereby, one chromosome consists of three sections. The first section represents a disassembly sequence, taking into account the precedence constraints. In the second section, all battery components are assigned a circular economy strategy dependent on their condition. The third section consists of only one gene, representing the disassembly depth.

- First chromosome section

A precedence-matrix (P) is used to derive feasible disassembly sequences. It describes the order of precedence of the disassembly steps and can be obtained from carrying out manual disassembly experiments or by using the computer-aided design (CAD) models of the product to be disassembled [28]. Another approach to determining possible precedence relationships in an automatic way based on computer vision was presented in [13]. This can avoid the disadvantages of the two mentioned methods: while manual pre-processing is error-prone and time-consuming, pre-processing using CAD data is often inaccurate. This is because CAD models can rarely still reliably describe the product at their EoL phase, for example, due to corrosion or changes that have been applied to the product during the usage phase. We used the disassembly precedence graph (DPG) to generate the P matrix for the use case in this paper. It allows a simple one-to-one comparison of all components. If a component i is the predecessor of the component j, the value P_{ij} gets the value 1, otherwise 0. The numerical method to derive feasible disassembly sequences from the P matrix is illustrated in Figure 5.

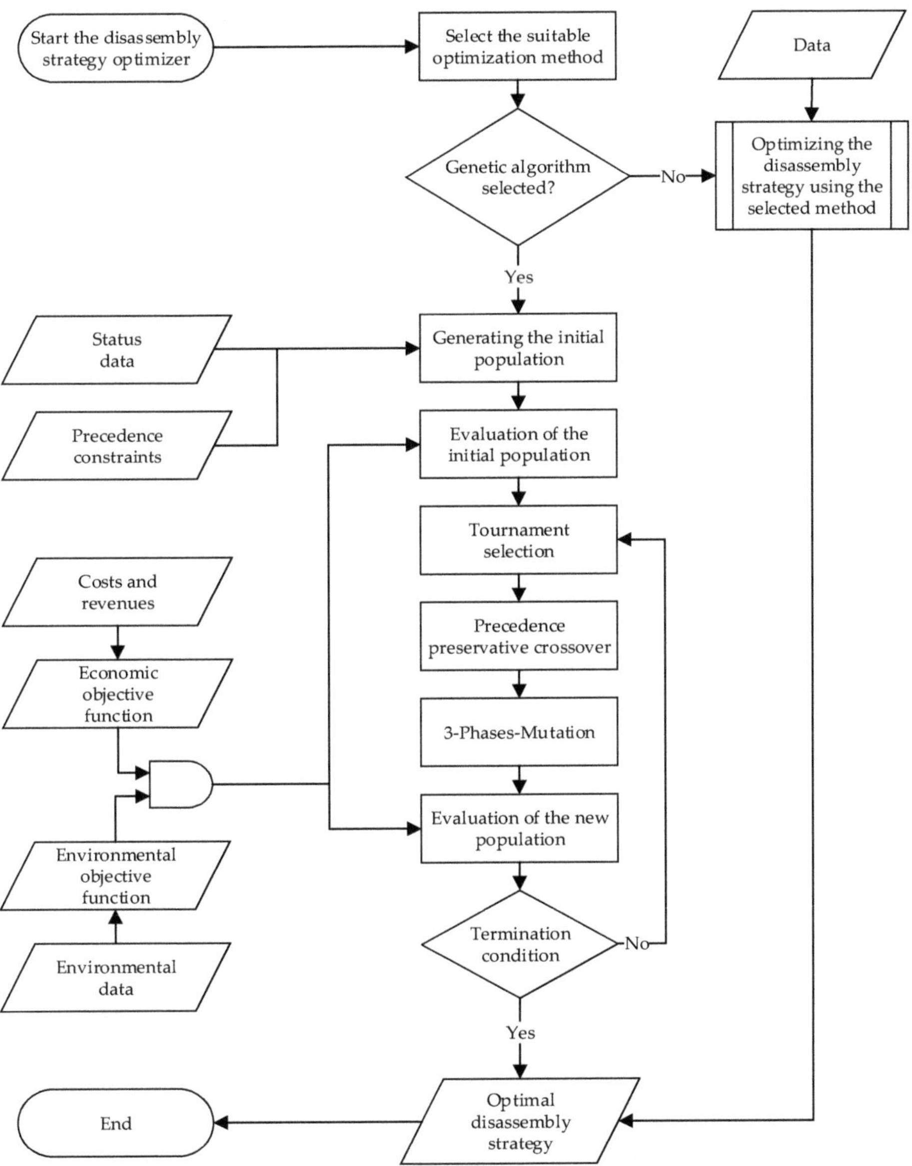

Figure 4. Disassembly strategy optimizer using a genetic algorithm.

```
Input: Precedence matrix P
Output: Matrix A^Npxn with feasible disassembly sequences
% Np: Population size
% n: Component number
j = 1;
While (j ≤ Np)
Find The columns of P where the sum is equal to 0
and store the column numbers in a vector x;
% components which can already be disassembled;
i = 1;
While (i ≤ n)
Select a random value from vector x to be stored
in the disassembly sequence yj at position i;
Delete the selected component in the vector x;
Go to the row with the index of the selected component
find the column elements that are equal to 1
and store their column index in the vector x;
i = i + 1;
End
j = j + 1;
End
```

Figure 5. Method to generate feasible disassembly sequences from a given precedence matrix.

- Second chromosome section

The transfer of battery components into a circular economy takes place via the implementation of circular economy strategies. Here, there are different strategies, which mainly differ in preparing the components and the application field. The waste hierarchy of the European Commission contains five priorities [29]. In this work, we focus on the priorities on the top, as they should be preferred in a circular economy. The highest one is prevention, for example, by extending the life of products through predictive maintenance or simple repair operations. The second priority is repair for reuse, followed by recycling. These priorities can be achieved by applying diverse circularity strategies. Potting et al. [30] identified ten strategies and divided them into three categories: (1) smarter product use and manufacture, (2) extend lifespan of product and its parts, and (3) useful application of materials. Here, we consider only the strategies, which can be applied at the EoL phase. Circularity solutions during the stages of design, manufacturing, and use are ignored. In this scope, the EoL strategies can be reduced to four strategies: reuse, remanufacturing, repurposing, and recycling. The integration of these strategies in the waste hierarchy and the material flows after their application are illustrated in Figure 6. In the literature, different definitions can be found for these strategies [24]. In this paper, we adopt the definitions of Potting et al. [30]. Reuse means the utilization of a discarded battery or a set of its components by another user in the automotive field. Remanufacturing is the treatment of battery parts so that they can at least meet the requirements of newly manufactured products and their utilization for the manufacturing of EVBs. If the battery or its parts are reconditioned and used in another application field, such as stationary energy storage, this is called repurposing. Recycling is the recovery of materials. In this case, pure and high-quality materials for use in the automotive sector should be targeted.

Figure 6 shows that disassembly is of central importance for the implementation of all these circularity strategies. The components of the EVBs can be allocated to different strategies depending on their condition. However, the selection of the optimal circular economy strategy does not depend exclusively on this. Other factors, such as the disassembly costs and the market constraints, such as the potential revenues, also play an important role. Therefore, we consider the selection of the EoL strategy as a part of the disassembly planning. This paper assumes that the feasibility of a circular economy strategy for a given component depends on its condition. That is why the second section of the chromosome must fulfill the condition constraints defined by a condition vector S containing the feasible circular economy strategies CES_i for every component i—see Equation (1). Here, the priority of the circular economy strategies is taken into account. If a component i is in excellent condition, it can be allocated to all strategies (reuse, remanufacturing, repurposing, and recycling). In this case, CES_i is assigned the value 1. If CES_i equals 2, the reuse option will be excluded. Part i can be neither reused nor remanufactured if CES_i equals 3. CES_i is assigned the value 4 if recycling is the only possible recovery option.

$$S = \begin{pmatrix} CES_1 \\ \dots \\ CES_n \end{pmatrix}; \; CES_i \in \{1; 2; 3; 4\} \tag{1}$$

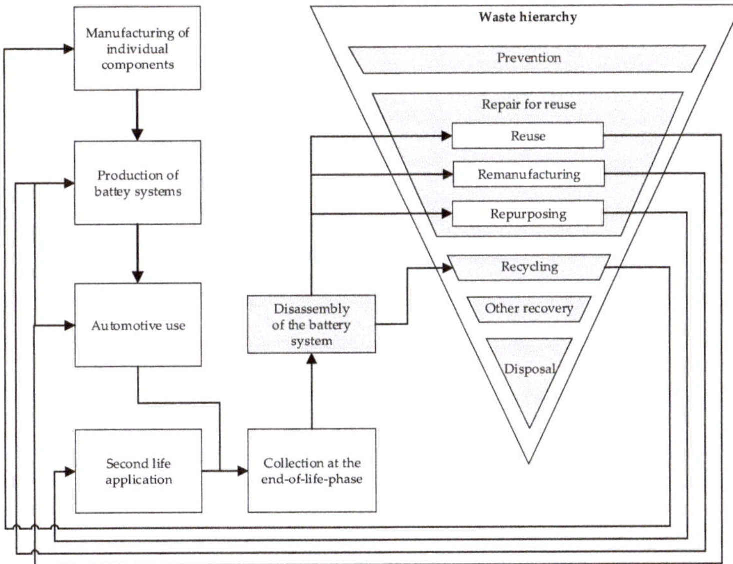

Figure 6. Overview about the considered end-of-life strategies.

The S vector cannot just be seen as a collection of testing results. Other factors can play a role in determining the potentially possible circular economy strategies for the different battery parts, such as the employees' experience in the disassembly factory. Testing results include the state of health (SoH) and state of charge (SoC) of battery cells and modules and additional parameters for the rest of the components.

- Third chromosome section

This section consists of a single gene and is used to define the disassembly depth. An EVB could be entirely disassembled by separating all its parts. However, this approach is neither economically nor environmentally practical in an industrial context [16]. Therefore, EVBs are more likely to be subject to incomplete disassembly. Here, there are two methods to perform incomplete disassembly: (1) the selective method and (2) the unrestricted method. The selective method means that specific components are selected to be disassembled. Subsequently, the disassembly planner needs to calculate a strategy for the optimal extraction of these parts. Here, the high-value strategy and the high-impact strategy can be distinguished [31]. For EVBs, the removal of the modules could present a high-value disassembly strategy. The high-impact strategy applies when, for example, a module with safety risks is identified and has to be replaced before reusing the battery. In contrast, no target components are selected in the unrestricted incomplete disassembly. The disassembly planner can freely calculate the optimal disassembly strategy based on an objective function. This method is considered in this paper. Thereby, the gene representing the disassembly depth is randomly generated with values between 0 (no disassembly) and the maximum number of components n while generating the initial population.

In Fehler! Verweisquelle konnte nicht gefunden werden, a disassembly precedence graph of a theoretical product, the associated precedence matrix, and the entire structure of a possible chromosome depending on a given condition vector are presented. In this case, one feasible disassembly sequence is 1-2-3-5-6-4, possible circular economy strategies at the component level are 1-2-4-1-2-4, and the disassembly is complete, since the last gene matches the component number.

3.2.2. Evaluation Method

The performance of the disassembly strategies, coded in chromosomes, has to be evaluated using an objective function, which depends on several parameters. Thereby, different evaluation criteria can be involved, such as the economic and social performance or the environmental impacts. In this paper, we focus on the economic performance of the disassembly strategies by implementing the following objective function to maximize the economic profit (see Equation (2)).

$$y = \left(\sum_{i=1}^{DL} RV_{i,j} - RC_{i,j} - OC_{i,j} - F \cdot DT_{i,j} \right) + \left(q \cdot \sum_{i=DL+1}^{n} RV_{i,4} - RC_{i,4} \right) \tag{2}$$

y: Economic profit
i: Index of components
j: Circular economy strategy
DL: Disassembly depth
$RV_{i,j}$: Revenues from component i while applying the circular economy strategy j
$RC_{i,j}$: Recovery costs for component i to apply the circular economy strategy j, for example, costs of cleaning, further mechanical treatment, replacement of elements, pyrometallurgical and hydrometallurgical treatment, etc.
$OC_{i,j}$: Overhead costs for component i to apply the circular economy strategy j
F: Machine and personnel hourly rate for the disassembly process
$DT_{i,j}$: Disassembly time of component i while applying the circular economy strategy j
q: Value reduction factor for the achievable yield from recycling in case of incomplete disassembly
n: Number of components.

It is assumed that the revenues, recovery costs, and overhead cost depend on the selected circular economy strategy. This also applies to the disassembly times because, in the case of recycling, disassembly operations can be performed faster due to destructive disassembly techniques.

3.2.3. Selection and Crossover

During the selection phase, parents are selected based on their fitness value. The candidates with higher fitness will subsequently mate to produce new generations. In the literature, there are several selection procedures. Ke et al. [12] used the roulette wheel selection method to select the fittest disassembly sequences for EVBs. In this work, a tournament selection technique is used. In a tournament, each chromosome competes twice against two random other chromosomes. The winners move into a mating pool consisting of parents of the same size as the initial population.

Afterward, the crossover phase takes place, usually with high probability (P_c). During this phase, children representing new solutions are generated using the genetic material of two parents. In the context of the disassembly planning task in this paper, it is essential to ensure that the created chromosomes during the crossover phase represent feasible solutions by not violating the precedence relationships and the condition constraints. Therefore, the precedence preservative crossover method described in [27] was chosen and adapted to the characteristics of our chromosome structure. Thereby, two parents generate two children whose chromosome structure is determined by two randomly generated masks. The first child is recombined by using mask 1, and the second one by mask 2. The chromosomes of the children are built up step by step. If the used mask has the value 1 at position i, parent 1 is used to specify the gene i of the child. Here, the leftmost element of parent 1 will be deleted from both parents and placed in the position i of the child. Otherwise, parent 2 is used to define the gene i. Figure 7 shows two possible disassembly sequences of the theoretical product presented in Figure 8, as well as two randomly generated masks, which were used to create new feasible disassembly sequences. The section of the chromosome, representing the circular economy strategies, is recombined simultaneously with the disassembly sequence using the same masks. However, the masks do not play any role in the definition of the disassembly depth. Here, child 1 gets the disassembly depth from parent 1 and child 2 from parent 2.

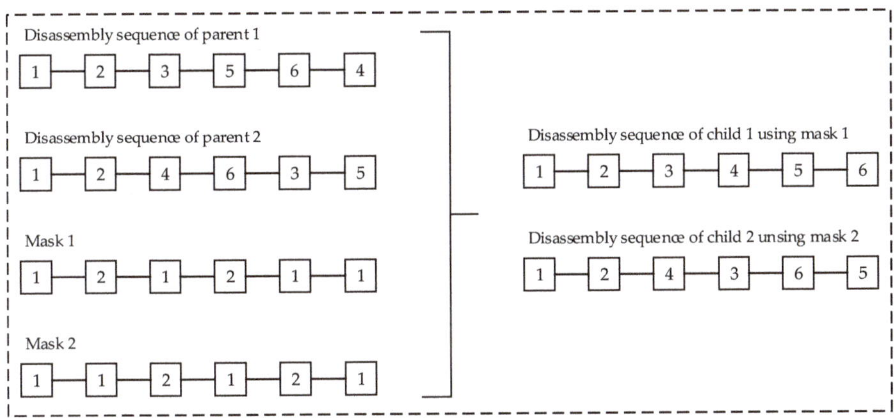

Figure 7. Using randomly generated masks to produce new feasible disassembly sequences during the crossover phase of the genetic algorithm.

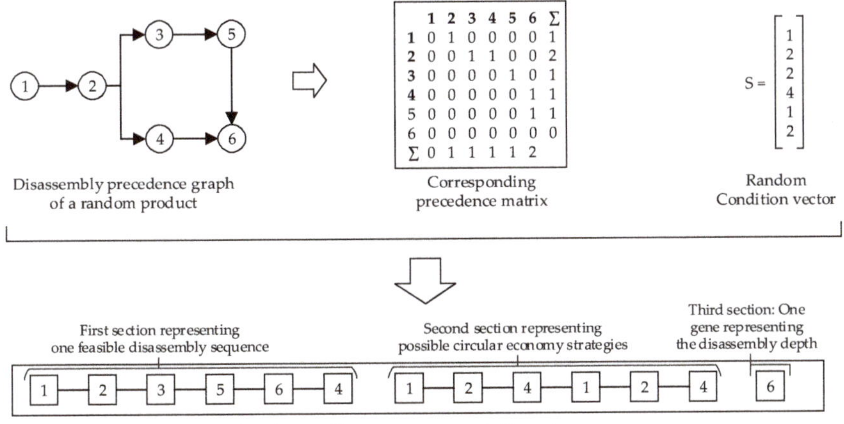

Figure 8. Structure of the chromosomes of the initial population.

3.2.4. Mutation

The mutation phase plays a crucial role in enhancing the quality of the solution. It increases the diversity in a population and the possibility of discovering new candidates with high performance [27]. In addition, mutation increases the robustness of genetic algorithms concerning local optima [15]. In the context of disassembly planning, the feasibility of mutated solutions must be guaranteed. In this work, the disassembly strategy involves three decisions: the disassembly sequence, the circular economy strategies, and the disassembly depth. Therefore, we chose a three-step mutation by applying the swap mutation method to sections 1 and 2 of the chromosome, representing the disassembly sequence and the circular economy strategies, respectively, and a random resetting of the third chromosome section describing the disassembly depth.

- Step 1: With a mutation probability $P_{m,1}$, two randomly selected genes in the first chromosome section are swapped. Thereby, the associated circular economy strategies in the second chromosome section must be swapped in the same manner to ensure the satisfaction of the condition constraints. The mutation is only accepted if the precedence relationships described by the precedence matrix P are not violated.
- Step 2: With a mutation probability $P_{m,2}$, two randomly selected genes in the second chromosome section are swapped. The mutation is only accepted if the condition constraints given by the condition vector S are not violated.
- Step 3: With a mutation probability $P_{m,3}$, the last gene of the chromosome is reset by randomly adding or subtracting up to 20% of the number of components.

Figure 9 shows an exemplary execution of the three-step mutation based on a disassembly strategy of the theoretical product presented in Figure 7. During the first step, the fifth and last genes of the first section, as well as the corresponding circular economy strategies, were changed. The mutation is accepted because the disassembly sequence 1-2-3-5-4-6 is feasible. The mutation in step 2 is rejected because part 3 cannot be reused (see condition vector S in Figure 7). The mutation during the last step consists of substracting 16.7% of the total number of components (one part).

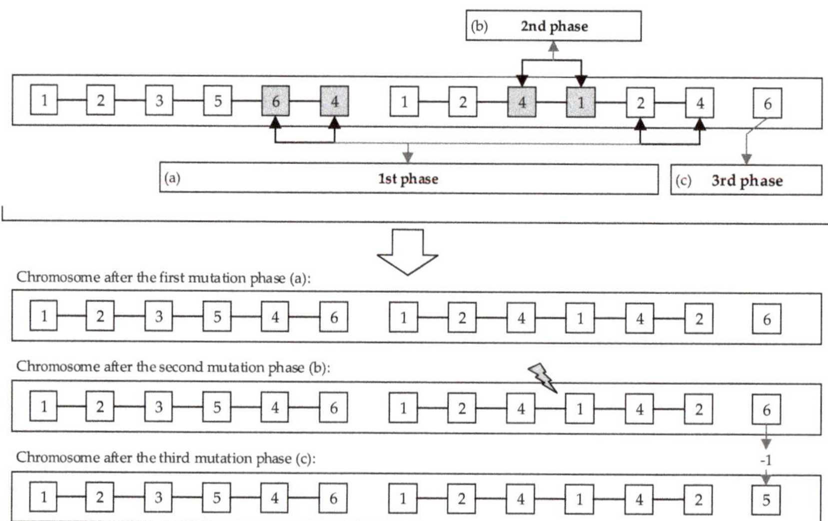

Figure 9. Three-step mutation in the chromosome sections of the defined theoretical product: (**a**) step 1: swap mutation in the disassembly sequence and the corresponding circular economy strategies; (**b**) step 2: swap mutation in the circular economy strategies; (**c**) step 3: random resetting mutation of the gene representing the disassembly depth DL (DL = DL $\pm$ random x; $x \in [0, 20\% \cdot n]$; $DL \in [0, n]$; n: number of components).

4. Results and Discussion

4.1. Case Study

The Audi A3 Sportback e-tron hybrid Li-ion battery pack was chosen as the use case in this paper to demonstrate our approach to planning disassembly strategies for battery systems. The selected battery was described in detail by Alfaro-Algaba and Ramirez [14]. The selected disassembly steps are presented, and most of the data required for our proposed disassembly strategy optimization method are available. The relevant assumptions for our use case are listed below:

- Currently, we cannot quantify how diverse disassembly techniques for implementing different circular economy strategies affect disassembly times. Therefore, we assume in the following that the disassembly time per component does not depend on the selected route.
- From interviews with two battery recyclers in Germany, we found out that two workers need 30 min on average to remove the modules manually from a medium-sized battery. This is consistent with the assumption made by Alfaro-Algaba et Ramirez [14]. However, the disassembly times for each component were not given in [14]. In this work, we have estimated the disassembly times for the selected battery based on the experience of Rallo et al. [32] during the disassembly of a similar battery. In Figure 10, the component-specific disassembly times are integrated into the precedence graph of the considered battery. Thereby, disassembly times are coded as follows: x[number of units]; [disassembly time per unit]; [total disassembly time].
- We only consider a disassembly station with two workers with 100% availability. This means that 3300 units can be disassembled per year (see Table 1).
- We used the same overhead costs and their allocation at the component level as in [14]. However, we assume that the overhead costs do not depend on the applied circular economy strategy.
- No components are disposed of. If components are in very poor condition, they must be recycled.
- Processing costs to repurpose the components are assumed to be 25% of the revenues.
- Repurposing revenues are 10% lower than remanufacturing revenues.
- In case of incomplete disassembly, the non-disassembled parts will be recycled. Here, the profit is reduced by 10% due to the missing separation resulting in impure material composition.
- It is assumed that the overhead cost per battery is EUR 96.97 on average (see Table 1). The total overhead costs (OC_T) are taken from [14].
- The economic input data and additional assumptions for the cost structure can be found in [14] in order to reproduce the results presented in the next subsection.

Table 1. Throughput of a disassembly station and overhead costs.

Variable	Formula	Parameters
Throughput a	$a = (n_d \cdot n_h)/t_B = 3300$	n_d = 220 days: Working day per year n_h = 7.5 h: Working hours per day t_B = 30 min: Average disassembly time per battery
Overhead costs per battery OC_B	$OC_B = OC_T/a = 96.97€$ data	OC_T = 320,00€: Total overhead costs per year data

In the following, we consider two scenarios for disassembly planning of the considered battery. The upper and lower housing shells and the cooling plates are in poor condition in the first scenario and can, therefore, only be recycled. In the second scenario, they can be assigned to all possible circular economy strategies. The eight modules are in different conditions: two modules must be recycled (CES_i = 4), two modules can be reprocessed for second-life applications (CES_i = 3), such as stationary energy storage, two modules can be remanufactured for automotive applications (CES_i = 2), and the last two modules can be directly reused in the automotive sector with little effort, for example for cleaning and packaging (CES_i = 1). In addition, we assume that all connecting elements cannot be reused and consequently have to be recycled in both scenarios. All other components of the considered have the same condition in both scenarios; see the condition vectors of both scenarios in Equations (3) and (4).

$$S_1^T = [4\,4\,1\,1\,1\,1\,1\,1\,1\,1\,1\,1\,4\,4\,4\,4\,1\,1\,2\,2\,3\,3\,4\,4\,4\,4\,4\,4\,4\,4\,4\,4] \tag{3}$$

$$S_2^T = [1\,4\,1\,1\,1\,1\,1\,1\,1\,1\,1\,1\,1\,4\,4\,1\,1\,1\,2\,2\,3\,3\,4\,4\,4\,4\,4\,4\,4\,4\,4\,4] \tag{4}$$

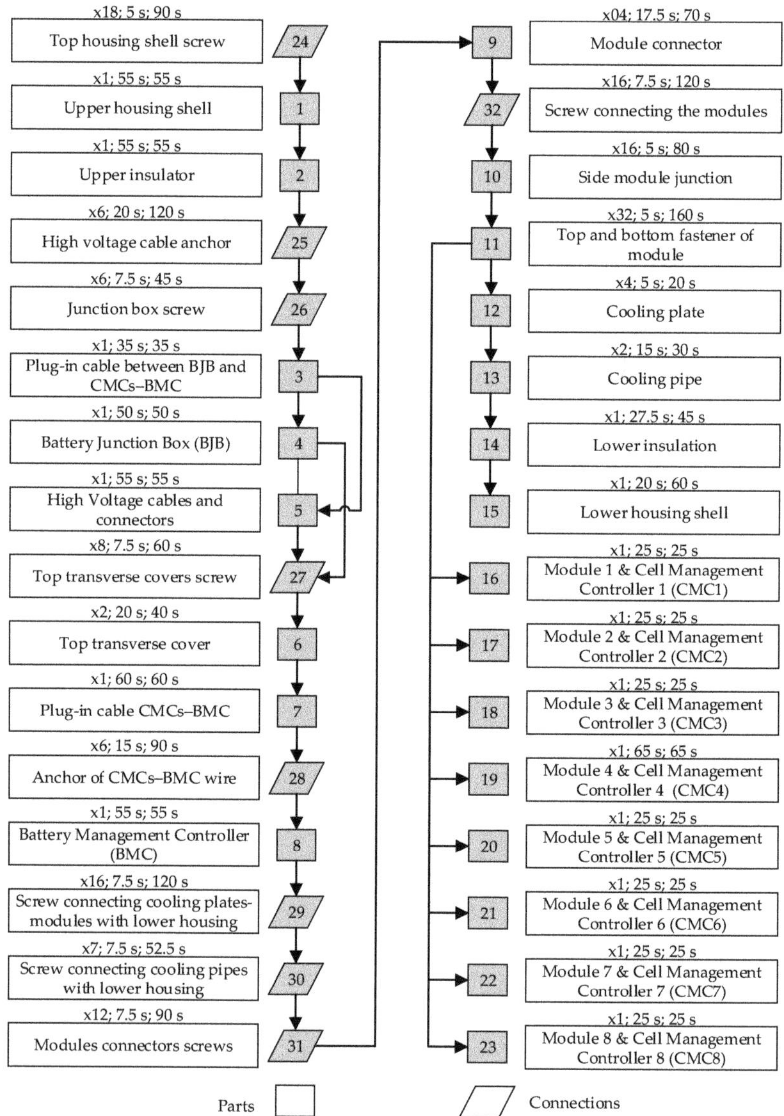

Figure 10. Li-ion battery of the Audi A3 Sportback e-tron hybrid: disassembly precedence graph and disassembly times at the component level. The disassembly times are specified by the following format: x[number of units]; [disassembly time per unit]; [total disassembly time].

4.2. Results

A disassembly strategy in this work consists of three decisions: (1) the optimal disassembly sequence, (2) the optimal circular economy strategy for each component, and (3) the optimal disassembly depth, which represents the stopping point of the disassembly process. These three decisions for both defined scenarios are shown in Figure 11. The used parameters for the initialization of the implemented genetic algorithm are listed Table 2. In the first scenario, the result is an incomplete disassembly with a disassembly depth of 90.63%. We see that the disassembly stops immediately after the modules are removed. In this case, two modules are reused, and two modules are recycled. This represents the best possible route due to the condition constraints. However, the remaining four modules are repurposed, although two modules could be remanufactured. This is because remanufacturing is not economically feasible in our use case. In the first scenario, an economic profit of EUR 553 can be achieved (see Figure 12). In the second scenario, it can be increased by 3.16% to EUR 570.5, although the disassembly costs are higher due to the complete disassembly. This is due to the fact that the upper and lower casing shells, as well as the cooling plates, are reused, and thus higher revenues can be realized. It is worth mentioning here that although the modules are the most valuable components on a battery, other parts contribute in a significant way to increasing the profitability of the disassembly. In the course of our research, we spoke with an EVB recycler in Germany. He said that recycling some EVB variants is particularly attractive because of massive busbars made of copper. Figure 12 shows the evolution of the objective function and the disassembly depth over the 100 generations. Thereby, the algorithm converges after few generations (<20), which means that the optimization can be terminated earlier and thus performed faster.

4.3. Discussion

Optimizing disassembly strategies for EVBs shows a key role in making the circularity of these systems more efficient by targeting higher priorities from the waste hierarchy presented in Figure 6. Here, selecting the optimal circular economy strategies at the component level must be considered as part of the disassembly process planning, since the chosen strategy influences the disassembly techniques and thus the disassembly times and costs.

Disassembly planning and optimization are becoming increasingly complex due to several factors: first, disassembly is subject to many uncertainties, which makes disassembly planning an adaptive and iterative process. Second, disassembly can be performed using different modes (sequential/parallel, complete/incomplete, destructive/non-destructive, automated/manual). Third, on the one hand, disassembly planning is a data-intensive process, and on the other hand, it must be ensured that disassembly is executed even when data are lacking. Fourth, there are multiple variables to optimize the defined objective function, such as the number of tool and direction changes and disassembly times, which depend on other factors such as the joining methods, the disassembly techniques, and the accessibility of the parts. Finally, disassembly strategies do not only consist of a disassembly sequence but also include other decisions, such as the disassembly depth and circular economy strategies for the different components. These three decisions are considered in our proposed adaptive disassembly planner with an integrated disassembly strategy optimizer.

Table 2. Used parameters for the genetic algorithm.

Parameter	Value
Population size N_p	300
Number of generations T	100
Crossover probability P_c	0.9
Mutation probability for the first chromosome section $P_{m,1}$	0.1
Mutation probability for the second chromosome section $P_{m,2}$	0.1
Mutation probability for the third chromosome section $P_{m,3}$	0.1

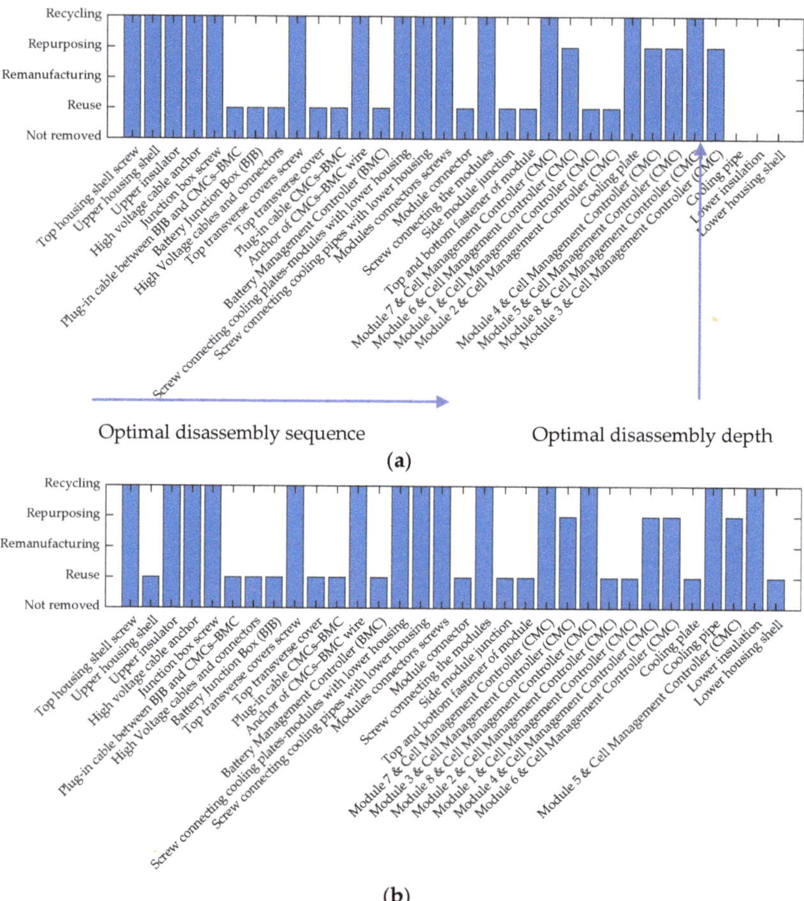

Figure 11. Optimal disassembly strategies consisting of three decisions (1—optimal disassembly sequence, 2—optimal disassembly depth, and 3—optimal circular economy strategies at the component level): (**a**) upper and lower housing shells and the cooling plates are in a bad condition and have to be recycled; (**b**) upper and lower housing shells and the cooling plates are in perfect condition and can be assigned to every circular economy strategy.

However, in the context of this paper, we only addressed sequential disassembly, since the aim of our current research is to develop an automated disassembly solution for battery packs down to the module level using a robot arm as a single manipulator. Our proposed disassembly strategy optimization method still needs to be extended to the following aspects: (1) including planning methods for cooperative disassembly, which can be applied by using at least two manipulators in fully automated disassembly solutions or by employing a human–machine collaboration, (2) taking into account the tool and direction changes, as they definitely influence the disassembly time, and (3) integrating adaptive methods for updating the condition and precedence constraints in case of complications during the disassembly process or when destructive disassembly steps are used. Furthermore, the adaptive planner should consider further factors, such as the configuration and the availability of the stations. On the one hand, the stations in a disassembly factory may have to be designed differently to be able to disassemble different battery variants and are therefore not suitable for carrying out all disassembly strategies and, on the other hand, a high-capacity utilization should be achieved. This means that EVBs have to be assigned to stations that cannot perform the best possible disassembly strategy in some cases. However, this measure can significantly improve capacity utilization and consequently contribute to establishing highly automated and flexible disassembly factories in the near future, which will become more and more profitable with increasing return volumes. In the literature, there are no concepts for highly flexible disassembly factories for EVBs. In the following publications, we will present several future layouts for disassembly factories under consideration of the presented building blocks of an automated disassembly station in section 0 and show potential challenges for the adaptive disassembly planner with respect to the proposed layouts.

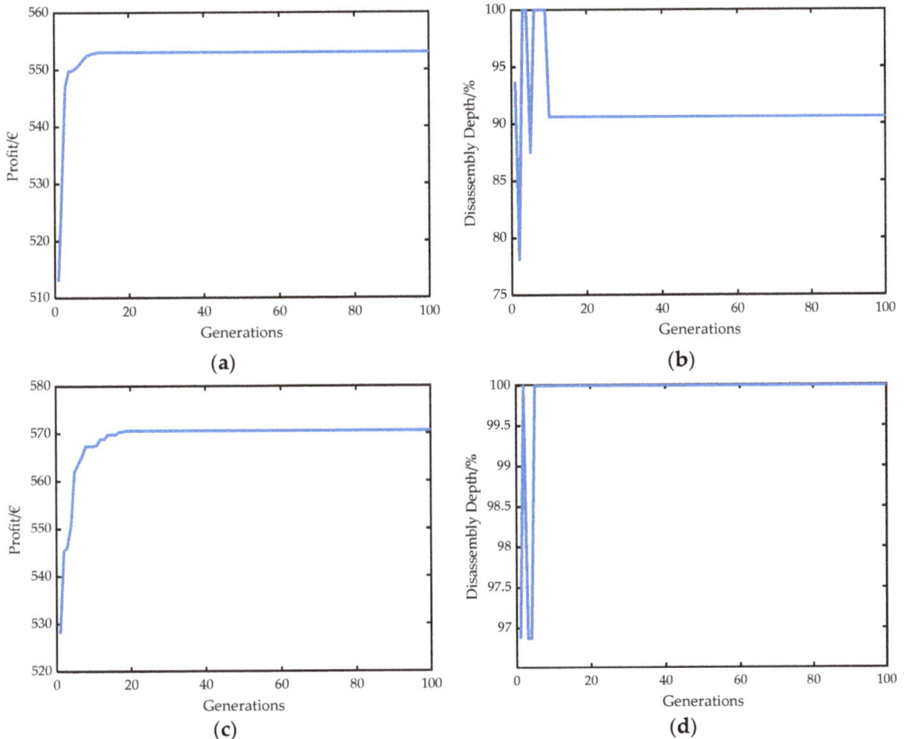

Figure 12. Model results: (**a,b**) optimal economic profit and disassembly depth for scenario 1; (**c,d**) same results for scenario 2.

Lastly, optimization of disassembly strategies, often described as disassembly sequence planning (DSP) in several literature sources, should be addressed in the product design phase. This will clearly contribute to achieving fully automated, cost-effective, and environmentally efficient disassembly for battery systems in the automotive sector. In particular, the modules, as the most valuable components in the battery, should be removable after only a few disassembly steps. This is obviously not the case for the battery considered in this paper.

5. Conclusions

An adaptive disassembly planner with an integrated disassembly strategy optimizer for electric vehicle batteries is presented in this paper. It serves to adaptively plan disassembly strategies and optimize them using heuristic optimization algorithms. A disassembly strategy consists of three decisions about the optimal disassembly sequence, disassembly depth, and circular economy strategy for each component. The disassembly strategy optimizer is implemented using a modified genetic algorithm and tested on a selected battery. Thereby, two condition scenarios were considered. In both scenarios, all modules are removed. The disassembly of the remaining components depends on their subsequent route. The presented optimization method is computationally efficient and can be further improved by applying a convergence termination condition. The introduced disassembly planning method can be used at the end-of-life phase to plan the disassembly depending on components' state and market conditions. Furthermore, our approach is also suitable for use in the begin-of-life stage to ensure the guidelines of "design for disassembly" in the design stage. Nowadays, there is a need for action in both application cases because, first, disassembly processes are mainly carried out based on experience, and second, battery treatment at the end-of-life phase is hardly considered when designing these systems.

Author Contributions: Conceptualization, S.B.; methodology, S.B.; software, S.B. and F.P.R.; writing—original draft preparation, S.B.; writing—review and editing, M.W., K.P.B. and F.P.R.; visualization, S.B.; project administration, M.W.; funding acquisition, M.W. and K.P.B. All authors have read and agreed to the published version of the manuscript.

Funding: The authors wish to thank the Ministry of the Environment, Climate Protection and the Energy Sector Baden-Wuerttemberg for funding this work under the funding code L7520101 as part of the accompanying research of the project "DeMoBat". The financial support is gratefully acknowledged.

Institutional Review Board Statement: Not applicable.

Informed Consent Statement: Not applicable.

Data Availability Statement: The data presented in this study are available within the article.

Conflicts of Interest: The authors declare no conflict of interest.

References

1. Acatech/Circular Economy Initiative Deutschland/SYSTEMIQ. Ressource-Efficient Battery Life Cycles: Driving Electric Mobility with the Circular Economy. 2020. Available online: https://www.acatech.de/publikation/ressourcenschonende-batteriekreislaeufe/ (accessed on 14 July 2021).
2. Helms, H.; Kämper, C.; Biemann, K.; Lambrecht, U.; Jöhrens, J.; Meyer, K. Klimabilanz Von Elektroautos: Einflussfaktoren Und Verbesserungspotenzial. 2019. Available online: https://www.agora-verkehrswende.de/fileadmin/Projekte/2018/Klimabilanz_von_Elektroautos/Agora-Verkehrswende_22_Klimabilanz-von-Elektroautos_WEB.pdf (accessed on 14 July 2021).
3. Karabelli, D.; Kiemel, S.; Singh, S.; Koller, J.; Ehrenberger, S.; Miehe, R.; Weeber, M.; Birke, K.P. Tackling xEV Battery Chemistry in View of Raw Material Supply Shortfalls. *Front. Energy Res.* **2020**, *8*, 331. [CrossRef]
4. Sato, F.E.K.; Nakata, T. Recoverability Analysis of Critical Materials from Electric Vehicle Lithium-Ion Batteries through a Dynamic Fleet-Based Approach for Japan. *Sustainability* **2020**, *12*, 147. [CrossRef]
5. Kotak, Y.; Marchante Fernández, C.; Canals Casals, L.; Kotak, B.S.; Koch, D.; Geisbauer, C.; Trilla, L.; Gómez-Núñez, A.; Schweiger, H.-G. End of Electric Vehicle Batteries: Reuse vs. Recycle. *Energies* **2021**, *14*, 2217. [CrossRef]
6. Glöser-Chahoud, S.; Huster, S.; Rosenberg, S.; Baazouzi, S.; Kiemel, S.; Singh, S.; Schneider, C.; Weeber, M.; Miehe, R.; Schultmann, F. Industrial disassembling as a key enabler of circular economy solutions for obsolete electric vehicle battery systems. *Resour. Conserv. Recycl.* **2021**, *174*, 105735. [CrossRef]
7. Yun, L.; Linh, D.; Shui, L.; Peng, X.; Garg, A.; Le, M.L.P.; Asghari, S.; Sandoval, J. Metallurgical and mechanical methods for recycling of lithium-ion battery pack for electric vehicles. *Resour. Conserv. Recycl.* **2018**, *136*, 198–208. [CrossRef]
8. Elwert, T.; Römer, F.; Schneider, K.; Hua, Q.; Buchert, M. Recycling of Batteries from Electric Vehicles. In *Behaviour of Lithium-Ion Batteries in Electric Vehicles*; Pistoia, G., Liaw, B., Eds.; Springer International Publishing: Cham, Switzerland, 2018; pp. 289–321, ISBN 978-3-319-69949-3.
9. Tan, W.J.; Chin, C.M.M.; Garg, A.; Gao, L. A hybrid disassembly framework for disassembly of electric vehicle batteries. *Int. J. Energy Res.* **2021**, *45*, 8073–8082. [CrossRef]
10. Gerlitz, E.; Greifenstein, M.; Hofmann, J.; Fleischer, J. Analysis of the Variety of Lithium-Ion Battery Modules and the Challenges for an Agile Automated Disassembly System. *Procedia CIRP* **2021**, *96*, 175–180. [CrossRef]
11. Schwarz, T.E.; Rübenbauer, W.; Rutrecht, B.; Pomberger, R. Forecasting Real Disassembly Time of Industrial Batteries Based on Virtual MTM-UAS Data. *Procedia CIRP* **2018**, *69*, 927–931. [CrossRef]
12. Ke, Q.; Zhang, P.; Zhang, L.; Song, S. Electric Vehicle Battery Disassembly Sequence Planning Based on Frame-Subgroup Structure Combined with Genetic Algorithm. *Front. Mech. Eng.* **2020**, *6*. [CrossRef]
13. Choux, M.; Marti Bigorra, E.; Tyapin, I. Task Planner for Robotic Disassembly of Electric Vehicle Battery Pack. *Metals* **2021**, *11*, 387. [CrossRef]
14. Alfaro-Algaba, M.; Ramirez, F.J. Techno-economic and environmental disassembly planning of lithium-ion electric vehicle battery packs for remanufacturing. *Resour. Conserv. Recycl.* **2020**, *154*, 104461. [CrossRef]

15. Ren, Y.; Zhang, C.; Zhao, F.; Xiao, H.; Tian, G. An asynchronous parallel disassembly planning based on genetic algorithm. *Eur. J. Oper. Res.* **2018**, *269*, 647–660. [CrossRef]
16. Wegener, K.; Andrew, S.; Raatz, A.; Dröder, K.; Herrmann, C. Disassembly of Electric Vehicle Batteries Using the Example of the Audi Q5 Hybrid System. *Procedia CIRP* **2014**, *23*, 155–160. [CrossRef]
17. Schäfer, J.; Singer, R.; Hofmann, J.; Fleischer, J. Challenges and Solutions of Automated Disassembly and Condition-Based Remanufacturing of Lithium-Ion Battery Modules for a Circular Economy. *Procedia Manuf.* **2020**, *43*, 614–619. [CrossRef]
18. Leitz, A. *Laserstrahlschweißen Von Kupfer-Und Aluminiumwerkstoffen in Mischverbindung*; Herbert Utz Verlag Wissenschaft: München, Germany, 2016; ISBN 978-3-8316-4549-7.
19. Kay, I.; Esmaeeli, R.; Hashemi, S.R.; Mahajan, A.; Farhad, S. Recycling Li-Ion batteries: Robotic disassembly of electric vehicle battery systems. In Proceedings of the International Mechanical Engineering Congress and Exposition (ASME 2019), Salt Lake City, UT, USA, 11–14 November 2019; Volume 6, p. 11112019, ISBN 978-0-7918-5943-8.
20. Li, L.; Zheng, P.; Yang, T.; Sturges, R.; Ellis, M.W.; Li, Z. Disassembly Automation for Recycling End-of-Life Lithium-Ion Pouch Cells. *JOM* **2019**, *71*, 4457–4464. [CrossRef]
21. Vongbunyong, S.; Chen, W.H. *Disassembly Automation*; Springer International Publishing: Cham, Switzerland, 2015; ISBN 978-3-319-15182-3.
22. Schmitt, J.; Haupt, H.; Kurrat, M.; Raatz, A. Disassembly automation for lithium-ion battery systems using a flexible gripper. In Proceedings of the 15th International Conference on Advanced Robotics (ICAR 2011), Tallinn, Estonia, 20–23 June 2011; pp. 291–297, ISBN 978-1-4577-1159-6.
23. Herrmann, C.; Raatz, A.; Andrew, S.; Schmitt, J. Scenario-Based Development of Disassembly Systems for Automotive Lithium Ion Battery Systems. *AMR* **2014**, *907*, 391–401. [CrossRef]
24. Kiemel, S.; Koller, J.; Kaus, D.; Singh, S.; Full, J.; Weeber, M.; Miehe, R.; Ehrenberger, S.; Österle, I.; Senzeybek, M.; et al. Untersuchung: Kreislaufstrategien für Batteriesysteme in Baden-Württemberg: KSBS BW, Stuttgart. 2020. Available online: https://www.ipa.fraunhofer.de/content/dam/ipa/de/documents/Kompetenzen/Nachhaltige-Produktion-und-Qualitaet/Endbericht_KSBS_offen.pdf (accessed on 16 November 2020).
25. Zhou, Z.; Liu, J.; Pham, D.T.; Xu, W.; Ramirez, F.J.; Ji, C.; Liu, Q. Disassembly sequence planning: Recent developments and future trends. *Proc. Inst. Mech. Eng. Part B J. Eng. Manuf.* **2019**, *233*, 1450–1471. [CrossRef]
26. Ren, Y.; Jin, H.; Zhao, F.; Qu, T.; Meng, L.; Zhang, C.; Zhang, B.; Wang, G.; Sutherland, J.W. A Multiobjective Disassembly Planning for Value Recovery and Energy Conservation From End-of-Life Products. *IEEE Trans. Automat. Sci. Eng.* **2021**, *18*, 791–803. [CrossRef]
27. Kheder, M.; Trigui, M.; Aifaoui, N. Disassembly sequence planning based on a genetic algorithm. *Proc. Inst. Mech. Eng. Part C J. Mech. Eng. Sci.* **2015**, *229*, 2281–2290. [CrossRef]
28. Zhou, Z.; Liu, J.; Pham, D. Disassembly Sequence Planning: Recent Developments and Future Trends. 2018. Available online: https://journals.sagepub.com/doi/full/10.1177/0954405418789975 (accessed on 1 May 2021).
29. Hernández Parrodi, J.C.; Höllen, D.; Pomberger, R. Potential and main technological challenges for material and energy recovery from fine fractions of landfill mining: A critical review. *Detritus* **2018**, *3*, 19–29. [CrossRef]
30. José, P.; Marko, H.; Ernst, W.; Aldert, H. Circular Economy: Measuring Innovation in the Product Chain; Policy Report. 2017. Available online: https://www.pbl.nl/sites/default/files/downloads/pbl-2016-circular-economy-measuring-innovation-in-product-chains-2544.pdf (accessed on 1 May 2021).
31. Ren, Y.; Tian, G.; Zhao, F.; Yu, D.; Zhang, C. Selective cooperative disassembly planning based on multi-objective discrete artificial bee colony algorithm. *Eng. Appl. Artif. Intell.* **2017**, *64*, 415–431. [CrossRef]
32. Rallo, H.; Benveniste, G.; Gestoso, I.; Amante, B. Economic analysis of the disassembling activities to the reuse of electric vehicles Li-ion batteries. *Resour. Conserv. Recycl.* **2020**, *159*, 104785. [CrossRef]

MDPI

Article

Investigation of a Novel Ecofriendly Electrolyte-Solvent for Lithium-Ion Batteries with Increased Thermal Stability

Marco Ströbel *, Larissa Kiefer and Kai Peter Birke

Electrical Energy Storage Systems, Institute for Photovoltaics, University of Stuttgart, Pfaffenwaldring 47, 70569 Stuttgart, Germany; st142567@stud.uni-stuttgart.de (L.K.), Peter.Birke@ipv.uni-stuttgart.de (K.P.B.)
* Correspondence: Marco.Stroebel@ipv.uni-stuttgart.de

Abstract: This study presents tributyl acetylcitrate (TBAC) as a novel ecofriendly high flash point and high boiling point solvent for electrolytes in lithium-ion batteries. The flash point ($T_{FP} = 217\,°C$) and the boiling point ($T_{BP} = 331\,°C$) of TBAC are approximately 200 K greater than that of conventional linear carbonate components, such as ethyl methyl carbonate (EMC) or diethyl carbonate (DEC). The melting point ($T_{MP} = -80\,°C$) is more than 100 K lower than that of ethylene carbonate (EC). Furthermore, TBAC is known as an ecofriendly solvent from other industrial sectors. A life cycle test of a graphite/NCM cell with 1 M lithium hexafluorophosphate ($LiPF_6$) in TBAC:EC:EMC:DEC (60:15:5:20 wt) achieved a coulombic efficiency of above 99% and the remaining capacity resulted in 90 percent after 100 cycles ($C/4$) of testing. As a result, TBAC is considered a viable option for improving the thermal stability of lithium-ion batteries.

Keywords: ecofriendly electrolyte for lithium-ion batteries; increased thermal stability of electrolytes; enhanced electrolyte safety based on high flash point; tributylacetylcitrate; acetyltributylcitrate

Citation: Ströbel, M.; Kiefer L.; Birke, K.P. Investigation of a Novel Ecofriendly Electrolyte-Solvent for Lithium-Ion Batteries with Increased Thermal Stability. *Batteries* **2021**, 7, 72. https://doi.org/10.3390/batteries7040072

Academic Editor: Catia Arbizzani

Received: 28 July 2021
Accepted: 12 October 2021
Published: 28 October 2021

Publisher's Note: MDPI stays neutral with regard to jurisdictional claims in published maps and institutional affiliations.

1. Introduction

Lithium-ion cells are the technological standard for portable devices such as smartphones, notebooks, and electric vehicles, and as a result, they are viewed as a key for the global transition to electro-mobility. In terms of thermal stability, the electrolyte of a lithium-ion cell is considered a critical component, which is responsible for ionic conductivity. The electrolyte is mainly composed of solvents, lithium conducting salts, and various additives [1,2]. The usage of conventional solvents for electrolytes with low boiling points and flash points (T_{FP}) like dimethyl carbonate (DMC, $T_{FP} < 20\,°C$ [3]), EMC ($T_{FP} \approx 22\,°C$ [4]), or DEC ($T_{FP} \approx 25\,°C$ [5]) bear an increased risk to ignite lithium-ion cells [6–8]. The low boiling point generates high pressure gradients at moderate temperatures ($<100°C$), which can lead to the explosion of the cell. The chemical products of burned fluorine-containing electrolytes are highly toxic [9–12]. Therefore, an expensive thermal management system and a massive casing for lithium-ion batteries are required. This heavy casing for battery packs for electric vehicles lowers the gravimetric density of the pack and increases the weight of the vehicles. Increasing the intrinsic thermal stability is a key factor to lower costs for cell protection and increase the gravimetric density of the battery pack. Investigations on new electrolyte formulations have been considered before, for example by using flame retardant additives like organic phosphates [13] or phosphonates [14]. However, using these additives to improve the thermal stability reduces the cell performance [15]. Investigations show that the use of ionic liquids can increase the flash point of electrolytes however, these ionic liquids are linked with high costs [16–19]. Another promising approach is to investigate co-solvents with higher flash points like adiponitrile (ADN, $T_{FP} \approx 163\,°C$ [20]) by Isken et al [21]. Co-solvents were able to increase the flash point significantly from $T_{FP,EC:DEC} \approx 36\,°C$ of the EC:DEC (3:7 wt) mixture to $T_{FP,EC:ADN} \approx 149\,°C$ of the EC:ADN (1:1 wt) mixture. This indicates that it is possible to formulate electrolytes with higher flash points by replacing volatile carbonates. However,

101

the melting points (MP) of EC $T_{\text{MP,EC}} \approx 36\,^{\circ}\text{C}$ and ADN $T_{\text{MP,ADN}} \approx 2\,^{\circ}\text{C}$ are extremely high for low temperature applications. To lower the working temperature applicability $T < 0\,^{\circ}\text{C}$ research has shown many different classes of solvents like sulfones [22,23]. Unfortunately, most of these solvents are ecologically harmful.

Therefore, we investigated tributyl acetylcitrate as a solvent to formulate an electrolyte composition with a high flash point and a wide operating range from very low to high temperature. TBAC has a high flash point of $T_{\text{FP}} \approx 217\,^{\circ}\text{C}$ [24], its melting point is very low with $T_{\text{MP}} \approx -80\,^{\circ}\text{C}$ and the boiling point is at $T_{\text{BP}} \approx 330\,^{\circ}\text{C}$. Another noteworthy advantage of TBAC is that it is ecofriendly [25], nontoxic [26], and therefore safe to handle. For example, it is well known as a plasticizer in the nail polish industry [27].

This study presents tributyl acetylcitrate as a novel solvent for lithium-ion cells. The combination of conventional solvents like EC and DEC with TBAC creates an electrolyte with improved thermal stability.

2. Materials and Methods

2.1. Materials

Tributyl acetylcitrate (TBAC, purity > 98% Sigma Aldrich, Germany) was dried over a molecular sieve (mesh size ≈ 0.3 nm) inside a glovebox exposed to argon atmosphere over night (moisture content of less than 0.5 ppm). Figure 1 shows the chemical structure of TBAC [24]. Ethylene carbonate (EC), ethyl methyl carbonate (EMC, purity > 99.9%), propylene carbonate (PC, purity > 99.7%), diethyl carbonate (DEC, purity > 99.9%), lithium tetrafluoroborate (LiBF$_4$ purity > 99%, all from Sigma Aldrich), lithium bis(trifluoromethanesulfonyl)imide (LiTFSI, purity > 99%), lithium bis(fluorosulfonyl)imide (LiFSI, both fromIOLITEC GmbH, Germany), and lithium hexafluorophosphate (LiPF$_6$, BASF) were opened in the glovebox and were used as received.

Graphite electrodes (3 mAh/cm^2, provided by Varta AG) and lithium nickel manganese cobalt oxide electrodes (NCM 622, 1.3 mAh/cm^2, provided by Münster Electrochemical Energy Technology) were punched to 18 mm coins and dried in a vacuum oven (Buechi Labortechnik AG B-585, $p < 50$ mbar, $T = 120\,^{\circ}\text{C}$). The separator Freudenberg 2190 (thickness $= 176$ µm) was punched to coins with 21 mm in diameter and dried in a vacuum oven ($p < 50$ mbar, $T = 120\,^{\circ}\text{C}$).

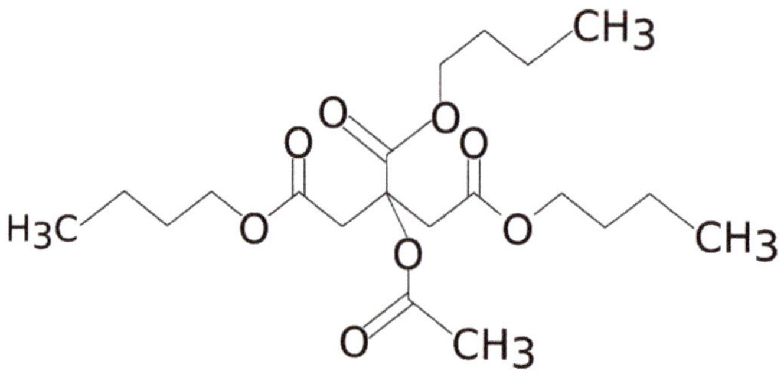

Figure 1. Chemical structure of tributyl acetylcitrate [24].

2.2. Electrolyte Preparation

The electrolytes were prepared in an argon-filled glovebox (H$_2$ and O$_2$ content lower than 0.5 ppm). TBAC was combined with different well-known solvents like EC and EMC. The formulated solvent mixtures and the amount of each solvent (in wt) is shown in Table 1a. Table 1b shows the combination of the solvent mixtures and the investigated lithium salts with a salt concentration of 1 M.

Table 1. (**a**) Formulated solvent-mixtures wt. (**b**) Investigated combinations of lithium salts and solvent-mixtures (concentration of 1 M).

(a)					
Mixture	**TBAC**	**EC**	**EMC**	**PC**	**DEC**
F1	85%	15%	0%	0%	0%
F2	80%	15%	5%	0%	0%
F3	60%	15%	5%	20%	0%
F4	60%	15%	5%	0%	20%

(b)				
Salt	**Solvent-Mixtures**			
LiBF$_4$		F2		
LiTFSI	F1	F2	F3	F4
LiPF$_6$		F2	F3	F4
LiFSI		F2		

2.3. Cell Preparation

Cells were assembled as coin cells in PAT-Cell or ECC-Ref cells from EL Cell. For constant current cycling and C-rate tests, cells were assembled with NCM as the working electrode and graphite as the counter electrode. The polypropylene separator (Freudenberg FS2190) was moistened on both sides with a total of 120 µL of the selected electrolyte before the electrodes were added. For cyclic voltammetry measurements, two electrode cells were assembled with stainless steel or aluminum as working electrode and stainless steel as counter and reference electrode.

2.4. Electrochemical Characterization

All tests were done in a Memmert IPP110 thermal chamber at a constant temperature of $T = 25 \pm 0.1$ °C. The cell performance was evaluated by cycling tests with constant current charge and discharge with the battery tester CTS from BaSyTec. At the beginning, three preliminary formation cycles were performed at $I = C/10$ in a potential range from $U = 2.5$ V to $U = 4.2$ V. Subsequently lifecycle tests were performed at $C/4$.

The cyclic voltammetry measurements were performed with a reference 3000 AE from Gamry. The electrochemical stability was measured by linear sweep voltammetry. The scan rate was set to 1 mVs^{-1}, the potential limits were set from -4.3 V to 4.3 V vs. stainless steel.

3. Results

Table 2 shows the melting, flash, and boiling point of TBAC, EC, EMC, PC, and DEC. The boiling point as well as the flash point $T_{FP} = 217$ °C of TBAC is almost 200 K higher than that of EMC and DEC. The melting point is more than 100 K lower than that of EC. The high boiling point and the high flash point are the main advantages of TBAC as a solvent for lithium-ion batteries. However, during cell preparation, TBAC showed a higher viscosity than the other solvents, which could cause a lower conductivity. Therefore, the solvent-mixtures shown in Table 1 were investigated to find a composition with adequate cycling behavior and a high amount of TBAC.

Table 2. Physical properties of TBAC [24], EC [28], EMC [4], PC [29], and DEC [5]. Symbols used: T_{MP}, melting point; T_{FP}, flash point; T_{BP}, boiling point.

	T_{MP} (°C)	T_{FP} (°C)	T_{BP} (°C)
TBAC	-80	217	331
EC	36	143	248
EMC	-55	23.9	101
PC	-49	123	243
DEC	-43	25	126

The following discussion is structured by presenting the results for each investigated conducting salt in combination with TBAC-based solvent mixtures from Table 1.

The main focus lies in combining TBAC with LiTFSI since other conducting salts are thermally less stable [30]. However, other possible combinations of salts are presented to give an overview of further possibilities.

3.1. TBAC Solvent-Mixtures with LiTFSI

The lithium salt concentration of 1 M LiTFSI seems to be completely dissolvable in F1 = TBAC:EC (85:15 wt). However, in a cell with NCM as positive electrode and graphite as a negative electrode with LiTFSI in F1, the coulombic efficiency of the life cycle test was only about $\eta_{coul} = 80\%$ and the ionic conductivity limits the current to a maximum of C/10 at $T = 25\,°C$. By adding EMC to the solvent-mixture, the coulombic efficiency increased above $\eta_{coul} = 99\%$. To increase the life cycle performance, DEC was added to the electrolyte. This resulted in an increase of conductivity that allowed to cycle the cells with C/4. Figure 2 shows the results of the cycling test of a graphite/NCM cell with the electrolyte formulation with 1 M LiTFSI in the solvent mixture F4 = TBAC:EC:EMC:DEC (60:15:5:20 wt) at a C-rate of C/4 at $T = 25\,°C$.

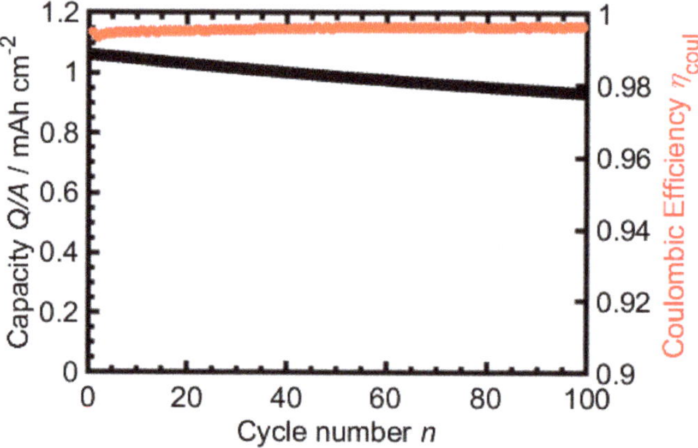

Figure 2. Cycling performance of a graphite/NCM cell with 1 M LiTFSI in TBAC:EC:EMC:DEC (60:15:5:20 wt) electrolyte at $T = 25\,°C$. The cell was charged and discharged at C−rate C/4.

The coulombic efficiency achieved was $\eta_{coul} = 99.6\%$. After 100 cycles there was about 88% of the initial capacity left. Different lithium salt concentrations between 0.8 M and 1.2 M were investigated. The best cycling performance was achieved with a concentration of 1 M. Therefore, all further tests were done with a salt concentration of 1 M.

C-rate tests were performed to investigate the electrode performance in combination with the novel electrolyte composition for different currents. Figure 3 shows the capacity obtained at different C-rates for a graphite/NCM cell with 1 M LiTFSI in the solvent-mixture F4.

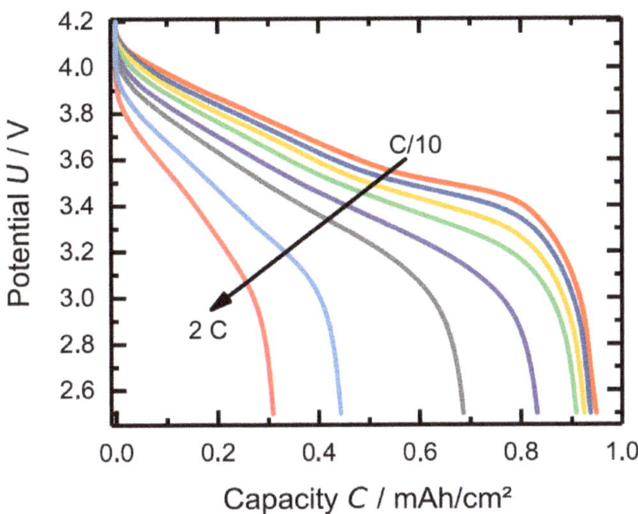

Figure 3. C-rate performance of a graphite/NCM cell with 1 M LiTFSI in TBAC:EC:EMC:DEC (60:15:5:20 wt) electrolyte at $T = 25\,°C$ in the potential range of 2.5–4.2 V.

The applied currents were: $C/10, C/5, C/3, C/2, 0.75C, 1C, 1.5C$, and $2C$. At a current rate of $1C$, the usable capacity dropped to 70% of the initial capacity. Higher C-rates decrease the usable capacity significantly.

Figure 4 shows the discharge capacity and the potential profiles at different temperatures between $T = 30\,°C$ and $T = 70\,°C$. Until 60 °C the voltage profile seems to be nearly independent of temperature. At 70 °C there was a small drop in voltage observed.

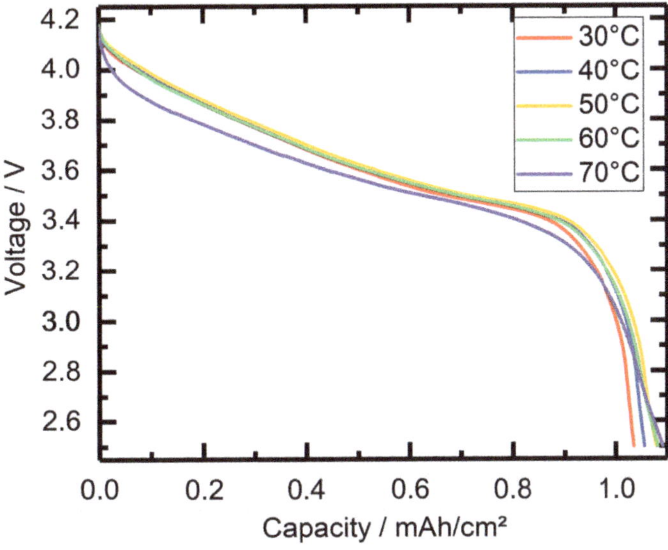

Figure 4. Voltage profiles at different temperatures of a graphite/NCM cell with 1 M LiTFSI in TBAC:EC:EMC:DEC (60:15:5:20 wt) electrolyte.

The solvent mixture F3 = TBAC:EC:EMC:PC (60:15:5:20 wt) was investigated in combination with LiTFSI. However, as expected from the literature, the combination of PC and LiTFSI is not compatible and it is not possible to cycle a lithium-ion cell with this electrolyte [31].

3.2. TBAC Solvent-Mixtures with LiBF$_4$

A total of 1 M LiBF$_4$ was investigated in TBAC:EC:EMC (80:15:5 wt). However, LiBF$_4$ did not dissolve. It was not possible to cycle the electrolyte formulation in a graphite/NCM cell. We assume that TBAC is incompatible with LiBF$_4$.

3.3. TBAC Solvent-Mixtures with LiFSI

The results for the electrolytes with LiFSI as conducting salt are comparable to LiTFSI. Graphite/NCM cells with 1 M LiFSI dissolved in F2 showed a coloumbic efficiency of about $\eta_{coul} = 98\%$. The results of the cycling tests are shown in Figure 5 with three formation cycles at $C/10$ and 100 cycles at $C/4$.

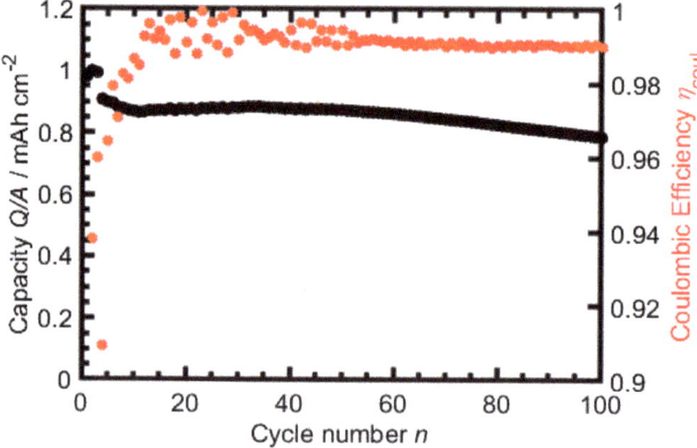

Figure 5. Cycling performance of a graphite/NCM cell with 1 M LiFSI in TBAC:EC:EMC (80:15:5 wt) electrolyte at $T = 25\,°C$. The cell was charged and discharged at C–rate $C/4$.

However, it was not possible to achieve a durable coulombic efficiency above $\eta_{coul} = 99\%$. Therefore TBAC-LiFSI electrolytes were not further investigated.

3.4. TBAC Solvent-Mixtures with LiPF$_6$

Similar to LiTFSI and LiFSI, the conducting salt LiPF$_6$ does dissolve in F2. The coulombic efficiency of 1 M LiPF$_6$ in F2 in a graphite/NCM cell was about $\eta_{coul} = 98\%$. It was necessary to add DEC or PC to improve the ionic conductivity to cycle cells at C-rate $C/4$. Figure 6 shows the cycling results for a graphite/NCM cell with 1 M LiPF$_6$ in F4 at $T = 25\,°C$ with C-rate of $C/4$. The results of the electrolyte formulation with PC (F3) are comparable to the DEC electrolyte. Therefore, we will discuss only the DEC containing electrolyte..

The coulombic efficiency after 100 cycles is above $\eta_{coul} = 99.5\%$, the remaining capacity is about 93% of the initial capacity. Compared to TBAC-LiTFSI electrolytes, the coulombic efficiency and the remaining total capacity are increased.

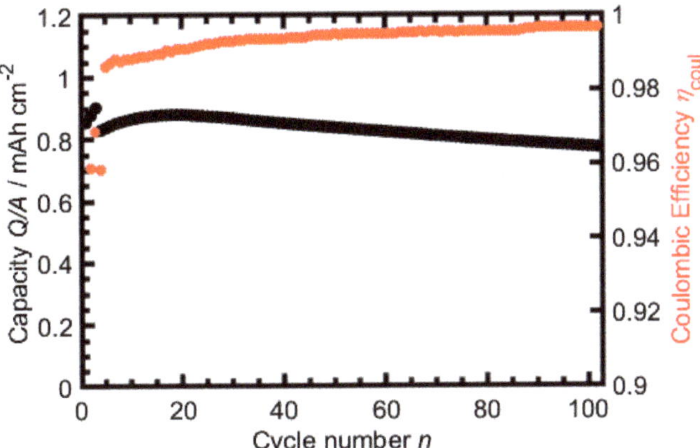

Figure 6. Cycling performance of a graphite/NCM cell with 1 M LiPF$_6$ in TBAC:EC:EMC:DEC (60:15:5:20 wt) electrolyte at $T = 25\ ^\circ$C. The cell was charged and discharged at C–rate $C/4$.

3.5. Cyclic Voltammetry

To determine the electrochemical stability window, CV measurements between electro-chemical inert stainless steel electrodes were observed. Figure 7a shows the electrochemical stability window of LP40 electrolyte as reference. Therefore, the potential of a stainless steel working electrode connected to the electrolyte LP40 against stainless steel was scanned at 1 mVs^{-1}. The electrolyte is stable up to 3.5 V vs. stainless steel.

The stability window of LiPF$_6$ dissolved in TBAC:EC:EMC:DEC (60:15:5:20 wt) is shown in Figure 7b. For the second cycle, the TBAC based electrolyte seems to be stable over the total potential range. Compared to the LP40 reference electrolyte the stability is improved. At 3.5 V vs. stainless steel, the current begins to rise but is still below 3 μA/cm^2.

Figure 7c shows the electrochemical stability window for LiTFSI dissolved in TBAC:EC:EMC:DEC (60:15:5:20 wt) between two stainless steel electrodes and Figure 7d be-tween aluminum as working electrode and stainless steel as the counter electrode. Between two stainless steel electrodes, the stability window of the LiTFSI-TBAC based electrolyte is comparable to the LP40 reference. The electrolyte decomposes above 3.5 V vs. stainless steel. However, after the second cycle the decomposition current is still below 4 μA/cm^2.

Since aluminum is used as a current collector in commercial lithium-ion cells the electrochemical stability window was investigated for an aluminum working electrode. CV measurements with LP40 and LiPF$_6$-TBAC based electrolytes in contact with aluminum were comparable to CV measurements without aluminum. Figure 7d shows the result for the LiTFSI-TBAC based electrolyte in contact with aluminum. Above 3 V vs. stainless steel, the decomposition current is increased by an order of one magnitude compared to the sample without aluminum. This effect is described in the literature as aluminum corrosion by the LiTFSI conducting salt [32,33].

LiFSI-TBAC based electrolytes were not further investigated since the tendency to allow aluminum current collector corrosion beyond 3.8 V vs. Li$^+$/Li^0 is also described in literature [34].

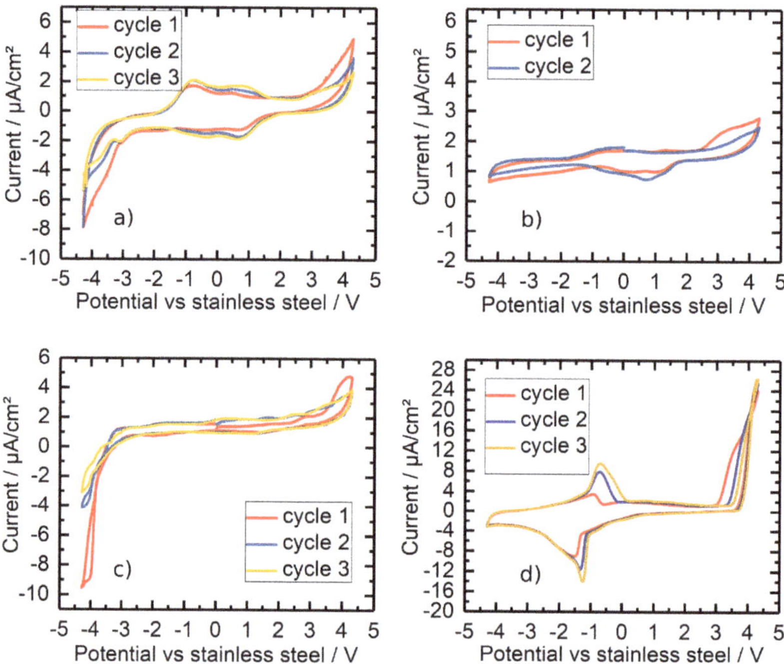

Figure 7. Electrochemical stability window of (**a**) LP40, (**b**) 1 M LiPF$_6$ dissolved in TBAC:EC:EMC:DEC (60:15:5:20 wt), (**c**) 1 M LiTFSI in TBAC:EC:EMC:DEC (60:15:5:20 wt) (all with stainless steel as working and counter electrodes), and (**d**) 1 M LiTFSI in TBAC:EC:EMC:DEC (60:15:5:20 wt) (with aluminum as working electrode). Scan rate 1 mVs^{-1}. The electrochemical stability of LiPF$_6$ in TBAC seems to be improved compared to LP40. LiTFSI–TBAC based electrolytes in contact with aluminum show higher decomposition rates, which is assumed to be linked to aluminum corrosion by LiTFSI.

4. Discussion

TBAC in combination with EC and EMC dissolves the conducting salts LiTFSI, LiFSI, and LiPF$_6$. The main focus of this study lays in the combination of TBAC with LiTFSI to create an electrolyte with high thermal stability and high flash point, since the thermal stability and the ionic conductivity of LiTFSI are improved compared to LiPF$_6$. By dissolving 1 M LiTFSI in TBAC:EC:EMC:DEC (60:15:5:20 wt) a coulombic efficiency of 99% at a C-rate of $C/4$ was achieved. However, the coulombic efficiency of LiTFSI in life cycle tests was slightly reduced compared to LiPF$_6$. To determine the reason for the efficiency reduction CV measurements were performed. They showed that the electrochemical stability window of LiPF$_6$-TBAC based electrolytes is improved compared to the LP40 reference. Furthermore, the electrochemical stability window of LiTFSI-TBAC based electrolyte between stainless steel is comparable to LP40. CV measurements of the LiTFSI-TBAC based electrolyte connected to an aluminum working electrode show decomposition reactions beyond 3.5 V vs. stainless steel. In literature this is referred to as aluminum corrosion by LiTFSI.

TBAC itself seems to be electrochemically stable in the same window as LP40.

5. Conclusions

In this paper we have investigated tributyl acetylcitrate as a novel and ecofriendly electrolyte-solvent for lithium-ion cells. Compared to conventional solvents, the flash point and the boiling point are almost 200 K higher than that of EMC or DEC. The melting point is more than 100 K lower than that of EC. In combination with LiPF$_6$ and other solvents (TBAC:EC:EMC:DEC, 60:15:5:20 wt), TBAC shows good electrochemical stability between -4 V and 3 V vs. stainless steel. In graphite/NCM cells the coulombic efficiency is above

$\eta_{coul} = 99\%$. After 100 cycles the remaining capacity is still above 90%. Therefore, TBAC when compared to other solvents is a promising alternative to improve the intrinsic thermal stability of lithium-ion cells.

Author Contributions: Conceptualization, M.S. and K.P.B.; methodology, M.S. L.K., and K.P.B.; investigation, M.S. and L.K.; writing—original draft preparation, M.S.; writing—review and editing, L.K. and K.P.B.; visualization, M.S. and L.K.; supervision, K.P.B.; project administration, M.S.; funding acquisition, K.P.B. All authors have read and agreed to the published version of the manuscript.

Funding: This research was funded by the Bundesministerium fuer Bildung und Forschung within the BCT—Battery Cell Technology project (03XP0109H).

Institutional Review Board Statement: Not Applicable.

Informed Consent Statement: Not Applicable.

Data Availability Statement: Not Applicable.

Acknowledgments: The authors thank Münster Electrochemical Energy Technology for providing the NCM622 electrode material and Varta AG for providing the graphite electrode material.

Conflicts of Interest: The authors declare no conflict of interest. The funders had no role in the design of the study; in the collection, analyses, or interpretation of data; in the writing of the manuscript, or in the decision to publish the results.

References

1. Xu, K. Nonaqueous liquid electrolytes for lithium-based rechargeable batteries. *Chem. Rev.* **2004**, *104*, 4303–4418. [CrossRef]
2. Xu, K. Electrolytes and interphases in Li-ion batteries and beyond. *Chem. Rev.* **2014**, *114*, 11503–11618. [CrossRef] [PubMed]
3. *Dimethyl Carbonate, CAS Number 616-38-6*; MERCK KGaA: Darmstadt, Germany, 2020.
4. *Ethyl Methyl Carbonate, CAS Number 623-53-0*; MERCK KGaA: Darmstadt, Germany, 2020.
5. *Diethyl Carbonate, CAS Number 105-58-8*; MERCK KGaA: Darmstadt, Germany, 2020.
6. Hofmann, A.; Hanemann, T. Novel electrolyte mixtures based on dimethyl sulfone, ethylene carbonate and LiPF6 for lithium-ion batteries. *J. Power Sources* **2015**, *298*, 322–330. [CrossRef]
7. Hess, S.; Wohlfahrt-Mehrens, M.; Wachtler, M. Flammability of Li-ion battery electrolytes: flash point and self-extinguishing time measurements. *J. Electrochem. Soc.* **2015**, *162*, A3084. [CrossRef]
8. Lisbona, D.; Snee, T. A review of hazards associated with primary lithium and lithium-ion batteries. *Process. Saf. Environ. Prot.* **2011**, *89*, 434–442. [CrossRef]
9. Lei, B.; Zhao, W.; Ziebert, C.; Uhlmann, N.; Rohde, M.; Seifert, H.J. Experimental analysis of thermal runaway in 18650 cylindrical Li-ion cells using an accelerating rate calorimeter. *Batteries* **2017**, *3*, 14. [CrossRef]
10. Hammami, A.; Raymond, N.; Armand, M. Runaway risk of forming toxic compounds. *Nature* **2003**, *424*, 635–636. [CrossRef]
11. Larsson, F. *Assessment of Safety Characteristics for Li-Ion Battery Cells by Abuse Testing*; Dempartment of Applied Physics, Chalmers University of Technology: Gothenborg, Sweden, 2014; p. 94.
12. Larsson, F.; Andersson, P.; Blomqvist, P.; Lorén, A.; Mellander, B.E. Characteristics of lithium-ion batteries during fire tests. *J. Power Sources* **2014**, *271*, 414–420. [CrossRef]
13. Xu, K.; Ding, M.S.; Zhang, S.; Allen, J.L.; Jow, T.R. Evaluation of fluorinated alkyl phosphates as flame retardants in electrolytes for Li-ion batteries: I. Physical and electrochemical properties. *J. Electrochem. Soc.* **2003**, *150*, A161. [CrossRef]
14. Dalavi, S.; Xu, M.; Ravdel, B.; Zhou, L.; Lucht, B.L. Nonflammable electrolytes for lithium-ion batteries containing dimethyl methylphosphonate. *J. Electrochem. Soc.* **2010**, *157*, A1113. [CrossRef]
15. Kalhoff, J.; Eshetu, G.G.; Bresser, D.; Passerini, S. Safer electrolytes for lithium-ion batteries: state of the art and perspectives. *ChemSusChem* **2015**, *8*, 2154–2175. [CrossRef]
16. Kim, G.; Jeong, S.; Joost, M.; Rocca, E.; Winter, M.; Passerini, S.; Balducci, A. Use of natural binders and ionic liquid electrolytes for greener and safer lithium-ion batteries. *J. Power Sources* **2011**, *196*, 2187–2194. [CrossRef]
17. Lux, S.F.; Schmuck, M.; Jeong, S.; Passerini, S.; Winter, M.; Balducci, A. Li-ion anodes in air-stable and hydrophobic ionic liquid-based electrolyte for safer and greener batteries. *Int. J. Energy Res.* **2010**, *34*, 97–106. [CrossRef]
18. Tsurumaki, A.; Agostini, M.; Poiana, R.; Lombardo, L.; Lufrano, E.; Simari, C.; Matic, A.; Nicotera, I.; Panero, S.; Navarra, M.A. Enhanced safety and galvanostatic performance of high voltage lithium batteries by using ionic liquids. *Electrochim. Acta* **2019**, *316*, 1–7. [CrossRef]
19. Navarra, M.A. Ionic liquids as safe electrolyte components for Li-metal and Li-ion batteries. *MRS Bull.* **2013**, *38*, 548. [CrossRef]
20. *Adiponitrile, CAS Number 111-69-3*; MERCK KGaA: Darmstadt, Germany, 2020.
21. Isken, P.; Dippel, C.; Schmitz, R.; Schmitz, R.; Kunze, M.; Passerini, S.; Winter, M.; Lex-Balducci, A. High flash point electrolyte for use in lithium-ion batteries. *Electrochim. Acta* **2011**, *56*, 7530–7535. [CrossRef]

22. Sun, X.G.; Angell, C.A. New sulfone electrolytes for rechargeable lithium batteries.: Part I. Oligoether-containing sulfones. *Electrochem. Commun.* **2005**, *7*, 261–266. [CrossRef]
23. Xu, K.; Angell, C.A. Sulfone-based electrolytes for lithium-ion batteries. *J. Electrochem. Soc.* **2002**, *149*, A920. [CrossRef]
24. *Tributyl Acetylcitrat, CAS Number 77-90-7*; MERCK KGaA: Darmstadt, Germany, 2020.
25. Bae, J.W.; Yeo, M.; Shin, E.J.; Park, W.H.; Lee, J.E.; Nam, B.U.; Kim, S.Y. Eco-friendly plasticized poly (vinyl chloride)–acetyl tributyl citrate gels for varifocal lens. *RSC Adv.* **2015**, *5*, 94919–94925. [CrossRef]
26. Johnson, W., Jr. Final report on the safety assessment of acetyl triethyl citrate, acetyl tributyl citrate, acetyl trihexyl citrate, and acetyl trioctyl citrate. *Int. J. Toxicol.* **2002**, *21*, 1–17.
27. Renard C. Tributyl Acetylcitrat as Component of Nail Polish. U.S. Patent 20040170584A1, 2 September 2004.
28. *Ethylcarbonate, CAS Number 96-49-1*; MERCK KGaA: Darmstadt, Germany, 2021.
29. *Propylencarbonate, CAS Number 108-32-7*; MERCK KGaA: Darmstadt, Germany, 2021.
30. Dahbi, M.; Ghamouss, F.; Tran-Van, F.; Lemordant, D.; Anouti, M. Comparative study of EC/DMC LiTFSI and LiPF6 electrolytes for electrochemical storage. *J. Power Sources* **2011**, *196*, 9743–9750. [CrossRef]
31. Pan, Y.; Wang, G.; Lucht, B.L. Cycling performance and surface analysis of Lithium bis (trifluoromethanesulfonyl) imide in propylene carbonate with graphite. *Electrochim. Acta* **2016**, *217*, 269–273. [CrossRef]
32. Krämer, E.; Schedlbauer, T.; Hoffmann, B.; Terborg, L.; Nowak, S.; Gores, H.J.; Passerini, S.; Winter, M. Mechanism of anodic dissolution of the aluminum current collector in 1 M LiTFSI EC: DEC 3: 7 in rechargeable lithium batteries. *J. Electrochem. Soc.* **2012**, *160*, A356. [CrossRef]
33. Morita, M.; Shibata, T.; Yoshimoto, N.; Ishikawa, M. Anodic behavior of aluminum current collector in LiTFSI solutions with different solvent compositions. *J. Power Sources* **2003**, *119*, 784–788. [CrossRef]
34. Abouimrane, A.; Ding, J.; Davidson, I. Liquid electrolyte based on lithium bis-fluorosulfonyl imide salt: Aluminum corrosion studies and lithium ion battery investigations. *J. Power Sources* **2009**, *189*, 693–696. [CrossRef]

Article

Thermal Runaway Modelling of Li-Ion Cells at Various States of Ageing with a Semi-Empirical Model Based on a Kinetic Equation

Mathilde Grandjacques *, Pierre Kuntz , Philippe Azaïs , Sylvie Genies and Olivier Raccurt

CEA, LITEN, DEHT, LMP, University Grenoble Alpes, F-38000 Grenoble, France; pierre.kuntz@cea.fr (P.K.); philippe.azais@cea.fr (P.A.); sylvie.genies@cea.fr (S.G.); olivier.raccurt@cea.fr (O.R.)
* Correspondence: mathilde.grandjacques@cea.fr; Tel.: +33-4387-840-83

Abstract: The thermal runaway model used is a semi-empirical model based on a kinetic equation, parametrised by three parameters (m, n, p). It is believed that this kinetic equation can describe any reaction. The choice of parameters is often made without justifications. We assumed it necessary to develop a method to select the parameters. The method proposed is based on data coming from an accelerating rate calorimeter (ARC) test. The method has been applied on data obtained for cells aged on different conditions. Thanks to a post-mortem analysis and simulations carried out using the parameters obtained, we have shown that the ageing mechanisms have a major impact on the parameters.

Keywords: Li-ion battery; thermal runaway; model; post-mortem analysis

Citation: Grandjacques, M.; Kuntz, P.; Azaïs, P.; Genies, S.; Raccurt, O. Thermal Runaway Modelling of Li-Ion Cells at Various States of Ageing with a Semi-Empirical Model Based on a Kinetic Equation. *Batteries* **2021**, *7*, 68. https://doi.org/10.3390/batteries7040068

Academic Editor: Kai Peter Birke and Duygu Kaus

Received: 31 July 2021
Accepted: 9 October 2021
Published: 18 October 2021

Publisher's Note: MDPI stays neutral with regard to jurisdictional claims in published maps and institutional affiliations.

1. Introduction

Lithium ion battery technology is more and more widespread due to its high energy density and good cyclability. However, this technology can suffer from safety issues. Most Li-ion battery electric vehicles catch fire due to thermal runaway of the battery. Therefore, safety is a key issue in the development of this technology.

Thermal runaway of the battery results from a chain of reactions, some of which are exothermic. Chemical reactions occur one after another, raising the temperature and triggering new reactions. The different exothermic reactions are as follows: first, the decomposition of the solid electrolyte interphase (SEI); then, the reaction of the anode active material and the electrolyte and the reaction between the anode active material and the binder. Next, the cathode active material decomposes and can release oxygen as the battery temperature goes higher. The released oxygen can oxidise the electrolyte or react with the anode active material bringing the battery temperature to a higher value. Combustion of the flammable gases, such as the gas-state solvents and the gases released from the chemical reactions, would happen at a high temperature if an ignition source occurred (short-circuit), resulting in fire and explosion [1].

Most of the studies concerning safety are conducted on new/fresh cells. However, different ageing mechanisms inside the cell can induce physical and chemical modifications of the internal components that could influence the initiation of possible combustion. Ageing can be an important factor that must be taken into account for safety. Few studies investigate the influence of ageing on safety and most of them are conducted from an experimental point of view. Cells are cycled and tested in order to exhibit different ageing mechanisms that impact the safety. For example, Friesen et al. [2] studied the impact of low-temperature cycling on safety, and Borner et al. [3] or Larson et al. [4] showed the correlation between ageing and thermal stability of batteries. All existing studies use the runaway temperature and the heat rate to determine the thermal stability. This article proposes to better understand the correlation between ageing and safety. Several

111

accelerating rate calorimeter (ARC) tests on cells ageing in different conditions have been realised. The results of the experiment have been crossed with a model.

Thermal runaway models are based on the kinetic equation parametrised by the temperature (T) and α ($0 \leq \alpha \leq 1$), the fractional conversion degree of the reactants. The kinetic model has different forms. Avrami–Erofeev introduced the semi-empirical model parametrised by three constants (m, n, p). Frequently, the more general Sestak and Berggren [5] model is used.

It is believed that this kinetic equation can describe any reaction. The choice of the parameters (m, n, p) is often made without justifications and the parameters ($m = 1, n = 1, p = 0$) are classically chosen.

We assumed it necessary to develop a method to select the parameters; the parameters are calculated from the ARC test data. The test protocol will be described in Section 2 and the model in Section 3. Section 4 will expose the results of the crossing of the test results with the model. The data-crossing proves that ageing mechanisms have a major impact on the parameters that is confirmed by the post-mortem analysis.

2. Test Protocol Description

The commercial Li-ion 18650 cell format was chosen for this protocol. Tests were performed on INR18650-30Q cells (from Samsung DI Co., Yongin, South Korea). The cell chemistry is a blend of graphite/silicon (approximately 2–3% Si) in the negative electrode and a blend of NCA ($LiNi_{0.89}Co_{0.10}Al_{0.01}O_2$) and NCO ($Ni_{0.8}Co_{0.2}$) in the positive electrode. Performance of each cell batch was measured at 1C: capacity was 3000 mAh and energy density was 230 Wh/kg. Cells were aged using a battery electric vehicle (BEV) protocol with representative ageing cycles at various temperatures and calendar ageings according to the international standard "IEC 62660-1". The temperature tests were ($-20\,°C, 0\,°C, 25\,°C, 45\,°C$) and the calendar ageing was realised at $45\,°C$ at constant voltage (CV) at 4.2 V and at $45\,°C$ in an open circuit voltage (OCV). Ageing profiles contained current pulses in charge or discharge that simulated car breaking and acceleration. For each condition 15 cells were aged on a battery tester ($PEC^{®}$ SBT 05250). The choice of these conditions was guided by the IEC 62660 standard.

During the test, the state of health (SOH) was followed via an electrical performance measurement realised at $25\,°C$ every 28 days for cells aged by cycling and every 6 weeks for calendar ageing on a battery cycler. This electrical test allowed us to track the evolution of some relevant characteristics of the cell: capacity, internal resistance [6,7], and nominal voltage. The internal resistance was evaluated by the current pulse method (10 s, I discharge max = 15 A).

In order to identify the favoured ageing mechanism in each condition, aged cells were dismantled in a glove box under argon (>99.999%) for post-mortem analyses. Scanning electronic microscopy (SEM) and electrochemical analyses were performed on both positive and negative electrodes recovered from the aged cells. Glow discharge-optical emission spectroscopy (GD-OES) and solid state ^7Li NMR were realised in addition on the negative electrodes to clearly identify the main ageing mechanisms. Indeed, different studies [8–10] have shown that the NCA positive electrode has no significant changes after ageing. The only suspect ageing mechanism on such NCA active material is the transition metal (Ni, Co, Al) dissolution during high-temperature ageing.

To help in the comprehension of mechanisms of ageing, electrochemical impedance spectra (EIS) measurements were carried out on cells at 60% SOC. EIS was performed using a VMP3 multi-potentiostat (Biologic Science Instruments) equipped with an EIS option. The frequency range used was from 1 mHz to 500 kHz, starting from high frequency to low frequency in a logarithmic scan. The alternating voltage amplitude was set to 5 mV at $25\,°C$.

Finally, an accelerated rate calorimetry (ARC) test was performed. An ARC test is a technique that allows measurement of the self-heating of a cell. The cell is heated in $5\,°C$ steps. After a 35 min break, if the heat rate β is not over $0.02\,°C/min$, a new step is started. As soon as the heat rate reaches over 0.02 °C/min, the so-called "onset temperature" is identified.

In this study, only the beginning of self-heating was studied. Tests were stopped at 90 °C in order not to damage the cell and allow further investigations. Two cells per ageing condition were tested. A fresh cell was also used as a reference in order to compare the different safety behaviours.

3. Model

The thermal runaway model is based on the kinetic equation parametrised by the temperature (T) and α ($0 \leq \alpha \leq 1$), the fractional degree of conversion of the reactants:

$$\frac{\partial \alpha}{\partial t} = k(T)f(\alpha) \tag{1}$$

The temperature dependence of the process rate is typically parametrised through the Arrhenius equation:

$$k(T) = A \exp^{-\frac{E_a}{RT}} \tag{2}$$

where A and E_a are kinetic parameters. A is the pre-exponential factor, E_a the activation energy, and R the universal gas constant.

The fractional degree of conversion of the reactants α is determined experimentally as the fraction of the overall change in the physical properties that go with the process (ex: loss of mass). The overall transformation can generally involve more than a single reaction or, in other words, multiple steps, each of which has its specific extent of conversion. This is why some authors have proposed to model all the different steps of reactions. This is the case for Coman et al. [11] or Hatchard et al. [12]. Hatchard et al. [12] built their model by distinguishing the reaction of the SEI, the anode, and the cathode, and also introduced the state of charge (SOC). A couple (A, E_a) was identified for each reaction.

Generally, models consider only the total reaction. First, because the various reactions are quite differentiated during the first steps, each reaction has its own kinetics. For example, a reaction can be described by:

$$\frac{\partial \alpha}{\partial t} = \begin{cases} k_1(T)f_1(\alpha_1) \text{ if } T \in [T_0, T_1] \\ k_2(T)f_2(\alpha_2) \text{ if } T \in [T_1, T_2] \end{cases} \tag{3}$$

α_1 and α_2 are the specific conversion rates of the two individual reactions and their sum is $\alpha = \alpha_1 + \alpha_2$ with:

$$\begin{cases} \alpha_1 = 0 \text{ if } T \in [T_1, T_2] \\ \alpha_2 = 0 \text{ if } T \in [T_0, T_1] \end{cases}$$

Second, even if the mechanism involves several steps, one of them can determine the overall kinetics. For instance, this would be the case for a mechanism composed of two consecutive reactions when the first reaction is significantly slower than the second. Then, the first process would determine the overall kinetics that would obey a single step, whereas the mechanism involves two steps.

In this study, ARC tests were stopped at 160 °C (for new cells) and 90 °C (for aged cells). The aged cells were subjected to post-abusive characterisation. The use of this temperature range guarantees keeping the cell intact. The principal reactions at this temperature are first the SEI and then anode reactions: consumption of Li in the anode when the tests stop at 90 °C, and the consumption of active materials in the anode for tests stopped at 160 °C. It seems reasonable to model the process as a single-step equation.

$f(\alpha)$ is the kinetic model. This function is based on physical-geometrical assumptions of regularly shaped bodies. We understand that these assumptions cannot describe heterogeneous systems. Therefore, the idea is to describe as easily as possible more complex reactions.

The first expression of $f(\alpha)$ comes from the experimental fitting of α through relation (1) under isothermal conditions:

$$f(\alpha) = \sum_k a_k \alpha^k \tag{4}$$

Sestak and Berggren [5] introduced a semi-empirical model (SB model) that estimates this:

$$f(\alpha) = l\alpha^m (1-\alpha)^n (-ln(1-\alpha))^p \tag{5}$$

Several variations of this model exist. The three principal ones are as follows:
The Avrami–Erofeev model:

$$f(\alpha) = n(1-\alpha)(-ln(1-\alpha))^{(n-1)/n} \tag{6}$$

The Prout–Tompkins model:

$$f(\alpha) = \alpha^m (1-\alpha)^n \tag{7}$$

and the Prout–Tompkins regular model, where $(p = 0, m = n = 1)$.

The lack of information regarding the estimation of these parameters led us to propose a way to estimate them. In the following sections, a method to build a thermal runaway model is proposed. To model it some parameters are required:

1. the onset temperature T_0 and so the degree of reactants converted α_0;
2. the parameters (l, m, n, p) of the kinetic model $f(\alpha)$;
3. the parameters of the Arrhenius law: E_a and A;
4. the time of simulation.

The following sections describe how to obtain these parameters from ARC tests.

3.1. Determination of Parameters T_0 and α_0

The first step is to determine the onset temperature. A new definition is proposed. Rather than defining the temperature onset as the temperature at the instant when the heat rate is over 0.02 °C/min, we will say that it is the temperature at the instant when the heat rate starts to increase.

The heat rate is defined as:

$$\beta(t) = \frac{\partial T(t)}{\partial t} \tag{8}$$

ans so the time onset t_0 is defined as:

$$\forall t \leq t_0, \frac{\partial \beta(t)}{\partial t} \geq 0 \tag{9}$$

This instant is denoted t_0. Different parameters can be evaluated at this instant:

- the temperature onset $T_0 = T(t_0)$;
- the degree of reactant converted $\alpha_0 = \alpha(t_0)$;
- the heat rate $\beta_0 = \beta(t_0) = \frac{\partial T(t)}{\partial t}(t_0)$.

In Figure 1, an example of heat rate measured during an ARC test on a new cell is given. On the left side (Figure 1a), the different heating levels (black peaks) can be observed. On the right side (Figure 1b), a zoom on the plot is chosen as the onset temperature. In orange is the trend.

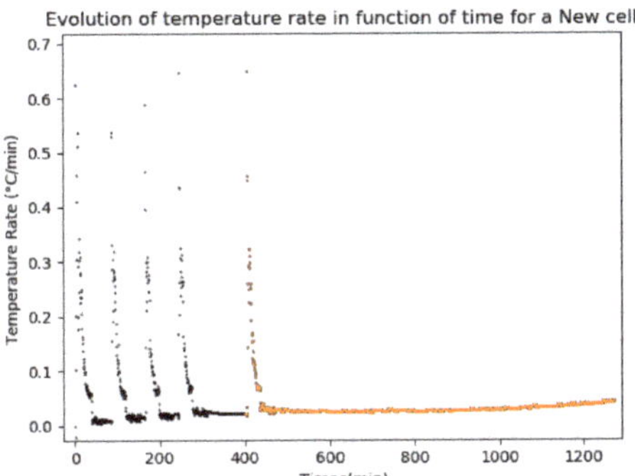

(**a**)

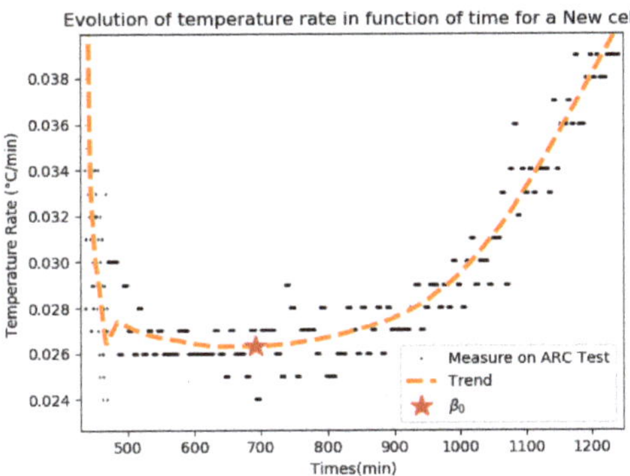

(**b**)

Figure 1. Location of onset temperature. (**a**) Example of heat rate measured during an ARC test on a new cell. (**b**) Example of situation of the onset temperature.

3.2. Estimation of Parameters (l, m, n, p)

Once the onset temperature of each cell is determined, the parameters of the model can be identified.

The demonstration is realised under adiabatic conditions. The energy conservation equation for an adiabatic reaction guarantees:

$$\frac{\partial T}{\partial t} = \frac{E_{Tot}}{C_{tot}} \frac{\partial \alpha}{\partial t} \tag{10}$$

where E_{Tot} (J) is the total heat that can be produced by the reaction and C_{tot} is the total capacity (JK^{-1}).

The heating rate is denoted by:

$$\beta(t) = \frac{\partial T}{\partial t} \tag{11}$$

With this notation the thermal runaway model (1) becomes:

$$\ln\left(\beta(t)\frac{C_{tot}}{E_{Tot}}\right) - \ln(f(\alpha)) = ln(A) - \frac{E_a}{RT} \tag{12}$$

Therefore:

$$\ln(f(\alpha)) = \frac{E_a}{RT} + \ln(\frac{\beta(t)}{A}\frac{C_{tot}}{E_{Tot}}) \tag{13}$$

At the start of the reaction $t = t_0$, $\alpha = 0$. The reactants begin to transform and we can consider that the heat rate is constant:

$$\beta(t) = \beta \tag{14}$$

Therefore, the SB kinetic model can be approached by a limited development:

$$ln(f(\alpha)) \underset{\alpha \to 0}{\sim} ln(l) + (m + p)ln(\alpha) - n\alpha \tag{15}$$

Therefore, when $\alpha \to 0$ Equation (13) is:

$$\frac{E_a}{RT} + \ln\left(\frac{\beta(t)}{A}\frac{C_{tot}}{E_{Tot}}\right) = ln(l) + (m + p)ln(\alpha) - n\alpha \tag{16}$$

By deriving Equation (16) by T:

$$-\frac{E_a\beta}{RT^2} = (m + p)\frac{\partial\alpha}{\partial t}\frac{1}{\alpha} - n\frac{\partial\alpha}{\partial t} \tag{17}$$

Due to the adiabatic condition $\frac{\partial\alpha}{\partial t}$ verifies (10) so:

$$-\frac{E_a E_{Tot}}{RC_{tot}T^2} = (m + p)\frac{1}{\alpha} - n \tag{18}$$

Three cases are possible at the beginning of the reaction ($\alpha \to 0$):

- $E_a = 0$. The reaction is instantaneous. Then:

$$\frac{m + p}{n} = 0$$

- $E_a > 0$ then:

$$\frac{m + p}{n} < 0$$

- $E_a < 0$. The phenomenon occurs in a chain reaction. Then:

$$\frac{m + p}{n} > 0$$

At the beginning of the reaction, it seems reasonable that E_a be positive or null. This implies:

$$\frac{m + p}{n} \leq 0 \tag{19}$$

Gorbatchev [13] has shown that no more than two kinetic exponents are necessary to describe any kinetic schema, so we choose $p = 0$.

n is considered as the order of the reaction, so $n > 0$, and therefore:

$$m \leq 0 \tag{20}$$

Let us study the Gibbs free energy of activation, defined as:

$$\Delta G = \Delta H - T\Delta S = -RT \ln\left(\frac{kh}{k_B T}\right) \tag{21}$$

where k_B is the Boltzmann constant and h the Planck constant and so $\frac{k_B}{h} = 2.084 \times 10^{10}$. k is the constant reaction.

Chemical thermodynamics explains that if

- $\Delta G < 0$: the reaction is thermodynamically favourable and will occur spontaneously;
- $\Delta G = 0$: the reaction is at its state of equilibrium;

- $\Delta G > 0$: additional energy must be input for the reaction to occur.

In ARC tests, the cell is heated for a few minutes and then there is a 35 min break: if the heat rate is not over 0.02 °C a new step is started. During these first steps additional energy is necessary to trigger the reactions. Therefore, ΔG must be positive. At the onset temperature: T_0, the thermal runaway starts and reactions become spontaneous so $\Delta G < 0$.

We consider that at t_0, when $T = T_0$ the reaction is at its state of equilibrium. The free enthalpy is then equal to 0.

Therefore, at $t = t_0$:

$$ln\left(\frac{kh}{k_B T_0}\right) = 0 \iff k_0 = \frac{k_B T_0}{h} \tag{22}$$

From (1) and (10):

$$\ln(k) = \ln\left(\frac{\beta(t)C_{tot}}{E_{Tot}}\right) - \ln(f(\alpha)) \tag{23}$$

At T_0:

$$\ln(k_0) = \ln\left(\frac{\beta(t_0)C_{tot}}{E_{Tot}}\right) - (\ln(l) + m\ln(\alpha_0) + n\ln(1 - \alpha_0)) \tag{24}$$

Then:

$$m\ln(\alpha_0) = \ln\left(\frac{\beta(t_0)C_{tot}}{E_{Tot}}\right) - \ln(k_0) - n\ln(1 - \alpha_0) - \ln(l) \tag{25}$$

$m \le 0$ and $\alpha_0 < 1$ so:

$$\ln\left(\frac{\beta(t_0)C_{tot}}{E_{Tot}}\right) - \ln(k_0) - n\ln(1 - \alpha_0) - \ln(l) \ge 0 \tag{26}$$

Therefore:

$$\begin{cases} \ln(l) \le \ln\left(\frac{\beta(t_0)C_{tot}}{E_{Tot}}\right) - \ln(k_0) - n\ln(1 - \alpha_0) = ln(l_{max}) \\ m = \frac{\ln\left(\frac{\beta(t_0)C_{tot}}{E_{Tot}}\right) - \ln(k_0) - n\ln(1 - \alpha_0) - \ln(l)}{\ln(\alpha_0)} \end{cases} \tag{27}$$

l will be chosen so that the Gibbs free energy is positive before T_0 and negative after, and so that $\ln(l) \le \ln(l_{max})$ is verified. Until these conditions are verified, $\ln(l)$ is reduced per step to 1.

For a fixed n, Algorithm 1 can be used to choose the parameter (l, m).

Algorithm 1 Calculate l.

Require: n, idx_0 the onset index and $\ln(l_{max})$
    Calculate m thanks $\ln(l_{max})$
    Calculate G thanks $(m, \ln(l_{max}))$
    while $G[0 : idx_0] \le 0$ **do**
    $\ln(l) = \ln(l) - 1$
    Calculate new m
    Calculate new G
    end while

3.3. Estimation of (E_a, A)

Energy activation (E_a) and pre-exponential factor A are calculated thanks to Equation (13) and the calculated parameters (l, m) (Figure 2).

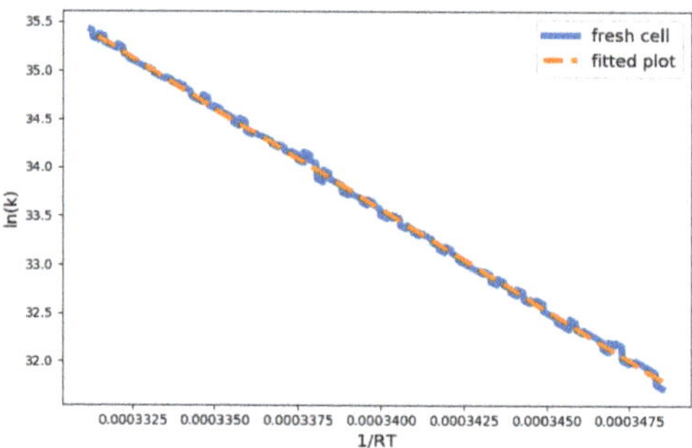

Figure 2. Estimation of E_a and A for fresh cell data: $ln(k(T)) = ln(A) - \frac{E_a}{RT}$.

The pre-exponential is defined as:

$$\ln(A_0) = \ln(A) + \ln(l) \tag{28}$$

3.4. Remarks about the Temperature Effects

The rate constant k is defined by the Arrhenius law, where energy of activation is considered a constant parameter (independent of the temperature). This approximation proves to be a very good one, at least over a moderate temperature range. In this case, kinetic phenomena can be interpreted as a set of elementary reactions. This theory is totally adapted for gas phases where chemical transformations take place in a series of isolated binary collisions of molecules, but less for solid or liquid phases or when the reactions are blended [5,14].

There is another way of calculating the rate constant. The expression comes from the transition state theory defined by Equation (21). In this case the energy of activation E_a and pre-exponential factor A are dependent on the temperature. It is this theory that is used to determined the values of (m, l). It seems well-adapted because in the system the reactions are blended and are in different phases. Pulses of heat impact the progress of the reaction (α) and so the transitory state of equilibrium. The level of energy of activation required to realise the reaction will change.

3.5. Remarks about the Precision of Measurements

Algorithm 1 updates l_{max} until all the values of G are negative before T_0. Sometimes, due to the precision of the measurement, few points are not under 0 during several iterations. For example, in Figure 3, 20 iterations more were performed because 3 points were still positive. l_{max} was then overvalued.

To avoid this kind of problem, instead of verifying $G[0 : idx_0] \leq 0$, a tolerance is added. In theory no value can be positive. In this case, we accepted a maximum of 10 positive values.

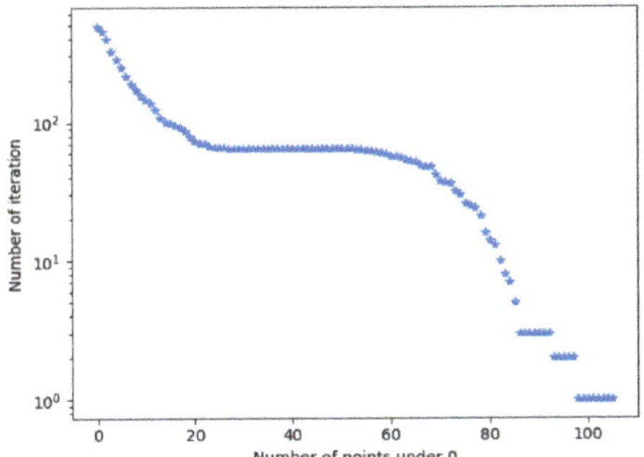

Figure 3. Number of negative G points as function of the number of iterations for a cell aged at 45 °C.

4. Results

All the parameters described were calculated on aged cells in order to appreciate the impact of the ageing on the kinetics model. The values were crossed with experimental results to verify the coherence of the values obtained.

4.1. Post-Mortem Analysis

The negative electrode is the site of the exchangeable lithium loss (SEI, Li-plating, ...). This is why the post-mortem analyses (GD-OES, Li-NMR and EIS) were focused on it. The results are presented in the next paragraph.

4.1.1. GD-OES and Li-NMR

Figure 4a presents the composition of the negative electrode surface after the different ageing conditions obtained by GD-OES. GD-OES allows us to study the chemical composition of the surface in the first 1.5 μm (total thickness of electrode is 44 μm) and allows us to understand which ageing mechanism takes place for each ageing condition.

For all aged cells, GD-OES measurements show that Si particles, initially well distributed in the negative electrode depth, are concentrated on the electrode surface. This migration should be related to the formation of Li-silicates [15], which consume cyclable Li.

Compared to fresh cells, the quantity of Li increases whereas the quantities of O (oxygen) and P decrease for cells aged at −20 °C. This shows that SEI does not grow on the electrode surface and it reveals the presence of metallic Li on its surface. Li-NMR measurements presented in Figure 4b confirm this observation.

When cells are aged at 25 °C and 45 °C, Li increases while O decreases and P increases, which reveals SEI growth by salt degradation [16].

At 45 °C calendar ageing, the observations are the same for both CV and OCV conditions. The increase in Li coupled with the increase in O and the decrease in P reveals that SEI grows by solvent degradation [16]. The mechanism is more pronounced for CV than OCV conditions.

Cells aged at 0 °C have no preponderant ageing mechanism considering the very small change in O and P quantities. ^{7}Li-NMR MAS (Figure 4b) shows that the observed Li is not in a metallic state. The Li detected here by the GD-OES analysis is not contained in the SEI layer and not in a metallic state.

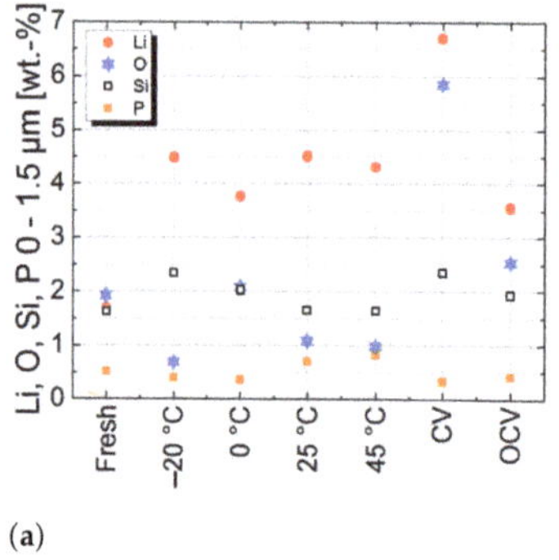

(a)

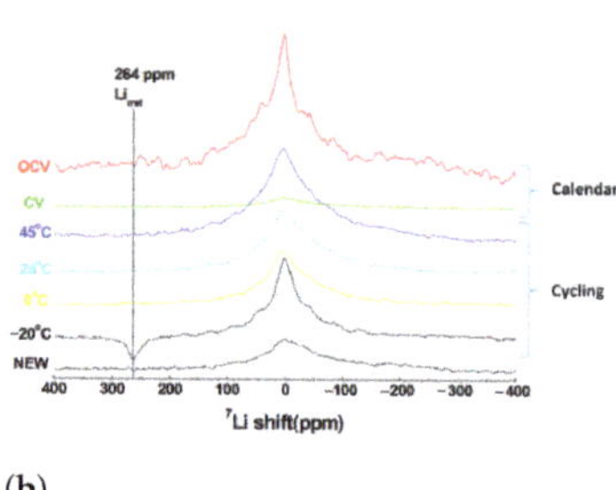

(b)

Figure 4. GD-OES and Li-NMR measurements. (**a**) Elementary composition of negative electrode surface by GD-OES analyses after ageing. (**b**) Li-NMR measurements of negative electrodes after ageing.

4.1.2. EIS

Figures 5 and 6 present the results of the impedance spectroscopy conducted on half of the cells built from the aged negative electrodes. The results are for a cell charged at 60%.

EIS spectra can be fitted by an equivalent circuit:

$$L_1 + R_1 + Q_2 // R_2 + Q_3 // R_3 \tag{29}$$

- R_1 is the resistance at high frequency associated to the ohmic resistance of the cell. It reflects the lithium ion transfers in the electrolyte.
- $Q_2 // R_2$ is linked to the first semi-loop. The frequency range of apparition is related to the migration of the lithium ion through SEI.
- $Q_3 // R_3$ is associated to the second semi-loop. The frequency range of apparition is related to the charge transfer through the electrode/electrolyte interface.

Values are summarised in Table 1.

Figure 5 represents the results in the Nyquist plan of the different cells.

Figure 6 presents the parameter values of the fitted models on each plot.

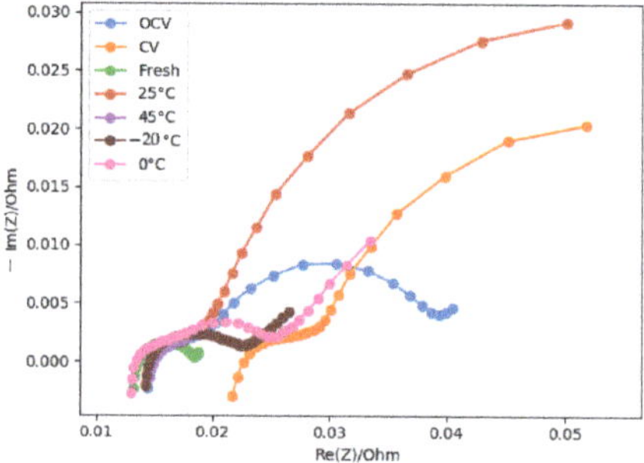

Figure 5. Nyquist plot of the different cells at SOC = 60%.

No ageing cell has the same resistance and capacitance, which reflects a difference in the morphology of the "electrodes".

The cells aged at −20 °C, 0 °C, and by calendar ageing at CV keep an interface resistance (R_1) in the same order as the fresh cells. These cells may not have developed extra SEI, or the thickness of the SEI has not increased.

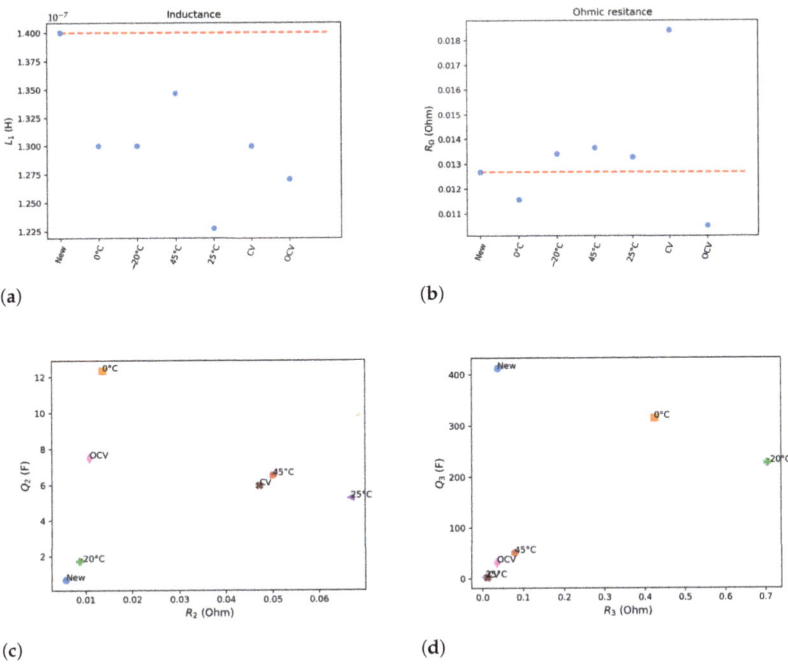

(a) (b)

(c) (d)

Figure 6. EIS results based on equivalent circuit $L_1 + R_1 + Q_2//R_2 + Q_3//R_3$ at SOC = 60%. (**a**) Comparison of inductance values. (**b**) Comparison of ohmic resistances. (**c**) Comparison of the interface resistance according to the interface capacitance. (**d**) Comparison of the charge transfer resistance according to the charge transfer capacitance.

Table 1. EIS results based on equivalent circuit $L_1 + R_1 + Q_2//R_2 + Q_3//R_3$ at SOC = 60%.

Test	R_1	R_2	R_3	Q_2	Q_3	a_2	a_3	L
Fresh	1.26×10^{-2}	0.037	0.58×10^{-2}	412.23	0.69	0.70	0.73	1.4×10^{-7}
0 °C	1.2×10^{-2}	0.036	1.37×10^{-2}	31.72	7.50	0.23	0.99	1.2×10^{-7}
−20 °C	1.34×10^{-2}	0.706	0.87×10^{-2}	227.35	1.71	0.42	0.62	1.3×10^{-7}
25 °C	1.33×10^{-2}	0.008	6.69×10^{-2}	3.34	5.25	0.54	0.91	1.2×10^{-7}
45 °C	1.2×10^{-2}	0.079	5.01×10^{-2}	50.93	6.52	0.19	0.97	1.2×10^{-7}
CV	1.84×10^{-2}	0.012	4.73×10^{-2}	2.30	5.94	0.42	0.90	1.3×10^{-7}
OCV	1.2×10^{-2}	0.036	1.08×10^{-2}	31.72	7.50	0.23	0.99	1.2×10^{-7}

The interface capacitance Q_2 of the fresh cells and the cells aged at −20 °C are quite the same, which confirms the results of GD-OES: no extra SEI has been formed.

The interface capacitance of cells aged at 0 °C and by calendar ageing at OCV is more important than that of the fresh cells, but the resistance is in the same order of value. It induces a transformation in the morphology of the SEI for these cells. The SEI is less compact.

EIS confirms that cells aged at 25 °C and 45 °C have the same kind of SEI morphology. The resistance and capacitance interfaces are closed (Figure 6c).

Cells aged by a calendar process at CV have a resistance R_2 and Q_2 close to the one of cells aged at 25 °C and 45 °C. GD-OES has shown that the extra SEI developed is not of the same nature. One growth is due to salt degradation, the other to solvent degradation. Compared to cells aged at OCV the resistance is higher, which means that the thickness developed by the CV process is greater than the one developed by an OCV process. This is confirmed by GD-OES.

Regarding charge transfer resistance R_3, cells aged at 25 °C, 45 °C, and by calendar ageing keep the same order of value as the fresh cells. However, charge transfer capacitances (Q_3) are lower than for the fresh cells. The interface film solution might be less porous certainly due to the growth of SEI.

The cells aged by calendar ageing at constant voltage tend to have the same morphology as cells aged at 25 °C and 45 °C regarding the values of R_3, R_2, Q_2, Q_3. Nevertheless, the ohmic resistance (Figure 6b) is more important, which means that the properties of the electrolyte are different. The GD-OES shows that the processes of SEI generation are not similar. One is based on the decomposition of salt and the other on the degradation of the solvents. Components present in the electrolyte might be different.

Cells aged at −20 °C and 0 °C have an important resistance charge transfer R_3. The growth of the resistance value, in the case of the cells aged at −20 °C, can be explained by the development of metallic lithium. In the case of the cells aged at 0 °C, the GD-OES analysis exhibits the presence of lithium (Figure 4a). The analyses show that lithium is not contained in the SEI layer and is not in a metallic state. It is as if lithium was blocked on the the film interface.

4.2. Model Results

Kinetic parameters have been calculated from the ARC tests. Results are presented in the following paragraphs and compared to the results of the post-mortem analysis summarised in Table 2.

Table 2. Summary of the conclusion of the tests.

Test	0 °C	−20 °C	25 °C	45 °C	CV	OCV
GD-OES	Li available	≃ Fresh	SEI by salt degradation	SEI by salt degradation	SEI by solvent degradation (+)	SEI by solvent degradation
Li-NMR		Li metal				
EIS R_1	≃ Fresh	≃ Fresh	≃ Fresh	≃ Fresh	+	≃Fresh
EIS interface/SEI	≃ Fresh	porous ++	morpho 1	morpho 1	thickness (+) than OCV	porous +
EIS charge transfer/film interface			porous − − −	porous − −	porous − − −	porous −
Kinetic parameters	G1	G1	G2	G2	G2	G2

4.2.1. Onset Temperature

The onset temperature was calculated for each cell (Figure 7) thanks to the criteria described before. The instant when T_0 is measured represents the start of the reaction. Once the beginning of the reaction is defined, the fractional degree of conversion of reactants α_0 and the temperature rate (β_0) can be evaluated.

Two groups appear. One group has a lower temperature onset than the fresh cells: cells aged at (0 °C, −20 °C). A second group has a higher onset temperature than the fresh cells: cells aged at (25 °C, 45 °C) and cells aged by a calendar process. The fractional degree of conversion of reactants follows the same order, which is totally normal in adiabatic conditions.

Notice that the dispersion of the fractional degree of conversion of reactants for cells aged at 45 °C is more pronounced even if the temperature onset is identical. One more temperature pulse has been applied to one of the cells, which explains a larger degree of conversion. Yet it has no impact on the measure of the onset temperature.

The temperature rate is quasi-identical at the beginning of the reaction for each test. Reactions start at the same speed.

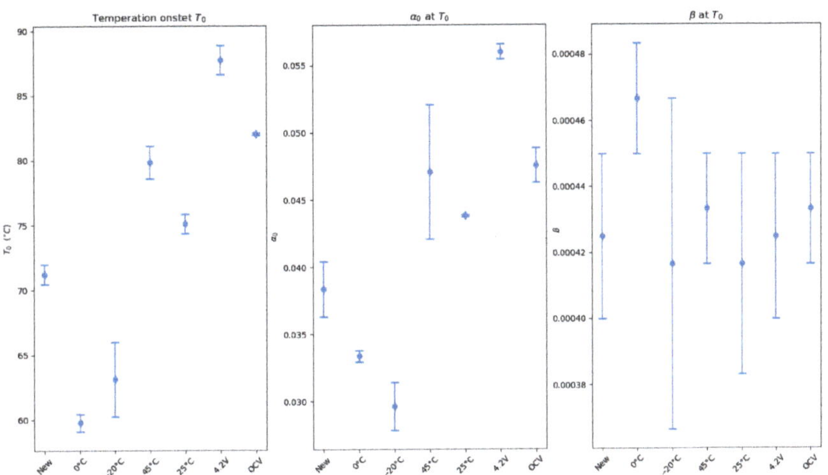

Figure 7. Onset temperature for different ageing conditions (new/fresh, 0 °C, −20 °C, 45 °C, 25 °C, at constant voltage (calendar ageing) at 4.2 V and OCV after charge at 4.2 V at room temperature). Fractional degree of conversion (α_0) of reactants at t_0 and temperature rate (β_0) at t_0.

4.2.2. Parameters (l, m)

Thanks to Algorithm 1, parameters (l_{max}, l, m) are calculated on all the aged cells with $n = 2$. The results are presented in Figure 8.

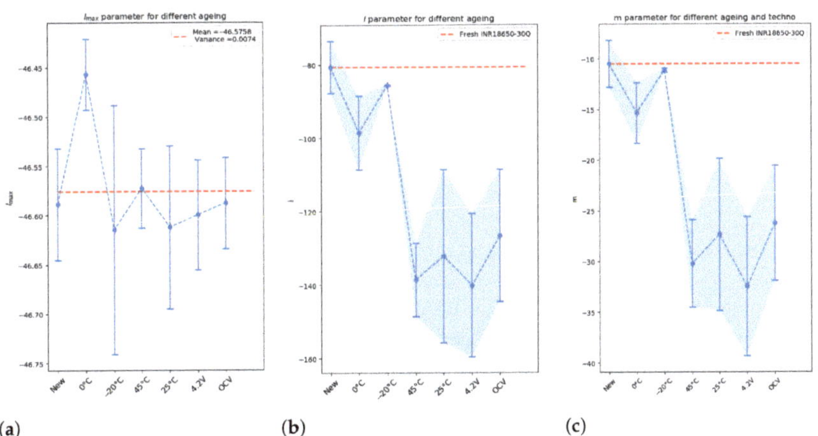

(a)   **(b)**   **(c)**

Figure 8. Kinetic parameters. (**a**) l_{max} parameter for different ageing conditions. (**b**) l parameter for different ageing conditions. (**c**) m parameter for different ageing conditions.

The dispersion of parameter l_{max} is very small: 0.007 (Figure 8a). Conversely, l (Figure 8b) does not have the same value depending on the ageing process.

Parameter m follows the same trend as l (Figure 8c). Two groups are formed:

- G1: cells aged at (-20 °C, 0 °C) and fresh cells;
- G2: cells aged at (25 °C, 45 °C), and cells aged by a calendar process at constant voltage and at OCV.

The dispersion per condition observed for parameters m and l is due to the quality of reproduction of the ARC test and the precision of the measurements previously explained.

Post-mortem analyses show that the main difference between the two groups formed here is the development of extra SEI. Group G1 has not developed extra SEI contrary to group G2. That is in accordance with the interpretation of the m parameter. The m parameter contains information regarding the basic geometrics of nuclei (spherical, prismatic...) and the physico-chemical aspects of processes (nucleation, diffusion, transport process...) [5]. The development of extra SEI changes the morphology of the cell.

Cells aged at 0 °C have the smallest m in group G1, although no ageing mechanism is preponderant. The EIS study shows that the property of interface capacitance is different. The morphology is different.

In group G2, the values of m are dispersed, which makes interpretation difficult.

4.2.3. Parameters (E_a, A_0)

Energy of activation E_a and pre-exponential factor A_0 were calculated thanks to Equation (13) and the calculated parameters (l, m) (Figure 9).

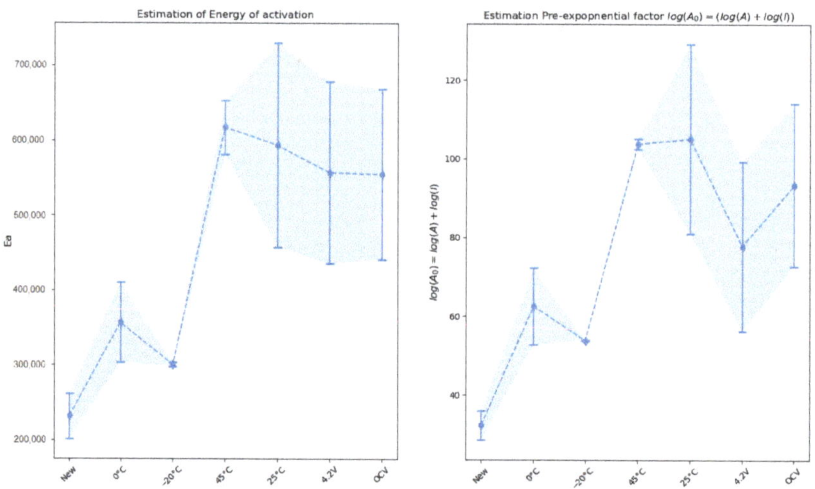

Figure 9. E_a and pre-exponential factor A_0 for different ageing conditions.

Group G1 has the smallest energy of activation. The energy required to start the reaction is smaller than the energy of activation of the other cells. No extra SEI has been formed in this group and it is well known than SEI plays a protector role in the thermal stability. However we can notice that the activation energy values within this group are not identical (Figure 9).

Cells aged at 0 °C have the highest energy of activation in group G1. A high energy of activation means that the reaction is slow. This can be explained by the fact that these cells have a different morphology (as shown by parameter m) and so require more energy to trigger the reaction. Post-mortem analyses confirm that these cells have been transformed. GD-OES exhibits that lithium is available at the surface of the electrode. EIS has shown that the interface capacitance has grown, revealing a more porous morphology. The porosity may create an imbalance in the concentration distribution of Li. SEI can be formed irregularly on the pores and so creates zones with Li and zones without Li [17].

In group $G2$, cells have quite the same energy of activation but the pre-exponential factor is different. The pre-exponential factor of the cell aged at CV is lower than for one the other cells of group $G2$. The pre-exponential factor includes many constants describing the initial state of the sample, such as three-dimensional shape factors of initial particles, molecular mass, density, stoichiometric factors of chemical reaction, active surface, number of lattice imperfections, etc. EIS shows that the structure of the interface is closer to the structure of cells aged at 45 °C and the cells aged at 25 °C. Extra SEI formed in these conditions is formed by solvent degradation like for the cells aged at the OCV. The EIS and the GD-OES analyses show that the quantity of SEI is more important in the case of the CV ageing process. This can explain the differences in the value A_0.

4.2.4. Simulation

It is often considered that the combination of all the parameters (m, n, l, E_a, A) describes the kinetics of the reaction.

To illustrate this, a simulation using the parameters estimated previously (m, n, l, E_a, A) was realised for each ageing process (Figure 10b). To do so, the initialisation of the temperature and α is required. Two algorithms are proposed.

The simulations in Figure 10 were performed with the true α_0 and T_0. In Figure 10b, the simulation performed with Algorithm 2 is compared to the true data. The parameters estimated previously (m, n, l, E_a, A) are representative of the kinetics. The kinetics seem to be well reproduced except for the cells aged at 0 °C. One possible explanation is that only one energy of activation is considered. In fact, several reactions occur at the same time. The kinetics can change over the time due to a new group of reactions, which will impose their kinetics. This is why the kinetics must be studied on a range of temperatures to improve the results, which verifies an isokinetic relationship.

By applying the second Algorithm 3, it is possible to evaluate the time of reaction. In this case no extra reaction occurs. The parameters (m, n, l, E_a, A) represent the whole reaction. This is the most optimistic case. The most reactive cells are the cells aged at 25 °C and cells aged in open circuit voltage (OCV). They are considered more thermally stable because the onset temperature is high and yet the time of reaction is shorter. Cells aged at 0 °C and −20 °C are less stable (onset temperature small) but their time of reaction is longer. Therefore, mitigation solutions can be set up more easily. The most thermally stable cells are the fresh cells and the cells aged at constant voltage.

Algorithm 2 Calculation of T.

Require: $n, m, E_a, A_0, T_{max}, T_0, \alpha_0, dt$
$\quad T_0 = T_0$
$\quad \alpha_0 = \alpha_0$
$\quad$ **while** $\mathrm{T} \leq T_{max}$ **do**
$\quad\quad \alpha_{i+1} = A_0 \exp(-E_a / RT_i)\alpha_i^m (1 - \alpha_i)^n * dt + \alpha_i$
$\quad\quad T_{i+1} = \frac{E_{Tot}}{C_{tot}}(\alpha_{i+1} - \alpha_i) + T_i$
$\quad$ **end while**

Algorithm 3 Calculation of T.

Require: $n, m, E_a, A_0, T_{max}, T_0, \alpha_0, dt$
$\quad T_0 = T_0$
$\quad \alpha_0 = \alpha_0$
$\quad$ **while** $\alpha_i \leq 1$ **do**
$\quad\quad \alpha_{i+1} = A_0 \exp(-E_a / RT_i)\alpha_i^m (1 - \alpha_i)^n * dt + \alpha_i$
$\quad\quad T_{i+1} = \frac{E_{Tot}}{C_{tot}}(\alpha_{i+1} - \alpha_i) + T_i$
$\quad$ **end while**

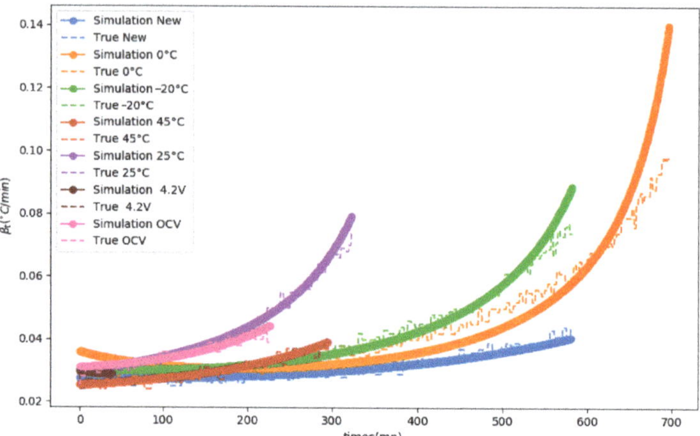

(**a**)

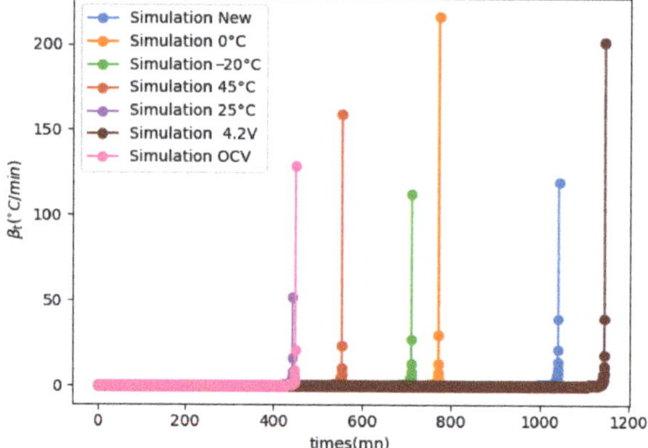

(**b**)

Figure 10. Simulation. (**a**) Simulation of the different kinetics for a time of simulation corresponding to the tests. Comparison with the test data. (**b**) Simulation of the different kinetics considering that no other reaction occurs until the end.

5. Conclusions

Thanks to the ARC tests realised on cells aged by different processes, we have shown that (m, l) parameters must be chosen carefully. Indeed these parameters vary according to ageing and what the cell experienced.

The values of (m, l) impact the energy of activation and the pre-exponential factor. The transition state theory shows that energy of activation and pre-exponential factor cannot be considered constant. They vary according to the history of the cell. First, the ageing changes the morphology of the cell and consequently the reaction that will be produced. Secondly, the rise in the temperature will change the equilibrium of the reaction, therefore the energy required to trigger the reaction will not be the same.

Once all the kinetic parameters are calculated, simulations can be used to reproduce the rise in temperature. Simulations confirm that the kinetic parameters must be estimated at different temperatures but with constant kinetics. One reaction can impose its kinetics but a new reaction can take over and in turn impose its own kinetics. In this case all the kinetic parameters must be re-evaluated, for each range of temperatures.

Simulations have exposed another problem: some cells are considered more stable thermally because the reaction starts at a higher temperature. However, in this case, the reaction will be realised faster. Other cells have a lower temperature onset but the reaction will be slower, which gives time to find solutions to a thermal runaway. Should we focus on stability in order to minimise the risk of thermal runaway, or should we focus on the time response in case of thermal runaway?

Author Contributions: Writing, modelling, and simulation: M.G.; post-mortem analysis and cell performance measurement (ARC, cycling, ...): P.K.; EIS and post-mortem analysis: S.G.; validation and supervision: O.R.; methodology and supervision: P.A. All authors have read and agreed to the published version of the manuscript.

Funding: This research received no external funding.

Conflicts of Interest: The authors declare no conflict of interest.

Abbreviations

The following abbreviations are used in this manuscript:

SOC	State of charge
SOH	State of health
BEV	Battery electric vehicle
ARC	Accelerating rate calorimeter
SEM	Scanning electronic microscopy
GD-OES	Glow discharge-optical emission spectroscopy
SB kinetic model	Sestak–Bergam kinetic model
EIS	Electrochemical impedance spectra
SEI	Solid electrolyte interphase
OCV	Open circuit voltage
CV	Constant voltage

References

1. Feng, X.; Ouyang, M.; Liu, X.; Lu, L.; Xia, Y.; He, X. Thermal runaway mechanism of lithium ion battery for electric vehicles: A review. *Energy Storage Mater.* **2018**, *10*, 246–267. [CrossRef]
2. Friesen, A.; Horsthemke, F.; Mönnighoff, X.; Brunklaus, G.; Krafft, R.; Börner, M.; Risthaus, T.; Winter, M.; Schappacher, F.M. Impact of cycling at low temperatures on the safety behavior of 18650-type lithium ion cells: Combined study of mechanical and thermal abuse testing accompanied by post-mortem analysis. *J. Power Sources* **2016**, *334*, 1–11. [CrossRef]
3. Börner, M.; Friesen, A.; Grützke, M.; Stenzel, Y.; Brunklaus, G.; Haetge, J.; Nowak, S.; Schappacher, F.; Winter, M. Correlation of aging and thermal stability of commercial 18650-type lithium ion batteries. *J. Power Sources* **2017**, *342*, 382–392. [CrossRef]
4. Larsson, F.; Bertilsson, S.; Furlani, M.; Albinsson, I.; Mellander, B.E. Gas explosions and thermal runaways during external heating abuse of commercial lithium-ion graphite-LiCoO2 cells at different levels of ageing. *J. Power Sources* **2018**, *373*, 220–231. [CrossRef]
5. Šesták, J.; Berggren, G. Study of the kinetics of the mechanism of solid-state reactions at increasing temperatures. *Thermochim. Acta* **1971**, *3*, 1–12. [CrossRef]
6. Schweiger, H.G.; Obeidi, O.; Komesker, O.; Raschke, A.; Schiemann, M.; Zehner, C.; Gehnen, M.; Keller, M.; Birke, P. Comparison of several methods for determining the internal resistance of lithium ion cells. *Sensors* **2010**, *10*, 5604–5625. [CrossRef] [PubMed]
7. Barai, A.; Uddin, K.; Dubarry, M.; Somerville, L.; McGordon, A.; Jennings, P.; Bloom, I. A comparison of methodologies for the non-invasive characterisation of commercial Li-ion cells. *Prog. Energy Combust. Sci.* **2019**, *72*, 1–31. [CrossRef]
8. Kabir, M.; Demirocak, D.E. Degradation mechanisms in Li-ion batteries: a state-of-the-art review. *Int. J. Energy Res.* **2017**, *41*, 1963–1986. [CrossRef]
9. Balakrishnan, P.; Ramesh, R.; Kumar, T.P. Safety mechanisms in lithium-ion batteries. *J. Power Sources* **2006**, *155*, 401–414. [CrossRef]
10. Zhang, S.S. A review on electrolyte additives for lithium-ion batteries. *J. Power Sources* **2006**, *162*, 1379–1394. [CrossRef]
11. Coman, P.T.; Darcy, E.C.; Veje, C.T.; White, R.E. Numerical analysis of heat propagation in a battery pack using a novel technology for triggering thermal runaway. *Appl. Energy* **2017**, *203*, 189–200. [CrossRef]
12. Hatchard, T.; MacNeil, D.; Basu, A.; Dahn, J. Thermal model of cylindrical and prismatic lithium-ion cells. *J. Electrochem. Soc.* **2001**, *148*, A755–A761. [CrossRef]
13. Gorbachev, V. Some aspects of Šesták's generalized kinetic equation in thermal analysis. *J. Therm. Anal. Calorim.* **1980**, *18*, 193–197. [CrossRef]

14. Espenson, J.H. *Chemical Kinetics and Reaction Mechanisms*; Citeseer: New York, NY, USA, 1995; Volume 102.
15. Richter, K.; Waldmann, T.; Kasper, M.; Pfeifer, C.; Memm, M.; Axmann, P.; Wohlfahrt-Mehrens, M. Surface Film Formation and Dissolution in Si/C Anodes of Li-Ion Batteries: A Glow Discharge Optical Emission Spectroscopy Depth Profiling Study. *J. Phys. Chem. C* **2019**, *123*, 18795–18803. [CrossRef]
16. Andersson, A. Surface Phenomena in Li-Ion Batteries. Ph.D. Thesis, Acta Universitatis Upsaliensis, Uppsala, Sweden, 2001.
17. Gnanaraj, J.; Thompson, R.W.; Iaconatti, S.; DiCarlo, J.; Abraham, K. Formation and growth of surface films on graphitic anode materials for Li-ion batteries. *Electrochem. Solid State Lett.* **2005**, *8*, A128. [CrossRef]

Article

Influence of Temperature and Electrolyte Composition on the Performance of Lithium Metal Anodes

Sanaz Momeni Boroujeni *, Alexander Fill, Alexander Ridder and Kai Peter Birke

Institute of Photovoltaics, Electrical Energy Storage Systems, University of Stuttgart, 70569 Stuttgart, Germany; alexander.fill@ipv.uni-stuttgart.de (A.F.); alexander.ridder@ipv.uni-stuttgart.de (A.R.); peter.birke@ipv.uni-stuttgart.de (K.P.B.)
* Correspondence: sanaz.momeni@ipv.uni-stuttgart.de

Abstract: Lithium metal anodes have again attracted widespread attention due to the continuously growing demand of cells with higher energy density. However, the lithium deposition mechanism and the affecting process of influencing factors, such as temperature, cycling current density, and electrolyte composition are not fully understood and require further investigation. In this article, the behavior of lithium metal anode at different temperatures (25, 40, and 60 °C), lithium salts, electrolyte concentrations (1 and 2 M), and the applied cell current (equivalent to 0.5 C, 1 C, and 2 C). is investigated. Two different salts were evaluated: lithium bis(fluorosulfonyl)imide (LiFSI) and lithium bis(trifluoromethanesul-fonyl)imide (LiTFSI). The cells at a medium temperature (40 °C) show the highest Coulombic efficiency (CE). However, shorter cycle life is observed compared to the experiments at room temperature (25 °C). Regardless of electrolyte type and C-rate, the higher temperature of 60 °C provides the worst Coulombic efficiency and cycle life among those at the examined temperatures. A higher C-rate has a positive effect on the stability over the cycle life of the lithium cells. The best performance in terms of long cycle life and relatively good Coulombic efficiency is achieved by fast charging the cell with high concentration LiFSI in 1,2-dimethoxyethane (DME) electrolyte at a temperature of 25 °C. The cell has an average Coulombic efficiency of 0.987 over 223 cycles. In addition to galvanostatic experiments, Electrochemical Impedance Spectroscopy (EIS) measurements were performed to study the evolution of the interface under different conditions during cycling.

Keywords: lithium battery; temperature dependency; ether based electrolyte, *insitu* deposited lithium-metal electrode; Coulombic efficiency; lithium deposition morphology

Citation: Momeni Boroujeni, S.; Fill, A.; Ridder, A.; Birke, K.P. Influence of Temperature and Electrolyte Composition on the Performance of Lithium Metal Anodes. *Batteries* **2021**, *7*, 67. https://doi.org/10.3390/ batteries7040067

Academic Editor: Seokheun Choi

Received: 30 July 2021
Accepted: 29 September 2021
Published: 14 October 2021

1. Introduction

1.1. Motivation

Lithium metal has always been one of the most attractive candidates for anode materials in lithium batteries. This is due to its potential to extend the energy density of conventional lithiumion batteries. State-of-the-art Li-ion cells, depending on the cell chemistry, can deliver a specific energy density of 130 Wh·kg^{-1} to 250 Wh·kg^{-1} [1]. This is already behind the U.S. Department of Energy's (DOE) target for advanced batteries for electric vehicles [2]. Lithium has a theoretical specific capacity of 3860 mAh·g^{-1} and a higher redox potential of $\phi_{Li||H_2} = -3.04$ V versus standard hydrogen electrodes in comparison to electrodes based on graphite. This means using lithium (Li) instead of conventional intercalating anode materials like graphite (LiC$_6$), which has a theoretical specific capacity of 372 mAh·g^{-1} that can increase the specific energy and volumetric energy density of cells significantly. In one study, researchers reported a 35% increase in specific energy and 50% increase in volumetric energy density when the graphite electrode is replaced with a Li metal electrode [3]. In addition to the aforementioned change in electrode material, they considered a solid electrolyte for the Li-metal cell and a liquid electrolyte for the conventional Li-ion cell in their estimates [3]. However, commercializing Li-metal batteries has been paused due to various challenges in both production and performance of lithium cells [4]. The issues with the production are that the Li surface is highly reactive and sensitive to humidity, oxygen and nitrogen, which are all present in the air atmosphere [5,6]. The challenges of the performance are given by the unstable and different Li growth morphology, low Coulombic efficiency, and considerable volume change during cycling [1]. Lithium metal cannot be utilized with known carbonate electrolytes because its electrochemical potential causes the electrolyte to continuously decompose until a passivating solid electrolyte interface (SEI) is built up [7]. One approach to utilizing the Li-metal is creating a stable and uniform SEI layer on the lithium surface that can withstand significant volumetric changes of Li during cycling. This is of critical concern to ensure safe and efficient lithium metal cells. Local variations in the SEI layer's composition might cause uneven Li deposition, resulting in changes in Li-ion conductivity across the electrode or SEI rupture, which can facilitate the creation of Li dendrites. Another approach is to use solid membranes, such as solid polymer electrolytes (SPE), which are less reactive to the Li [8] and their soft nature could withstand the extensive volume change of the Li anode [9]. In this study we focus on compatible liquid electrolytes in lithium cells. More experiments investigating the influence of different electrolyte compositions and the cycling conditions on effective SEI layer formation are much needed.

1.2. Relevant Literature

The high reactivity of metallic Li, which causes difficulty during the production process, can also be a source of performance challenges. Potential corrosive reactions at the surface of Li metal often lead to an increase in interfacial resistance, a reduction in Coulombic efficiency (CE) and a poor lifetime. Additionally, the large volume expansion of the electrode during repeated Li deposition/dissolution will seriously deteriorate the interfacial stability and in general increase the gap between theoretical and practical energy density of the Li cells. Continuous interface reactions, together with the surface enlargement due to new depositions, consume the fresh Li more and more during the cycle life of an Li metal anode. This means that an excess of lithium and electrolytes are strongly needed to increase the cycle life and improve the stability of Li metal cells [10]. There are studies implying that 20% is the optimum excess of lithium; a greater excess of lithium will increase the possibility of side reactions and consequently shorten the cycle life of the cells [11]. This required excess of Li and electrolytes is another limitation to increasing practical energy density in Li-metal cells.

The function and properties of Li cells are strongly dependent on the growth morphologies. However, predicting the kinetic structures is difficult as they are influenced by different parameters. There are studies [12–15] investigating the parameters that have an influence on the shape, morphology and growth of metal particles. Tao Yang et al. [12] proposed three different growth modes: reaction limited, diffusion limited and the so-called reaction–diffusion balance mode. The reaction limited mode is dominated by a slow reaction rate. The reaction in the case of Li cells, which is the focus of this paper, is the nucleation and transformation of an Li ion to deposited Li metal:

$$Li^+ + e^- \rightarrow Li. \tag{1}$$

The slow reaction rate combined with a sufficient mass transport diffusion in the electrolyte from the cathode to the Li metal anode leads to a rich concentration of reactants (Li ions) around the nucleus which further leads to a classic crystallization process. The diffusion limited mode is dominant if the reaction rate is much faster compared to the mass transport diffusion, which causes a lack of reactants around the nucleus. In both mentioned conditions—reaction limited and diffusion limited—the particles' growth is slow. However, in the reaction–diffusion balance mode, the morphology is dendritic and the growth speed is fast. In this region, there is a concentration gradient around the nucleus. Yang and his group [12] examined the morphology evolution on silver, gold and copper, coming to the same conclusion that the relation between diffusion and reaction rate is the key factor in predicting the shape of particle growth. With this as the background, the importance of operation temperature and applied current, as the key kinetics definers for the performance of lithium metal anodes, should be emphasized.

There are studies showing that a local temperature rise can have healing effects on Li dendrites at different current densities [13,14]. By increasing the temperature, the diffusion of Li ions in the bulk of the electrolyte will be faster [15]; therefore, based on Tao Yang's model, in comparison to the cell performed at $T_{Cell} = 25\,°C$, elevated temperature should move the cells either in the direction of reaction limited or to a balanced region. Yehu et al. [13] investigated the temperature influence on the stability and efficiency of the Li metal anode. They found out that in their studied system, temperature rise has a positive influence on the efficiency and life time of the cells. They also studied the impact of different temperatures and electrolyte compositions on the morphology of Li depositions. They found out that bigger Li spheres form at the initial stage of deposition at elevated temperatures, leading to a lower specific area of plated Li and consequently to reducing the probability of dendritic formation [13].

At the same time, there are studies showing that the temperature rise leads to more unstable lithium deposition [16–18]. The structural uniformity and mechanical strength of the Solid Electrolyte Interface (SEI) play important roles in defining the type of deposition as they directly influence the dynamic of Li plating and stripping [19]. As the SEI layer consists of reduced and decomposed electrolyte components, different electrolytes induce totally different SEI layers. One influencing component in the electrolyte is the used Li salt. The salt lithium hexafluorophosphate $LiPF_6$ exhibits poor thermostability [20]; Lithium perchlorate $LiClO_4$ can strongly oxidize the Li metal [21] and causes low safety. An alternative salt given by lithium bis(fluorosulfonyl)imide (LiFSI) is reported to form a robust SEI protecting layer [22]. In this work, the influence of LiFSI and lithium bis(trifluoromethanesulfonyl)imide (LiTFSI) and their concentrations in electrolytes on the cycle life of Li metal cells is studied.

1.3. Structure and Technical Contribution

In this article, the influences of temperature, C-rate, type, and concentration of Li salt in the electrolyte on the lithium deposition were studied. A variety of electrolytes including LiFSI in 1,2-dimethoxyethane (DME) with a molar concentration of $c = 1\,M$, LiFSI in DME with $c = 2\,M$ and lithium LiTFSI in DME, also with a concentration of $c = 1\,M$, were investigated. In addition to the Coulombic efficiency, EIS measurements were carried out to evaluate the stability and aging behavior. The best performance is achieved by a high concentration of LiFSI-DME electrolyte at $25\,°C$. The remainder of this article is structured as follows. Section 2, presents the materials and methods for evaluating the degradation of the cells. The cell preparation, the measurement procedure and matrix are discussed in detail. The results of the conducted experiments, including the data on the Coulombic efficiency and the Electrochemical Impedance Spectroscopy (EIS) measurements, are shown and discussed in Sections 3 and 4. Finally, an outlook of the future work and a conclusion are given.

2. Materials and Methods

In this section, the preparation of the cell structure consisting of the electrolyte and the electrodes will be discussed first. Thereafter, the measurement procedure of the cycling tests and the boundary conditions of the conducted experiments will be presented.

2.1. Material and Cell Preparation

Electrolytes preparation: LiTFSI and LiFSI (with a chemical purity of $x_{LiTFSI,LiFSI} > 99\%$, manufactured by Ionic Liquid Technologies GmbH (Heilbronn, Germany), were purchased and used as received. The electrolyte DME with a chemical purity of $x_{DME} > 99\%$ manufactured by Thermo Fisher GmbH (Kandel, Germany) was used as the solvent. Three different ether-based electrolytes were prepared as follows: (1) LiTFSI 2M in DME, (2) LiFSI 2M in DME and (3) LiFSI 1M in DME.

Electrode preparation: An oxygen-free copper (Cu) foil from SCHLENK Metal Foils GmbH & Co. KG. (Roth, Germany) with a thickness of $h_{Cu} = 6$ µm was purchased and used as received. Current collector (CC) disks with a diameter of $d_{CC} = 18$ mm were punched out of the foil to be used as working electrodes. Lithium metal foil manufactured by Sigma Aldrich (St. Louis, MO, USA) with a thickness of $h_{Li\,foil} = 320$ µm was used as the counter electrode. The lithium surface was cleaned of an oxide layer before punching. Disks with a diameter of $d_{Li\,foil} = 18$ mm were made. All handling was performed inside an argon-filled glove box from M. Braun Inertgas-Systeme GmbH (Garching, Germany) with an oxygen (O_2) and water (H_2O) concentration $c_{O_2,H_2O} < 0.5$ ppm. Both counter and working electrodes had an effective surface area of $A = 2.54$ cm^2.

2.2. Cell Degradation Experiments

Cell degradation experiments were conducted using EL-CELL PAT-Cell cases manufactured by EL-Cell GmbH (Hamburg, Germany). Before cell assembly, Cu current collectors were dried under vacuum for $t_{dry} = 16$ h at $T_{dry} = 120$ °C and were transferred into the glove box. Celgard 2500 from Celgard (North Carolina, USA) was used as a separator and was also dried under vacuum for $t_{dry} = 16$ h at $T_{dry} = 80$ °C prior to use. Prepared electrolytes were poured on both sides of the separator and the amount was determined depending on the measurement temperature and applied current density. For the measurements performed at $T_{Cell} = 25$ °C, an electrolyte amount of $V_{elec} = 50$ µL was used for the current densities of $J = 0.5$ mAh · cm^{-2} and $J = 1$ mAh·cm^{-2}. The amount of electrolyte was increased to $V_{elec} = 100$ µL for the experiments with an applied current density of $J = 2$ mAh·cm^{-2}. Regardless of current density, $V_{elec} = 100$ µL was used for the measurements at $T_{Cell} = 40$ °C and $T_{Cell} = 60$ °C.

In order to evaluate the degradation behavior, cycling tests were conducted. A segment of the measurement procedure is visualized in Figure 1 with a C-rate of $I_{Cell} = 1$ C. The current is presented in Figure 1a and the corresponding voltage in Figure 1b.

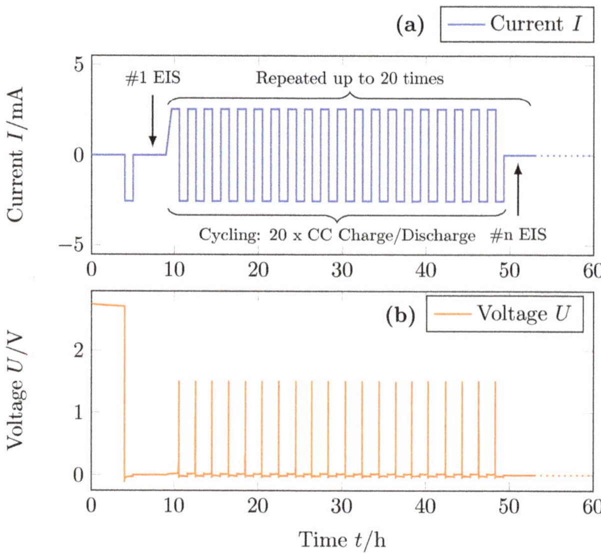

Figure 1. Segment of the measurement procedure of the conducted cycling tests example at $I_{Cell} = 1$ C. The current data are presented in (**a**) and the corresponding voltage data are visualized in (**b**). Before the first cycle and after each 20 full charge/discharge cycles, EIS measurements were conducted.

The tests were conducted using a battery cell tester from Basytec GmbH (Asselfingen, Germany) in combination with a Reference 3000 from Gamry Instruments (Warminster, PA, USA) in a climate chamber manufactured by Memmert GmbH (Schwabach, Germany). Prior to cycling, cells were relaxed at the considered measurement temperature for $t = 4\,\text{h}$ to achieve a homogeneous electrolyte distribution and a steady state temperature. The Cu/Li cells showed an open circuit potential (OCP) of $U_{\text{OCP}} \approx 2.7\,\text{V}$ which is observable during the relaxation period in the first 4 h of the procedure.

The cells were continuously charged and discharged at the considered C-rate with the maximum voltage limit of $U_{\text{max}} = 1.5\,\text{V}$. In this set of experiments, no minimum voltage limit (U_{min}) was defined and instead the specified time period based on the applied current density limited the discharge process. The cells were not relaxed between the charge and discharge cycles. After each 20th full charge/discharge cycle the impedance of the cells was evaluated via EIS measurements. To consider the initial impedance behavior of the cells, an EIS was also conducted after the first deposition period (discharge process) prior to 20 full cycles repetition. The segment including the 20 full cycles and the EIS was repeated up to 20 times depending on the degradation level of the cells.

The cell current and the cell voltage were controlled and captured by the mentioned battery tester. The EIS measurements were conducted using the Reference 3000 from Gamry. The frequency of the EIS was varied between $f_{\text{EIS, min}} = 0.1\,\text{Hz}$ and $f_{\text{EIS, max}} = 100\,\text{kHz}$.

The purpose of this article is to investigate the influence of the C-rate, cell temperature, used salt and salt concentration on the aging behavior of the cells. The set boundary conditions of the conducted experiments are given by the measurement matrix in Figure 2. The different salts used are separated by color; orange measurement points correspond to experiments with LiTFSI and for the measurement points colored blue, LiFSI was used as the salt.

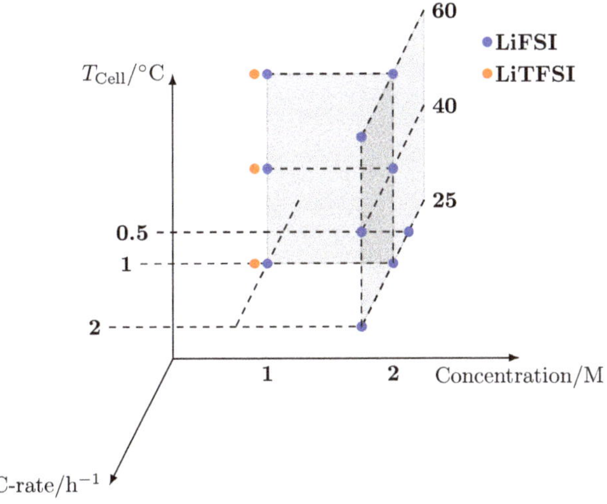

Figure 2. Matrix of the conducted measurements. For the salt LiFSI the c-rate was varied with $I\epsilon\{0.5, 1, 2\}$ C at a salt concentration of $c = 2$ M and a cell temperature of $T_{\text{Cell}} = 25\,^{\circ}\text{C}$. The temperature was varied with $T_{\text{Cell}} = \{25, 40, 60\}\,^{\circ}\text{C}$ at the same concentration and $I\epsilon\{1, 2\}$ C. At a concentration of $c = 1$ M measurements at $I = 1$ C and $T_{\text{Cell}} = \{25, 40, 60\}\,^{\circ}\text{C}$ for two salts: LifTSI and LiFSI.

As presented in the measurement matrix, the C-rate was varied with $I_{\text{Cell}}\epsilon\{0.5, 1, 2\}$ C in order to evaluate the influence of the C-rate. For the current stress of $I_{\text{Cell}}\epsilon\{1, 2\}$ C the cell temperature was varied with $T_{\text{Cell}} = \{25, 40, 60\}\,^{\circ}\text{C}$. All mentioned measurements were conducted with LiFSI and a salt concentration of $c = 2$ M. To evaluate the impact of the salt and salt concentration experiments at a molar concentration of $c = 1$ M and two different salts—LiFSI (blue) and LiTFSI (orange)—were performed. Some measurement points were repeated in order to check the reproducibility and to determine the standard deviation of the experimental data due to the measurement setup, procedure and cell production. The number of measurements for each measurement point is listed in Table 1.

Table 1. Number of measurements for each condition considered.

		LiFSI			LiTFSI
		2 C	1 C	0.5 C	1 C
2 M	25 °C	#1	#3	#3	
	40 °C	#1	#2		
	60 °C	#1	#2		
1 M	25 °C	-	#1	-	#1
	40 °C	-	#2	-	#1
	60 °C	-	#2	-	#2

3. Results

In this section, the results of the experimental data will be presented. First, the reproducibility will be analyzed. Thereafter, the influence of the C-rate, temperature and electrolyte composition will be shown.

3.1. Reproducibility of Measurements

To validate the reproducibility of the galvanostatic cycling measurements on Cu/Li cells for selected points in the measurement matrix, up to three different EL-CELLs with exactly the same cell chemistry were assembled and the cycling results of the conducted measurements were compared (see Table 1). As an example, the cycling results of three cells at one specific measurement point are illustrated in Figure 3, with the areal capacity in Figure 3a and the Coulombic efficiency in Figure 3b. The cells contained LiFSI based electrolyte with a concentration of $c = 2$ M in DME and the measurements were performed at a current density of $J = 1$ mAh·cm^{-2} and a temperature of $T_{Cell} = 25$ °C.

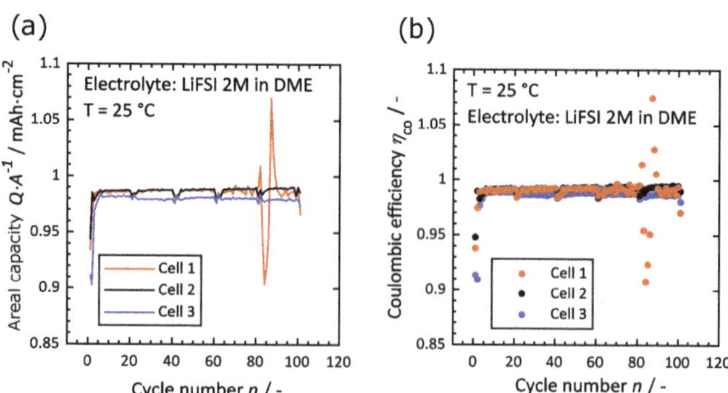

Figure 3. Galvanostatic cycling performance of three identical cells with Cu/Li structure. Electrolyte used in all cells is LiFSI 2M in DME. (a) areal capacity and (b) Coulombic efficiency over life time is shown for cells running at $T_{Cell} = 25$ °C with the current density of $J = 1$ mA·cm^{-2}.

Both the areal capacity (Figure 3a) and the Coulombic efficiency (Figure 3b) of all three cells show a high comparability at the first 100 cycles. This is indicated by the calculated standard deviation of $\sigma_{CE} \pm 0.005$ for the Coulombic efficiency and the corresponding value of $\sigma_{A.Cap} \pm 0.011$ mAh · cm^{-2} for the areal capacity. Only cell 1 shows an instability from cycle number 80 onward, which is not the case for cell 2 and cell 3. The ideal discharge capacity that the cells could reach is 1 mAh·cm^{-2}. The average Coulombic efficiency of $\eta_{CE-mean} \approx 0.99$ indicates the average areal capacity with a value of 0.99 mAh·cm^{-2} as well. For cell 1 there were four cycles which had a areal capacity and Coulombic efficiency above the theoretical maximum: $\eta_{CE} > 1$. It has been reported that micro shorts can happen during the Li deposition period using Li metal electrodes, which could be seen as tiny fluctuations in the voltage behavior of the cell [23]. The micro shorts happen when local dendrites can penetrate through the separator and contact the other electrode. In these cases, the cell locally experiences the occurrence of charge and discharge processes simultaneously. Due to this effect, the cell could show a longer deposition period, as under normal conditions, causing values above the theoretical maximum. In the following, only the Coulombic efficiency is presented and discussed as the areal capacity and the Coulombic efficiency correlate to each other.

3.2. Influence of C-Rate

In Figure 4, the influence of the C-rate on the Coulombic efficiency is presented for the measurements using LiFSI with a concentration of $c = 2$ M and varying C-rates of $I_{Cell} \epsilon \{0.5, 1, 2\}$ C at $T_{Cell} = 25\,^{\circ}$C.

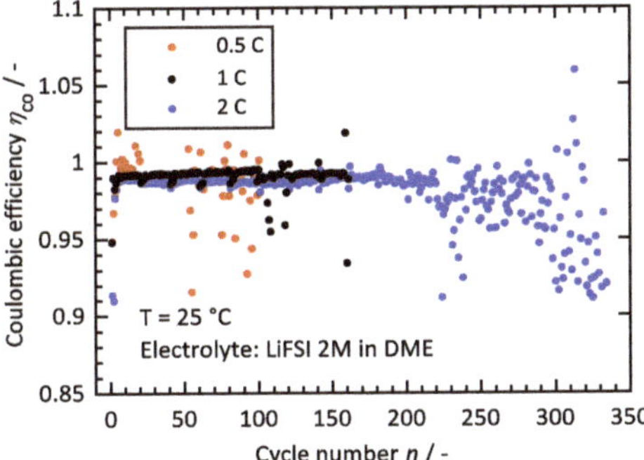

Figure 4. Galvanostatic cycling performance of cells with Cu/Li structure using LiFSI 2M in DME as the electrolyte. Cells are run each with an individual C-rate of $I_{Cell} \epsilon \{0.5, 1, 2\}$ C and are shown with orange, black and blue colors, respectively. The cell temperature for all three experiments was set to $T_{Cell} = 25\,^{\circ}$C.

Interestingly, the cycling results presented in Figure 4 show that increasing the current density of charging and discharging positively influence the cyclability of *insitu* deposited lithium electrodes. All three cells during the initial cycles have a relatively low Coulombic efficiency which increases gradually with increasing cycle number for all three cells until a Coulombic efficiency of $\eta_{CE} \approx 0.99$ is reached. This could be caused by a rapid SEI formation during the initial cycles. Interface reactions consume electrolytes and a part of the cyclable Li, which negatively influences the Coulombic efficiency during the formation cycles, as can be seen in Figure 4. The cell with the lowest C-rate ($I_{Cell} = 0.5$ C), which is highlighted with the orange color, illustrates an instability at the early cycles ($\eta_{CE} > 1$). Further, a fast drop of the Coulombic efficiency was found and already after 50 cycles the cell reached an efficiency below the set threshold value of $\eta_{CE} < 0.95$. Due to this reason, the cycling of the cell was stopped after 100 cycles. The cell with the medium C-rate ($I_{Cell} = 1$ C) presented in the black color is more stable at the first cycles and the Coulombic efficiency of the cell is very high with $\eta_{CE} \approx 0.99$ for the first 160 cycles. This leads to an increase of the maximum number of cycles by a factor of 3.2 compared to the cell cycled at $I_{Cell} = 0.5$ C. Finally, the cell with the highest C-rate ($I_{Cell} = 2$ C, blue) shows the most stable behavior. This is visible in the Coulombic efficiency, which drops below the set reference value of $\eta_{CE} < 0.95$ not until the 220th cycle.

3.3. Influence of Temperature

In order to investigate the effect of temperature on the kinetics of Li deposition and consequently the stability and cyclability of *insitu* deposited lithium electrodes, experiments were performed at three different temperatures with $T_{Cell} = \{25, 40, 60\}$ °C on the Cu/Li cells using LiFSI with a concentration of $c = 2$ M in DME as the electrolyte. The Coulombic efficiency results obtained from cycling tests are shown in Figure 5. Two different C-rates of $I_{Cell} \epsilon \{1, 2\}$ C were investigated, and the corresponding results are displayed in Figure 5a,b, respectively. The cells cycled at a temperature of $T_{Cell} = 25$ °C are presented in the orange color, the results at $T_{Cell} = 40$ °C and $T_{Cell} = 60$ °C are visualized in black and blue colors, respectively.

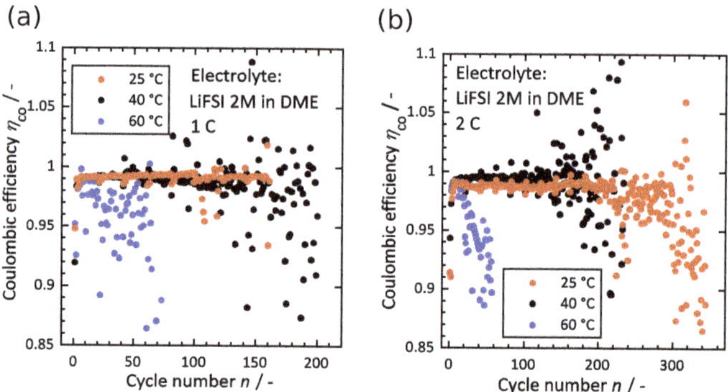

Figure 5. Cycle performance of cells with Cu/Li structure having LiFSI 2M in DME as the electrolyte. (a) Cells are running with the C-rate of $I_{Cell} = 1$ C and each at the different temperatures of $T_{Cell} = \{25, 40, 60\}$ °C shown in orange, black and blue colors, respectively. (b) Cells are running with the C-rate of $I_{Cell} = 2$ C and each at the different temperatures of $T_{Cell} = \{25, 40, 60\}$ °C shown with orange, black and blue colors, respectively.

In both investigated C-rates, the most stable cycling results were reached at a temperature of $T = 25$ °C. At this temperature, the cell cycled with $I_{Cell} = 1$ C shows 160 smooth cycles with a high average Coulombic efficiency of $\eta_{CE} > 0.99$. After 160 cycles the Coulombic efficiency drops below the threshold value of $\eta_{CE} < 0.95$. The cell cycled at $T_{Cell} = 40$ °C shows an instability at earlier cycles and reached a Coulombic efficiency of $\eta_{CE} < 0.95$ already after 130 cycles. The worst cycling performance was obtained at a temperature of $T_{Cell} = 60$ °C. The instability starts at the initial cycles and no smoothing behavior, as was found at $T_{Cell} = 40$ °C, could be achieved during the 70 cycles for which the cell was running. The same trend is observable for the cells running with a C-rate of $I_{Cell} = 2$ C. The positive influence of higher C-rates is reflected in an increase of the number of cycles reached for all three cells at the considered temperatures.

3.4. Influence of Salt, Concentration and Temperature

Motivated by the enhanced cycle life at lower temperatures of the cells containing LiFSI with a concentration of $c = 2$ M in the DME electrolyte, the influence of the salt concentration and the used salt on the degradation was investigated. Therefore, the concentration was reduced from $c = 2$ M to $c = 1$ M and the used salt was changed from LiFSI to LiTFSI at a concentration of $c = 1$ M. In Figure 6 the impact of the electrolyte composition is visualized. The influence of the concentration is presented in Figure 6a,c,e. The graphs differ for the considered cell temperatures. In Figure 6b,d,f, the effects of the used salt are also given at different temperatures.

Independent of the operating temperature, decreasing the LiFSI concentration from $c = 2$ M to $c = 1$ M in the electrolyte drastically impaired the cell's performance. This is clearly evident in the reduced cycle life of the cell with lower concentrations. The limiting value of $\eta_{CE} > 0.95$ is reached after 50 cycles for the experiments with $c = 1$ M at $T_{Cell} = 25\,°C$. This is three times less than the lifetime of an equivalent cell at a concentration of $c = 2$ M. Similar results were found at a cell temperature of $T_{Cell} = 40\,°C$. The maximum number of cycles increased from $n_{max} = 40$ to $n_{max} = 130$ as a consequence of the increased concentration. The cells performing at $T_{Cell} = 60\,°C$ showed, in general, the worst stability. The cell with 1 M LiFSI reached a cycle number of $n_{max} = 20$. The maximum number of cycles was minimally increased to $n_{max} = 60$ by increasing the concentration to $c = 2$ M.

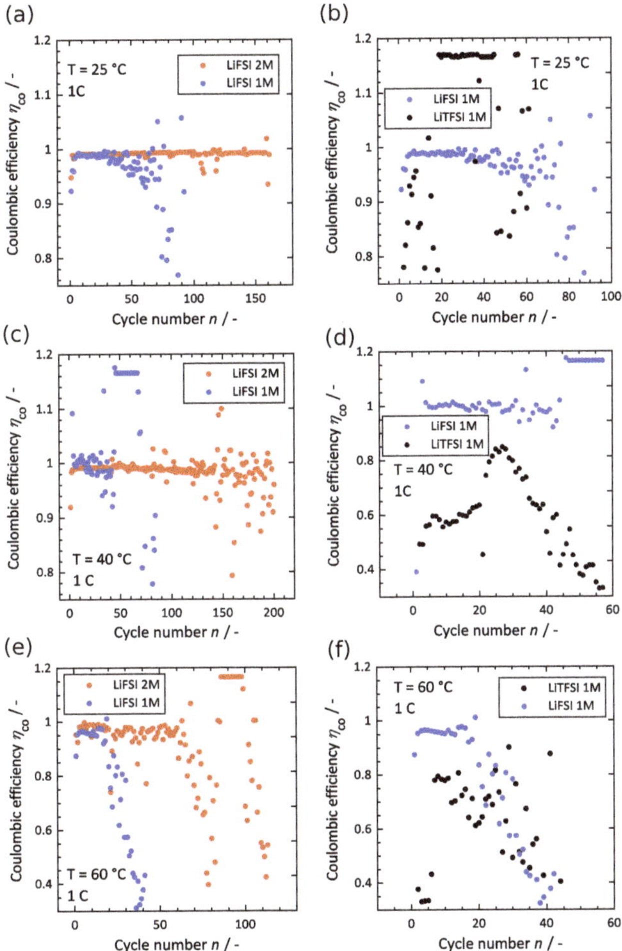

Figure 6. Cycle performance of cells with Cu/Li structure. All cells ran with the C-rate of $I_{Cell} = 1$ C. (**a,c,e**) Both cells use LiFSI Li salt in electrolyte with the different concentrations of $c = \{1,2\}$ M represented by orange and blue colors, respectively. (**b,d,f**) Both cells use a salt concentration of $c = 1$ M. Blue points represent the cell using LiFSI salt, and black points show the results of a cell using LiTFSI. (**a,b**), cell temperature is set to $T_{Cell} = 25\,°C$. (**c,d**), cell temperature is set to $T_{Cell} = 40\,°C$. (**e,f**), cell temperature is set to $T_{Cell} = 60\,°C$.

Replacing LiFSI at a concentration of $c = 1$ M with LiTFSI at the same concentration affects the cells' performances drastically. Similar to the effects seen by reducing the salt concentration, the use of LiTFSI negatively influences the cell degradation. At none of the investigated temperatures did the experiments using LiTFSI show stable behavior, not even for a small number of cycles. Independent of the applied temperature, the cells show low and random values for the CE and no trend in the degradation behavior is evident. The cells with LiTFSI also frequently show values above the theoretical maximum value of $\eta_{CE} > 1$, which is a sign of the inhomogeneous, poor and weak SEI formation potential of LiTFSI salt [24]. This might be because of the lower LiF content formed during the degradation of LiTFSI, which plays a major role in stabilizing the cell performance, resulting in a longer cycle life [25]. A general trend of CE development observed in Figures 4–6 is that the random behavior (noise-like) is a sign of instability. The longer cells run smoothly, the better the cycling performance and lifetime get. A CE value of higher than one could be a sign of micro Li plating, while a CE value of lower than one could be caused by the loss of deposited lithium in the form of SEI or dead lithium. A common behavior in all cells is that the cell cyclability reduces significantly as soon as noises start.

4. Discussion

It has been realized that the type of lithium salt as well as its concentration can strongly influence the performance and cycle life of Li-metal cells. To further investigate this issue, EIS measurements were performed on the cells with different electrolytes of (1) LiFSI 2M in DME, (2) LiFSI 1M in DME, (3) LiTFSI 1M in DME, different measurement temperatures of $T_{Cell} \epsilon \{25, 40, 60\}$ °C and different C-rates of $I_{Cell} \epsilon \{0.5, 1, 2\}$ C.

As explained in Figure 1, the first EIS measurement was carried out after the first Li deposition on Cu and then was repeated every 20 cycles until the Coulombic efficiency of the cell reached the value of 0.95. The spectra of cells with different electrolytes after the first Li plating performed at $T_{Cell} = 25$ °C and an applied current density of $j = 1$ mAh·cm^{-2} (C-rate = 1 C) are presented in Figure 7a. The EIS spectra of cells after the first Li plating performed at different measurement temperatures, having LiFSI 2M in DME electrolyte and a C-rate of $I_{Cell} = 1$ C, are presented in Figure 7b. The influence of aging on the EIS spectra of a Cu/Li cell with LiFSI 2M in DME electrolyte, performed at $T_{Cell} = 25$ °C and a C-rate of $I_{Cell} = 1$ C, is visualized in Figure 7c. The spectrum #1 is the first EIS performed after the first Li deposition, #2 is assigned to the second EIS performed after 20 full cycles, #3 is the third EIS after 40 full cycles and so on.

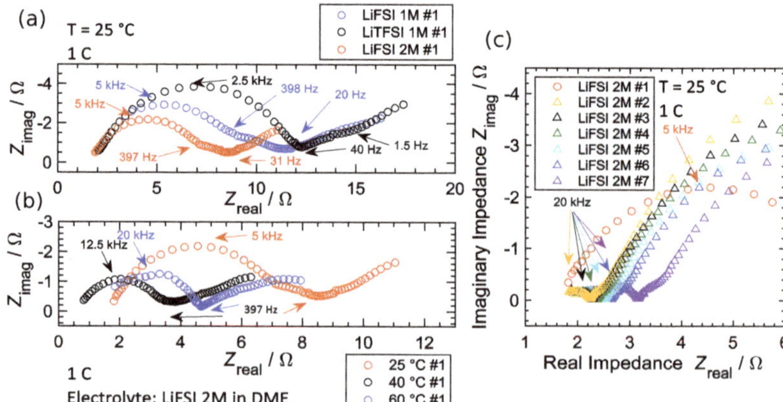

Figure 7. EIS spectra of Cu/Li cells: (**a**) Three cells with different electrolytes after the first Li plating. Orange represents LiFSI 2M in DME, black is LiTFSI 1M in DME, and blue shows LiFSI 1M in DME at $T_{Cell} = 25$ °C and $j = 1$ mAh·cm^{-2}. (**b**) Three cells at different temperatures $T_{Cell} \epsilon \{25, 40, 60\}$ °C after the first Li plating. Electrolyte: LiFSI 2M in DME with an applied current density of $j = 1$ mAh·cm^{-2}. (**c**) EIS spectra of a single cell, performed every 20th cycle during the degradation test with LiFSI 2M in DME, $T_{Cell} = 25$ °C and $j = 1$ mAh·cm^{-2}. EIS # 1 is after first plating, EIS # 2 is after 21 cycles and EIS # 7 is after 121 cycles.

The correlation of EIS measurements with cycling results could be better realized by considering the induced overpotentials of Li deposition nucleation $\mu_{nucleation}$ and particle growth μ_{growth} during one full cycle. The initial voltage drop at the beginning of plating on Cu is considered to be nucleation overpotential and the steady potential during the rest of deposition period is considered to be growth overpotential [13]. These values are illustrated in Figure 8. The influence of the electrolyte variation is presented in Figure 8a, the influence of temperature variation is visualized in Figure 8b, and at last the impact of the applied current density is displayed in Figure 8c. The cells presented in Figure 8a,b are identical to the cells shown in Figure 7a,b.

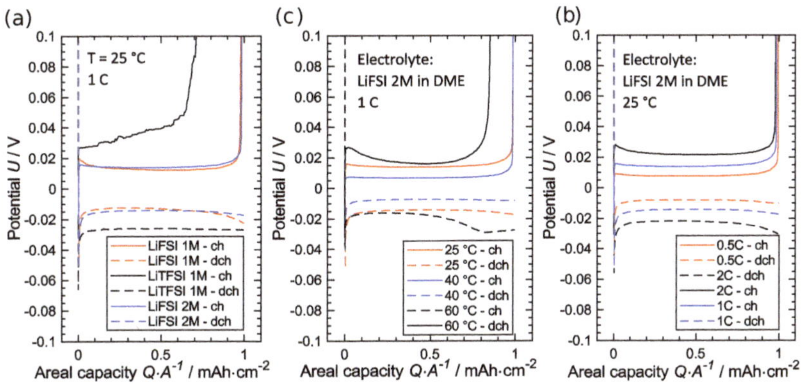

Figure 8. Potential–Capacity profile at cycle number 10 for Li/Cu cells, (**a**) with different electrolytes of LiFSI 1M in DME (orange), LiTFSI 1M in DME (black), and LiFSI 2M in DME (blue). Measurements are performed at $T_{Cell} = 25\,°C$ and with an applied current density of $j = 1\ \text{mAh·cm}^{-2}$. (**b**) performed with different C-rates of $I_{Cell} = 0.5\ C$ (orange), $I_{Cell} = 1\ C$ (black) and $I_{Cell} = 2\ C$ (blue), with LiFSI 2M in DME as electrolyte at $T_{Cell} = 25\,°C$. (**c**) performed at different temperature of $T_{Cell} = 25\,°C$ (orange), $T_{Cell} = 40\,°C$ (black) and $T_{Cell} = 60\,°C$ (blue), using LiFSI 2M in DME as electrolyte and a current density of $j = 1\ \text{mAh·cm}^{-2}$.

All three cells presented in Figure 7a show a comparable ohmic resistance (R_{ohmic}), which is mainly correlated to electrolyte and contact resistances. This was expected as the cells consist of the same electrodes and the cells are still too fresh to be influenced by different aging rates due to different used electrolytes. The typical semi-circle is easily noticeable in the EIS data of all cells. Additionally, all three cells in Figure 7a show a second semi-circle at lower frequencies, which are partly overlapped with the first ones. The semi-circles at higher frequencies are formed in a similar frequency range for the cells having an LiFSI based electrolyte (see Figure 7a). The frequency values are different for the cells containing the LiTFSI based electrolyte.

These results indicate that the first semi-circle is based on the interface or SEI related impedance. The second semi-circle shows the impedance related to the charge transfer R_{ct}. The consequence of a different impedance behavior of the cells due to the used electrolyte can also be seen in the overpotential of cells during cycling. The voltage behavior of one charge and discharge process (at cycle number #10) is illustrated in Figure 8a; these are the cells identical to those presented in Figure 7a. As expected, the cell with the LiTFSI based electrolyte shows the highest overpotentials ($\mu_{nucleation} = -6.5$ mV and $\mu_{growth} = -2.6$ mV) and irreversible capacity among the rest. The two LiFSI based cells show comparable overpotentials of $\mu_{nucleation} = -5.2$ mV and $\mu_{growth} = -1.4$ mV for LiFSI 2M and of $\mu_{nucleation} = -4.5$ mV and $\mu_{growth} = -1.5$ mV for LiFSI 1M. The high concentrated LiFSI based cell shows the minimum of irreversible capacity.

By varying the temperature it can be noticed that the cells have the most stable performance at $T = 25\,°C$ (see Figure 7b). By increasing the temperature, the frequency which corresponds to a maximum of the semi-circle moves to higher values (from 5 kHz at $T_{Cell} = 25\,^{\circ}C$ to 12.5 kHz at $T_{Cell} = 40\,^{\circ}C$ and to 20 kHz at $T_{Cell} = 60\,^{\circ}C$). The second semi-circle at elevated temperatures is not distinct anymore and is hardly noticeable at $T_{Cell} \geq 40\,^{\circ}C$. This effect can be explained by the fact that with increasing temperature the charge transfer resistance decreases and consequently the correlated semi-circle is more overlapped to the SEI based semi-circle. The ohmic resistances (R_{ohmic}) are comparable in both cells at elevated temperatures and are smaller compared to the resistance of the cell performed at $T_{Cell} = 25\,^{\circ}C$. The cell cycled at $T_{Cell} = 40\,^{\circ}C$ has the smallest length of the SEI related semi-circle. This is a consequence of the improved kinetic at $T_{Cell} = 40\,^{\circ}C$ in comparison to $T_{Cell} = 25\,^{\circ}C$. This is in line with the seen voltage profile of the cells presented in Figure 8b. The data at $T_{Cell} = 40\,^{\circ}C$ show the lowest overpotentials of $\mu_{nucleation} = -4$ mV and $\mu_{growth} = -0.7$ mV. The cell cycled at $T_{Cell} = 60\,^{\circ}C$ also has a low nucleation overpotential of $\mu_{nucleation} = -4$ mV; however, μ_{growth} in contrast to the rest of the cells is increasing during the plating period and approaches bigger values ($\mu_{growth} = -2.8$ mV). Cycling at a temperature of $T_{Cell} = 60\,^{\circ}C$ has the maximum growth overpotential and the worst cycling stability among all considered temperatures.

Another influencing parameter on the kinetics of Li deposition is the applied current density. We have noticed that cells with a higher C-rate achieve a longer cycle life. This can be seen in Figure 8c, which shows the data extracted from three Li/Cu cells containing LiFSI 2M in DME electrolyte at $T_{Cell} = 25\,^{\circ}C$ with different current densities. As expected, the overpotential increases with increasing current density which causes a higher deviation from equilibrium. However, the better cycling performance of the cells cycled at a higher C-rate could be due to fewer interfacial side reactions as the cycling time is shorter. On the other hand, however, the higher current density could be a trigger to side reactions. More interfacial investigation (EIS and *insitu* observation) is needed for a better understanding of this process. EIS measurements on the cell with LiFSI 2M in the DME electrolyte performed at $T_{Cell} = 25\,^{\circ}C$ show that, after 20 cycles, the interface resistance significantly decreased. This could be due to the SEI layer not being formed homogeneously and completely after the first deposition. By continuing the cycling, however, the layer becomes denser and more uniform and therefore after 20 cycles the semi-circle is significantly smaller than that after the first deposition. By continuing the cycling, the second semi-circle at low frequencies is still noticeable. This was not the case for the cells cycled at higher temperatures. By cycling the frequency corresponding to the maximum of the first, the semi-circles move towards higher values (similar to the high temperature behavior). The R_{ct} increases by cycling but is still smaller than that of the very first cycle. The ohmic resistance is also slightly increased by cycling. A relatively sharp increase of R_{ohmic} is noticeable between cycle 100 and cycle 120. It is worth mentioning that, based on cycle performance results, the cell starts to show slight instabilities in the CE trend from the 110th cycle. It can be concluded that both R_{ohmic} and R_{ct} influence the cycle behavior and the cycle life of the Li-metal cells.

5. Future Work

Another method for investigating lithium deposition during cycling is *insitu* observation using special cell holders. *Insitu* observations in the ECC-Opto Stud, manufactured by EL-Cells GmbH (Hamburg, Germany), on battery cells were made in [26–33] using Raman spectroscopy, X-ray diffraction (XRD) and optical analytics. Observations were made using Raman spectroscopy [27,31,32] and XRD [26,28,33] to observe changes in electrodes and SEI during aging. Merryweather et al. [29] used the above-mentioned cell housing for optical interferometric scattering measurements to detect single-particle ion dynamics, and Rittweger et al. [30] observed the reflectivity of cathodes during charge and discharge.

In future work, the above-mentioned cell holder ECC-Opto-Std cell, manufactured by EL-Cell, will be used to conduct *insitu* measurements. The cell setups, like those described in this article, consisting of Cu/Li, will be used. The experiments will also investigate the influence of C-rate, temperature, used salt and salt concentration on the morphology of lithium deposition. Additionally, in order to better understand the behavior of the lithium at different conditions, a fully developed model will be included.

6. Conclusions

In summary, the electrochemical performance of Cu/Li cells was investigated with the motivation of understanding the kinetics of the deposition mechanism of Li metal electrodes. The influence of temperature $T_{Cell} \epsilon \{25, 40, 60\}$ °C and C-rate $I_{Cell} \epsilon \{0.5, 1, 2\}$ C was examined. Additionally, the variety of electrolytes, including LiFSI 2M in DME, LiFSI 1M in DME, and LiTFSI 1M in DME, were utilized to address the impact of type and concentration of Li salt on the electrolyte. Coulombic efficiency and induced depositions overpotential, as well as EIS measurements, were used to evaluate the aging behavior of cells under different conditions. Based on our results, it is confirmed that cycling at a temperature of $T_{Cell} = 40$ °C has the best kinetics in comparison to the cycling data with $T_{Cell} = 25$ °C and $T_{Cell} = 60$ °C, as it shows minimum deposition overpotentials and impedance. However, the best performance regarding the stability and long cycle life is achieved at $T_{Cell} = 25$ °C. The LiFSI in general showed better cyclability in Cu/Li cells compared to LiTFSI and the best performance could be gained by a high concentration of the LiFSI-DME electrolyte.

Author Contributions: Conceptualization, S.M.B., A.F. and A.R.; methodology, S.M.B.; validation, S.M.B.; investigation, S.M.B.; data curation, S.M.B.; writing—original draft preparation, S.M.B.; writing—review and editing, A.F.; visualization, S.M.B., A.F.; supervision, K.P.B.; project administration, K.P.B.; funding acquisition, K.P.B. All authors have read and agreed to the published version of the manuscript.

Funding: This research was partly funded by Robert Bosch GmbH.

Institutional Review Board Statement: Not applicable.

Informed Consent Statement: Not applicable.

Data Availability Statement: The data presented in this study are available on request from the corresponding author

Acknowledgments: The authors thankfully acknowledge Leonie Wolf for helping with the preparation of the cell assemblies. The authors would also like to thank Robert Bosch GmbH for their support during the Bosch Doctoral College.

Conflicts of Interest: The authors declare no conflict of interest.

Abbreviations

The following abbreviations are used in this manuscript:

A	Effective surface area
c	Molar concentration
c_{H_2O}	water concentration
c_{O_2}	oxygen concentration
CC	Current Collector
CE	Coulombic efficiency
Cu	Copper
d_{CC}	Diameter of the current collector
$d_{Li\ foil}$	Diameter of the Lithium foil
DME	1,2-dimethoxyethane
DOE	US Department of Energy
e^-	electron
EIS	Electrochemical Impedance Spectroscopy
$f_{EIS,\ max}$	Maximum frequency of EIS
$f_{EIS,\ min}$	Minimum frequency of EIS
H_2	Hydrogen
H_2O	Water

$h_{Cu} = 6$ — Thickness of the Copper foil
$h_{Li\,foil}$ — Thickness of the Lithium foil
I_{Cell} — Cell current
j — Current density
Li — Lithium
Li^+ — Lithium Ion
LiFSI — Lithium bis(fluorosulfonyl)imide
LiTFSI — Lithium bis(trifluoromethanesul-fonyl)imide
n — Number of cycles
n_{max} — Maximum number of cycles
O_2 — Oxygen
OCP — Open Circuit Potential
R_{ct} — Charge transfer resistance
R_{ohmic} — Ohmic resistance
SEI — Solid Electrolyte Interface
SPE — Solid Polymer Electrolyte
t — Time
T_{Cell} — cell temperature
t_{dry} — Time of drying
T_{dry} — Temperature of drying
U_{min} — Lower voltage limit
U_{max} — Upper voltage limit
V_{elec} — Amount of electrolyte
$x_{LiTFSI,LiFSI}$ — Purity of LiTFSI and LiFSI
XRD — X-ray diffraction
Z_{imag} — Imaginary part of the impedance
Z_{real} — Real part of the impedance
η_{CE} — Coulombic efficiency
$\eta_{CE\text{-}mean}$ — Average Coulombic efficiency
$\mu_{nucleation}$ — overpotential of Li deposition nucleation
μ_{growth} — overpotential of particle growth
$\phi_{Li\,|\,H_2}$ — Potential of Li versus H_2
σ_{CE} — Standard deviation of the Coulombic efficiency
$\sigma_{A.Cap}$ — Standard deviation of the areal capacity

References

1. Ding, Y.; Cano, Z.P.; Yu, A.; Lu, J.; Chen, Z. Automotive Li-ion batteries: Current status and future perspectives. *Electrochem. Energy Rev.* **2019**, *2*, 1–28. [CrossRef]
2. *USABC Goals for Advanced Batteries for EVs—CY 2023 Commercialization (USABC, 2017)*; 2018.
3. Albertus, P.; Babinec, S.; Litzelman, S.; Newman, A. Status and challenges in enabling the lithium metal electrode for high-energy and low-cost rechargeable batteries. *Nat. Energy* **2018**, *3*, 16–21. [CrossRef]
4. Brandt, K. Historical development of secondary lithium batteries. *Solid State Ionics* **1994**, *69*, 173–183. [CrossRef]
5. Wang, H.; Zhang, W.D.; Deng, Z.Q.; Chen, M.C. Interaction of nitrogen with lithium in lithium ion batteries. *Solid State Ionics* **2009**, *180*, 212–215. [CrossRef]
6. Furukawa, T.; Hirakawa, Y.; Kondo, H.; Kanemura, T.; Wakai, E. Chemical reaction of lithium with room temperature atmosphere of various humidities. *Fusion Eng. Des.* **2015**, *98*, 2138–2141. [CrossRef]
7. Peled, E. The electrochemical behavior of alkali and alkaline earth metals in nonaqueous battery systems—The solid electrolyte interphase model. *J. Electrochem. Soc.* **1979**, *126*, 2047. [CrossRef]
8. Sergi, F.; Arista, A.; Agnello, G.; Ferraro, M.; Andaloro, L.; Antonucci, V. Characterization and comparison between lithium iron phosphate and lithium-polymers batteries. *J. Energy Storage* **2016**, *8*, 235–243. [CrossRef]
9. Mindemark, J.; Lacey, M.J.; Bowden, T.; Brandell, D. Beyond PEO—Alternative host materials for Li+-conducting solid polymer electrolytes. *Prog. Polym. Sci.* **2018**, *81*, 114–143. [CrossRef]
10. Nagpure, S.C.; Tanim, T.R.; Dufek, E.J.; Viswanathan, V.V.; Crawford, A.J.; Wood, S.M.; Xiao, J.; Dickerson, C.C.; Liaw, B. Impacts of lean electrolyte on cycle life for rechargeable Li metal batteries. *J. Power Sources* **2018**, *407*, 53–62. [CrossRef]
11. Christensen, J.; Albertus, P.; Sanchez-Carrera, R.S.; Lohmann, T.; Kozinsky, B.; Liedtke, R.; Ahmed, J.; Kojic, A. A critical review of Li/air batteries. *J. Electrochem. Soc.* **2011**, *159*, R1. [CrossRef]
12. Yang, T.; Liu, J.; Dai, J.; Han, Y. Shaping particles by chemical diffusion and reaction. *CrystEngComm* **2017**, *19*, 72–79. [CrossRef]

13. Han, Y.; Jie, Y.; Huang, F.; Chen, Y.; Lei, Z.; Zhang, G.; Ren, X.; Qin, L.; Cao, R.; Jiao, S. Enabling stable lithium metal anode through electrochemical kinetics manipulation. *Adv. Funct. Mater.* **2019**, *29*, 1904629. [CrossRef]
14. Li, L.; Basu, S.; Wang, Y.; Chen, Z.; Hundekar, P.; Wang, B.; Shi, J.; Shi, Y.; Narayanan, S.; Koratkar, N. Self-heating–induced healing of lithium dendrites. *Science* **2018**, *359*, 1513–1516. [CrossRef]
15. Aryanfar, A.; Brooks, D.J.; Colussi, A.J.; Merinov, B.V.; Goddard, W.A., III; Hoffmann, M.R. Thermal relaxation of lithium dendrites. *Phys. Chem. Chem. Phys.* **2015**, *17*, 8000–8005. [CrossRef] [PubMed]
16. Zhu, Y.; Xie, J.; Pei, A.; Liu, B.; Wu, Y.; Lin, D.; Li, J.; Wang, H.; Chen, H.; Xu, J.; et al. Fast lithium growth and short circuit induced by localized-temperature hotspots in lithium batteries. *Nat. Commun.* **2019**, *10*, 1–7. [CrossRef]
17. Adair, K.R.; Banis, M.N.; Zhao, Y.; Bond, T.; Li, R.; Sun, X. Temperature-Dependent Chemical and Physical Microstructure of Li Metal Anodes Revealed through Synchrotron-Based Imaging Techniques. *Adv. Mater.* **2020**, *32*, 2002550. [CrossRef]
18. Geng, Z.; Lu, J.; Li, Q.; Qiu, J.; Wang, Y.; Peng, J.; Huang, J.; Li, W.; Yu, X.; Li, H. Lithium metal batteries capable of stable operation at elevated temperature. *Energy Storage Mater.* **2019**, *23*, 646–652. [CrossRef]
19. Cheng, X.B.; Zhang, R.; Zhao, C.Z.; Wei, F.; Zhang, J.G.; Zhang, Q. A review of solid electrolyte interphases on lithium metal anode. *Adv. Sci.* **2016**, *3*, 1500213. [CrossRef]
20. Andersson, A.; Edström, K. Chemical composition and morphology of the elevated temperature SEI on graphite. *J. Electrochem. Soc.* **2001**, *148*, A1100. [CrossRef]
21. Aurbach, D.; Weissman, I.; Zaban, A.; Chusid, O. Correlation between surface chemistry, morphology, cycling efficiency and interfacial properties of Li electrodes in solutions containing different Li salts. *Electrochim. Acta* **1994**, *39*, 51–71. [CrossRef]
22. Kim, H.; Wu, F.; Lee, J.T.; Nitta, N.; Lin, H.T.; Oschatz, M.; Cho, W.I.; Kaskel, S.; Borodin, O.; Yushin, G. In situ formation of protective coatings on sulfur cathodes in lithium batteries with LiFSI-based organic electrolytes. *Adv. Energy Mater.* **2015**, *5*, 1401792. [CrossRef]
23. Dornbusch, D.A.; Hilton, R.; Lohman, S.D.; Suppes, G.J., Experimental validation of the elimination of dendrite short-circuit failure in secondary lithium-metal convection cell batteries. *J. Electrochem. Soc.* **2014**, *162*, A262. [CrossRef]
24. He, Y.; Zhang, Y.; Yu, P.; Ding, F.; Li, X.; Wang, Z.; Lv, Z.; Wang, X.; Liu, Z.; Huang, X. Ion association tailoring SEI composition for Li metal anode protection. *J. Energy Chem.* **2020**, *45*, 1–6. [CrossRef]
25. Jurng, S.; Brown, Z.L.; Kim, J.; Lucht, B.L. Effect of electrolyte on the nanostructure of the solid electrolyte interphase (SEI) and performance of lithium metal anodes. *Energy Environ. Sci.* **2018**, *11*, 2600–2608. [CrossRef]
26. Bärmann, P.; Mohrhardt, M.; Frerichs, J.E.; Helling, M.; Kolesnikov, A.; Klabunde, S.; Nowak, S.; Hansen, M.R.; Winter, M.; Placke, T. Mechanistic Insights into the Pre-Lithiation of Silicon/Graphite Negative Electrodes in "Dry State" and After Electrolyte Addition Using Passivated Lithium Metal Powder. *Adv. Energy Mater.* **2021**, *11*, 2100925. [CrossRef]
27. Blanchard, D.; Slagter, M. In operando Raman and optical study of lithium polysulphides dissolution in Lithium-Sulfur Cells with carrageenan binder. *J. Phys. Energy* **2021**, *3*, 044003. [CrossRef]
28. Chladil, L.; Kunický, D.; Vanýsek, P.; Čech, O. In-Situ X-Ray Study of Carbon Coated LiFePO$_4$ for Li-Ion Battery in Different State of Charge. *ECS Trans.* **2018**, *87*, 107–114. [CrossRef]
29. Merryweather, A.J.; Schnedermann, C.; Jacquet, Q.; Grey, C.P.; Rao, A. Operando optical tracking of single-particle ion dynamics in batteries. *Nature* **2021**, *594*, 522–528. [CrossRef] [PubMed]
30. Rittweger, F.; Modrzynski, C.; Roscher, V.; Danilov, D.L.; Notten, P.H.L.; Riemschneider, K.-R. Investigation of charge carrier dynamics in positive lithium-ion battery electrodes via optical in situ observation. *J. Power Sources* **2021**, *482*, 228943. [CrossRef]
31. Vinayan, B.P.; Diemant, T.; Lin, X.-M.; Cambaz, M.A.; Golla-Schindler, U.; Kaiser, U.; Behm, R.J.; Fichtner, M. Nitrogen Rich Hierarchically Organized Porous Carbon/Sulfur Composite Cathode Electrode for High Performance Li/S Battery: A Mechanistic Investigation by Operando Spectroscopic Studies. *Adv. Mater. Interfaces* **2016**, *3*, 1600372. [CrossRef]
32. Vinayan, B.P.; Euchner, H.; Zhao-Karger, Z.; Cambaz, M.A.; Li, Z.; Diemant, T.; Behm, R.J.; Gross, A.; Fichtner, M. Insights into the electrochemical processes of rechargeable magnesium–sulfur batteries with a new cathode design. *Adv. Mater. Chem. A* **2019**, *7*, 25490–25502. [CrossRef]
33. Zou, J.; Sole, C.; Drewett, N.E.; Velický, M.; Hardwick, L.J. In Situ Study of Li Intercalation into Highly Crystalline Graphitic Flakes of Varying Thicknesses. *J. Phys. Chem. Lett.* **2016**, *7*, 4291–4296. [CrossRef] [PubMed]

 *batteries*

Article

Power and Energy Rating Considerations in Integration of Flow Battery with Solar PV and Residential Load

Purnima Parmeshwarappa, Ravendra Gundlapalli and Sreenivas Jayanti *

DST-Solar Energy Harnessing Center, Department of Chemical Engineering, IIT Madras, Chennai 60036, India; purnimagp@gmail.com (P.P.); ravendra.gundlapalli@gmail.com (R.G.)
* Correspondence: sjayanti@iitm.ac.in; Tel.: +91-44-2257-4168

Abstract: Integration of renewable energy sources such as solar photovoltaic (PV) generation with variable power demand systems like residential electricity consumption requires the use of a high efficiency electrical energy system such as a battery. In the present study, such integration has been studied using vanadium redox flow battery (VRFB) as the energy storage system with specific focus on the sizing of the power and energy storage capacities of the system components. Experiments have been carried out with a seven-day simulated solar insolation and residential load characteristics using a 1 kW VRFB stack and variable amounts of electrolyte volume. Several scenarios have been simulated using power and energy scale factors. Battery response, in terms of its power, state of charge and efficiency, has been monitored in each run. Results show that the stack power rating should be based on peak charging characteristics while the volume of electrolyte should be based on the expected daily energy discharge through the battery. The PV source itself should be sized at about 25% more energy rating than the average daily load. The ability to design a VRFB with a high power-to-energy ratio makes it particularly attractive for PV-load integration.

Keywords: solar photovoltaic energy; redox flow battery; residential load; renewable energy integration; battery sizing; battery efficiency

Citation: Parmeshwarappa, P.; Gundlapalli, R.; Jayanti, S. Power and Energy Rating Considerations in Integration of Flow Battery with Solar PV and Residential Load. *Batteries* **2021**, *7*, 62. https://doi.org/10.3390/batteries7030062

Academic Editor: Kai Peter Birke

Received: 30 July 2021
Accepted: 2 September 2021
Published: 8 September 2021

Publisher's Note: MDPI stays neutral with regard to jurisdictional claims in published maps and institutional affiliations.

1. Introduction

Use of renewable energy sources is one of the best solutions to reduce society's dependence on fossil fuels so as to reduce emission of both conventional pollutants and greenhouse gases. Amongst various renewable energy sources, solar photovoltaic (PV) panels are the most commercially viable due to the ease in installation, cost and scalability, especially in countries close to the equator [1]. Wide adoption of renewable energy as a power generation source is only possible if the power can be delivered on and as per the demand of the consumer. In case of solar or wind energy, there is uncertainty with the amount of energy available at any given time as the exploitable energy depends on a number of factors that include location of the PV panel or wind turbine, time of the day, and seasonal and weather conditions [1,2]. Energy storage systems coupled with these energy sources can reduce the impact of their natural fluctuations and can provide power needed by the consumer. An energy storage system used for solar PV applications requires specific properties to aid this integration while maintaining the good performance and long life of the system. High cycle life, high capacity appreciation at slow rate of discharge, good reliability under cyclic discharge conditions, low equalizing and boost charging requirements, high watt-hour (round-trip energy) efficiency and ampere-hour (coulombic) efficiency at different states of charge (SOC) levels, low self-discharge, wide operating temperature range, robust design and low maintenance and cost effectiveness are some of the necessary characteristics anticipated from a battery for storing solar energy [3,4]. Energy storage integrated with a renewable energy source can serve as a stand-alone power generation system for a variety of applications [5,6].

145

Lead acid or lithium ion type batteries have been conventionally used as battery storage devices in integration studies. These batteries are best operated in their safe operating range; in the context of solar PV applications, the life of these batteries degrades considerably due to varying input from solar energy and due to their limited range of operating depth of discharge (DOD) and operational conditions. Lead acid battery is economical, sustainable and has good operational safety and quick response time, but it suffers from low efficiency, low energy density (30–50 Wh/kg) and has limitations on low power rating and DOD for assured life. Lithium ion battery has higher energy density and higher efficiency than lead acid battery but is expensive and prone to thermal runaway and can thus be a fire hazard [7]. In addition, solid state batteries have a fixed power to energy (P/E) ratio due to fixed volume of electrolyte.

A number of studies have been reported recently on improving the characteristics and performance of integrated renewable energy source-energy storage systems. Angenendt et al. [8] studied stringent control strategies based on forecast for a PV-battery system to improve the system performance and battery life. PV self-consumption improvement using frequency restoration reserves was analyzed by Litjens et al. [9] for both residential and commercial applications. Hybridization is another way of improving the system robustness and enhancing its performance with improved life. Single energy source with multiple storage systems (for example, PV with lead acid battery and supercapacitor for improved battery life) has been studied by Jing et al. [10]. Multiple energy sources, for example, solar PV, wind turbine, diesel generator coupled with single or multiple storage systems such as lead acid batteries, lithium ion batteries, flow batteries, reversible fuel cells, etc. have been investigated extensively and various algorithms and optimization models have been employed for better energy management and reduced system cost [11–18].

The present work focuses on redox flow batteries as energy storage system. A redox flow battery (RFB) is an electrochemical energy storage device that has several attractive features especially for large-scale stationary storage, such as independent scalability in energy and power levels, large number of life cycles and absence of fire hazard [19–23]. The energy storage capacity of a typical redox flow battery is determined by the volume of the electrolyte taken, while the power at which the energy can be delivered or absorbed is controlled by combination of active area and number of cells in a stack [20–22]. The positive (catholyte) and negative (anolyte) electrolyte species are stored in separate reservoirs. Each electrolyte is circulated through the respective electrodes of a stack for either charging or discharging of the battery. There is broad consensus that RFBs can be highly cost-competitive when used in large scale power applications such as microgrids, power islands, peak shaving and renewable energy applications [24]. The integration of solar cell and redox flow battery offers a unique advantage, namely, the liquid electrolytes of redox flow battery system can also be used as a coolant for the photovoltaic panels and the battery stacks so as to have integrated thermal management capabilities. Vanadium based redox flow batteries have gained significance and market penetration compared to other flow battery systems in view of its same chemical species on positive and negative redox couples and ease of recyclability [3,20]. Although the VRFB has considerable capacity fade induced by cross-over of vanadium species through membrane, either the cross-over can be reduced using operating protocols or the active state can be reversed back by remixing schedules [25–27]. Compared to a lithium-ion battery, it suffers from relatively low efficiency, low energy density and high electrolyte cost. It is considered is to be cost-competitive for GWh-scale energy storage applications [24] and several studies have recently reported on its integration with renewable energy sources. Garcia-Quismondo et al. [23] reported on a nine-month performance analysis of a 5 kW/5kWh VRFB system coupled to a PV system. Bhattacharjee and Saha [28] designed an electrical equivalent model of 1 kW/6 kWh VRFB system, validated it in MATLAB/SIMULINK and later integrated with solar PV for residential application. Zhang et al. [29] studied an integrated solar PV- VRFB system for residential applications using MATLAB and brought out the importance of battery sizing considering cost and battery efficiency. Sarkar et al. [30] designed an integrated solar

PV, wind turbine, biomass and VRFB system and studied its performance using Homer software. A virtual power plant was designed by Behi et al. [31] using solar PV (810 kW) with VRFB (700 kWh, 350 kW) to cater to 67 dwelling power requirements.

Several design/ simulation studies [12–17,29] have highlighted the importance of optimal sizing of battery energy storage system to ensure uninterrupted energy availability, improved life span of battery, less maintenance and cost, etc. The present work on simulation of the integrated system is primarily experimental and is focused on dealing with natural fluctuations that arise both from supply side (solar PV) and demand side (residential load) in a solar PV-VRFB integrated system. By running different scenarios with a VRFB stack over a seven-day solar PV-load profile, the study brings out how the power and energy characteristics influence the sizing of VRFB and the PV systems in terms of power and energy ratings.

2. System Lay-Out and Experimental Details

2.1. Conceptual System and Experimental Set-Up

The system being studied consists of a set of PV panels, residential electrical load, a vanadium flow battery system and a charge controller power electronics module to link the three. The conceptual model is shown in Figure 1a. At every instant of time, the charge controller determines the residential load, assesses the available power output from the PV panels and the state of charge of the battery and aims to serve the load demand from PV to the extent possible. When this is not feasible, it attempts to source power from the battery to meet the demand may be partly or fully. Any excess power from the PV will be sent to the battery to the extent possible. The charge controller will, in practice, regulate the power drawn from the PV panels to meet the combined demand from residential load and battery charging. It will also regulate the power going to (and coming from) the battery by operating the latter within its safe operating voltage window. The present study deals with the VRFB system only, and the roles of load, PV panels and power electronics module are taken over by a battery charger (Bitrode Model), as shown in Figure 1b. The time-wise continuous load and PV output profiles are replaced by discrete time steps (of 11 min) during which the power going to or coming from the battery is held constant, subject to the battery lying within its pre-set voltage limits. The battery power is the difference between the PV output and the load, both averaged over the 11-min interval. The battery power may be positive (when the PV output is higher than the load) or negative (when the load is higher than the PV). At the end of the time step (of 11 min), the power is held at zero for one minute so as to measure the open circuit voltage (OCV) of the battery which is used to monitor its state of charge (SoC).Each experiment simulated the continuous running of the VRFB system over a seven-day period with natural variations of solar irradiations and residential power demand incorporated into the profiles. Setting up of these profiles is discussed below.

2.2. Construction of PV Output Profile

A number of factors such as time of the day, location of PV, season of the year, atmospheric conditions, cell temperature, etc. influence solar insolation at a particular point of time and location. In order to get a realistic simulation of these natural factors in the PV output, solar insolation was simulated using the NREL software SAM for the year 2019 and was compared with global horizontal irradiation (GHI) data for a solar park site in Tumkur, Karnataka, India obtained from Meteonorm 7.0 database [32]. Weekly variation from the first method was found to be between 6.7 and 3.3 kWh/m^2 while that from the latter was between 7.7 and 1.98 kWh/m^2 over a week in the month of June. In view of the fairly good agreement between the two, seven successive days of insolation data from Meteonorm 7.0 database was chosen to construct the PV output profile for the present study. These give only the total PV output over the day. In order to get a minute-by-minute variation, measurements were carried out over several days using a set of WAAREE WS-325 solar PV panels. These profiles were then normalized by dividing the instantaneous output power by the maximum daily output over all these days. These normalized profiles would then show how insolation might vary over the day so as to give a total daily output which would be less than the expected (clear day) value. These normalized profiles were then used to create a minute-by-minute profile of PV output over seven consecutive days such that their daily output agreed with that obtained from the Meteonorm database [32]. Figure 2a shows the modeled instantaneous PV output over seven consecutive days in the month of June. Salient numerical values of this output profile are listed in Table 1. It can be seen that the peak to average ratio of solar power is rather high at about 3.

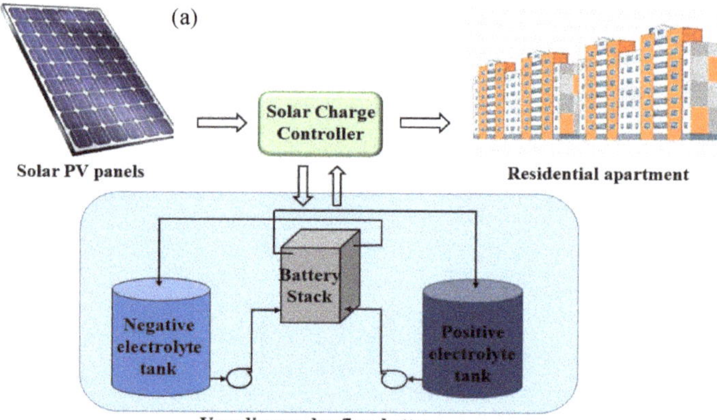

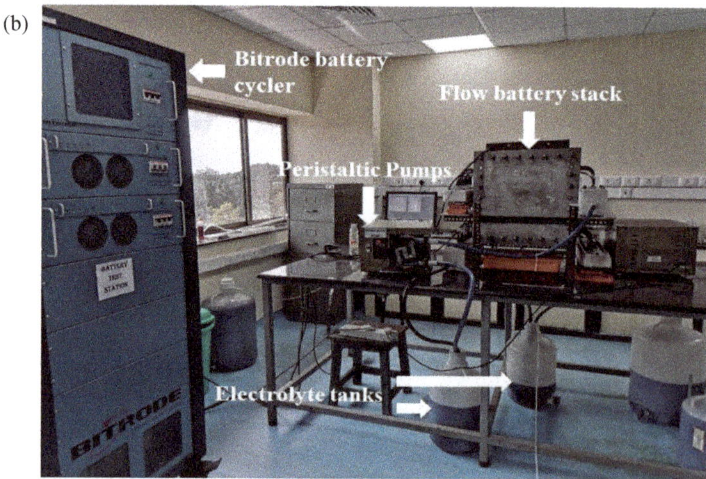

Figure 1. (**a**) Schematic diagram of the solar PV-flow battery-residential load integrated system. (**b**) Modelled system for the experimental study.

2.3. Construction of Residential Load Profile

Typical residential load demand variation in India has two noticeable time periods where the load requirement is high. The morning time of 6 a.m. to 10 a.m. where the residents are getting ready for school and office is a period of relatively high power consumption as is the time period from around 6 p.m. to 11 p.m. wherein the family comes back home to regroup. The power requirement during weekends is also more than that during weekdays. According to the weather conditions prevailing in India, the average household electricity consumption increases further during summer seasons as air conditioning is required. There will be a noticeable dip in the power consumption in the night during the winter months [33,34].

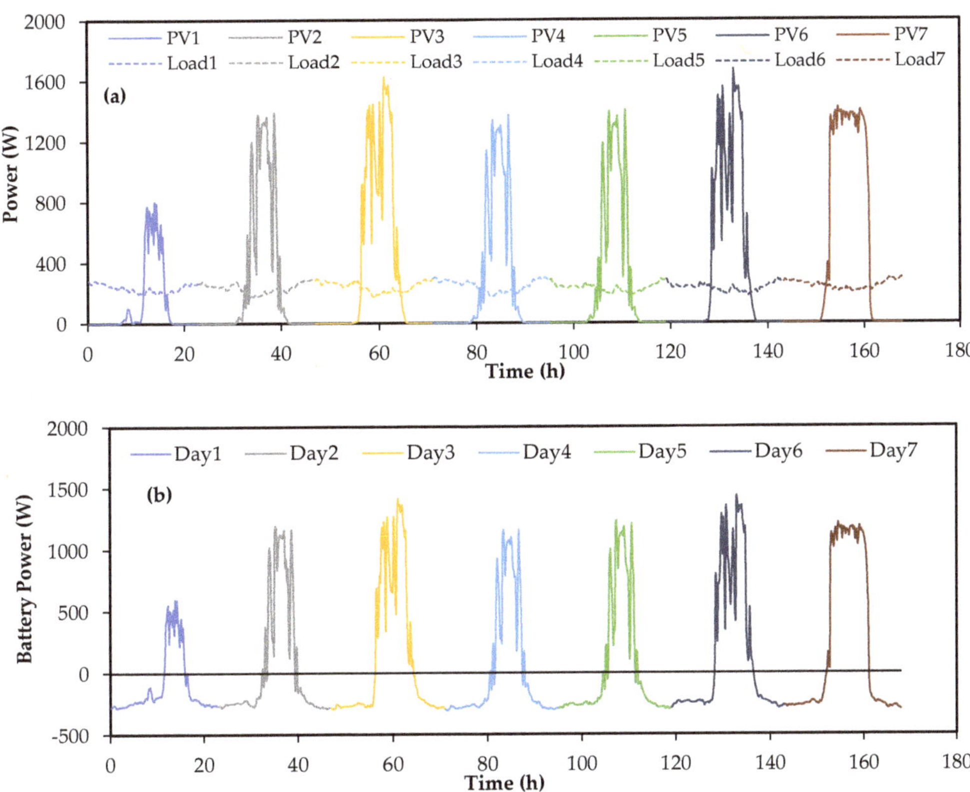

Figure 2. Seven day profile of June 2019 (**a**). Solar PV variation and load demand (**b**). Expected battery charge and discharge cycle.

Table 1. Summary of June 2019 profile (no scale down, $S_E = 1$).

PV	Load	Battery	
Power (W)	Power (W)	Charge Power (W)	Discharge Power (W)
Max–1670	Max–303	Max–1432	Max–303
Min–0	Min–156	Min–2.9	Min–5.9
Avg–640	Avg–238	Avg–769	Avg–238
Energy per day (Wh)	Energy per day (Wh)	Charge energy per day (Wh)	Discharge energy per day (Wh)
Max–11,110	Max–5944	Max–9240	Max–4316
Min–2600	Min–3716	Min–1720	Min–1840
Avg–7144	Avg–5713	Avg–5450	Avg–4029
Solar insolation to load (%)	23.4	Solar insolation to battery (%)	76.6%

In the present work, realistic variation of electricity consumption has been incorporated by sourcing data from real-time measurements (https://.prayaspune.org/ accessed on 15 January 2020 of the load requirement of a residential apartment complex of

120–145 dwellings in Pune, India [34,35]. Prayas energy group is a non-governmental and not-for-profit Indian organization. As part of one of its projects, the organization recorded real time power demand of residential apartment based in Pune for various appliances, weekends and weekdays over an extended period. The data collected at fifteen-minute intervals over a full week in June 2019 was used to construct the power demand profile. The data were numerically interpolated to generate minute-by-minute variation which is shown in Figure 2a in the form of average value per household. Because of aggregation over a large number of dwellings, the average consumption of power per dwelling does not show sudden spikes associated with the switching on or off of large power consuming devices such as the air conditioner or the washing machine [36].

Salient features of the load variation over the week are listed in Table 1. It can be seen that the maximum load per household over the seven-day duration is 303 W while the minimum is 156 W, that is, a ratio of about 2. This is in contrast to the PV output which has a maximum power of 1670 W, i.e., more than five times the maximum load. This wide disparity between source output and load is characteristic of solar PV-residential load applications. The battery needs to be capable of tackling these wide variations.

2.4. Salient Features of the Redox Flow Battery

An in-house vanadium redox flow battery stack has been used as the battery in the present study. The stack is rated at 1 kW power and has 8 cells of 900 cm^2 active area in each cell and features a specially designed flip-flop serpentine flow field [37] on the graphite plates. Nafion 117 membrane is used between the porous electrodes (SGL GFD4.6 4.6 mm thickness) with 35% compression as separator between anode and cathode. Full details of the stack construction can be found in [38–40]. Furthermore, 1.6 M vanadium electrolyte, procured from Oxkem, UK in an oxidation state of 3.5 is used as the electrical energy storage medium. In all the experiments, a constant flow rate of electrolyte, which corresponds to a stoichiometric factor of 9 at a current density of 60 mA/cm^2, was maintained on each side [22,39]. Comprehensive testing over a range of flow conditions and state of charge (SoC) shows that, over a wide range of SoC, the stack can deliver 1200 W in charging and 750 W in discharging when operated at a current density of 100 mA/cm^2. The maximum amount of energy stored in the battery depends on the electrolyte volume. Experiments have been conducted with electrolyte volumes of 30, 35 and 40 liters on each side. At a rated discharge energy of 25 Wh/litre, these electrolyte volumes amount to energy rating of 750 to 1000 Wh and about 50% more in charging.

A couple of runs have also been done with a commercially available lead-acid battery having a rating of 150 Ah, 12 V with a recommended C-rating of 10. Two of these batteries were connected in series thus giving a nominal power and energy rating of 360 W and 3.6 kWh, respectively for the lead-acid battery system. Compared to the VRFB, the lead-acid battery has one-third power and about four times energy storage capacity. Thus, the P/E ratio of the two storage systems varies by more than an order of magnitude.

2.5. Estimation of State of Charge (SoC) of the Battery

State of charge (SoC) of the battery, which represents the fractional energy remaining in the battery for discharge, is an important parameter in assessing battery integration with upstream and downstream equipment. In the present study, SoC has been estimated from measurements of the open circuit voltage (OCV) of the stack at frequent intervals during each run. A single cell using the same materials and methods used to construct the stack was assembled and charge-discharge cycling was done at constant current densities for a single VRFB cell with 1 liter of vanadium electrolyte each as positive and negative electrolyte. OCV of the cell was measured with specified amount of change in SoC intermittently during charge and discharge. At moderate current densities, the overpotential during charge and discharge at a particular SoC value was found to be nearly the same in the SoC range of 20 to 80%. An average of the two curves is taken to be OCV and these data have been used to obtain a correlation between SoC and OCV which then has been used to estimate the SoC dynamically using measured value of the OCV.

2.6. Set-Up of the Experimental Study

A comparison of the load and supply profiles over a seven-day period is shown in Figure 2a. Figure 2b shows the difference between the two; this is the power that would be drawn from or sent to the battery. As remarked earlier, the residential load is fairly constant compared to the large spikes in solar output. There is also considerable variation in the day-to-day energy output from the PV. The integrated PV-battery-load system should be able to work with these typical variations and still deliver as much power as demanded by the consumer at every moment. For the integrated system to perform effectively, the individual systems should be properly scaled so that the systems are not too oversized nor too undersized. The performance of the integrated system—in terms of being able to meet the load demand at any instant, being able to use PV output for either meeting the load or charging the battery and for the battery to operate at high charging /discharging efficiencies—and its relation to the sizing of the components is of interest to the present study.

Examination of the load and supply profiles shows that on the whole, there is considerable overlap between the two. In energy terms, this PV self-consumption (PVSC) amounts to about 23.4% of PV output that can go directly to the load during the sunshine hours; the rest of the 76.6% of load needs to be met through battery which is charged by the PV, assuming the system to be working independently of the grid. Keeping these numbers in view, the sizing of the PV and the battery is done as follows. Over the seven-day period in consideration, the residential load amounts to 39.9 kWh. Assuming the PV to be oversized by a factor of 25% (given the rather wide variations in the day-to-day energy output from the PV), the PV system is nominally rated at 50 kWh. The energy output per day is scaled as per Figure 2a which reflects measured day-to-day variation over a week in June 2019. The battery power profile is simulated as the difference between the power from PV and load demand and is shown in Figure 2b. This data (see also Table 2), with a scale factor, constitutes the input to Bitrode battery cycler to simulate the dynamic changes in PV-battery-load integrated system.

Table 2. Daily energy variations for PV, load and battery in Wh with an energy scaling factor, S_E, of 1/4 (C-charge, DC-discharge).

Cycle	PV	Load	Battery
1 half DC	47	697	653
1 C	650	232	−419
2 DC	51	1004	957
2 C	1545	373	−1173
3 DC	35	1065	1035
3 C	2066	421	−1648
4 DC	70	1090	1025
4 C	1494	354	−1142
5 DC	66	1095	1034
5 C	1521	317	−1205
6 DC	36	1005	973
6 C	2082	425	−1660
7 DC	52	979	931
7 C	2775	486	−2291
8 half DC	13	454	443

The experiments were designed to operate the battery with scaled down factors and varying electrolyte volumes to understand the stack performance for varying energy and power demand. Keeping in mind different power and energy characteristics of the two battery systems used in the present study, a power scale factor, S_P, an energy scale factor, S_E, and a time scale factor, S_T, have been varied independently while maintaining the relation that $S_E = S_P \times S_T$. Thus, having $S_P = 1/2$, $S_E = 1/2$ and $S_T = 1$ will simulate an integrated system with a total weekly solar energy input of $50/2 = 25$ kWh. If S_T is changed to $1/2$ while keeping S_P to be $1/2$, the energy input to the system will be reduced to 12.5 kWh over the week. Given that the VRFB system has a high power to energy ratio, for VRFB studies, experiments were conducted with S_P of 1 (full scale) or $1/2$, S_T of $1/2$ or $1/3$ or $1/5$ resulting in S_E of $1/4$ or $1/5$ or $1/6$. The lead-acid battery has a rated P/E of $1/10$. Therefore, it has been operated at S_P of $1/6$ and $1/8$ with S_T of 1 giving an S_E of $1/6$ or $1/8$. It may be noted that these scale factors refer only to traded energy; the inherent maximum energy storage capacity or power delivery capacity depend on the state of the battery and the operating conditions, including the volume of electrolyte and stack size and cell characteristics in the case of VRFB and nominal factory rating in case of the lead-acid battery.

In the experiments, the actual power vs. time profiles of the battery (shown typically as in Figure 2b) are imposed on the battery through a battery cycler such that the pre-determined power is held constant for 11 min in case of $S_T = 1$ (and 5.5 min in case of $S_T = 1/2$). The experiment is designed such that when the battery is not able to deliver the intended power, it goes to rest for a time period of 1 min. Open circuit voltage (OCV) of battery was also monitored over a one-minute rest period after every power step, whether or not the power step was a failure. The OCV data was later used to evaluate the battery efficiency characteristics. Such experiments have been conducted for different scale factors and electrolyte volumes. Results from these simulation runs, together with initial redox flow battery characterization study, are discussed in the next Section 3.

3. Results and Discussion

3.1. Battery Performance for the Baseline Case

As a baseline case, the VRFB stack was operated at power scale factor (S_P) of $1/2$ and time scale factor (S_T) of $1/2$ with electrolyte volume of 30 L in each tank. The 7-day battery profile shown in Figure 2b was discretized into constant power time steps of 5.5 min and was imposed on the stack using the battery cycler. OCV was measured after every time step over a period of 1 min during which the battery power was set to zero. The battery cycler was programmed in such a way that if the intended power during a particular time step could not be delivered to or drawn from the battery, then that power step would be aborted, and the testing would continue to the next step which would be the OCV measuring step. At the end of the OCV step, the next power step would be imposed. Thus, battery failure would be indicated by a power step that did not last the full duration of 5.5 min. During the whole experiment, the voltage and current across the battery were measured at 5 s intervals. Prior to each run, the battery was charged until the stack voltage reached the pre-set value of 1.7 V per cell.

The response of the battery to the imposed power steps is depicted in Figure 3a which shows the variation of power and SoC over the seven-day period. (It may be noted that the total duration of each experiment would depend on the time scale factor as well as on the number of failures and the number of OCV steps and would therefore change from run to run. For S_T of $1/2$, if the OCV measurement steps are discounted and if there are no charging or discharging failures, then the duration would be 84 h). The experiment started at midnight of the first day, and thus, with a discharge step and would then go continuously over a series of charging and discharging steps and would finish with a partial discharge of the 7th day. The SoC obtained using measured OCV is also plotted in the figure. One can see that the power variation follows the expected variation and that the steep changes in power are easily handled by the battery. The repeated variation of SoC over a wide range without lasting ill effects on the battery is also noteworthy. The full range of SoC of the battery is thus being utilized and this is an important feature of the vanadium flow battery.

One can also see in Figure 3a several instances of failure of the stack to deliver or absorb the required amount of power. Charging failures occur when the SoC is very high and discharging failures occur at very low SoC values. An expanded view of one such discharging failure is shown in Figure 3b which shows repeated failure to execute discharge power step over the simulation period of 13.6 h to 14.3 h. During this period, the SoC is particularly low (~9%) and the stack was unable to discharge even 120 W of power. (More power could have been extracted from the battery by increasing the electrolyte circulation rate or decreasing the power; however, such active management of the electrolyte was not done in the present study). An instance of battery failure during charging is shown in close-up view in Figure 3c which shows charging failure towards the end of charging on the 7th day over the period 74.3 to 75 h. One can see that the SoC is rather high at around 92%. Figure 3d shows the nomenclature used in the present study to characterize battery failure to deliver by time and energy. Typically, the stack will be able to meet the power demand for part of a power step and it may then fail for the rest of the duration of the time step. Cumulative amounts of time during which the power demand has not been met is used to determine battery failure by time. The shaded areas in Figure 3d represent the amount energy demand that has not been met. This is used to determine the cumulative battery failure % by energy.

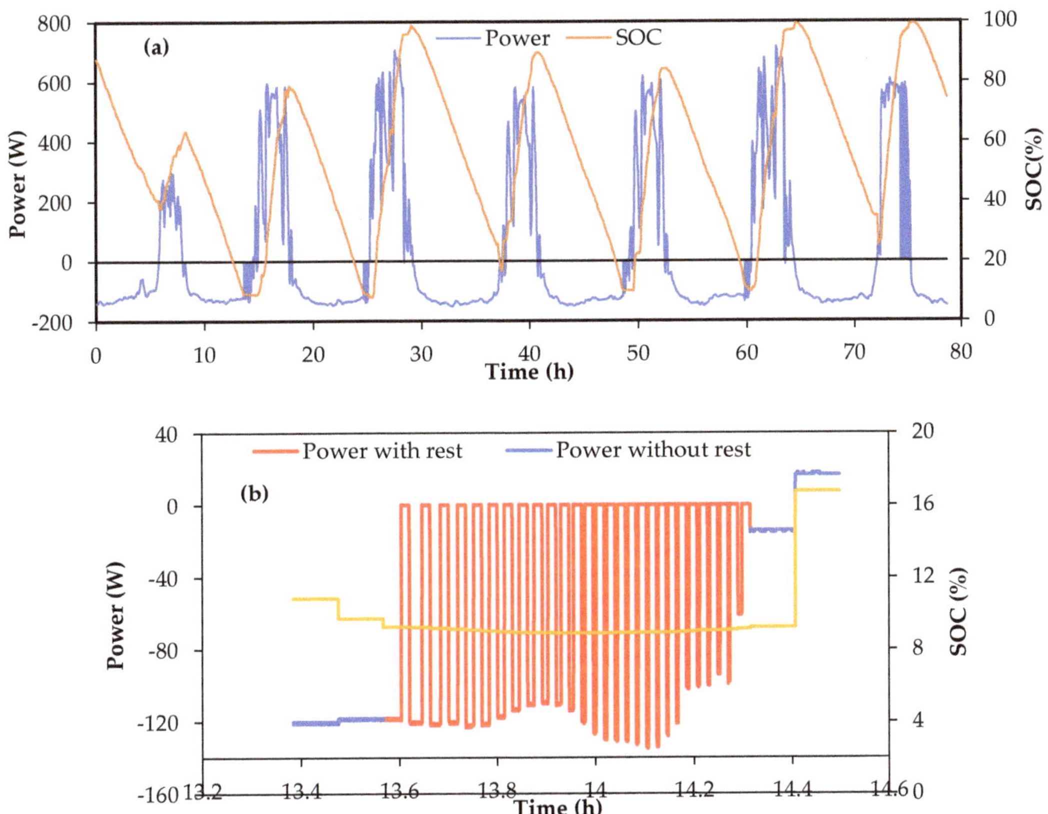

Figure 3. *Cont.*

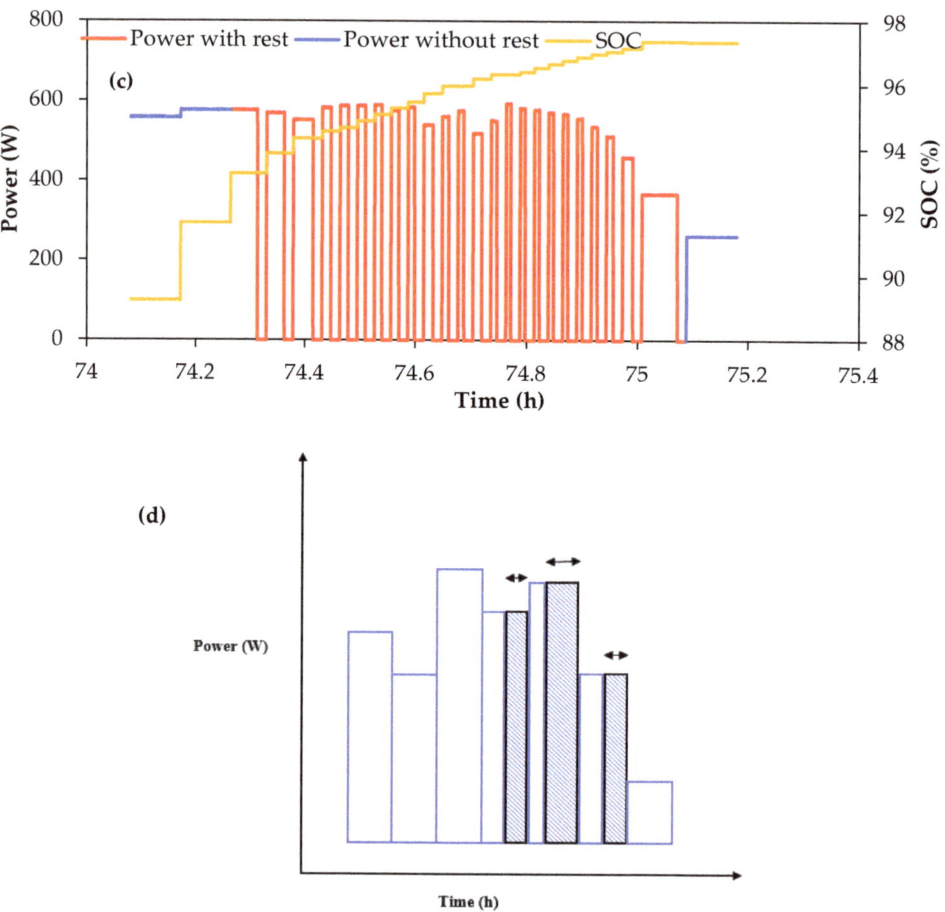

Figure 3. VRFB performance for $S_P = 1/2$ and $S_T = 1/2$ showing with 30 L electrolyte: (**a**) Complete 7-day profile with SoC variation. (**b**) Discharge failure in 1st discharge cycle. (**c**) Charge failure in 7th day charge. (**d**) Schematic diagram of battery with failure to deliver power and without failure.

The daily changes in energy flows during charge and discharge are listed in Table 2. One can see the amount of energy output from the battery during the 1st charge is only 419 Wh and the profile requires the battery to discharge up to 1100 Wh on certain days. Due to depletion of stored energy, its SOC becomes so low that it has been unable to deliver the load demand requirement. The battery is able to accommodate the charge energy from PV during the 2nd day of insolation but is not able to deliver again during the early morning of the 2nd night discharge as the intended discharge energy is 10% higher than what the battery has been able to absorb during the 2nd charge while the amount of energy put in during the 2nd day charging is only 10% higher than what battery is expected to discharge in the night. Third day's charging is of high energy content and the battery fails briefly towards the end of the charging time because the SoC has reached 95% and it is unable to accommodate the amount of energy coming in at the high power of 570 W. The high energy input during the third day means no discharge failures on the 3rd night. The relatively lower amounts of charge on the 4th and the 5th day of charging lead to discharge failures on the succeeding discharging events. The amount of energy charged on the 6th day is the highest among the seven-day profile; as a result, the battery gets fully charged resulting in no discharge failure on the 6th day. The last day's charge too is considerable and due to the high SoC at the beginning of charge (~25%), the battery soon gets fully charged leading to extensive charging failure on the last day and no discharge failure during the last night.

Table 3 shows the details of energy delivered and charged for various charge and discharge cycles. Energy-based % failure and time-based % failure for each case is also tabulated. Minor (1–2%) energy-based failures without corresponding time-based failures can be neglected as errors due to fluctuations in battery power supplied by the battery cycler. One can see that significant discharge failure has occurred in 1st, 2nd, 4th, and 5th discharges and significant charge failures in the 3rd, 6th, and 7th charges. The first discharge failure may be attributed to the exceptionally low insolation on the first day from an already depleted battery and is thus caused by initial conditions of the set-up. (In view of this, subsequent runs have been done with second day's half discharge.) The 3rd and the 4th discharge failures may be attributed to the large amount of discharge that needed to be done. The more severe 5th discharge failure is somewhat surprising as the amount of discharge energy is smaller and the amount of charge energy in the preceding charge is higher. However, as can be seen in Figure 3a, the SoC at the beginning of discharge is lower than that at the corresponding stage of the 3rd and 4th discharges. Thus, discharge failures are associated with low SoC rather than high power under typical residential load conditions. In contrast, charge failures are a combination of high power and high SoC (compare the conditions of the 3rd, 6th, and 7th day charge failures in Figure 3a) as the charging power is significantly higher than the average discharging power in solar PV-residential load integration applications.

Table 3. VRFB energy during charge/discharge cycle 30 L for $S_P = 1/2$, $S_T = 1/2$.

Cycles	Intended Wh	Experimental Wh	Energy Based % Failure	Time Based % Failure
0th Discharge	653	643	2	0
1st Charge	419	415	1	0
1st Discharge	957	677	29	28
2nd Charge	1173	1172	0	0
2nd Discharge	1035	850	18	16
3rd Charge	1648	1606	3	1
3rd Discharge	1025	1011	1	0
4th Charge	1142	1136	1	0
4th Discharge	1034	977	5	5
5th Charge	1205	1199	0	0
5th Discharge	973	877	10	8
6th Charge	1660	1550	7	4
6th Discharge	931	912	2	0
7th Charge	2291	1288	44	40
7th Discharge	443	437	1	0

3.2. Power and Energy Scaling Studies

In order to study the influence of electrolyte volume, and hence the battery system's energy storage capacity, on the dynamic performance of the integrated system, experiments have been carried out with the electrolyte volume increased from 30 liters to 35 liters on each side and then once again from 35 liters to 40 liters. The same stack operated with the same voltage limits and electrolyte circulation rates was used in these simulations. Thus, compared to the baseline case, the power scaling remains the same but energy storage capacity has been increased by 16.6% and 33%, respectively. As mentioned earlier, the ordering of the days has been changed from 1-2-3-4-5-6-7 to 2-3-4-5-6-7-1 to remove the anomalous influence of the starting day. The battery response in terms of power and SoC in these two cases is compared in Figure 4. It can be seen the first day discharging failures have disappeared but mid-week discharge failures are still there, though with reduced intensity, especially in the case with 40 L electrolyte volume on each side. In both cases, the mid-week discharge failures are associated with the battery reaching very low SoC despite increased storage capacity compared with the baseline case. Thus, increasing the electrolyte volume is beneficial in reducing discharge failures. Charge failures too have come down significantly, especially in the 40 L case in which instances of high SoC have reduced considerably.

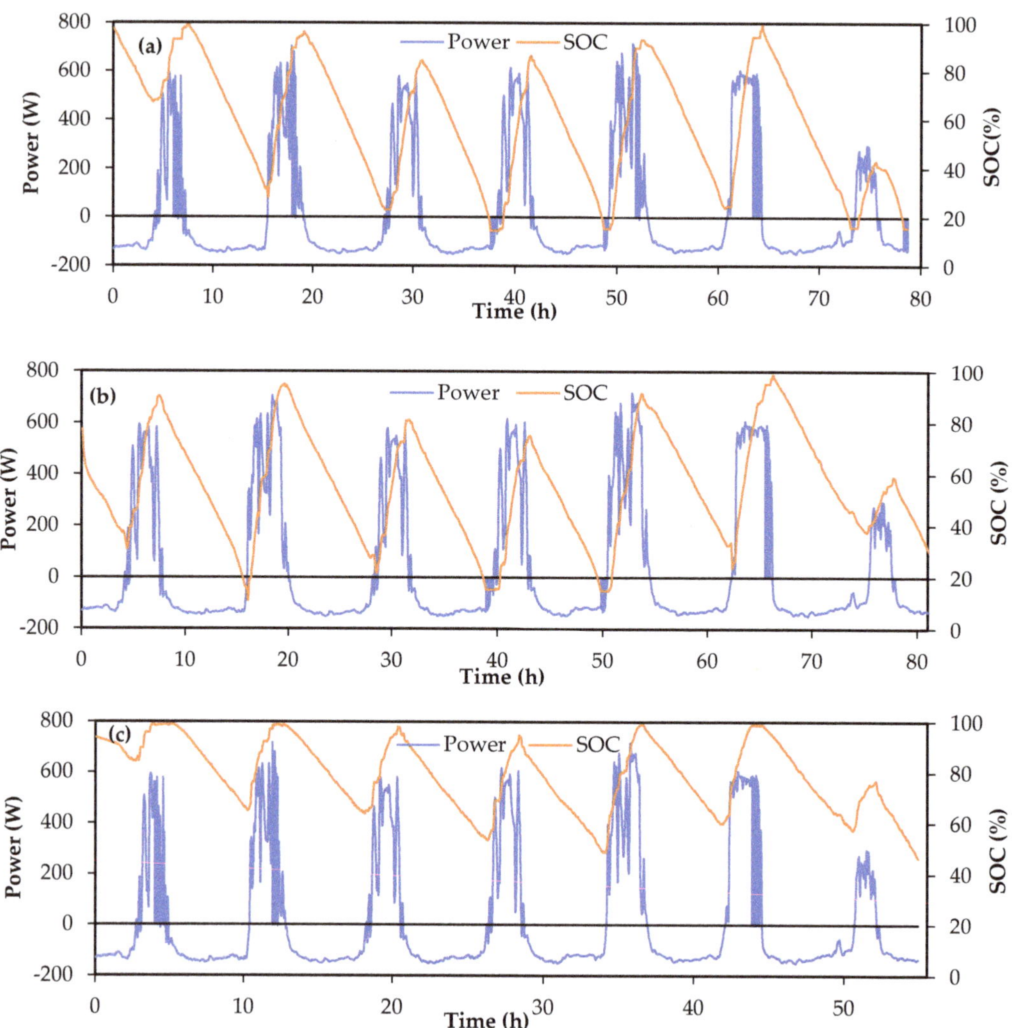

Figure 4. Battery performance with SoC variation for $S_P = 1/2$ and $S_T = 1/2$ for (**a**) 35 L electrolyte (**b**) 40 L electrolyte and (**c**) with 40 L electrolyte with SOC variation for $S_P = 1/2$ and $S_T = 1/3$.

In the above two cases, the total energy intended to be traded is the same as that in the baseline case; only the energy storage capacity of the battery system is increased by increasing the electrolyte volume. Figure 4c shows the response for a stack with 40 L of electrolyte which is run on the same power scaling of $S_P = 1/2$ but with a time scaling of $S_T = 1/3$. This system has the same energy storage capacity as that of Figure 4b but the energy traded is reduced by 1/3rd. Thus, this experiment is equivalent to $S_P = 1/2$ with $S_T = 1/2$ scaling with 60 L of electrolyte at initial SoC of around 60% for the same battery stack. As can be seen from Figure 4c, there are no discharge failures as the SoC remains comfortably high (>40%) throughout the equivalent seven-day cycle. However, the battery has more charging failures because the SoC reaches high values (>90%) repeatedly. However, this failure to charge does not translate to failure to meet load demand because the system has high storage capacity (twice as much as that of the reference case).

Figure 5 compares the energy traded (charged or discharged) in every charge/discharge cycle over the seven days for the three cases shown in Figure 4. In addition, given here is the intended amount of energy to be traded in each cycle. All the three cases (of 35 L with $S_P = 1/2$ and $S_T = 1/2$, 40 L $S_P = 1/2$ and $S_T = 1/2$ and 40 L with $S_P = 1/2$ and $S_T = 1/3$) have been non-dimensionalized by dividing the energy by the maximum energy traded in that seven-day period (this corresponds to the charging energy on the 6th day). It can be seen that the combination of (40 L, $S_P = 1/2$ and $S_T = 1/3$) is able to deliver discharge energy as intended over the entire period; however, there is considerable wastage of PV output. This is borne out by the SoC variation which is mostly in the range of 60 to 99% showing underutilization of the electrolyte. The intermediate case of (40 L, $S_P = 1/2$ and $S_T = 1/2$) is better at utilizing the PV output with occasional discharge failure while the case with (35 L, $S_P = 1/2$ and $S_T = 1/2$) may be said to be undersized with respect to energy storage capacity. Proper energy sizing of the energy system is necessary to optimally use its storage capacity.

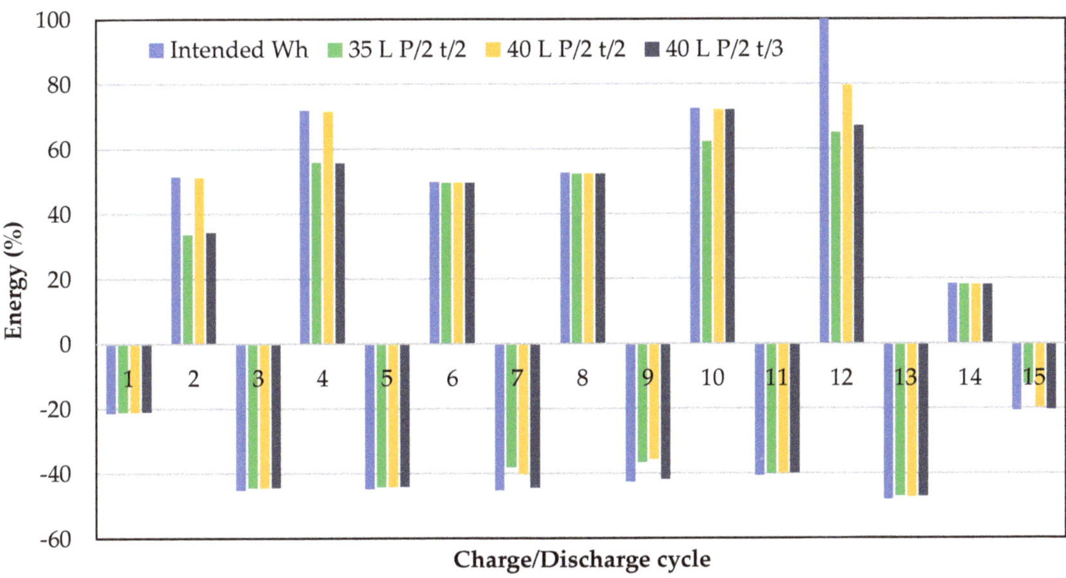

Figure 5. Comparative plot showing traded energy (%) in each charge/discharge cycle over a seven-day period for a VRFB with (electrolyte volume, S_P and S_T) combinations of (35 L, 1/2, 1/2), (40 L, 1/2, 1/2) and (40 L, 1/2, 1/3).

In order to bring out the effect of stack sizing, a further simulation has been carried with an electrolyte volume of 40 L but with $S_P = 1$. In view of the limited energy storage capacity associated with 40 L electrolyte volume, the time scale factor is reduced such that $S_T = 1/5$ leading to S_E of 1/5 compared to that of 1/4 for the baseline case. The reduced energy scaling is consistent with anticipated higher efficiency losses associated with higher powers of charge and discharge due to doubling of power compared to the baseline case.

The response of the battery system for the first two days is summarized in Figure 6. One can see that high power charging leads to significantly higher overpotential so that charging failure occurs on the first day itself even for SoC < 80% and the SoC remains less than 90% at the end of first day's charging. As a result, there is a hint of failure at the far end of the first full discharge. Due to high charging losses, the SoC does not even reach 80% at the end of 2nd day's charging. This is in contrast to the corresponding case in Figure 4b where the SoC is well in excess of 90% at the corresponding stage. Coupled with this low SoC, higher discharge losses due to higher average discharge power leads to extensive failure in the next discharge cycle (not shown) which has a knock-on effect on subsequent charge and discharge cycles, each of which suffers from extended periods of battery failure to meet power demand. This case, which is representative of an undersized stack, thus illustrates the importance of power sizing of the integrated system. The battery should be able to tackle high charging powers; otherwise, subsequent discharge and charge cycles will suffer.

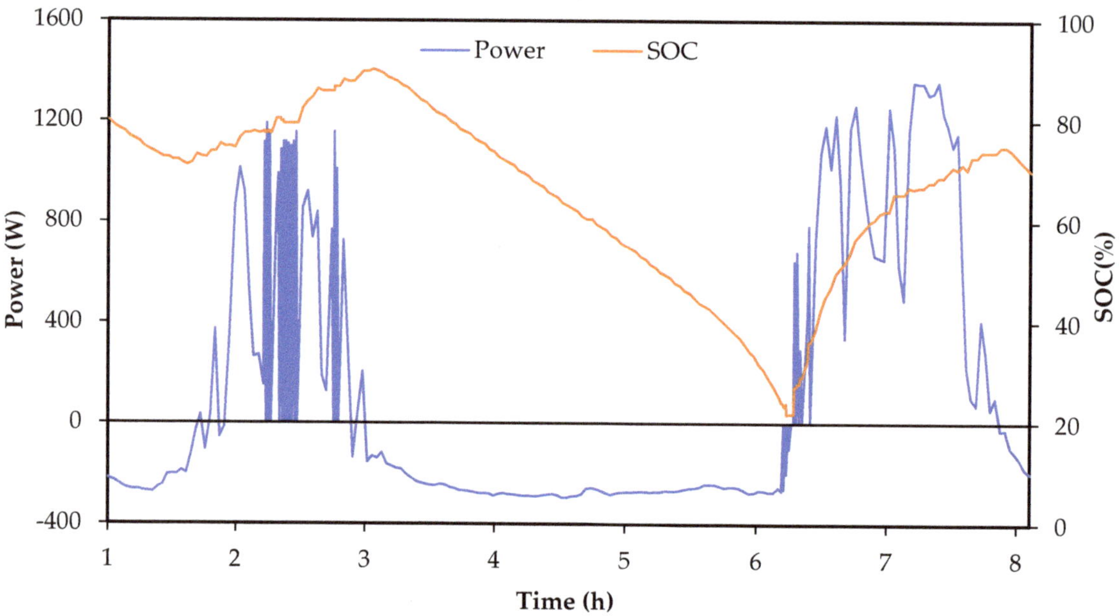

Figure 6. Battery performance with 40 L electrolyte with SoC variation for $S_P = 1$ and $S_T = 1/5$.

3.3. Charging and Discharging Efficiency

From measured OCV at the end of every power step, the average overpotential for every step could be determined. From this, the charging efficiency and discharging efficiency could be computed over the entire seven-day period for a given integrated system. For a given VRFB stack operating at a constant electrolyte circulation rate, both efficiencies depend primarily on the power at which charge or discharging takes places and the SoC of the electrolyte, both of which vary dynamically in a given situation. Figure 7 shows the data of charging and discharging efficiencies over the seven-day period wherein data from for all the three cases shown in Figure 4 are plotted together. One can see that the discharge efficiency is not strongly influenced by power (probably because the discharging power is low and varies in a rather narrow range) and remains relatively high (~0.95) for SoC > 30%. On the contrary, the charging efficiency is a strong function of power and almost varies linearly with charging power. Since charging power is about three times higher than the discharging power, the average charging efficiency is lower at about 0.9.

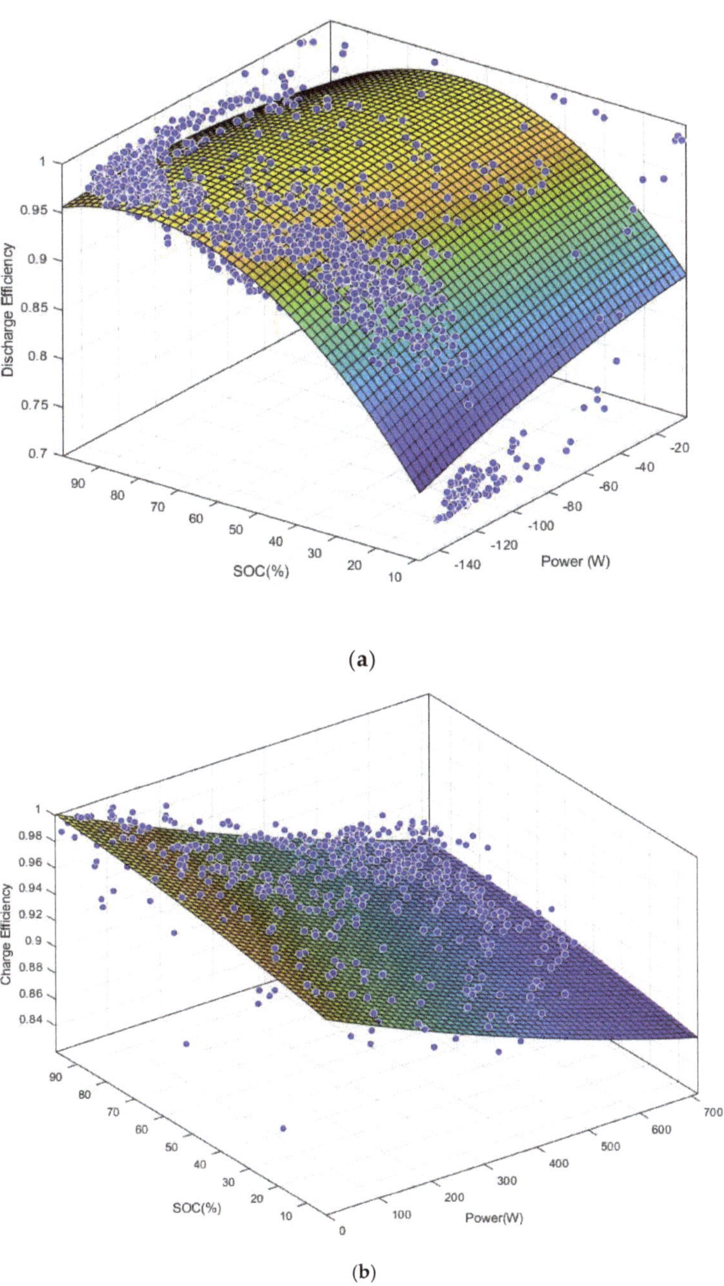

Figure 7. Efficiency variation with SoC and operating power of VRFB for (**a**) discharging and (**b**) charging for the three runs shown in Figure 4.

3.4. Lead Acid Battery Performance

The same 7-day profile is operated using a lead-acid battery for understanding its performance to dynamically changing requirements. A new lead-acid battery of 150 Ah capacity at 12 V was used for these experiments. Since the commercial lead acid battery has a power rating of 180 W and energy rating of 1800 Wh, a power scaling factor, S_P, of 1/8 and a time scale factor, S_F, of 1 were used giving energy scaling factor, S_E, of 1/8. Since $S_T = 1$, each power step lasted 11 min; this was followed by a one-minute OCV measurement step. Figure 8 shows the battery response for the seven-day profile in terms of power step by power step response (Figure 8a) and in terms of normalized cumulative energy traded in each charge/discharge step (Figure 8b) where the data for the VRFB case of 40 L electrolyte volume, $S_P = 1/2$ and $S_T = 1/2$ are also given. One can see from Figure 8a that the lead-acid battery suffers charging failures on each of the first six days; however, due to its high energy storage capacity (and favorable power scaling), it shows no failures in discharge. Figure 8b shows that in comparison with VRFB, the energy charged in each cycle is significantly less in charging cycle. On the other hand, it outperforms the VRFB in discharge as it is able to meet the discharge demand in every cycle. However, it must be kept in mind that the lead acid battery case has an energy scaling factor of 1/8 whereas S_E for the VRFB is 1/4. Given that the rated energy storage capacity of the former is 1800 Wh compared to about 1000 Wh for the latter, the lead acid battery system is grossly underutilizing its energy storage capacity.

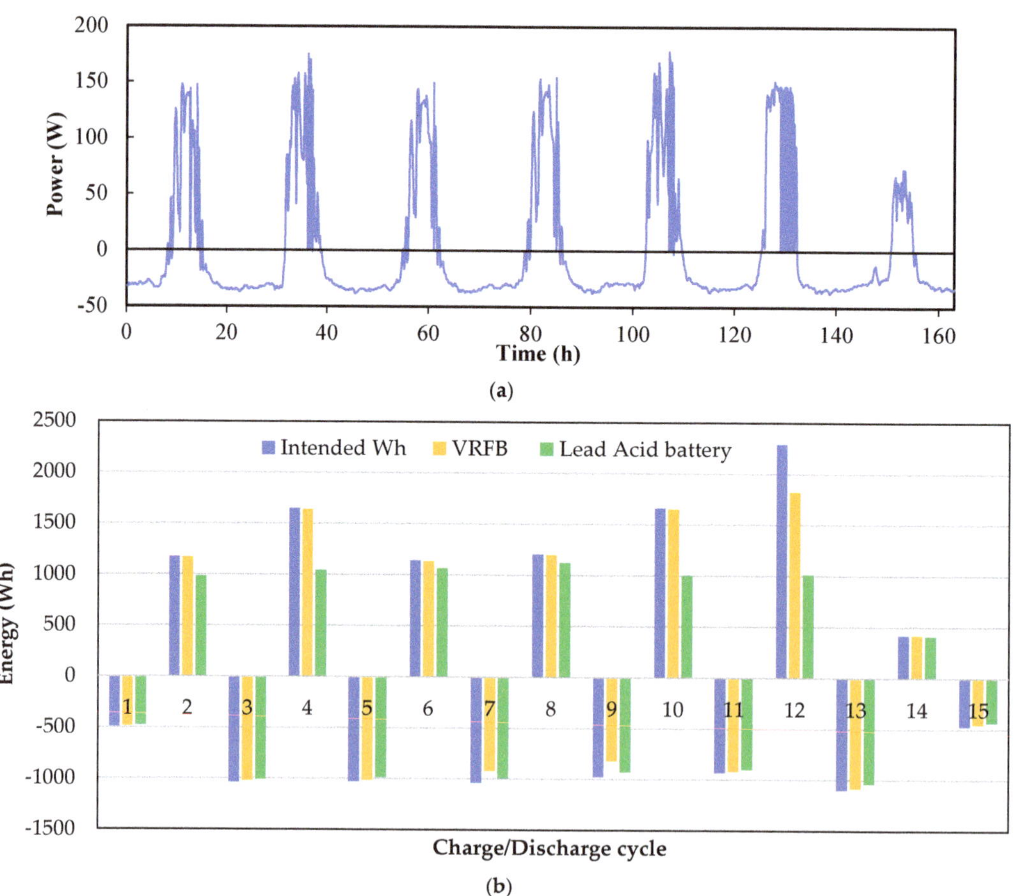

Figure 8. (**a**) Lead acid battery performance for $S_P = 1/8$ and $S_T = 1$. (**b**) Energy comparison for VRFB with 40 L with $S_P = 1/2$ and $S_T = 1/2$ with 2 lead acid battery performance.

It may further be noted that the VRFB system worked well for about 50 consecutive charge + discharge cycles without noticeable degradation in performance whereas the new lead acid battery showed signs of rapid degradation. After three seven-day profile experiments were conducted with S_P of 1/8, 1/6 and 1/4, the battery was found to have retained a capacity of only 50 Ah. When the seven-day profile was repeated with S_P of 1/8, the reduced capacity was found to have led to extensive failures in both charging and discharging. Although this cannot be considered as a well-controlled study of degradation behavior of the lead acid battery (which was procured off-the-shelf from a commercial store), the VRFB system does have the advantage of long life under repeated cycles which is necessary for solar PV applications.

4. Conclusions

A 1 kW-rated 8-cell stack of 900 cm^2 cell active area vanadium redox flow battery has been investigated to serve as electrical energy storage system in a conceptually integrated PV-battery-residential load system intended to run as a stand-alone system. A series of seven-day profile runs of battery power have been made to simulate different energy storage and power delivery capacities. The following conclusions can be drawn from the study:

1. Typical insolation and load profiles for an integrated solar PV-battery-residential load system have about a quarter of the PV output going directly to meet the load demand during sunshine hours. This means that nearly three-quarters of energy flow to the load occurs through the battery. Account must therefore be taken of charging and discharging energy efficiencies.
2. The seven-day power profiles for the integrated system show the vast differences between charging and discharging conditions for the battery of such an integrated system. Average charging power is about three times that of average discharge power. Further, due to the different nature of variations of solar insolation and aggregated residential load, ratio of peak power in charging to that in discharging of the battery is 4.7. Thus, charging conditions are much more severe than discharging. The stack power rating should therefore be based on the charging condition during peak solar insolation. Too low stack power rating can lead to considerable charging failures which can subsequently translate into discharging failures.
3. For residential load applications, the daily energy variation is not highly variable. The energy rating of the VRFB system, i.e., electrolyte volume, should be based on the maximum daily discharge load and the range of operable SoC.
4. Due to the rather mild discharging conditions of the battery, favorable discharging efficiencies can be maintained in a properly-sized stack for SoC variations in the range of 20 to 90%. A stack operating over this range may be expected to have, given that it is designed for harsh charging conditions, net discharge efficiency of 90% or higher.
5. The sizing of the PV plant should be based on both charging and discharging efficiencies of the battery. Given that the stack should be designed for peak charging power, and given that the peak-to-average charging power is nearly two, the stack charging efficiency is likely to between 85 to 90%. With discharge efficiency being in the range of 90 to 95%, a net round trip energy loss of about 80% may be expected.
6. Considering the fact that about 75% energy flow occurs through the battery (with a round-trip efficiency of 80%) with the rest going through to load directly with a considerably higher efficiency, and making some allowance for failure to charge, the PV plant should be rated at about 25% more energy than the average daily energy demand from the load.

The present analysis does not include transmission and distribution losses and consideration of these may increase about 10% power and energy storage requirements. In summary, it can be concluded that the ability to design a VRFB with a high ratio of power-to-energy makes it particularly attractive for PV-load integration.

Author Contributions: P.P.: conceptualization, methodology, software, data curation, visualization, writing—original draft. R.G.: conceptualization, methodology, visualization, data curation. S.J.: conceptualization, data curation, validation, writing—review and editing, supervision, funding acquisition. All authors have read and agreed to the published version of the manuscript.

Funding: This research was funded by grants from MHRD (Grant reference no. F.NO.41-2/2015-T.S.-I (Pt.)) and DST-Solar Energy Harnessing Centre (Grant reference no. DST/TMD/SERI/HUB/1(C)), both from the Government of India.

Institutional Review Board Statement: Not applicable.

Informed Consent Statement: Not applicable.

Data Availability Statement: Not applicable.

Acknowledgments: Financial supports from the Ministry of Human Resource Department and Department of Science and Technology, both from the Government of India, are gratefully acknowledged. The authors also acknowledge the load data made available by eMARC, Prayas Energy group for case studies and analysis performed in the present work.

Conflicts of Interest: The authors declare no conflict of interest.

References

1. Global Renewables Outlook: Energy Transformation 2050, International Renewable Energy Agency. 2020. Available online: https://www.irena.org/publications/2020/Apr/Global-Renewables-Outlook-2020 (accessed on 20 June 2020).
2. Duffie, A.J.; Beckman, A.W. *Solar Engineering of Thermal Processes*; John Wiley & Sons: Hoboken, NJ, USA, 2013.
3. Gur, T.M. Review of electrical energy storage technologies, materials and systems: Challenges and prospects for large scale grid storage. *Energy Environ. Sci.* **2018**, *11*, 2696–2967. [CrossRef]
4. Sufyan, M.; Rahim, N.A.; Aman, M.M.; Tan, C.K.; Raihan, S.R.S. Sizing and applications of battery energy storage technologies in smart grid system: A review. *J. Renew. Sustain. Energy* **2019**, *11*, 014105. [CrossRef]
5. Liu, J.; Chen, X.; Cao, S.; Yang, H. Overview on hybrid solar photovoltaic-electrical energy storage technologies for power supply to buildings. *Energy Convers. Manag.* **2019**, *187*, 103–121. [CrossRef]
6. Huggins, R.A. *Energy Storage: Fundamentals, Materials and Applications*; Springer: Berlin/Heidelberg, Germany, 2016.
7. Whitehead, A.H.; Rabbow, T.J.; Trampert, M.; Pokorny, P. Critical safety features of vanadium redox flow battery. *J. Power Sources* **2017**, *351*, 1–7. [CrossRef]
8. Angenendt, G.; Zurmühlen, S.; Axelsen, H.; Sauer, D.U. Comparison of different operation strategies for PV battery home storage systems including forecast-based operation strategies. *Appl. Energy* **2018**, *229*, 884–899. [CrossRef]
9. Litjens, G.B.M.A.; Worrell, E.; van Sark, W.G.J.H.M. Assessment of forecasting methods on performance of photovoltaic-battery systems. *Appl. Energy* **2018**, *221*, 358–373. [CrossRef]
10. Jing, W.; Lai, C.H.; Wong, W.S.; Wong, M.D. A comprehensive study of battery-supercapacitor hybrid energy storage system of standalone PV power system in rural electrification. *Appl. Energy* **2018**, *224*, 340–356. [CrossRef]
11. Gharavi, H.; Ardehali, M.; Ghanbari-Tichi, S. Imperial competitive algorithm optimization of fuzzy multi-objective design of a hybrid green power system with considerations for economics, reliability, and environmental emissions. *Renew. Energy* **2015**, *78*, 427–437. [CrossRef]
12. Hadidian-Moghaddam, M.J.; Arabi-Nowdeh, S.; Bigdeli, M. Optimal sizing of a stand-alone hybrid photovoltaic/wind system using new grey wolf optimizer considering reliability. *J. Renew. Sustain. Energy* **2016**, *8*, 035903. [CrossRef]
13. Hakimi, S.M.; Moghaddas-Tafreshi, S.M. Optimal sizing of a stand-alone hybrid power system via particle swarm optimization for Kahnouj area in south-east of Iran. *Renew. Energy* **2009**, *34*, 1855–1862. [CrossRef]
14. Ekren, O.; Ekren, B.Y. Size optimization of a PV/wind hybrid energy conversion system with battery storage using simulated annealing. *Appl. Energy* **2010**, *87*, 592–598. [CrossRef]
15. Askarzadeh, A. A discrete chaotic harmony search-based simulated annealing algorithm for optimum design of PV/wind hybrid system. *Sol. Energy* **2013**, *97*, 93–101. [CrossRef]
16. Moghaddam, M.J.H.; Kalam, A.; Nowdeh, S.A.; Ahmadi, A.; Babanezhad, M.; Saha, S. Optimal sizing and energy management of stand-alone hybrid photovoltaic/ wind system based on hydrogen storage considering LOEE and LOLE reliability indices using flower pollination algorithm. *Renew. Energy* **2019**, *135*, 1412–1434. [CrossRef]
17. Fetanat, A.; Khorasaninejad, E. Size optimization for hybrid photovoltaic–wind energy system using ant colony optimization for continuous domains based integer programming. *Appl. Soft Comput.* **2015**, *31*, 196–209. [CrossRef]
18. Ajlan, A.; Tan, C.W.; Abdilahi, A.M. Assessment of environmental and economic perspectives for renewable-based hybrid power system in Yemen. *Renew. Sustain. Energy Rev.* **2017**, *75*, 559–570. [CrossRef]
19. Lu, P.; Leung, P.; Su, H.; Yang, W.; Xu, Q. Materials, performance, and systems design for integrated solar flow batteries—A mini review. *Appl. Energy* **2021**, *282*, 116210. [CrossRef]

20. Minke, C.; Turek, T. Materials, System designs and modelling approaches in techno-economic assessment of all-vanadium redox flow batteries—A review. *J. Power Sources* **2018**, *376*, 66–81. [CrossRef]
21. Akter, P.; Li, Y.; Bao, J.; Skyllas-Kazacos, M.; Rahman, M.F. Optimal Charging of Vanadium Redox Flow Battery with Time-Varying Input Power. *Batteries* **2019**, *5*, 20. [CrossRef]
22. Gundlapalli, R.; Jayanti, S. Comparative Study of Kilowatt-Scale Vanadium Redox Flow Battery Stacks Designed with Serpentine Flow Fields and Split Manifolds. *Batteries* **2021**, *7*, 30. [CrossRef]
23. García-Quismondo, E.; Almonacid, I.; Martinez, M.A.C.; Miroslavov, V.; Serrano, E.; Palma, J.; Salmerón, J.P.A. Operational Experience of 5 kW/5 kWh All-Vanadium Flow Batteries in Photovoltaic Grid Applications. *Batteries* **2019**, *5*, 52. [CrossRef]
24. Mongird, K.; Viswanathan, V.; Alam, J.; Vartanian, C.; Sprenkle, V.; Baxter, R. *2020 Grid Energy Storage Technology Cost and Performance Assessment*; Technical Report No. DOE/PA-0204; US Department of Energy: Washington, DC, USA, 2020.
25. Zhang, Y.; Liu, L.; Xi, J.; Wu, Z.; Qiu, X. The benefits and limitations of electrolyte mixing in vanadium flow batteries. *Appl. Energy* **2017**, *204*, 373–381. [CrossRef]
26. Wang, K.; Liu, L.; Xi, J.; Wu, Z.; Qiu, X. Reduction of capacity decay in vanadium flow batteries by an electrolyte-reflow method. *J. Power Sources* **2017**, *338*, 17–25. [CrossRef]
27. Sun, C.; Negro, E.; Nale, A.; Pagot, G.; Vezzù, K.; Zawodzinski, T.A.; Meda, L.; Gambaro, C.; Di Noto, V. An efficient barrier toward vanadium crossover in redox flow batteries: The bilayer [Nafion/(WO3)x] hybrid inorganic-organic membrane. *Electrochim. Acta* **2021**, *378*, 138133. [CrossRef]
28. Bhattacharjee, A.; Saha, H. Design and experimental validation of a generalized electrical equivalent model of Vanadium redox flow battery for interfacing with renewable energy sources. *J. Energy Storage* **2017**, *13*, 220–232. [CrossRef]
29. Zhang, X.; Li, Y.; Skyllas-Kazacos, M.; Bao, J. Optimal Sizing of Vanadium Redox Flow Battery Systems for Residential Applications Based on Battery Electrochemical Characteristics. *Energies* **2016**, *9*, 857. [CrossRef]
30. Sarkar, T.; Bhattacharjee, A.; Samanta, H.; Bhattacharya, K.; Saha, H. Optimal design and implementation of solar PV-wind-biogas-VRFB storage integrated smart hybrid microgrid for ensuring zero loss of power supply probability. *Energy Convers. Manag.* **2019**, *191*, 102–118. [CrossRef]
31. Behi, B.; Baniasadi, A.; Arefi, A.; Gorjy, A.; Jennings, P.; Pivrikas, A. Cost-benefits analysis of a virtual power plant including solar PV, flow battery, heat pump, and demand management: A western Australia case study. *Energies* **2020**, *13*, 2614. [CrossRef]
32. Solar Park 2000 MW Capacity in the State of Karnataka, Detailed Project Report. 2015. Available online: https://cdkn.org/wp-content/uploads/2017/07/Up-Karnataka-Solar-park-DPR-1.pdf (accessed on 10 November 2019).
33. Ali, S. The Future of Indian Electricity Demand: How Much, by Whom, and under What Conditions, Brooking India. 2018. Available online: https://www.brookings.edu/wp-content/uploads/2018/10/The-future-of-Indian-electricity-demand.pdf (accessed on 20 December 2019).
34. Available online: http://emarc.watchyourpower.org/energycurve.php (accessed on 20 December 2019).
35. Chunekar, A.; Sreenivas, A. Towards an understanding of residential electricity consumption in India. *Build. Res. Inf.* **2018**, *47*, 75–90. [CrossRef]
36. Walker, A.; Kwon, S. Analysis on impact of shared energy storage in residential community: Individual versus shared energy storage. *Appl. Energy* **2020**, *282*, 116172. [CrossRef]
37. Gundlapalli, R.; Jayanti, S. Effective splitting of serpentine flow field for applications in large-scale flow batteries. *J. Power Sources* **2021**, *487*, 229409. [CrossRef]
38. Jayanti, S.; Gundlapalli, R.; Chetty, R.; Jeevandoss, C.R.; Ramanujam, K.; Monder, D.S.; Rengaswamy, R.; Suresh, P.V.; Swarup, K.S.; Varadaraju, U.V.; et al. Characteristics of an Indigenously Developed 1 KW Vanadium Redox Flow Battery Stack. In *Proceedings of the 7th International Conference on Advances in Energy Research, Mumbai, India, 10–12 December 2019*; Springer: Singapore, 2021.
39. Gundlapalli, R.; Kumar, S.; Jayanti, S. Stack Design Considerations for Vanadium Redox Flow Battery. *INAE Lett.* **2018**, *3*, 149–157. [CrossRef]
40. Gundlapalli, R.; Jayanti, S. Effect of electrode compression and operating parameters on the performance of large vanadium redox flow battery cells. *J. Power Sources* **2019**, *427*, 231–242. [CrossRef]

Article

Non-Uniform Circumferential Expansion of Cylindrical Li-Ion Cells—The Potato Effect

Jessica Hemmerling [1,*], Jajnabalkya Guhathakurta [2], Falk Dettinger [1], Alexander Fill [1] and Kai Peter Birke [1]

[1]   Institute for Photovoltaics, Electrical Energy Storage Systems, University of Stuttgart, Pfaffenwaldring 47, 70569 Stuttgart, Germany; falkdett@web.de (F.D.); alexander.fill@ipv.uni-stuttgart.de (A.F.); Peter.Birke@ipv.uni-stuttgart.de (K.P.B.)

[2]   CT-Lab Stuttgart, Nobelstraße 15, 70569 Stuttgart, Germany; guhathakurta@ct-lab-stuttgart.de

*   Correspondence: Jessica.Hemmerling@ipv.uni-stuttgart.de

Abstract: This paper presents the non-uniform change in cell thickness of cylindrical Lithium (Li)-ion cells due to the change of State of Charge (SoC). Using optical measurement methods, with the aid of a laser light band micrometer, the expansion and contraction are determined over a complete charge and discharge cycle. The cell is rotated around its own axis by an angle of $\alpha = 10°$ in each step, so that the different positions can be compared with each other over the circumference. The experimental data show that, contrary to the assumption based on the physical properties of electrode growth due to lithium intercalation in the graphite, the cell does not expand uniformly. Depending on the position and orientation of the cell coil, there are different zones of expansion and contraction. In order to confirm the non-uniform expansion around the circumference of the cell in 3D, X-ray computed tomography (CT) scans of the cells are performed at low and at high SoC. Comparison of the high resolution 3D reconstructed volumes clearly visualizes a sinusoidal pattern for non-uniform expansion. From the 3D volume, it can be confirmed that the thickness variation does not vary significantly over the height of the battery cell due to the observed mechanisms. However, a slight decrease in the volume change towards the poles of the battery cells due to the higher stiffness can be monitored.

Keywords: lithium-ion battery cell; volumetric expansion; mechanical degradation; state of charge dependency; cell thickness; mechanical aging; non-uniform volume change

Citation: Hemmerling, J.; Guhathakurta, J.; Dettinger, F.; Fill, A.; Birke, K.P. Non-Uniform Circumferential Expansion of Cylindrical Li-Ion Cells—The Potato Effect. *Batteries* **2021**, 7, 61. https://doi.org/10.3390/batteries7030061

Academic Editor: Manickam Minakshi

Received: 26 July 2021
Accepted: 31 August 2021
Published: 6 September 2021

Publisher's Note: MDPI stays neutral with regard to jurisdictional claims in published maps and institutional affiliations.

1. Introduction

The mechanical properties of commercial Li-ion cells are increasingly coming into focus, especially considering the steadily growing requirements. Higher energy and power densities, less space consumption, and longer service life—these are the challenges that need to be overcome. However, many promising material combinations are limited by their mechanical properties or are not suitable for real applications. For example, silicon has a significantly higher energy density and specific capacity ($Q_{Si} = 4200\,\text{mAh g}^{-1}$ [1]) than graphite ($Q_{C_6} = 372\,\text{mAh g}^{-1}$ [1]), is available in sufficient quantities in nature, and is reasonably priced [1,2]. However, the volume change of $100\% \leq \Delta V_{SoC} \leq 300\%$ compared to the initial volume due to lithium intercalation is problematic. This can lead to cracking in the lattice structure and thus to delamination of the active material from the current collector [3], which leads to a faster aging. For this reason, an alloy or a compound of different materials is usually used to combine the most favorable properties [1–5].

In order to understand which underlying mechanical processes take place inside the battery cell, suitable measurement methods are necessary for recording and analyzing parameters such as electrode thickness and volume change of the cell components. Various methods of measuring electrode thickness have already been implemented and show a volume change of electrode materials and battery cells caused by the intercalation and deintercalation of Li-ions [6–9]. Depending on the material, the electrodes expand differently with lithium migration. Compared to silicon, graphite as an anode material is significantly inferior when considering the specific capacity, but it expands only up to $\Delta V_{\text{SoC}} = 12\%$ [6,7,10] of the initial volume during lithiation and delithiation and thus exerts significantly less mechanical pressure on the cell components. Cathode materials, metal oxides frequently, also show a volume change due to lithiation and delithiation. In this process, Nickel Cobalt Manganese Oxide (NCM) expands by up to $1\% \leq \Delta V_{\text{SoC}} \leq 2\%$ [7,8,10] and Nickel Cobalt Aluminium Oxide (NCA) by up to $\Delta V_{\text{SoC}} = 5\%$ [10] of the initial volume.

Other factors influencing the cell volume are the Solid Electrolyte Interface (SEI) layer thickness growth due to electrolyte decomposition products forming a covering layer on the electrodes [11,12] and Li-plating (especially for Lithium Metal cells) [8,11]. These effects lead to a reduction in volume due to irreversible layer thickness growth and, together with the gassing that takes place due to side reactions [13,14], leads to an increase in pressure inside the cell on the housing, resulting in a measurable change of the cell thickness. This was measured using dilatometry [7,8], imaging techniques such as computed tomography [6] and neutron imaging [15], or strain gauges glued to the cells [9], among other methods. In some cases, besides reversible expansion due to the migration of lithium ions, irreversible expansion was also shown, which also correlates with the loss of cyclizable lithium.

In addition, measurement of the internal gas pressure has already been realized for different cell formats [13,14,16]. Aiken et al. [13] performed tests according to the Archimedes principle on clamped and unclamped pouch cells and demonstrated reversible volume expansion as a function of SoC for initial cycling of battery cells with different electrolytes. In unstrained cells, the volume change built up via charge could be dissipated upon discharge and was thus almost completely reversible. In the case of clamped cells, the pressure no longer decreased completely, since the gas collects on the outer sides of the electrode stack due to the static pressure caused by the clamping. The results obtained from the measurements suggest that, depending on the electrolyte, reversible gas formation is also possible, with the gas formed as the reactant in the chemical reactions taking place. Schmitt et al. [14] showed for the first time that internal gas pressure measurement is possible with commercial pressure sensors on large-sized cells. The results show that during formation the gas pressure in the cell increases irreversibly. Afterwards, a clear SoC dependence can be seen, from which a correlation between the expansion and the graphite stages (pressure increase correlates with the calculated volume change due to lithiation of the graphite and stagnates with restructuring of the crystal lattice) can also be established. Schmitt et al. [14] also showed the irreversible increase in internal gas pressure with decreasing State of Health (SoH). This is also evident in work by Schiele et al. [16], which shows an irreversible increase of internal pressure with increasing number of cycles because of gas evolution attributed to the thermal decomposition of the conducting salt $LiPF_6$ using a multichannel in situ pressure measurement system.

While measurable changes in cell thickness are obvious for pouch cells, having a flexible aluminum composite foil housing, reversible, and irreversible cell thickness growth can also be observed for cylindrical cells, although to a much lesser extent due to the rigid housing [9]. Considering all mentioned effects and investigations, as well as due to the rigid housing and the cylindricity of the commercial cylindrical battery cells, the assumption of a homogeneous expansion of the battery cell over the entire surface is reasonable. The irreversible cell thickness growth due to the mentioned causes and the increase of the internal gas pressure due to side reactions would thus have to lead to an increasing load on the cell housing. Figure 1a schematically shows the assumed cell thickness growth of a cylindrical cell with increasing SoC as a result of the lithiation of the anode, which expands much more than the cathode. The increase in the radius of the cell is shown in blue. During discharge, the cell completely returns to its original shape as the lithium delithiates completely from the graphite. In Figure 1b, the assumed cell thickness growth of the cylindrical battery cell with decreasing SoH (marked in green) is visualized. The red area marks the aging, which increases with higher number of cycles over lifetime. The increase in battery cell radius Δr_{SoH} is directly related on the residual capacity, as already proven by Willenberg et al. [9]. Thus, by correlating with the SoC and the SoH, a state variable estimation can also be attempted by measuring the battery cell thickness change. However, a requirement for this is the uniform expansion of the battery cell over the entire circumference.

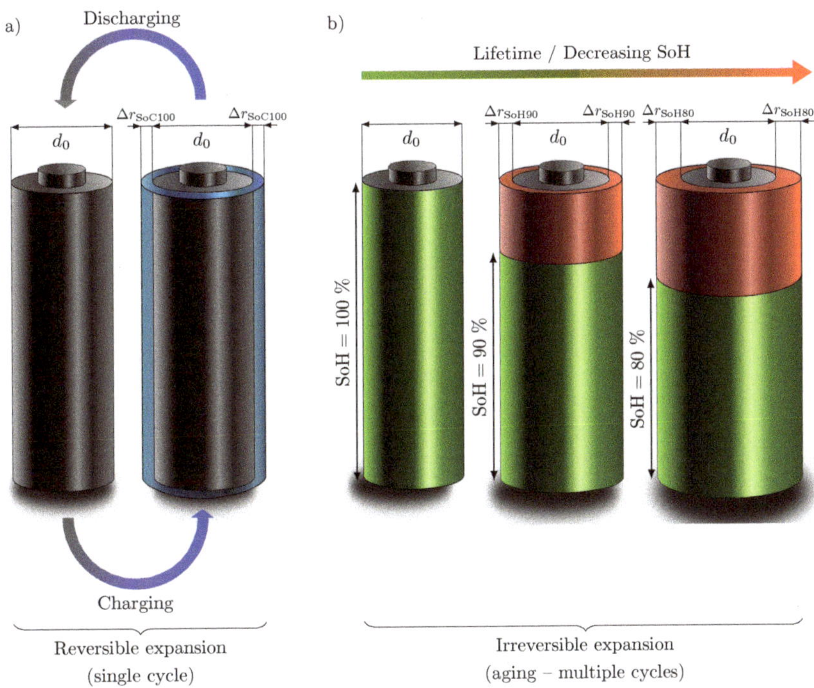

Figure 1. Theoretical schematic representation of (**a**) reversible and (**b**) irreversible cell thickness growth (Δr_{SoC}) compared to the initial diameter (d_0). (**a**) As a result of the volume expansion of graphite with increasing lithiation, the radius of the battery cell increases over every single cycle. The battery cell returns to its original shape when discharged. (**b**) Due to SEI layer thickness growth, defects in graphite, and pressure rise due to side reactions over the lifetime of a cylindrical Li-ion cell, the radius of the battery cell increases irreversibly by Δr_{SoH} with decreasing SoH (green) and increasing irreversible aging (red) until the end-of-life (EoL) of the cell.

The central question of this work was whether the cell really expands uniformly as expected. Both manufacturing and geometric variables are important here – the position and thickness of the current collector tabs that connects the electrodes, the number of layers of electrodes, and the position in the housing. How these influence and what difficulties can arise in measuring battery cell thickness growth are discussed below. It remains to be tested whether cell thickness growth can be used for cell state estimation.

2. Materials and Methods

The study of battery cell thickness growth is performed on a commercial cylindrical Li-Ion cell, LG INR18650 M29. The battery cell has a positive electrode made of nickel-rich NCM active material and a negative electrode made of graphite. The specifications of the battery cell are listed in Table 1. The battery cell has a usable voltage range of $2.5 \text{ V} \leq U \leq 4.2 \text{ V}$ according to the producer's datasheet [17].

Table 1. Product specification of the cylindrical Li-ion battery cell applied [17].

	LG INR18650 M29
Nominal Specification	
Nominal Capacity	2850 mAh
Charging Voltage	4.20 V
Nominal Voltage	3.67 V
Discharge Cut-off Voltage	2.50 V
Max. Charge Current	1.0 C (2.850 A)
Max. Discharging Current	6 A (for continuous discharge)
	10 A (not for continuous discharge)
Operating Temperature	Charge: 0 °C to 50 °C
	Discharge: −30 °C to 60 °C
Dimensions	
Diameter	18.3 + 0.1/−0.3 mm
Height	65.0 ± 0.2 mm
Weight	45.00 g

2.1. Experimental Setup

To investigate the punctual radial expansion of the LG INR18650 M29 battery cell, an optoCONTROL ODC2600 light band micrometer with a high-resolution CCD camera for measuring geometric quantities from Micro-Epsilon Messtechnik GmbH & Co. KG (Ortenburg, Germany) is used. The light band micrometer is shown schematically in Figure 2. The measuring range (red band between laser and acquisition unit) has a total height of $l_{\mathrm{range}} = 40\,\mathrm{mm}$. The linearity l, i.e., the deviation between an ideal straight characteristic curve and the real characteristic curve, is max. $l = 3\,\mu\mathrm{m}$. The resolution d_{res} of the measurement signal is $d_{\mathrm{res}} = 0.1\,\mu\mathrm{m}$, at a measurement rate f of up to $f = 2.3\,\mathrm{kHz}$. The micrometer operates on the principle of shading or light quantity measurement and thus detects the dimension and position of the object to be measured.

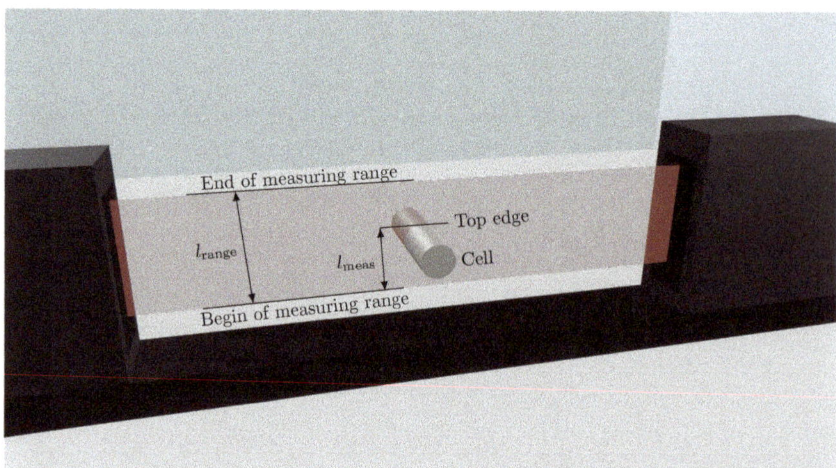

Figure 2. Schematic measurement setup with the light band micrometer from Micro-Epsilon Messtechnik GmbH & Co. KG for investigating the punctual expansion of the cylindrical Li-Ion cell (LG INR18650 M29) at the top edge of the cell located in the l_{range} high light band marked in red. In fact, the distance between the LG INR18650 M29 and the end of the measuring range is measured (l_{meas}). As the cell expands, the distance decreases and vice versa. The cell is cycled in the light band micrometer and rotated 10° after charge and discharge, until the complete circumference has been measured with a total of 36 cycles.

To examine the expansion of the battery cell, the cell is clamped at the poles in a holder for contacting so that it is free and orthogonal to the light band in the measuring area marked in red, as shown in Figure 2. The light band micrometer has different setting modes. The transition between the shading by the cell and the adjacent light edge of the laser is used as a reference point. Either the path between the upper edge of the battery cell and the lower edge of the measuring range, or the path between the lower edge of the battery cell and the upper edge of the measuring range can be measured. As the battery cell expands, the top edge of the cell shifts, resulting in increased shading and thus increased measurable distance. In addition, the diameter of the shading through the cell or a gap between two shadings can be determined. However, this mode is not suitable for the following investigations because the expansion of one position and a contraction at the opposite position can cancel each other out. Therefore, the change of the battery cell radius is always indicated in the following. The acquisition can only be performed for one position or one diameter at a time. To examine the cell thickness growth over the entire circumference, the battery cell must be rotated manually.

2.2. Measurement Procedure

The cylindrical Li-ion battery cell is subjected to the same measurement procedure at each position around the circumference. An ACT0550 battery tester from PEC Corp. (Leuven, Belgium) is used to cycle the battery cell. The battery cell tester can provide currents up to 50 A per channel and apply a voltage of 5 V maximum. All measurements are performed in a climatic chamber of Memmert GmbH & Co. KG (Schwabach, Germany) at a temperature of $T = 25\,^\circ\text{C}$. To investigate the effect of temperature, a temperature sensor PT100 is attached to the outside of the cell and connected to the battery tester.

The cell is always clamped in the cell holder with the position to be examined facing upwards. The battery cell is first fully charged with a current of $I = 1\text{C}$ ($1\text{C} \equiv 2.85\,\text{A}$) and also fully discharged again with $I = 1\text{C}$ in the voltage range of $2.5\,\text{V} \leq U \leq 4.2\,\text{V}$. Charge and discharge period consist of a CC phase and a CV phase as listed in Table 2. The termination criterion for the CV phase is when the current drops below $1/100\,\text{C}$, which corresponds to a current of $I_{\text{lim}} = 28.5\,\text{mA}$. For the following investigations, only this one current rate is used. Willenberg et al. [9] showed that expansions also occur at lower current strengths. The influence of the current strength on the uniformity of the expansion has yet to be shown in subsequent work.

The cell is rotated manually by $\alpha = 10^\circ$ after each charge and discharge cycle and measured again. During the measurement, voltage, current, and temperature are logged with one data point per second, averaged from the last 1000 recorded values at a frequency of $f = 1000\,\text{Hz}$. After a total of 36 cycles, the individual positions can be compared with each other.

Table 2. Consistent measurement procedure for each measured position around the circumference of the battery cell.

	Step	Parameter	Termination Criterion
1	CC-Charging	$I_{\text{ch}} = 2.85\,\text{A}$	$V = 4.2\,\text{V}$
2	CV-Charging	$U_{\text{const}} = 4.2\,\text{V}$	$I \leq 28.5\,\text{mA}$
3	CC-Discharging	$I_{\text{dch}} = -2.85\,\text{A}$	$V = 2.5\,\text{V}$
4	CV-Discharging	$U_{\text{const}} = 2.5\,\text{V}$	$I \geq -28.5\,\text{mA}$

2.3. Computed Tomography

In order to verify the measurements with the light band micrometer, a 3D non-destructive measurement was needed with high spatial resolution. X-ray computed tomography is one such method with the capability of providing high resolution 3D scans of objects. The combination of non-destructive methodology and high resolution makes it apt for analyzing the non-uniform expansions of Li-ion cells. A cone-beam X-ray computed tomography is used specifically in our experiment as shown in Figure 3. In this setup, at one end is an X-ray source radiating a cone shaped beam and on the other end there is a flat-panel X-ray detector. The object to be scanned is placed on a high precision rotation stage between the source and the detector. During the scan, the object is rotated slowly over 360° and, in this duration, several thousand 2D images called the projections are captured by the detector. This stack of 2D projections of the object to be measured is then fed into a tomographic reconstruction algorithm to generate the 3D volume of the scanned object. The resolution of the scan depends on how close the object is to the source. Like a shadow approach, the closer the object is to the source, the higher is the scan resolution; however, the field of view is smaller. If we move away from the source, we have a greater field of view but at the cost of resolution.

Figure 3. Computed tomography setup for 3D measurement of the battery cell for different loading conditions to compare the mechanical structure of the cylindrical cell.

The Li-ion cell is scanned in two states, once with SoC = 0 and once with SoC = 1. In both of the cases, the resolution of the 3D reconstructed volume was 10 μm. This was followed by a surface determination on the 3D volume data to determine the surface of the cell in the two states. These generated surfaces from both scans were then used to calculate the non-uniform expansion of the cell walls in 3D. In order to have an absolute co-relation between the two states, a reference metal tube was fixed to the setup, which was used to register the 3D volumes of both scans. This would ensure that there is no geometric shift or rotation influencing the measurement of the non-uniform expansion of the cells. Although CT has been used to study various effects in batteries [9,18–20], to the best of the author's knowledge, a 3D non-uniform expansion of Li-ion cells measured with CT is investigated for the first time in this work. Figure 3 shows the setup of the computed tomography for the 3D measurement of the battery cell with the source for the X-rays and the detector.

In order to be able to evaluate which local expansions belong to which geometrical influences within the cell later on also during the measurements with the light band micrometer, a picture of the cell examined in the light band micrometer was also taken with the computed tomography. In addition, the position of the anode current collector was marked on the housing of the battery cell to assign the corresponding positions.

3. Results

3.1. Position-Dependent Volume Change

Figure 4a shows the schematic illustration of a Li-ion cell as well as the examined positions over the circumference of the cell, starting at pos. 0, where the negative current collector (marked A in Figure 4b) is located inside the cell. Each further position is offset by $\alpha = 10°$ from the previous one (pos. 1 $\widehat{=}$ 10°, pos. 2 $\widehat{=}$ 20°, ...).

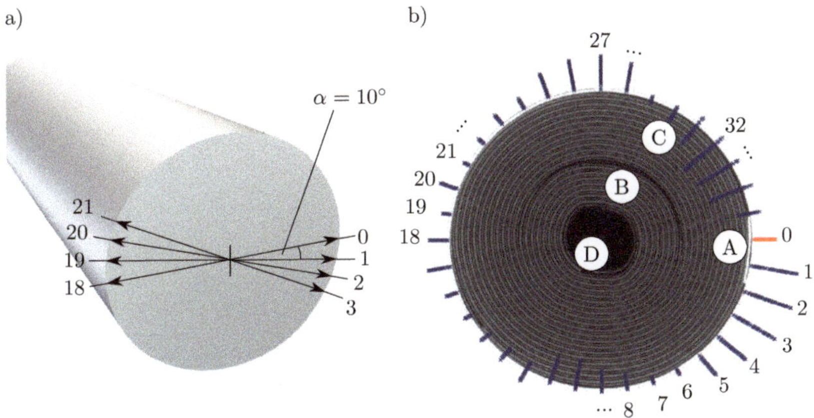

Figure 4. (**a**) Schematic drawing of a cylindrical Li-Ion cell, marked with the points measured in the light band micrometer, which are located at an angle of $\alpha = 10°$ to each other. The start position from the measurement (position 0) corresponds to the position of the negative current collector tab, which was marked in the CT (marked A in Figure 4b; (**b**) 100-fold amplified expansion of the cell thickness in steps of $\Delta SoC = 2.5\%$ at the corresponding positions including a CT image of the cell to assign the corresponding positions on the cell circumference. A marks the negative current collector tab, B the positive current collector tab. C highlights one of two spots, where a small gap between the jelly role and the housing is located, where the jelly role is not pressed against the housing. D marks the end of the jelly roll.

Figure 4b shows the change of the radius of the respective position Δr_{Pos} over the circumference of the battery cell per SoC change of $\Delta SoC = 2.5\%$ amplified by a factor of 100. Between positions 0 and 5, a clear expansion of the cell is evident, which continues to flatten and reverses at position 7. The battery cell also shows significant expansion directly opposite the negative current collector tab and at the position of the positive current collector tab (marked B). The marks C (associated measurement positions: 30–35) and D (associated measurement positions: 8–11) each highlight locations where the jelly roll is not as tight as at the other locations and thus does not press against the housing. Both locations show a contraction of the cell over a wide range. Whether the cell wrap is more densely or less densely packed is determined by the distance between the cell wrap and the housing. If the cell coil lies directly against the housing, a denser packing can be assumed then if there is still a cavity between the cell coil and the housing (as in mark C in Figure 4b).

The results of the change in radius over the circumference of the battery cell suggest that, in particular, expansion occurs at those locations where the packing of the jelly roll is significantly tighter than normal due to the current collector tabs. Contraction occurs mostly at locations where the jelly roll is significantly looser. This ultimately leads to a potato-shaped cell form, also visualized in Figure 5.

Figure 5a shows the radial change Δr (amplified by a factor of 100) over time t added to a fixed cell radius r_{mean} to visualize the change in cell volume of each circumferential position over time. Each line reflects the recorded expansion or contraction of the battery cell Δr in the light band micrometer at that point. The contraction at the bottom of the cell and the bulge near the negative current collector tab can be seen particularly clearly. The plane in gray shows the section plane applied at different states of charge to study the volume change of the battery cell for different states of charge.

Figures 5b–e show the sectional views at different states of charge (SoC = 0.25 ... 0.5 ... 0.75 ... 1), which once more clearly show the potato-shaped form of the battery cell.

The change in radius at the position of the negative current collector tab is marked in green. The top view of the battery cell in gray reflects the initial shape of the cell.

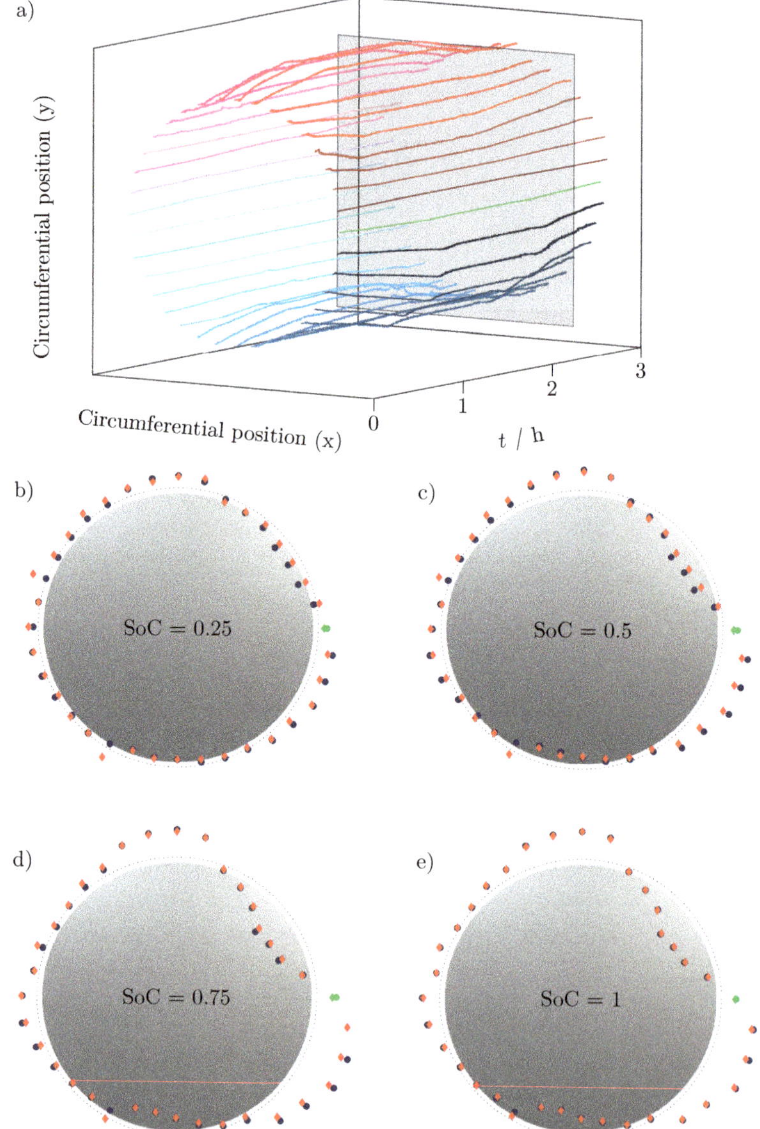

Figure 5. (**a**) 100-fold amplified change of cell thickness growth over the time *t* in h of charge and discharge cycle in $\alpha = 10°$ steps added to a fixed cell radius (mean diameter of the cell with SoC = 0) for each position measured in the light band micrometer over the circumference of LG INR18650 M29. The gray area is the intersection for plots (**b**–**e**), showing sectional views of relevant SoCs (SoC = 0.25 ... 0.5 ... 0.75 ... 1). The current collector position is marked by a green line/dot in every plot. The blue dots belong to the charge of the battery cell, the red marks to the discharge, revealing a visible hysteresis of the volume change of the cell.

The change in radius during charging is shown in blue, and the change when the corresponding SoC is reached during discharging is shown in red.

An increase in the extreme points over the SoC can be clearly seen: the expansion becomes larger as the charge increases, but the contraction also becomes stronger as other regions expand. The hysteresis of the voltage due to the applied load has already been investigated many times [21–23]. However, the expansion also has a hysteresis, recognizable by the non-overlapping points especially at the extreme points.

3.2. Detailed Radial Change and Influencing Factors

For a better understanding of the volume change, a detailed consideration of the alteration of the radius is useful. This also allows for identifying any influencing factors. Figure 6a shows the change in radius Δr over time t for the entire cycle of the position that expands the most, as well as current (I), voltage (U), and temperature (T) characteristics of the test.

The change in radius over the CC charge phase is about $\Delta r = 27$ μm. It decreases somewhat as the cell enters the CV charging phase. At the start of the discharge, there is a very brief increase in radius until the decrease in cell thickness due to the mechanical processes predominates and the cell contracts again. At the end of the CC discharge phase, respectively at the beginning of the CV discharge phase, a kink in the curve is visible.

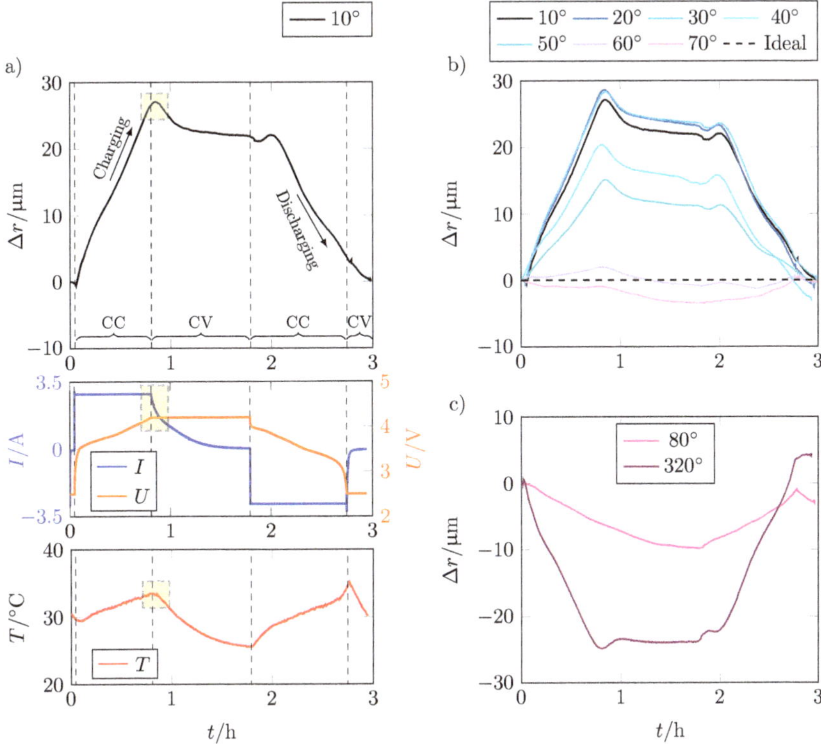

Figure 6. (**a**) Radial expansion Δr in μm of the LG INR18650 M29, the corresponding current I in A and voltage U in V, and the temperature T in °C over the time t in h for charge and discharge cycle for position 1 ($\alpha = 10°$ from the current collector tab). The area highlighted in yellow marks the end of the CC phase or the beginning of the CV phase at which the temperature inside the cell continues to increase despite the falling current, resulting in a change of the cell thickness; (**b**) radial expansion Δr in μm of position 1–7 ($\alpha \in \{10, 20, ..., 70\}°$) from the current collector tab; (**c**) comparison of the radial expansion of position 8 and 32 with negative expansion over the charge and discharge cycle.

If only the voltage was decisive for the expansion of the battery cell, the change in the radius would correspond to the voltage curve. Looking at the voltage curve and the expansion of the cell in Figure 6a, this means that other effects besides lithiation have an influence on the expansion. Particularly noticeable is the sharp peak between the CC and CV charging phases, which is marked in yellow here. Although the applied current drops and the voltage remains constant, the expansion continues to rise. Looking at the temperature curve, the peak of the expansion correlates with the temperature increase of the cell. This is due to the ideal gas law, according to which the gas pressure increases with increasing temperature while the volume remains constant. The increase in gas pressure also causes the battery cell to expand further, although no further electrode growth takes place. By decreasing temperature and thus also gas pressure, the expansion also decreases to an almost constant level until the end of the CV discharge phase.

At the beginning of the discharge phase, the expansion initially decreases analogously to the voltage due to the delithiation of the graphite, which leads to a contraction of the electrode. Shortly after the start of the CC discharge phase, the radius increases again despite progressive delithiation. The reason is presumably the increasing temperature, which causes the internal gas pressure to rise again and superimposes the reduction in the radius, thus leading to expansion again until the effect of delithiation predominates and the radius of the battery cell decreases again. Due to the temperature effect, the slope of the radius change is smaller during discharge than during charge.

Figure 6b,c show that the above effects also apply to lower expansions and even to locations where contraction of the battery cell occurs. At points where the cell contracts, an additional contraction occurs as the temperature rises. This is due to the strong expansion at other points (see positions 1–7 in Figure 6b), which causes the spaces between the bulges to contract due to the mechanical rigidity of the housing.

As already indicated in Figure 5, the expansion exhibits a hysteresis. Figure 7a shows the voltage for charging U_{ch} and discharging U_{dch} as well as the corresponding change in radius Δr_{ch} and Δr_{dch} plotted versus SoC.

As expected, the voltage exhibits hysteresis due to structural changes in graphite and non-uniform distribution of lithium ions [24,25]. The expansion also shows this hysteresis, with the curves of the radius change for charge and discharge intersecting at SoC = 0.08.

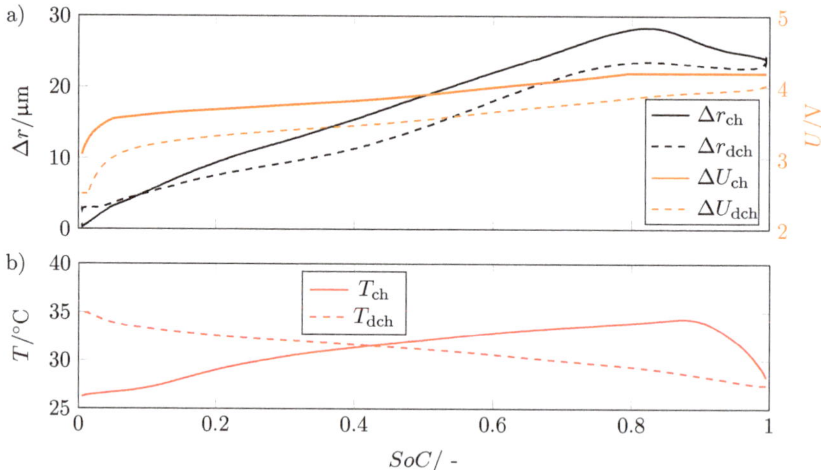

Figure 7. (a) Change in radius Δr in μm for charge and discharge and the associated voltage U in V versus State of Charge SoC showing the hysteresis between charging and discharging; (b) temperature curve T in °C during charging and discharging versus the state of charge SoC.

This hysteresis shows that the lithiation and delithiation of the electrodes results in an asymmetric volume change.

The overlap of the curves of the radius progression can be explained with regard to the temperature. The temperature increases with decreasing SoC due to the high applied current at a discharge rate of 1C to a temperature of $T > 33\,°C$ at the corresponding SoC. This causes the internal gas pressure to increase, which results in the radius not decreasing as fast as it increased during charging. This causes the two curves to meet. This is also the reason why the hysteresis becomes smaller for SoCs close to 1.

3.3. Investigations with the Help of Computed Tomography

To validate the measurements with the light band micrometer, computed tomography was used to determine battery cell thickness growth, as shown in Figure 8.

Therefore, images of the cell in a fully discharged state and in a fully charged state are generated, creating two three-dimensional models of the battery cell.

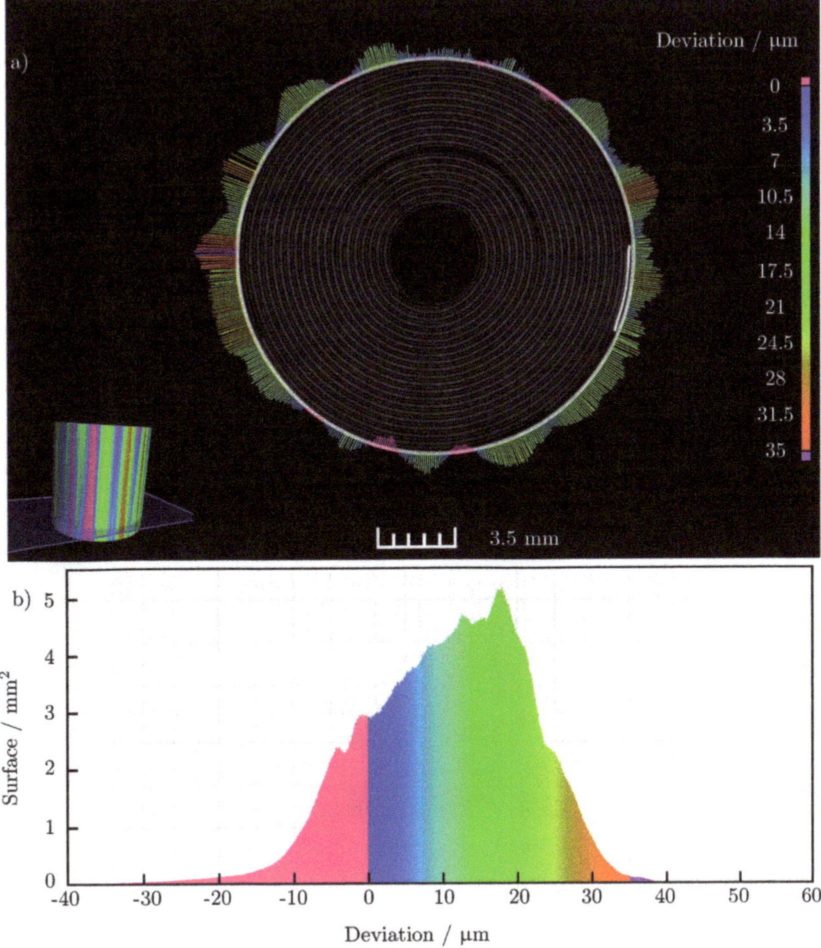

Figure 8. (**a**) Cross−section of the 3D CT image of the battery cell with magnified extent. Contraction of the battery cell is colored in pink; (**b**) representation of the absolute surface of the recorded battery volume, which changes over the charge according to the indicated expansion. The largest proportion of the battery surface expands between 10μm and 20 μm while charging.

By superposing these models, the difference in the shape of the outside of the casing between the two states of charge of the battery cell can be visualized. Figure 8a shows a sectional view of the three-dimensional scan of the cell and the 50-fold amplified change on the outer side of the cell. The figure confirms the clear expansion of the cell housing in the area of the anode current collector tab and at the opposite position of the cell (180° offset). The maximum absolute radius change at expanding regions of the cell is also of the same order of magnitude. In contrast to the measurements with the micrometer, the expansion in the area of the anode current collector tab does not go directly into a contraction of the surrounding positions around the circumference of the battery cell. Overall, there is a smaller number of contraction positions seen around the circumference of the cell. The absolute change in radius also lags well behind the maximum contraction that occurred in the micrometer. While the averaged radius change of the cell in the light band micrometer measurement is only about $\Delta r_{\mathrm{LBM}} = 1\,\mu m$, it is about $\Delta r_{\mathrm{CT}} = 12\,\mu m$ in the computed tomography measurements. Nevertheless, the inhomogeneous expansion of the cell is clearly emphasized. The potato shape due to the increased expansion is also clearly recognizable. In addition to the differently expanding areas, a sinusoidal shape of the change of the cell housing can be seen in the images. It is possible that this is due to the manufacturing process of the cell housing. Confirmation of this effect by subsequent measurements is still to be conducted. Figure 8b shows how much area of the outer housing expands or contracts. It is clear that the largest proportion of the surface of the housing exhibits a radius change of $\Delta r = 20\,\mu m$ during charging compared to the discharged state of the cell.

Figure 9a shows a representation of the 3D image of the cell including the analysis. The coloring clearly shows an area of contraction on the left side and areas of varying degrees of expansion on the remaining surface. The color scale can be found in Figure 8. Figure 9b shows the expansion of a cross-section over the height of the cell, proving that there is no significant change in cell thickness growth over height. Only the absolute expansion decreases in the direction of the poles. In addition, the left side of the plot shows that, even after the end of the current collector tab, there is a reduction in the absolute expansion of the cell over the height of the battery cell.

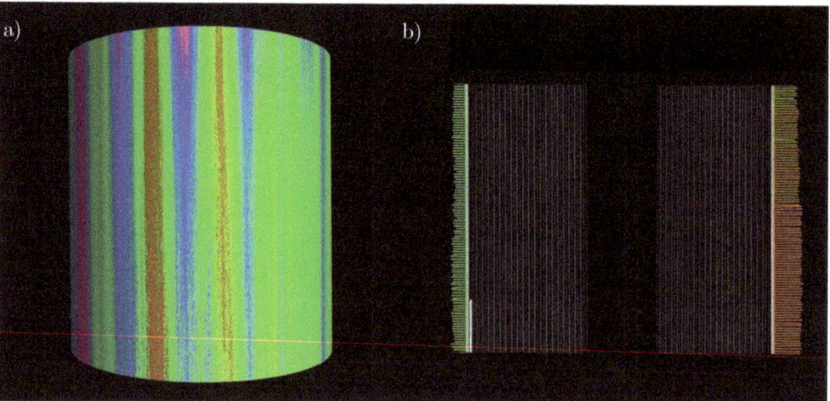

Figure 9. (a) 3D representation of the recorded battery volume analyzed for expansion; (b) longitudinal section through the battery cell over the recorded height with the corresponding expansion analysis.

4. Discussion

As shown in the section Results, the use of a light band micrometer is a stable and reliable method to study cell thickness growth. During the cycling of LG INR18650 M29, the change in radius of a single position could be measured with an accuracy of $a_{\mathrm{acc}} = 99.67\%$. The operating temperature of the light band micrometer is between $0\,°C \leq T \leq 50\,°C$, so, when measuring in a climatic chamber at a constant temperature of $T = 25\,°C$, a temperature effect of the environment on the measuring device and the battery cell can be neglected.

The measurement of the volume change over the complete circumference of the battery cell shows that a non-uniform change in the shape of the cell takes place. Instead of the expected uniform expansion, the cell takes on a potato shape, putting it simply. The roundness of the cell decreases and bulges and valleys form around the circumference depending on the position and orientation of the jelly roll inside the housing. Where the negative current collector tab is located, the cell expands at an above-average rate, while it contracts more than average at points where the jelly roll is not directly against the housing and a cavity is created.

The measured change in radius, caused by the intercalation and deintercalation of Li-ions in the electrodes, is up to a maximum of $\Delta r_{max,ex} = 27$ µm on expansion of the cell (as seen in Figure 6a) and up to $\Delta r_{max,con} = -25$ µm on contraction (as seen in Figure 6c). A SoC dependent volume change behavior was observed at all locations of the battery cell. The averaged diameter of the battery cell with a low SoC is $d_{SoC=0} = 18.15$ mm. The change in radius over charge averages $\Delta r_{mean} = 1.1$ µm for each position by the circumference of the cell, respectively $\Delta d_{mean} = 2.2$ µm, resulting in an averaged diameter of the battery cell in the fully charged state of $d_{SoC=1} = 18.152$ mm. As shown by the small difference between the average diameter when fully discharged and the average diameter when fully charged, the volume change of the battery cell is non-uniform because the areas of expansion and the areas of contraction compensate each other. The actual change of radius of different zones around the circumference of the cell is up to 12 times higher than the average expansion. Since the expansion of the individual positions was investigated in 36 consecutive cycles, deviations due to forming processes, aging processes, and possibly existing influences due to different relaxation of expansion and pressure are to be expected. The absolute values are therefore to be considered critically and must be validated elsewhere with an extended setup that allows the investigation of all of the different positions within a single cycle for multiple samples and different cell chemistries. In addition, the influence of the current and the ambient temperature on the uniformity of the expansion still need to be considered in future studies.

The CT scan could confirm the non-uniform change in the shape of the cell and the potato-shaped expansion. The absolute maximum radius change of both measurement methods is between 29 µm $< \Delta r_{max} < 37$ µm. However, there is a significant difference between the absolute surface area expanding and the absolute surface area contracting. It should be noted that the resolution of computer tomography is much higher than that of manual measurement or rotation by $\alpha = 10°$ in the light band micrometer. In addition, there is a risk that peaks or valleys near the area of interest will falsify the measurement due to the arrangement of the measurement area with the light band micrometer. Therefore, optimization of the measurement setup is mandatory for further investigations.

By analyzing a section of a battery cell with a height of $h = 1$ cm, computed tomography is able to confirm the initial assumption that the change in radius of the cell does not change significantly with height. Only the absolute value of the expansion or contraction decreases in the direction of the poles.

As shown in Figures 6 and 7, a correlation of the radius change with the SoC is very well possible. This correlation between the lithiation of the electrodes and the change in radius can be established regardless of whether the cell expands or contracts at the corresponding position. The characteristic points, for example, the end of the CC charging and discharging phase, as well as the influence of the temperature rise and the relaxation at the respective CV phases, can be seen over almost all of the examined positions. However, this correlation is much clearer for positions where the absolute value of the change in radius is higher. Figure 7 also shows that the intercalation and deintercalation of Li-ions result in different swelling and contraction, which leads to a visible hysteresis. The variation in the slope of the radius change in Figure 6 is probably related to the stage formation during the lithiation of graphite. Further investigation of this will be reported elsewhere.

In addition, investigation of the influence of the SoH on the expansion at various locations is still outstanding.

5. Conclusions

It has been shown that spatially resolved measurement of the expansion of battery cells is limited with a light band micrometer and very feasible with the aid of computed tomography. The investigated battery cells show significant deviations from the expected uniform expansion and form areas of expansion and contraction over their circumference, depending on the orientation of the jelly roll in the case.

Hence, the use of volume change to study aging or predict SoC, as indicated by different authors [9,12], requires special caution, since the expansion varies significantly depending on the measurement position and the measurement results may be biased depending on the measurement method.

Author Contributions: Conceptualization, J.H.; methodology, J.H., F.D., and J.G.; investigation, J.H., F.D., and J.G.; CT images, J.G.; writing—original draft preparation, J.H.; writing—original draft preparation CT part, J.G.; writing—review and editing, J.G., A.F., F.D., K.P.B., and J.H.; visualization, J.H. and A.F.; supervision, K.P.B.; All authors have read and agreed to the published version of the manuscript.

Funding: This research was funded by Bundesministerium für Wirtschaft und Energie (BMWi - ZF4370703LT9).

Data Availability Statement: The data presented in this study are available on request from the corresponding author.

Acknowledgments: The authors are thankful to Tina Kreher, Maike Lambarth, and Martin Kühnemund for reviewing the paper.

Conflicts of Interest: The authors declare no conflict of interest.

Abbreviations

The following abbreviations are used in this manuscript:

CC	Constant Current
CV	Constant Voltage
CT	Computed Tomography
LBM	Light Band Micrometer
NCA	Nickel Cobalt Aluminium Oxide
NCM	Nickel Cobalt Manganese Oxide
SEI	Solid Electrolyte Interface
SoC	State of Charge
SoH	State of Health

References

1. Kamali, A.R.; Fray, D.J. Review on Carbon and Silicon Based Materials as Anode Materials for Lithium Ion Batteries. *J. New Mater. Electrochem. Syst.* **2010**, *13*, 147–160.
2. Park, C.-M.; Kim, J.-H.; Kim, H.; Sohn, H.-J. Li-alloy based anode materials for Li secondary batteries. *Chem. Soc. Rev.* **2010**, *39*, 3115–3141. [CrossRef]
3. Dimov, N.; Kugino, S.; Yoshio, M. Carbon-coated silicon as anode material for lithium ion batteries: Advantages and limitations. *Electrochim. Acta* **2002**, *48*, 1579–1587. [CrossRef]
4. Andersen, H.F.; Foss, C.E.L.; Voje, J.; Ragner, T.; Mokkelbost, T.; Vullum, P.E.; Ulvestad, A.; Kirkengen, M.; Mæhlen, J.P. Silicon-Carbon composite anodes from industrial battery grade silicon. *Sci. Rep.* **2019**, *9*, 14814. [CrossRef]
5. Khomenko, V.G.; Barsukov, V.Z.; Doninger, J.E.; Barsukov, I.V. Lithium-ion batteries based on carbon-silicon-graphite composite anodes. *J. Power Sources* **2007**, *165*, 598–608. [CrossRef]
6. Michael, H.; Iacoviello, F.; Heenan, T.M.M.; Llewellyn, A.; Weaving, J.S.; Jervis, R.; Brett, D.J.L.; Shearing, P.R. A Dilatometric Study of Graphite Electrodes during Cycling with X-ray Computed Tomography. *J. Electrochem. Soc.* **2021**, *168*, 010507. [CrossRef]
7. Rieger, B.; Schlueter, S.; Erhard, S.V.; Schmalz, J.; Reinhart, G.; Jossen, A. Multi-scale investigation of thickness changes in a commercial pouch type lithium-ion battery. *J. Energy Storage* **2016**, *6*, 213–221. [CrossRef]

8. Bitzer, B.; Gruhle, A. A new method for detecting lithium plating by measuring the cell thickness. *J. Power Sources* **2014**, *262*, 297–302. [CrossRef]

9. Willenberg, L.K.; Dechent, P.; Fuchs, G.; Sauer, D.U.; Figgemeier, E. High-Precision Monitoring of Volume Change of Commercial Lithium-Ion Batteries by Using Strain Gauges. *Sustainability* **2020**, *12*, 557. [CrossRef]

10. Louli, A.J.; Li, J.; Trussler, S.; Fell, C.R.; Dahn, J.R. Volume, Pressure and Thickness Evolution of Li-Ion Pouch Cells with Silicon-Composite Negative Electrodes. *J. Electrochem. Soc.* **2017**, *164*, A2689–A2696. [CrossRef]

11. Pender, J.P.; Jha, G.; Youn, D.H.; Ziegler, M.; Andoni, I.; Choi, E.J.; Heller, A.; Dunn, B.S.; Weiss, P.S.; Penner, R.M.; et al. Electrode Degradation in Lithium-Ion Batteries. *Am. Chem. Soc. Nano* **2020**, *14*, 1243–1295. [CrossRef]

12. Deich, T.; Hahn, S.L.; Both, S.; Birke, K.P.; Bund, A. Validation of an actively-controlled pneumatic press to simulate automotive module stiffness for mechanically representative lithium-ion cell aging. *J. Energy Storage* **2020**, *28*, 101192. [CrossRef]

13. Aiken, C.P.; Xia, J.; Wang, D.Y.; Stevens, D.A.; Trussler, S.; Dahn, J.R. An Apparatus for the Study of In Situ Gas Evolution in Li-Ion Pouch Cells. *J. Electrochem. Soc.* **2014**, *161*, A1548–A1554. [CrossRef]

14. Schmitt, J.; Kraft, B.; Schmidt, J.P.; Meir, B.; Elian, K.; Ensling, D.; Keser, G.; Jossen, A. Measurement of gas pressure inside large-format prismatic lithium-ion cells during operation and cycle aging. *J. Power Sources* **2020**, *478*, 228661. [CrossRef]

15. Siegel, J.B.; Stefanopoulou, A.G.; Hagans, P.; Ding, Y.; Gorsich, D. Expansion of Lithium Ion Pouch Cell Batteries: Observations from Neutron Imaging. *J. Electrochem. Soc.* **2013**, *160*, A1031–A1038. [CrossRef]

16. Schiele, A.; Hatsukade, T.; Berkes, B.; Hartmann, P.; Brezesinski, T.; Janek, J. High-Throughput In Situ Pressure Analysis of Lithium-Ion Batteries. *Anal. Chem.* **2017**, *89*, 8122–8128. [CrossRef]

17. Product Specification of the Rechargeable Lithium Ion Battery Model: INR 18650 M29 2850 mAh. Available online: http://akkuplus.de/mediafiles/Datenblatt/LG/LG_INR18650M29.pdf (accessed on 7 July 2021).

18. Robinson, J.B.; Darr, J.A.; Eastwood, D.S.; Hinds, G.; Lee, P.D.; Shearing, P.R.; Taiwo, O.O.; Brett, D.J.L. Non-uniform temperature distribution in Li-ion batteries during discharge—A combined thermal imaging, X-ray micro-tomography and electrochemical impedance approach. *J. Power Sources* **2014**, *252*, 51–57. [CrossRef]

19. Yufit, V.; Shearing, P.; Hamilton, R.W.; Lee, P.D.; Wu, M.; Brandon, N.P. Investigation of lithium-ion polymer battery cell failure using X-raycomputed tomography. *Electrochem. Commun.* **2011**, *13*, 608–610. [CrossRef]

20. Pfrang, A.; Kersys, A.; Kriston, A.; Sauer, D.U.; Rahe, C.; Käbitz, S.; Figgemeier, E. Geometrical Inhomogeneities as Cause of Mechanical Failure in Commercial 18,650 Lithium Ion Cells. *J. Electrochem. Soc.* **2019**, *166*, A3745–A3752. [CrossRef]

21. Gerschler, J.B.; Sauer, D.U. Investigation of Open-Circuit-Voltage Behaviour of Lithium-Ion Batteries with Various Cathode Materials under Special Consideration of Voltage Equalisation Phenomena. In Proceedings of the International Battery, Hybrid and Fuel Cell Electric Vehicle Symposium EVS24, Stavanger, Norway, 13–16 May 2009.

22. Grimsmann, F.; Gerbert, T.; Brauchle, F.; Gruhle, A.; Parisi, J.; Knipper, M. Hysteresis and current dependence of the graphite anode color in a lithium-ion cell and analysis of lithium plating at the cell edge. *J. Energy Storage* **2018**, *15*, 17–22. [CrossRef]

23. Roscher, M.A.; Bohlen, O.; Vetter, J. OCV Hysteresis in Li-Ion Batteries including Two-Phase Transition Materials. *Int. J. Electrochem.* **2011**, *2011*, 984320. [CrossRef]

24. Zheng, T.; Dahn, J.R. Hysteresis observed in quasi open-circuit voltage measurements of lithium insertion in hydrogen-containing carbons. *J. Power Sources* **1997**, *68*, 201–203. [CrossRef]

25. Dreyer, W.; Jamnik, J.; Guhlke, C.; Huth, R.; Moskon, J.; Gaberseck, M. The thermodynamic origin of hysteresis in insertion batteries. *Nat. Mater.* **2010**, *9*, 448–453. [CrossRef] [PubMed]

batteries

MDPI

Article

Versatile AC Current Control Technique for a Battery Using Power Converters

S. M. Rakiul Islam * and Sung-Yeul Park

Electrical and Computer Engineering Department, University of Connecticut, Storrs, CT 06269, USA; sung_yeul.park@uconn.edu
* Correspondence: s.islam@uconn.edu; Tel.: +1-860-486-0915

Abstract: Although a battery is a DC device, AC current is often necessary for testing, preheating, impedance spectroscopy, and advanced charging. This paper presents a versatile control technique to inject AC current to a battery. Synchronous buck and H-bridge topologies are operated in bidirectional mode and controlled by uni-polar and bi-polar pulse width modulation techniques for the AC current injection. The input and output passive circuits are specially designed considering AC current and the properties of the battery. A controller is proposed considering a small internal impedance, small AC ripple voltage, and variable DC offset voltage of a battery. The controller is capable of maintaining stable operation of AC current injection in two power quadrant within a small DC voltage boundary of a battery. The controller is comprised of a feedback compensator, a feedforward term, and an estimator. The feedback gain is designed considering the internal impedance. The feedforward gain is designed based on estimated open circuit battery voltage and input voltage. The open circuit voltage estimator is designed based on filters and battery model. For validation, AC current is injected to a Valence U-12XP battery. The battery is rated for 40 Ah nominal capacity and 13.8 V nominal voltage The controller successfully injected AC current to a battery with +10 A, 0 A, and −10 A DC currents. The magnitude and frequency of the AC current was up to 5 A and 2 kHz respectively.

Keywords: lithium ion battery; AC current injection; bi-directional control; charger

Citation: Islam, S.M.R.; Park, S.-Y. Versatile AC Current Control Technique for a Battery Using Power Converters. *Batteries* **2021**, *7*, 47. https://doi.org/10.3390/batteries7030047

Academic Editors: Kai Peter Birke and Duygu Kaus

Received: 12 May 2021
Accepted: 9 July 2021
Published: 15 July 2021

Publisher's Note: MDPI stays neutral with regard to jurisdictional claims in published maps and institutional affiliations.

1. Introduction

Batteries are DC electrochemical energy storage devices which are mainly utilized as DC source or load in applications. However, AC current injection in a battery is necessary to preheat in cold weather, to measure the internal impedance, to balance the capacity based on impedance, and to charge by sinusoidal ripple current method [1–4]. Conventionally, AC current is injected to rechargeable batteries using linear circuits which work offline (the batteries are disconnected from the applications) [5,6]. Recently, power converters are being used to inject AC current to rechargeable batteries which can be operated either offline or online (the battery is connected to the applications) [1–4,7–13]. A buck converter topology was used to inject AC current on top of DC charging current in [9]. AC current was imposed on top of discharging current in [10]. The methods available in the existing literature for AC current injection using power converters can be operated either offline or online (either while charging or discharging). A versatile AC current control technique is proposed in this manuscript which can be operated in both offline and online conditions for a broad frequency spectrum.

The AC current injection using power converters for AC system (grid tied inverter) is an well established technology where the power flows unidirectionally and has a significant voltage variation corresponding to current [14,15]. The unidirectional power flow of a single phase inverter is depicted in Figure 1a. V_{out} and I_{out} are the output voltage and current. The arrow line depicts the steady state operating condition. The tilted arrow indicates the AC signal. Power converters can operate beyond the steady state for transient conditions which is depicted as control range by shaded area. Compared to AC systems, AC current injection to a DC system (rechargeable battery) using a power converter is a recent idea [16,17]. Power can flow bidirectionally in a rechargeable battery. The power flow of a conventional bidirectional converter is shown in Figure 1b, which can control DC current for a constant voltage. A controller for conventional bidirectional converter is not enough to inject AC current into a battery. This is because a battery has a small internal impedance and a variable DC voltage offset. A significant amount of AC current causes a small voltage ripple in a battery. The AC injector for rechargeable batteries should be able to control large current within a small voltage window as shown in Figure 1c. Therefore, AC current control for rechargeable batteries is more challenging than conventional bidirectional converters.

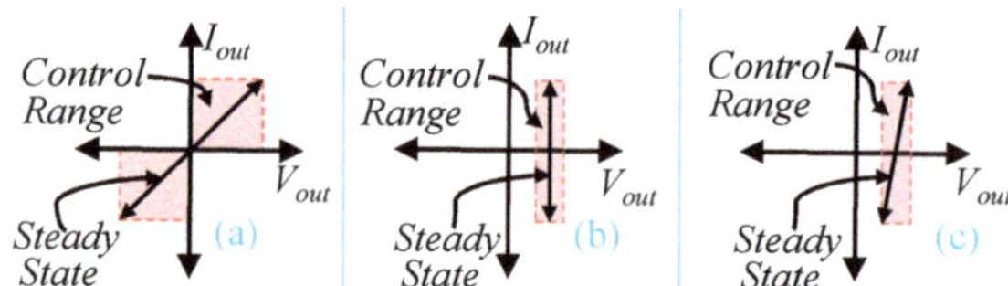

Figure 1. Power converters output: (**a**) AC current injector using an inverter for AC systems (**b**) fixed voltage bi-directional converter for DC bus, and (**c**) AC current injector required for batteries.

A versatile control technique to inject AC current into a rechargeable battery is presented in this manuscript which can inject AC current up to a magnitude of 5 A and a frequency of 2 kHz for a 40 Ah 13.8 V battery. The controller is capable of injecting AC current using two topologies: synchronous buck and H-bridge. In addition, the key contributions in this manuscript are listed as follows:

- A versatile controller is proposed which can inject AC current to a battery for zero, positive, and negative DC current i.e., it works for both offline and online (while charging and discharging).
- A controller gain selection method and controller architecture is developed for AC current injection considering small internal impedance and variable DC offset voltage of a battery.
- A procedure is developed to select topology, modulation, and passive components for AC current injection.

The modes of AC current injection to a battery is discussed in Section 2. Power converter topologies, small signal analysis, and controller architecture are in Sections 3–5. Finally, results are presented in Section 6.

2. Operational Modes for AC Current Injection

AC current can be injected to a rechargeable battery on top of DC charging and discharging current. AC current can also be injected solely without DC current. The modes of AC current injection in a battery are expressed by (1).

$$I_{out}(t) = I_{bat}(t) = \begin{cases} I_{dc} + I_m sin(\omega t), & \text{if Mode 1} \\ 0 + I_m sin(\omega t), & \text{if Mode 2} \\ -I_{dc} + I_m sin(\omega t), & \text{if Mode 3} \end{cases} \tag{1}$$

where, t is the time, I_{bat} is the battery current, I_{dc} is the DC offset current, I_m is the magnitude of AC signal, and ω is the angular frequency for AC signal. I_{dc} is positive for mode 1 where AC current is injected to the battery while charging. I_{dc} is negative for discharging condition. I_{dc} is zero for offline AC injection. The waveforms of each mode of AC current injection are shown in Figure 2a. The existing methods available in the literature for AC current injection using power converters can operate in only one mode at a time. All three modes can be operated using proposed controller which is unique in this manuscript.

The voltage vs. current of a battery during different modes of operation shown in Figure 2b. The power converters output is bounded by the rated operating range of the battery (shaded in area in Figure 2b). The rated maximum and minimum operating voltage of a battery is V_{max} and V_{min}. The rated currents are I_{max} and I_{min}. The control range for different modes are also shaded and has dashed boundary line. The steady state operating conditions are within the control range and shown as tilted arrow lines. The tilted arrow indicates AC voltage and current for different modes. The magnitude of AC current in Figure 2b is corresponding to the magnitude of of current in Figure 2a. To perform all modes of operation, the power converter should be capable of operating within the right half plane of the power quadrant (two quadrant as in Figure 2b).

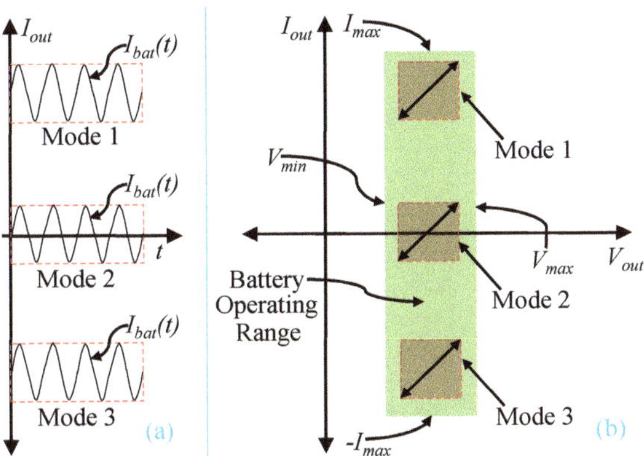

Figure 2. Modes of operation for AC current injector: (**a**) waveform of converter current and (**b**) operating conditions and control ranges.

3. Topologies and Modulation

An AC current injector for batteries was implemented using buck, synchronous buck, boost, and resonant power converters [1–4,7–10]. A synchronous buck converter for AC current injection to battery was used in this paper as shown in Figure 3. In addition, a technique is proposed to use H-bridge topology to inject AC current as shown in Figure 4. Q_1, Q_2, Q_3, and Q_4 are SiC MOSFETs with anti-parallel diodes. An input network was used to deploy three modes of operation for the versatile control which consists of a DC voltage source V_{in}, a resistor R_{in}, and a capacitor C_{in}. I_{in} is input current which could have three parts i_{vin}, i_{rin}, and i_{cin}. S_1 and S_2 are used to control the network as source and/or load which determines online charging/discharging mode or offline mode of operation for AC current injection. The AC injection modes and corresponding possible switch states are in Table 1.

Table 1. Operating modes and input switches.

Switches	$S_1 S_2$	$S_1 S_2$	$S_1 S_2$	$S_1 S_2$
State	00	10	01	11
Mode	2	1	3	1, 2, 3

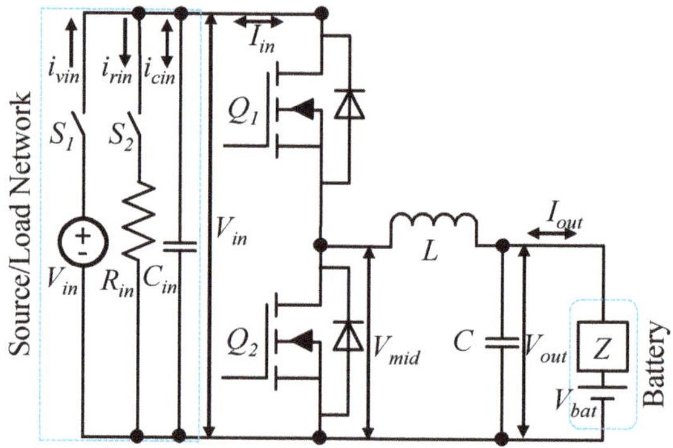

Figure 3. AC injector using synchronous buck converter (2 MOSFTETs).

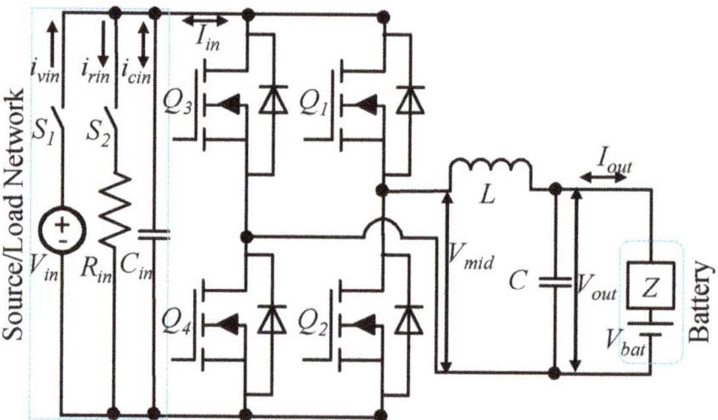

Figure 4. AC injector using H-bridge converter (4 MOSFTETs).

3.1. Design

V_{in} depends on the available input power source. For the buck mode of operation, V_{in} must be greater than open circuit battery voltage, V_{bat}. For maximum control flexibility, the duty cycle is 50% and V_{in} can be selected based on (2).

$$V_{in} = 2V_{bat\_n} \tag{2}$$

where, V_{bat_n} is the nominal battery voltage from datasheet.

R_{in} depends on the value of DC load current in mode 3. For maximum control flexibility, R_{in} can be selected by (3).

$$R_{in} = \frac{V_{in}}{i_{rin}} = \frac{V_{in}}{I_{in\_dc}} = \frac{2V_{bat\_n}}{I_{dc}/2} = \frac{4V_{bat\_n}}{I_{dc}} \tag{3}$$

The input capacitor should be capable to store the charge and energy for half cycle of the injected AC current and discharge in rest of the half cycle. The value of C_{in} is determined based on the allowable input ripple voltage Δv_{cin} and the magnitude, I_m, and frequency, ω, of AC current. The charge storage, sinusoidal ripple current, and ripple voltage across input capacitor is depicted in Figure 5. C_{in} is used to bypass most of the AC current. Therefore, AC current through R_{in} and V_{in} should be negligible. Input capacitor current, i_{cin}, can be represented by (4).

$$i_{cin} \approx I_m sin(\omega t) \tag{4}$$

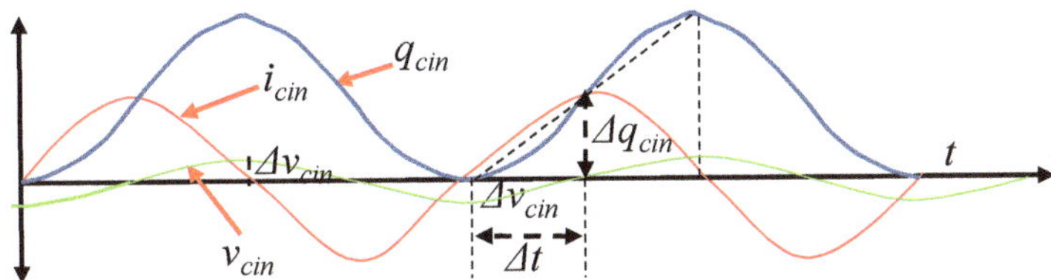

Figure 5. AC states at input capacitor: current, charge, and voltage.

The voltage and the charge of the input capacitor is determined based on the initial conditions and AC current. The charge of the input capacitor, q_{cin}, can be presented by (5).

$$q_{cin} = q_{in} + \int i_{cin} dt \tag{5}$$

where, q_{in} is the steady state stored charge in capacitor. The designed nominal input voltage, V_{in} can be considered is initial condition for capacitor voltage, v_{cin}. Therefore, v_{cin} can be represented by (6).

$$v_{cin} = V_{in} + \Delta v_{cin} sin(\omega t + \pi/2) \tag{6}$$

By the definition of capacitance, the relationship between voltage and charge can be presented by (7).

$$i_{cin} = \frac{dq_{cin}}{dt} = C_{in}\frac{dv_{cin}}{dt} \tag{7}$$

Equation (7) can be rewritten as (8) by linearization. The linearization is shown in Figure 5.

$$\frac{\Delta q_{cin}}{\Delta t} = C_{in}\frac{\Delta v_{cin}}{\Delta t} \tag{8}$$

For 1/4th of the period, Δq_{cin} is determined by (9).

$$\frac{\Delta q_{cin}}{\Delta t} = \frac{\int_0^{T/4} i_{cin} dt}{T/4} = \frac{\int_0^{T/4} I_m sin(\omega t) dt}{T/4} = \frac{I_m}{8\pi} \tag{9}$$

where, T is period. Therefore, (8) can be represented by (10).

$$C_{in}\frac{\Delta v_{cin}}{T/4} = \frac{I_m}{8\pi} \tag{10}$$

Rearranging (10), C_{in} can be calculated from (11).

$$C_{in} = \frac{I_m T}{32\pi \Delta v_{cin}} = \frac{I_m}{32\pi f \Delta v_{cin}} = \frac{I_m}{16\omega \Delta v_{cin}} \tag{11}$$

where, f is the frequency of the AC current and $\omega = 2\pi f$.

The output inductor, L, is selected based on the standard design as in (12) [18].

$$L = \frac{V_{out} \times (V_{in} - V_{out})}{\Delta I_L \times f_{SW} \times V_{in}} \tag{12}$$

where, ΔI_L is the allowable inductor switching ripple current, and f_{SW} is the pulse width modulation switching frequency. Switching ripple current, ΔI_L, is different than controlled AC injection current. The frequency of ΔI_L, is much higher than AC injection current and the magnitude should be almost negligible. A larger inductor gives better attenuation to switching ripple current. However, it reduces the control bandwidth. Considering (2) and control bandwidth, the design guideline for L can be presented in (13).

$$L = \frac{V_{bat\_n} \times (2V_{bat\_n} - V_{bat\_n})}{\Delta I_L \times f_{SW} \times 2V_{bat\_n}} = \frac{V_{bat\_n}}{2 \times \Delta I_L \times f_{SW}} \tag{13}$$

The output capacitor, C, reduces the switching ripple voltage. C can be selected by the standard design as in (14) [18].

$$C = \frac{\Delta I_L}{8 \times f_{SW} \times \Delta V_{out}} \tag{14}$$

A higher value of C provides lower switching ripple. However, the battery itself has very high capacitance. Therefore, a smaller value of C can be used.

The passive components for the topologies are calculated using (2), (3), (11), (13), and (14) considering $V_{bat\_n} = 13.8$ V, $I_m = 5$ A, and $f = 20$ Hz, $f_{SW} = 100$ kHz, $\Delta V_{out} = 7.5$ mV, and ΔI_L is 5% i of I_m. Available standard components are selected for experiments. The calculated and selected value of components are in Table 2.

Table 2. Passive Components for Topologies.

	V_{in}	R_{in}	C_{in}	L	C
Calculated	27.6 V	5.5 Ω	1800 μF	276 μH	41 μF
Selected	27.6 V	4.5 Ω	2200 μF	198 μH	24 μF

3.2. Modulation

The midpoint voltage, V_{mid}, is the voltage across L and C shown in Figures 3 and 4. V_{mid} determines the direction and magnitude of output current I_{out}. V_{mid} is controlled by pulse width modultion (PWM) of the gate driver signal. The output voltage, V_{out}, depends on internal impedance, Z, output current, I_{out} and open circuit battery voltage, V_{bat} as in (15).

$$V_{out} = V_{bat} + I_{out}Z \tag{15}$$

The PWM signals to drive the MOSFETs are shown in Figure 6. The gating signal and midpoint voltage for synchronous buck converter is in Figure 6a. Q_1 and Q_2 are complementary. Deadtime is considered between on states of Q_1 and Q_2. The estimated average midpoint voltage is expressed by (16).

$$\overline{V_{mid}} \approx dV_{in} \approx \overline{V_{out}} \tag{16}$$

where, d is the duty cycle. The effect of synchronous operation and deadtime are excluded from calculation in (16).

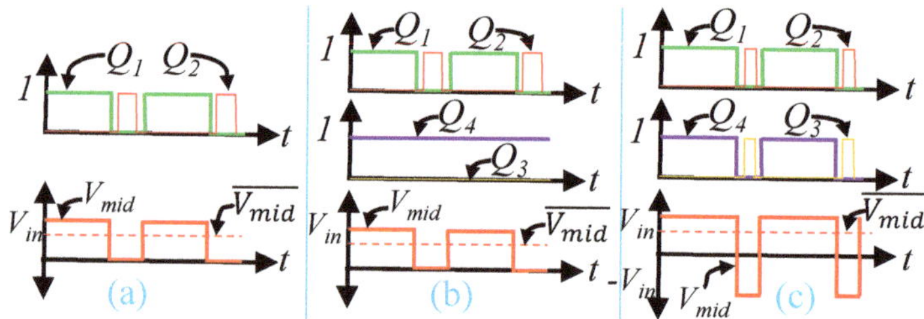

Figure 6. Gate driver PWM signal for power converters: (**a**) synchronous buck (**b**) H-bridge with unipolar switching (**c**) H-bridge with bi-polar switching.

The unipolar PWM signal for H-bridge topology is shown in Figure 6b. V_{out} is always positive. Therefore, Q_3 is always off and Q_4 is always on for unipolar switching. Q_1 and Q_2 are complementary with deadtime. The estimated average voltage for unipolar switching is approximately the same as the synchronous buck converter in (16).

$$\overline{V_{mid}} \approx d_{uni} V_{in} \approx \overline{V_{out}} \tag{17}$$

where, d_{uni} is the duty cycle of Q_1 for unipolar PWM. The relationship between d_{uni} and d is represented by (18).

$$d_{uni} = d \tag{18}$$

The bipolar PWM signal for the H-bridge is in Figure 6c. The estimated average midpoint voltage for bipolar PWM is expressed by (19).

$$\overline{V_{mid}} \approx d_{bi} V_{in} - (1 - d_{bi}) V_{in} \approx \overline{V_{out}} \tag{19}$$

where, d_{bi} is the duty cycle of Q_1 for bipolar switching. $1 - d_{bi}$ is applied to Q_3. Q_1 and Q_2 are complementary with deadtime. Q_3 and Q_4 are also complementary. $d_{bi} V_{in}$ represents positive parts of V_{mid} and $(1 - d_{bi}) V_{in}$ represents the negative part. V_{out} is always positive, therefore $d_{bi} V_{in}$ is always significantly higher than $(1 - d_{bi}) V_{in}$. The relationship between d_{bi} and d is represented by (20)–(22).

$$d = d_{bi} - (1 - d_{bi}) \tag{20}$$

$$d_{bi} = \frac{1}{2} + \frac{d}{2} \tag{21}$$

$$1 - d_{bi} = \frac{1}{2} - \frac{d}{2} \tag{22}$$

PWM techniques are selected based on the topologies used to inject AC current injection. Regardless of the topology chosen, V_{mid} is controlled by the duty cycle, d. d is converted to d_{uni} and d_{bi} for the H-bridge using unipolar and bi-polar switching. For both topologies anti-parallel diode ensures the reverse current flow i.e., discharging condition of the battery.

4. Small Signal Analysis

The output current of synchronous buck and H-bridge converters can be analyzed by a simplified average mode small signal model and transfer functions.

4.1. Simplified Model and Feedforward

Both synchronous buck and H-bridge converters can be represented by a simplified form as shown in Figure 7a. Modulation techniques are used to get the desired value of the midpoint voltage, $\overline{V_{mid}}$. The direction of average output current, $\overline{I_{out}}$, depends on value of $\overline{V_{mid}}$ and $\overline{V_{bat}}$. $\overline{I_{out}}(t)$ determines the mode of operation for the controller. In the steady state equilibrium condition, $\overline{I_{out}}=0$ and $\overline{V_{mid}}=V_{out}=V_{bat}$. A feedforward duty can maintain the power converters in equilibrium. The feedforward duty, d_{ff}, in averaged control mode can be expressed by (23) and (24).

$$\overline{V_{mid}} = V_{in} d_{ff} = V_{bat} \tag{23}$$

$$d_{ff} = \frac{V_{bat}}{V_{in}} \tag{24}$$

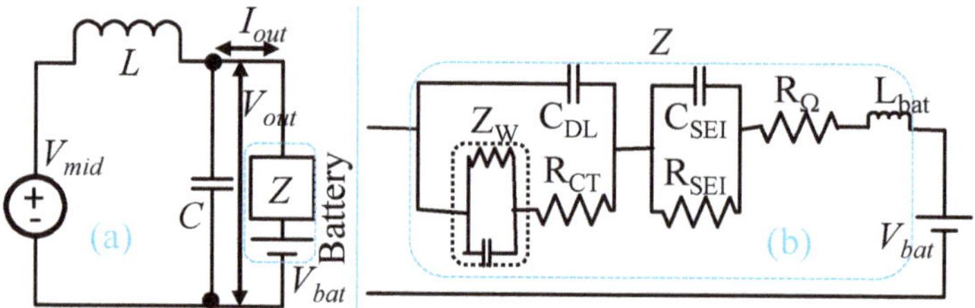

Figure 7. Small signal modeling: (**a**) simplified model of power converters, and (**b**) AR-ECM model of a battery.

4.2. Transfer Function of Switching Power Pole

The average midpoint voltage for a PWM cycle at any instant can be expressed by (25).

$$\overline{V_{mid}}(t) = V_{in}\overline{d}(t) \tag{25}$$

where, $\overline{d}$ is average duty for that cycle and the value of $\overline{d}$ is updated for every cycle based on control requirement. A small perturbation to steady state duty causes perturbation to average midpoint voltage. The small perturbations can be expressed by (26) and (27).

$$\overline{d}(t) = d_{ff}(t) + \tilde{d}(t) \tag{26}$$

$$\overline{V_{mid}}(t) = V_{in}d_{ff}(t) + \widetilde{V_{mid}}(t) \tag{27}$$

Considering the average mode control perturbation and steady input voltage, frequency domain interpretation of the midpoint voltage can be expressed by (28).

$$\overline{V_{mid}}(s) = V_{in}d(s) \tag{28}$$

where, $s = j\omega$, and ω is angular frequency.

4.3. Battery Model

A Li-Ion battery has a small AC voltage ripple for large AC current ripple due to small internal impedance. The adaptive Randle equivalent circuit model (AR-ECM) of a Li-Ion battery can explain the internal impedance. AR-ECM is shown in Figure 7b. It consists of an open circuit voltage source, V_{bat}, battery inductance, L_{bat}, Ohmic resistance, R_Ω, charge transfer resistance, R_{CT}, double layer capacitance, C_{DL}, Warburg impedance, Z_W, solid electrolytic interface resistance, R_{SEI}, and capacitance, C_{SEI}. The overall internal battery impedance in frequency domain, $Z(s)$ is expressed by (29) [19].

$$Z(s) = sL_{bat} + R_\Omega + \cfrac{1}{\frac{1}{R_{SEI}} + sC_{SEI}} + \cfrac{1}{\cfrac{1}{R_{CT}+\sigma\sqrt{\frac{2}{s}}} + sC_{DL}} \tag{29}$$

where, σ is Warburg coefficient, and Warburg impedance $Z_W = \sigma\sqrt{2/s}$. $Z(s)$ is taken into account for versatile current controller design.

For a Valence U-12XP 40 Ah 13.8 V Li-Ion battery, the model components are $L = 0.34\ \mu H$, $R_\Omega = 5.65\ m\Omega$, $C_{DL} = 4.29\ F$, $R_{CT} = 1.23\ m\Omega$ and $\sigma = 2.05 \times 10^{-3}$ at 25% state of charge [9]. The bode plot of the battery impedance is shown in Figure 8. The magnitude is calculated by $20log10(|Z|/R_{base})$, where R_{base} is 1 Ω. The magnitude of impedance is ≈ -40 dBΩ i.e., a small voltage perturbation will cause large current perturbation. The phase plot indicates that at low frequency perturbation the battery behaves in capacitive manner. However, at high frequency the inductive part becomes dominant.

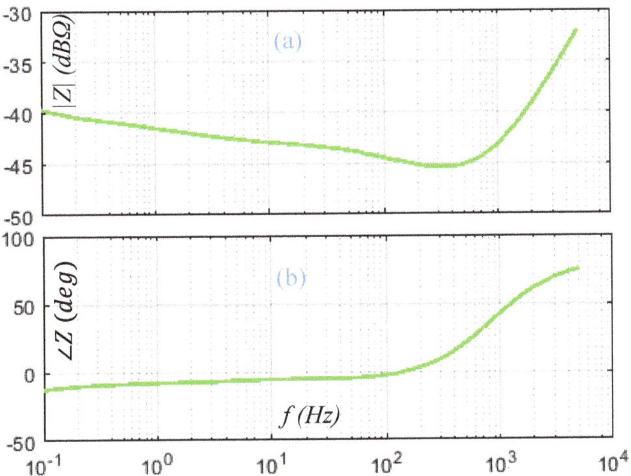

Figure 8. Bode-plot of internal impedance: (**a**) magnitude, (**b**) phase [9].

4.4. Open Loop Transfer Function

The output current is controlled by duty perturbation. Therefore, output current to duty transfer function is defined by G_{id} as in (30).

$$G_{id} = \frac{I_{out}(s)}{d(s)} \tag{30}$$

G_{id} can be determined by circuit analysis from Figure 7a. The overall impedance at the midpoint can be expressed by (31).

$$Z_{mid}(s) = sL + \frac{\frac{1}{sC}Z(s)}{\frac{1}{sC} + Z(s)} \tag{31}$$

The impedance at the output node can be expressed by (32).

$$Z_{out}(s) = \frac{\frac{1}{sC}Z(s)}{\frac{1}{sC} + Z(s)} \tag{32}$$

The inductance current, $I_L(s)$, can be expressed by capacitor current, $I_C(s)$, and output current, $I_{out}(s)$, based on Kirchhoff's current law as in (33).

$$I_L(s) = I_C(s) + I_{out}(s) \tag{33}$$

The equation for output voltage to midpoint voltage can be derived by (34) and (35).

$$H_v(s) = \frac{V_{out}(s)}{V_{mid}(s)} = \frac{I_L(s)Z_{out}(s)}{I_L(s)Z_{mid}(s)} \tag{34}$$

$$H_v(s) = \frac{1}{1 + \frac{sL}{Z(s)} + s^2 LC} \tag{35}$$

The battery ripple voltage transfer function, G_{vd}, can be expressed by (36)

$$G_{vd} = \frac{V_{out}(s)}{d(s)} = \frac{1}{1 + \frac{sL}{Z(s)} + s^2 LC} V_{in} \tag{36}$$

The $V_{out}(s)$ can be expressed by (37).

$$V_{out}(s) = I_{out}(s)Z(s) \tag{37}$$

Using (34) and (37), we can write (38).

$$H_v(s) = \frac{I_{out}(s)Z(s)}{V_{mid}(s)} \tag{38}$$

For average mode control, (38) can be expressed by (39)

$$H_v(s) = \frac{I_{out}(s)Z(s)}{V_{in}d(s)} = \frac{Z(s)}{V_{in}}G_{id} \tag{39}$$

Using (35) and (38), G_{id} can be determined by (40)

$$G_{id} = \frac{1}{Z(s) + sL + s^2LCZ(s)}V_{in} \tag{40}$$

The bode plot of G_{id} is shown in Figure 9 for the proposed system. The bode plot is based on selected parameters of Table 2 and $Z(s)$ from Figure 8. The LC resonance peak is not visible in bode plot due to small internal impedance. The internal parameters of a battery changes with state of charge and aging. Therefore, the nominal value of internal resistance can be used as an alternative to $Z(s)$. At lower frequency, G_{id} has a very high gain (≈ 70 dB). This means a very small duty perturbation causes a very high current perturbation which leads to instability i.e., 1% duty perturbation at 5 Hz would cause 31 A current perturbation whereas recommended current is only 20 A. This instability happens because of very low magnitude of internal impedance (≈ -40 dBΩ). The instability is removed by designing a proper feedback and feedforward controller.

4.5. Feedback Compensator Design

Stability, steady state error, and response time are three important criteria for feedback compensator design. The feedback compensator is designed based on the following steps:

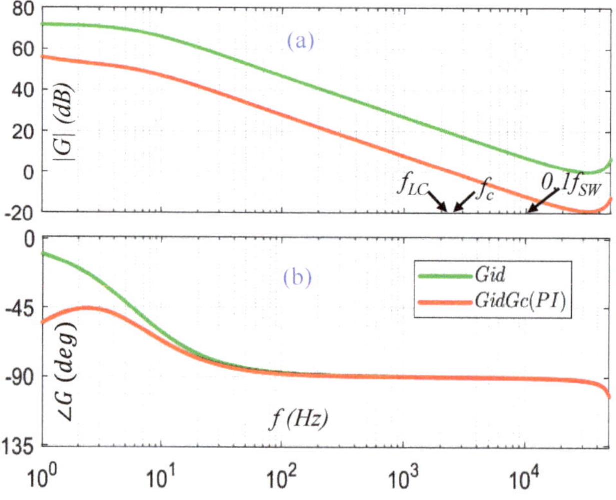

Figure 9. Bode-plot of open loop and compensated loop transfer functions: (**a**) magnitude, (**b**) phase.

4.5.1. Crossover Frequency Selection

The first step of controller design is to select crossover frequency, f_c. To make a stable controller f_c must be slightly higher than the resonance frequency of LC filter, f_{LC}. The resonance frequency for the LC filter is defined by (41).

$$f_{LC} = \frac{1}{2\pi\sqrt{LC}} \tag{41}$$

The rule of thumb to design an efficient and stable controller is to choose crossover frequency less than 1/10th of switching frequency f_{SW}. The crossover frequency selection criterion can be expressed by (42).

$$f_{LC} < f_c \leq \frac{1}{10}f_{SW} \tag{42}$$

Considering the value of f_{LC} and f_{SW} the value of f_c = 2.5 kHz is selected.

4.5.2. Gain Adjustment

The compensated loop gain at cross over frequency, $|G_{id}G_c|_{fc}$, should be 0 dB which can be expressed by (43).

$$|G_{id}G_c|_{fc} = 1 \tag{43}$$

where, G_c is controllers transfer function. (43) can be re-written as (44)

$$k = |G_c|_{fc} = \frac{1}{|G_{id}|_{fc}} \tag{44}$$

where, k is the factor for gain adjustment.

The value of $|G_{id}|_{fc}$ is 18.96 dB i.e., 8.87, therefore the value of k is 0.11. This system is reducing gain instead of boosting. Gain reduction is necessary to improve stability.

4.5.3. Phase Adjustment

Phase adjustment is done through controller selection. A proportional integral (PI) controller is selected for that purpose. The transfer function of the PI controller is expressed by (45).

$$G_c = k_p + \frac{k_i}{s} \tag{45}$$

The value of k_p and k_i are selected in such a way so that $|G_c|$ is 0 dB at f_c and frequency can be swept for all the frequency less than f_c. The value of k_p could be $\approx k$ and the value of k_i can be determined by (46).

$$k_i = k_p\omega_z \tag{46}$$

where, ω_z is the angular form of zero frequency f_z. The value of f_z should be less than the desired control frequency, f, of injected AC current. The relationship can be expressed by (47).

$$f_z < f < f_c \tag{47}$$

The parameters used for phase and gain adjustment are in Table 3. The loop controller transfer functions are shown in Figure 9.

Table 3. Parameters for proposed controller.

f_{LC}	f_c	f_{SW}	$|G_{id}|_{fc}$	$|G_{id}G_c|_{fc}$	f_z	k_p	k_i
2.3 kHz	2.5 kHz	100 kHz	18.96 dB	0 dB	1 Hz	0.11	0.7

5. Controller Architecture

The proposed controller is designed to inject AC current for all three modes as described in Section 2. The proposed versatile controllers architecture is shown in Figure 10. The overall proposed closed loop control system consists of references, feedback controller, sensing gain (H), feedforward term, estimator, PWM modulation transfer function, G_{PWM}, output current transfer function, G_{id}, and voltage transfer function, G_{vi}. The current reference, I_{out}^*, is selected based on desired mode of operation and frequency as in (48).

$$I_{out}^*(t) = I_{dc}^* + I_m^* sin(\omega^* t) \tag{48}$$

where, I_{dc}^* is the DC level of the current based on the desired mode, I_m^* is the desired amplitude of the AC signal, and ω^* is the desired angular frequency of the AC signals.

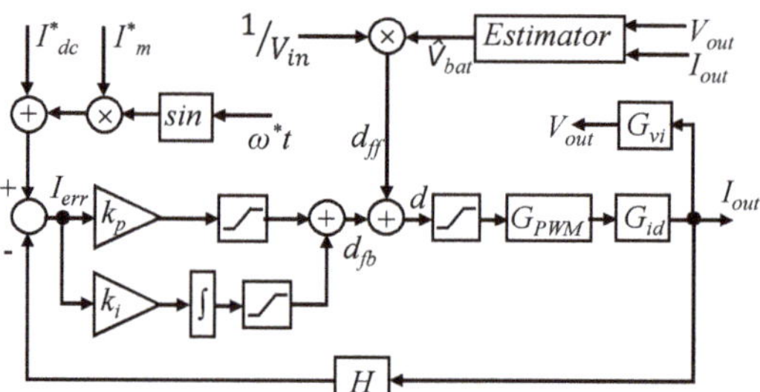

Figure 10. Structure of the proposed controller for AC current injector.

The value of reference current, I_{out}^*, is compared with the output current I_{out} and the current error, I_{err}, is calculated by (49). The error is compensated by a PI feedback controller. The feedback controller gives feedback duty, d_{fb}, by (50).

$$I_{err}(t) = I_{out}^*(t) - I_{out}(t) \tag{49}$$

$$d_{fb}(t) = k_p I_{err}(t) + k_i \int I_{err}(t)dt \tag{50}$$

The saturation blocks of the proposed controller keep the value of d_{fb} within the range of -1 to $+1$. The feedforward duty, d_{ff}, is calculated in (51) using the estimated open circuit voltage, $\widehat{V}_{bat}$, and input voltage, V_{in}.

$$d_{ff}(t) = \frac{\widehat{V}_{bat}(t)}{V_{in}} \tag{51}$$

The feedback and feedforward terms are combined by (52).

$$d = d_{ff} + d_{fb} \tag{52}$$

The value of $\widehat{V}_{bat}$ is estimated using an estimator as shown in Figure 11. The estimator takes V_{out} and I_{out} as input. It uses low pass filters to estimate $\widehat{V}_{dc}$ and $\widehat{I}_{dc}$. A high pass filter is used to estimate $\widehat{v}_{ac}$ and $\widehat{i}_{ac}$. From the value of $\widehat{v}_{ac}$ and $\widehat{i}_{ac}$, the magnitude of battery impedance $|\widehat{Z}|$ is calculated. The saturation block is used to limit the unexpected values. Using $\widehat{I}_{dc}$, $|\widehat{Z}|$ and $\widehat{V}_{dc}$, the value of $\widehat{V}_{bat}$ is calculated by (53).

$$\widehat{V}_{bat} = \widehat{V}_{dc} - \widehat{I}_{dc}|\widehat{Z}| \tag{53}$$

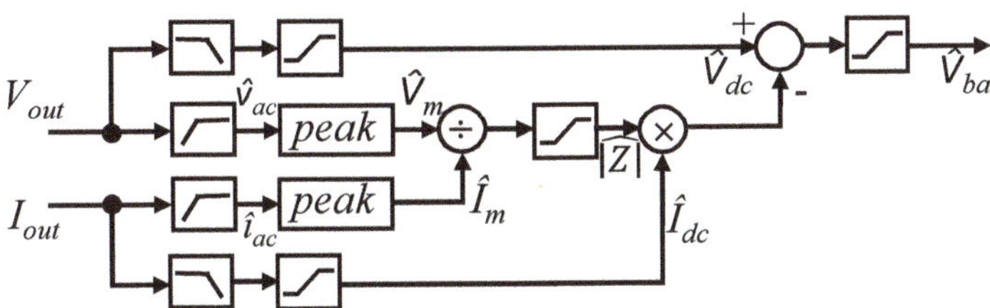

Figure 11. Proposed estimator for the open circuit voltage of a battery.

6. Experimental Results

Experiments were conducted for all three modes using Synchronous buck and H-bridge topologies for a Valence U-12XP 40 Ah 13.8 V Li-Ion battery [20].

6.1. Experimental Setup

A re-configurable test bed was set up for both synchronous buck and H-bridge topologies as shown in Figure 12. The topologies are shown in Figures 3 and 4. The switch, S_H was used to configure the options between synchronous buck and H-bridge topologies. S_H connected or disconnected the Q_3Q_4 leg of H-bridge. The switches S_1 and S_2 were to connect and disconnect the input source, V_{in} and input resistor, R_{in}. The value of V_{in}, C_{in}, R_{in}, L, and C were selected based on the discussion of III.A and Table 2. These MOSFETs were controlled by 100 kHz PWM signal generated by the Simulink based Opal-RT controller. The Opal-RT controller gets voltage and current as feedback through Op-Amp based offset clipping and scaling interface circuits. The proposed controller was implemented with a sample time of 20 µS using Opal-RT (OP4510).

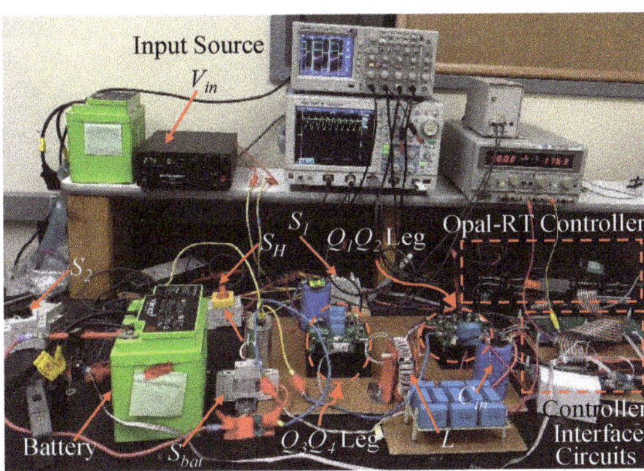

Figure 12. The re-configurable testbed of AC current injector for synchronous buck and H-bridge topologies.

6.2. Waveforms and Analysis

The steady state battery voltage and current for different modes of operation using the synchronous buck converter are shown in Figure 13. The DC current of the battery for different modes were +10 A, 0 A, and −10 A. The same operation was verified using the H-bridge converter also. The amplitude of the AC injection current was 5 A and the frequency was 100 Hz. The battery voltage was 13.5 V as DC average. The battery had a very small internal impedance (maximum 15 mΩ). Therefore, small battery voltage ripple due to AC injection was not visible by DC coupling in an oscilloscope. The battery voltage ripple is shown in Figure 14 using AC coupling. In this case the battery was operated in mode 1 for 100 Hz. The AC current peak was changed from 2 A to 5 A to observe the transient response and the effect of AC current to battery voltage. The controller was successfully able to regulate the current.

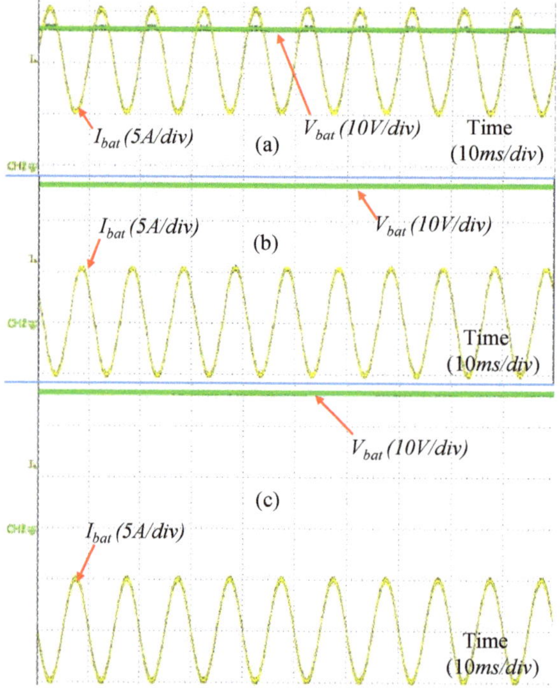

Figure 13. Voltage and current of the battery in steady state for: (**a**) mode 1, (**b**) mode 2, and (**c**) mode 3.

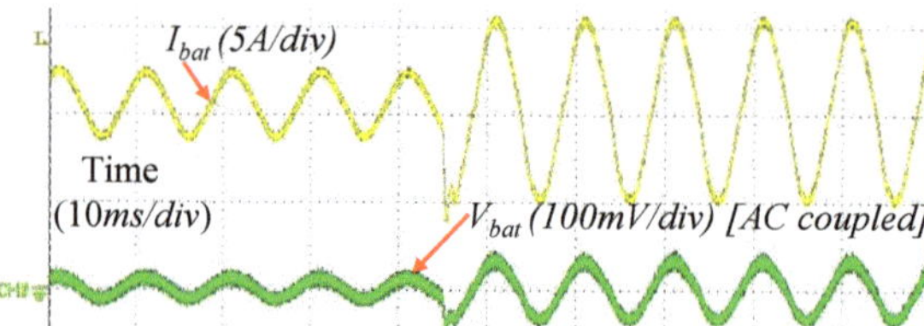

Figure 14. Transient response: magnitude change (from 2 A AC peak to 5 A) of battery current in mode 1 at 100 Hz.

The AC part of the battery current was changed from 0 to 2 kHz for all three modes of operation using both synchronous buck and H-bridge converters. The battery voltage and current for 10 Hz, 100 Hz and 1 kHz for mode 1 using the synchronous buck converter is shown in Figure 15. The dynamic response for mode change using the proposed controller is shown in Figure 16. In these cases, both S_1 and S_2 were turned on. The proposed versatile controller was validated by successfully testing the additional conditions listed in Table 4.

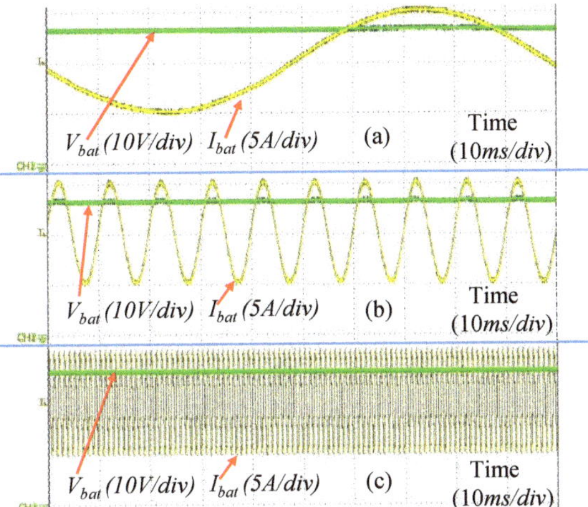

Figure 15. Voltage and current of the battery in steady state in mode 1 for: (**a**) 10 Hz, (**b**) 100 Hz, and (**c**) 1000 Hz.

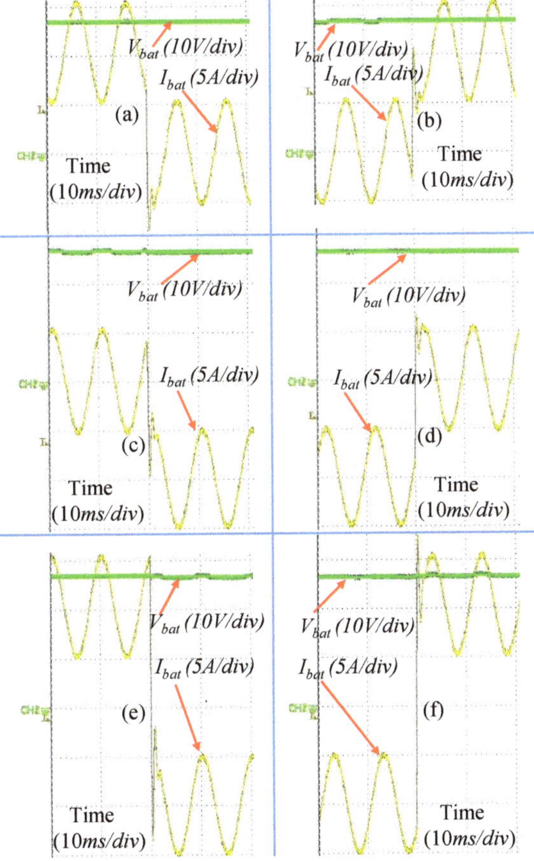

Figure 16. Dynamic response: mode change (for 100 Hz 5A AC peak) of battery current: (**a**) mode 1 to 2, (**b**) mode 2 to 1, (**c**) mode 2 to 3, (**d**) mode 3 to 2, (**e**) mode 1 to 3, and (**f**) mode 3 to 1.

Table 4. The test conditions for controller validation.

Condition	Sync-Buck			H-Bridge		
	Mode 1	Mode 2	Mode 3	Mode 1	Mode 2	Mode 3
Steady State	✓	✓	✓	✓	✓	✓
	0, 2, 3, & 5 A AC peak 0, 0.1, 10, 50, 100, 300, 500, 1k, 1.5k, & 2k Hz AC					
Dynamic Response	Mode change: 1-2, 2-1, 2-3, 3-2, 1-3, & 3-1. Maximum mode transition time: 2 mS Maximum overshoot current for mode change: 2.5 A					
	Magnitude change of AC current: ✓ transient response for 0-5 A, 5-0 A, 2-5 A, & 5-2 A					
	Frequency change of AC current: ✓ frequency transition for 0–50 Hz, 50–100 Hz, 100–50 Hz, & 0–2 kHz					

7. Discussion

A passive component selection guideline and controller design method for AC current injection to battery is presented in this manuscript. The novelty in this manuscript is to operate power converters for batteries in two quadrant applying the proposed guideline and method. The novelty is justified by comparing the AC current injection method for batteries in the recent journals as in Table 5.

Table 5. Comparison of AC current injection method for batteries.

Reference	Year	Power Quadrant	DC Current	AC Waveform	Topology	Control	Notes
[10]	2017	1	-	0.1–1.9 kHz multisine	DC/DC boost/buck	discharge, feedback	voltage control
[9]	2020	1	+	0.1 Hz–5 kHz	DC/DC	charge	feedback
[21]	2021	1	+	20 mHz–2 kHz	DC/DC,	charge	feedback
[22]	2021	2	±	<60 Hz sine	AC/DC/DC grid-tied	charge/discharge feedback	distort if $f > 60$ Hz
This work	2021	2	±, 0	0.1 Hz–2 kHz sine	DC/DC, Sync-buck, H-bridge	charge/discharge, feedback, feedforward	passive component selection

8. Conclusions

A complete controller design method for AC current injection to a battery is developed in this manuscript. This method includes the passive component selection, controller gain selection, and controller architecture. The controller was able to operate for both synchronous buck and H-bridge converters. The proposed versatile controller successfully controlled the AC battery current for offline and online modes (while charging/discharging) over a range of frequencies and magnitudes as expected from the bode plot of the simulation. The proposed controller and design method will be useful for internal impedance measurement, advanced charging, and preheating applications. The advantage of the proposed technique is that it does not require any changes in the control architecture for different mode of operation for various topologies over a range of frequencies. The disadvantage of the proposed controller is that it requires voltage feedback for optimized operation. The performance of this controller with AC/DC/DC grid tied application is not verified yet and will be investigated in the future.

Author Contributions: All the authors contributed substantially to the manuscript. Contributions of each author are as follows: conceptualization, S.M.R.I.; methodology, S.M.R.I.; software, S.M.R.I.; validation, S.M.R.I. and S.-Y.P.; formal analysis, S.M.R.I.; investigation, S.M.R.I.; resources, S.M.R.I. and S.-Y.P.; data curation, S.M.R.I.; writing—original draft preparation, S.M.R.I.; writing—review and editing, S.M.R.I. and S.-Y.P.; visualization, S.M.R.I.; supervision, S.-Y.P.; project administration, S.-Y.P.; funding acquisition, S.-Y.P. All authors have read and agreed to the published version of the manuscript.

Funding: This research was funded by the National Science Foundation of under Grant No. 1454578.

Institutional Review Board Statement: Not applicable.

Informed Consent Statement: Not applicable.

Data Availability Statement: Not applicable.

Acknowledgments: This work was supported by the National Science Foundation under grant no. 1454578. However, any opinions, findings, conclusions, or recommendations expressed herein are those of the authors and do not necessarily reflect the views of the National Science Foundation.

Conflicts of Interest: The authors declare no conflict of interest.The funders had no role in the design of the study; in the collection, analyses, or interpretation of data; in the writing of the manuscript, or in the decision to publish the results.

References

1. Shang, Y.; Liu, K.; Cui, N.; Zhang, Q.; Zhang, C. A Sine-Wave Heating Circuit for Automotive Battery Self-Heating at Subzero Temperatures. *IEEE Trans. Ind. Inform.* **2020**, *16*, 3355–3365. [CrossRef]
2. Din, E.; Schaef, C.; Moffat, K.; Stauth, J.T. A Scalable Active Battery Management System With Embedded Real-Time Electrochemical Impedance Spectroscopy. *IEEE Trans. Power Electron.* **2017**, *32*, 5688–5698. [CrossRef]
3. Varnosfaderani, M.A.; Strickland, D. A Comparison of Online Electrochemical Spectroscopy Impedance Estimation of Batteries. *IEEE Access* **2018**, *6*, 23668–23677. [CrossRef]
4. Lee, Y.D.; Park, S.Y. Electrochemical State-Based Sinusoidal Ripple Current Charging Control. *IEEE Trans. Power Electron.* **2015**, *30*, 4232–4243. [CrossRef]
5. Lasia, A. *Electrochemical Impedance Spectroscopy and Its Applications*; Springer: New York, NY, USA, 2014.
6. Orazem, M.E.; Tribollet, B. *Electrochemical Impedance Spectroscopy*; John Wiley & Sons: Hoboken, NJ, USA, 2011; Volume 48.
7. Lee, Y.D.; Park, S.Y.; Han, S.B. Online Embedded Impedance Measurement Using High-Power Battery Charger. *IEEE Trans. Ind. Appl.* **2015**, *51*, 498–508. [CrossRef]
8. Chen, J.; Lee, Y.; Park, S. Adaptive PI gain control to realize sinusoidal ripple current charging. In Proceedings of the 2015 9th International Conference on Power Electronics and ECCE Asia (ICPE-ECCE Asia), Seoul, Korea, 1–5 June 2015; pp. 2582–2589. [CrossRef]
9. Islam, S.M.R.; Park, S. Precise Online Electrochemical Impedance Spectroscopy Strategies for Li-Ion Batteries. *IEEE Trans. Ind. Appl.* **2020**, *56*, 1661–1669. [CrossRef]
10. Qahouq, J.A.A.; Xia, Z. Single-Perturbation-Cycle Online Battery Impedance Spectrum Measurement Method With Closed-Loop Control of Power Converter. *IEEE Trans. Ind. Electron.* **2017**, *64*, 7019–7029. [CrossRef]
11. Hajisadeghian, H.; Motie Birjandi, A.A.; Nahavandi, R. Sliding Mode Controller (SMC) For Sinusoidal Ripple Current (SRC) Charge of Li-ion Battery. In Proceedings of the 2019 10th International Power Electronics, Drive Systems and Technologies Conference (PEDSTC), Shiraz, Iran, 12–14 February 2019; pp. 600–605. [CrossRef]
12. Sadeghi, E.; Zand, M.H.; Hamzeh, M.; Saif, M.; Alavi, S.M.M. Controllable Electrochemical Impedance Spectroscopy: From Circuit Design to Control and Data Analysis. *IEEE Trans. Power Electron.* **2020**, *35*, 9933–9942. [CrossRef]
13. Bayati, M.; Abedi, M.; Gharehpetian, G.B.; Farahmandrad, M. Sinusoidal-Ripple Current Control in Battery Charger of Electric Vehicles. *IEEE Trans. Veh. Technol.* **2020**, *69*, 7201–7210. [CrossRef]
14. Teodorescu, R.; Liserre, M.; Rodriguez, P. *Grid Converters for Photovoltaic and Wind Power Systems*; John Wiley & Sons: Hoboken, NJ, USA, 2011; Volume 29.
15. Li, J. Design and Control Optimisation of a Novel Bypass-embedded Multilevel Multicell Inverter for Hybrid Electric Vehicle Drives. In Proceedings of the 2020 IEEE 11th International Symposium on Power Electronics for Distributed Generation Systems (PEDG), Dubrovnik, Croatia, 28 September–1 October 2020; pp. 382–385. [CrossRef]
16. Chen, L.; Wu, S.; Shieh, D.; Chen, T. Sinusoidal-Ripple-Current Charging Strategy and Optimal Charging Frequency Study for Li-Ion Batteries. *IEEE Trans. Ind. Electron.* **2013**, *60*, 88–97. [CrossRef]
17. Koch, R.; Kuhn, R.; Zilberman, I.; Jossen, A. Electrochemical impedance spectroscopy for online battery monitoring—Power electronics control. In Proceedings of the 2014 16th European Conference on Power Electronics and Applications, Lappeenranta, Finland, 26–28 August 2014; pp. 1–10. [CrossRef]
18. Instruments, T. *Basic Calculation of a Buck Converter's Power Stage*; Application Report-SLVA477B, Rev; Richtek Technology Corporation: Zhubei City, Taiwan, 2015.
19. Islam, S.; Park, S.Y.; Balasingam, B. Unification of Internal Resistance Estimation Methods for Li-Ion Batteries Using Hysteresis-Free Equivalent Circuit Models. *Batteries* **2020**, *6*, 32. [CrossRef]

20. Lithium Werks. Valence U1-12XP Batteries. Available online: https://www.lithionbattery.com/products/modules/u-charge-xp/ (accessed on 13 July 2021).
21. Koseoglou, M.; Tsioumas, E.; Papagiannis, D.; Jabbour, N.; Mademlis, C. A Novel On-Board Electrochemical Impedance Spectroscopy System for Real-Time Battery Impedance Estimation. *IEEE Trans. Power Electron.* **2021**, *36*, 10776–10787. [CrossRef]
22. Tang, C.Y.; Chen, P.T.; Jheng, J.H. Bidirectional Power Flow Control Integrated With Pulse and Sinusoidal-Ripple-Current Charging Strategies for Three-Phase Grid-Tied Converters. *IEEE Access* **2021**, *9*, 42151–42160. [CrossRef]

Article

Comparison of Aqueous- and Non-Aqueous-Based Binder Polymers and the Mixing Ratios for Zn//MnO$_2$ Batteries with Mildly Acidic Aqueous Electrolytes

Oliver Fitz [1,*], Stefan Ingenhoven [1], Christian Bischoff [1], Harald Gentischer [1], Kai Peter Birke [2], Dragos Saracsan [3] and Daniel Biro [4]

[1] Fraunhofer Institute for Solar Energy Systems ISE Battery Cell Technology, Department of Electrical Energy Storage, 79110 Freiburg, Germany; stefan.ingenhoven@ise.fraunhofer.de (S.I.); christian.bischoff@ise.fraunhofer.de (C.B.); harald.gentischer@ise.fraunhofer.de (H.G.)

[2] Chair for Electrical Energy Storage Systems, Institute for Photovoltaics, University of Stuttgart, 70569 Stuttgart, Germany; peter.birke@ipv.uni-stuttgart.de

[3] Faculty of Mechanical and Process Engineering, Offenburg University of Applied Sciences, 77652 Offenburg, Germany; dragos.saracsan@hs-offenburg.de

[4] Fraunhofer Institute for Solar Energy Systems ISE, Head of Department of Electrical Energy Storage, 79110 Freiburg, Germany; daniel.biro@ise.fraunhofer.de

* Correspondence: oliver.fitz@ise.fraunhofer.de

Citation: Fitz, O.; Ingenhoven, S.; Bischoff, C.; Gentischer, H.; Birke, K.P.; Saracsan, D.; Biro, D. Comparison of Aqueous- and Non-Aqueous-Based Binder Polymers and the Mixing Ratios for Zn//MnO$_2$ Batteries with Mildly Acidic Aqueous Electrolytes. *Batteries* **2021**, *7*, 40. https://doi.org/10.3390/batteries7020040

Academic Editor: Claudio Gerbaldi

Received: 20 May 2021
Accepted: 15 June 2021
Published: 18 June 2021

Publisher's Note: MDPI stays neutral with regard to jurisdictional claims in published maps and institutional affiliations.

Abstract: Considering the literature for aqueous rechargeable Zn//MnO$_2$ batteries with acidic electrolytes using the doctor blade coating of the active material (AM), carbon black (CB), and binder polymer (BP) for the positive electrode fabrication, different binder types with (non-)aqueous solvents were introduced so far. Furthermore, in most of the cases, relatively high passive material (CB+BP) shares ~30 wt% were applied. The first part of this work focuses on different selected BPs: polyacrylonitrile (PAN), carboxymethyl cellulose (CMC), styrene butadiene rubber (SBR), cellulose acetate (CA), and nitrile butadiene rubber (NBR). They were used together with (non-)aqueous solvents: DI-water, methyl ethyl ketone (MEK), and dimethyl sulfoxide (DMSO). By performing mechanical, electrochemical and optical characterizations, a better overall performance of the BPs using aqueous solvents was found in aqueous 2 M ZnSO$_4$ + 0.1 M MnSO$_4$ electrolyte (i.e., BP *LA133*: 150 mAh·g^{-1} and 189 mWh·g^{-1} @ 160 mA·g^{-1}). The second part focuses on the mixing ratio of the electrode components, aiming at the decrease of the commonly used passive material share of ~30 wt% for an industrial-oriented electrode fabrication, while still maintaining the electrochemical performance. Here, the absolute CB share and the CB/BP ratio are found to be important parameters for an application-oriented electrode fabrication (i.e., high energy/power applications).

Keywords: zinc ion batteries; stationary energy storage; polymer binder; solvent; doctor blade coating; manganese dioxide; mixing ratio; electrochemical impedance spectroscopy; SEM+EDX; electrode fabrication

1. Introduction

The research on aqueous battery technologies for stationary applications such as the aqueous rechargeable zinc-ion battery (ARZIB) is getting more and more attention. The ARZIB technology combines inherent safety, environmental friendliness, material abundance, low active material costs, and promising cycling stabilities. This publication focuses on the Zn//MnO$_2$ chemistry with zinc on the negative electrode side and manganese dioxide on the positive electrode side, together with an acidic ZnSO$_4$ + MnSO$_4$-based electrolyte.

Recent publications show different ways of fabrication procedures for the positive electrode with doctor blade coating on a current collector foil, electrodeposition of the active material or pasting of the electrode material (summary in Supplementary Figure S1) [1–24].

As the literature shows, the doctor blade coating is the most prominent fabrication procedure. For this, an electrode slurry with active material (MnO_2, AM), a conductive agent such as carbon black (CB), and a binder polymer (BP) in different fractions, most commonly 70/20/10 wt% (AM/CB/BP), is coated on the current collector sheet using a doctor blade [4,5,8–10,13,14,25–28].

Different polymer binders such as *LA133* (based on polyacrylonitrile (PAN)), carboxymethyl cellulose (CMC), styrene butadiene rubber (SBR), polytetrafluoroethylene (PTFE), and polyvinylidene fluoride (PVDF) based on different solvents such as DI-water or non-aqueous N-methyl-2-pyrrolidone (NMP) were used in the literature so far (summary in Supplementary Figure S1) [3,23]. Generally, an aqueous slurry processing enables inexpensive, safe, and environmentally friendly fabrication of the positive electrode. Still, a better understanding of the relation between the binder solvent (aqueous/non-aqueous) and the aqueous electrolyte, together with the reaction mechanism of ARZIBs, can enable a targeted electrode fabrication and an improved cell performance.

For a reasonable and application-oriented electrode fabrication, the consideration of the underlying reaction mechanism of the ARZIB cell chemistry is of high importance: So far, the reaction mechanism for the positive electrode seems to be a combination of multi-step chemical reactions such as a Zn^{2+} intercalation [12,26–36], a H^+/Zn^{2+}-Co-intercalation [17,20,37–39], a dissolution of the MnO_2 active material loading and their re-deposition on the positive electrode surface together with the deposition of pre-dissolved Mn^{2+} ions from the electrolyte (as far as $MnSO_4$ is pre-dissolved in the electrolyte) [19,33,37,40,41]. The reactions are often accompanied by a pH-dependent zinc hydroxide sulfate (ZHS) formation [14,19,28,33,37,39–44]. Furthermore, the formation of inert $Zn_xMn_yO_z$-species (i.e., $ZnMn_2O_4$) on the positive electrode were introduced in the recent literature [11,39,40]. For the negative electrode, there is a reversible Zn plating/stripping at the zinc electrode [9,14,20,42,45,46] as a consequence of the acidic pH value in accordance to the potential-pH diagrams for zinc [36,47]. This is in contrast to the alkaline $Zn//MnO_2$ batteries with the formation of irreversible Zn phases [36,48,49].

In consideration of the different characteristics (intercalation and conversion reactions) of the reaction mechanism, there are various requirements for the positive electrode. For example:

- Stability of the binder polymer in aqueous electrolyte (no peeling of the coating from the current collector due to dissolution or strong swelling of the binder polymer).
- Porosity of the coating for high specific surface area to provide a high (electrochemically active) surface area.
- Wetting of the coating by the aqueous electrolyte to enable deposition and intercalation processes of the dissolved components of the electrolyte.
- Advantageous transport characteristics of the porous electrode for the ions and electrons.

Herein, selected BPs (polyacrylonitrile (PAN), carboxymethyl cellulose (CMC), styrene butadiene rubber (SBR), cellulose acetate (CA), nitrile butadiene rubber (NBR)) with aqueous and non-aqueous solvents (DI-water, methyl ethyl ketone (MEK), and dimethyl sulfoxide (DMSO), for details see Section 4.1. Materials) are compared by applying a mechanical stress test (MST) on the positive electrode sheet and rate capability tests (RCT), together with electrochemical impedance spectroscopy (EIS) measurements, on experimental battery cells. The experimental results are compared with SEM+EDX images of cross-sections of pristine and cycled electrode sheets, giving further insights into the homogeneity, material distribution and morphology of the coatings. Furthermore, the mixing ratio of the electrode components is systematically investigated by using a ternary plot visualization to evaluate the influence of the different shares of the positive electrode ingredients on the mechanical stability of the coating and the cycling performance. Finally, recommendations for the selection of the binder polymer and the mixing ratios for the utilization in ARZIBs are made.

This study, besides well-known binder polymers such as CMC, SBR and *LA133*, also considers new BP/solvent combinations such as PAN+DMSO, NBR+MEK and CA+MEK for the utilization in ARZIBs.

As there are only few publications dealing with a comparison of different binder polymers for ARZIBs in literature (to our knowledge, only [4]), this publication is intended to provide a basis for comparing different BP/solvent combinations as well as different mixing ratios for the positive electrode fabrication. Hereby, the focus is set on NMP-free solvents, as NMP-free processing enables lower safety precautions (note: reproductive toxicity of NMP) and a clean-room-free processing, lowering the production costs and being a decisive advantage over the lithium-ion battery technology.

2. Results and Discussion

2.1. Binder Polymer Variation

For the binder polymer (BP) variation, different BPs based on aqueous solutions or dispersions, as well as non-aqueous solutions, were investigated. Since an aqueous electrolyte is used in the cell, the mechanical stability and structural integrity, especially of those electrodes produced with an aqueous binder preparation, were of interest.

In the following, the BPs based on aqueous solvent preparations are summarized as *aqueous electrodes*, whereas the BPs based on non-aqueous solvent preparations are summarized as *non-aqueous electrodes*.

2.1.1. Mechanical Stress Test (MST)

For the evaluation of the stability and structural integrity of the electrode coating with different BPs, the electrode coins were first mechanically tested by applying a mechanical stress test (MST).

Interestingly, after the MST, the aqueous electrode coatings are washed off and dissolved in DI-water (see Table 1a–d) leading to a turbidity of the solution (see Figure 1a). For the electrolyte solution, the MST shows no dissolution of the coating. Only the PVP binder flakes off the current collector in the electrolyte solution, but still does not dissolve. In contrast, the non-aqueous electrode coatings are stable in both DI-water and electrolyte solutions (see Table 1e–g and Figure 1b). The results of the MST follow the observations seen in Chang et al. (2020) for the non-aqueous binder (here: PVDF+NMP) and the aqueous CMC binder [4].

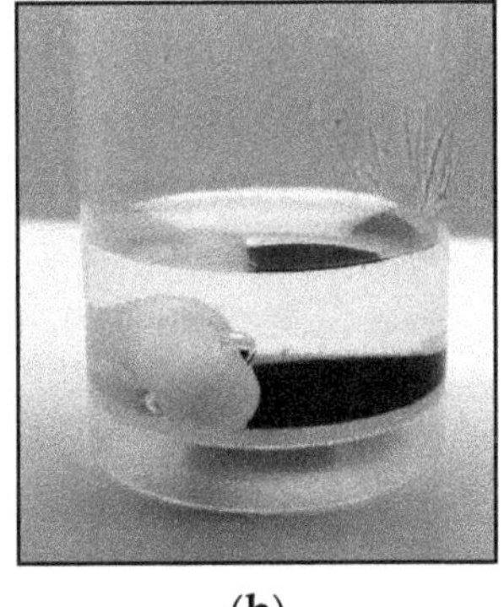

(a)   (b)

Figure 1. Example side views of the beakers after the MST with an aqueous (here: PAN-based LA133, (**a**) and non-aqueous (here: CA (MEK), (**b**) electrode coating in DI-water, respectively. For the aqueous coatings, the MST leads to a dissolution of the coatings and a turbidity of the solutions (the stirring bar is still inside the beaker but not visible anymore), whereas the non-aqueous coatings show no damages.

Table 1. Mechanical stress test of the electrode coatings with different binder polymers in DI-water and electrolyte (2 M $ZnSO_4$ + 0.1 M $MnSO_4$), respectively.

Binder (aq)	DI-Water	Electrolyte	Binder (non-aq)	DI-Water	Electrolyte
(a) PAN	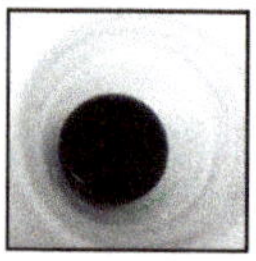		**(e)** PAN (DMSO)	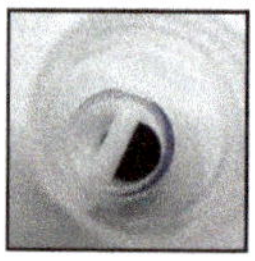	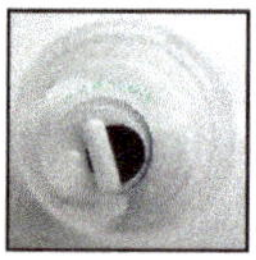
(b) CMC/SBR	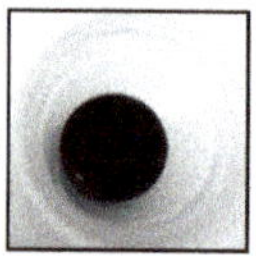	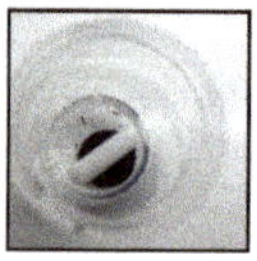	**(f)** NBR (MEK)	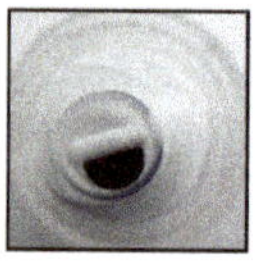	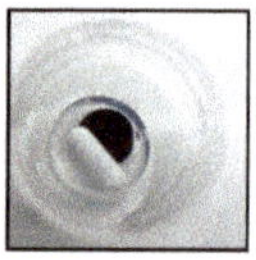
(c) CMC	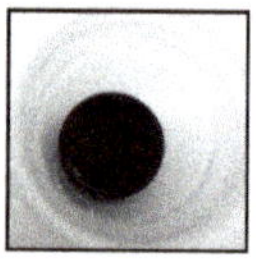	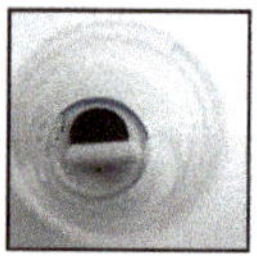	**(g)** CA (MEK)	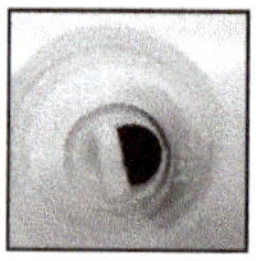	
(d) PVP	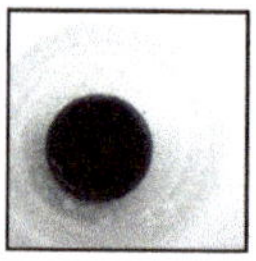	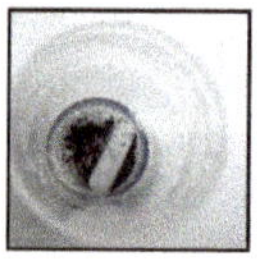			

For the aqueous binders, the different behaviour in DI-water and electrolyte solution could be referred to the reduced solvation free energy of the water molecules in consequence of the $ZnSO_4$ and $MnSO_4$ addition in the electrolyte ([50] and Supplementary Figure S12), retarding and reducing the BP dissolution: Instead of a dissolution, the BP could only swell while still maintaining the structural integrity of the coating.

For the non-aqueous binders, as was expected, both the DI-water and the aqueous electrolyte solution are not affecting the structural integrity of the coating during the MST. Furthermore, in consequence of the non-aqueous processing of the coating, no swelling of the coating should occur here, in contrast to the previous aqueous binder coatings.

Based on the observations of the MST, a different behaviour during the cycling of the aqueous and non-aqueous electrodes in the aqueous $ZnSO_4$/MnSO4 electrolyte is expected and shown in the following section.

2.1.2. Rate Capability Test (RCT)

For the evaluation of the electrochemical performance of the electrode coatings with different BPs and solvents (see Table 2), a rate capability test (RCT) was performed under various current rates. The current rates are given in $mA \cdot g^{-1}$ based on the active material mass loading. The RCT results are shown by plotting both the specific discharge (DCH) capacity as well as the specific discharge energy, together with the efficiency values, over the cycle number, which allows a more in-depth analysis of the effects of the different binder types.

The results of the RCT with specific capacities (see Figure 2a) shows significantly higher capacity values for all the aqueous binders compared to the non-aqueous binders.

Table 2. Overview of the slurry preparations with aqueous- and non-aqueous-based binder polymer (BP), active material (AM), and carbon black (CB).

#	Ratio AM/CB/BP	Binder #1	Binder #2	Solvent	Comment
1	70/20/10	PAN (*LA133*)		DI-water	aqueous suspension
2	70/20/10	CMC		DI-water	aqueous solution
3	70/20/10	50 wt% CMC	50 wt% SBR	DI-water	aqueous solution/suspension
4	70/20/10	PAN		DMSO	solution
5	70/20/10	NBR		MEK	solution
6	70/20/10	CA		MEK	solution

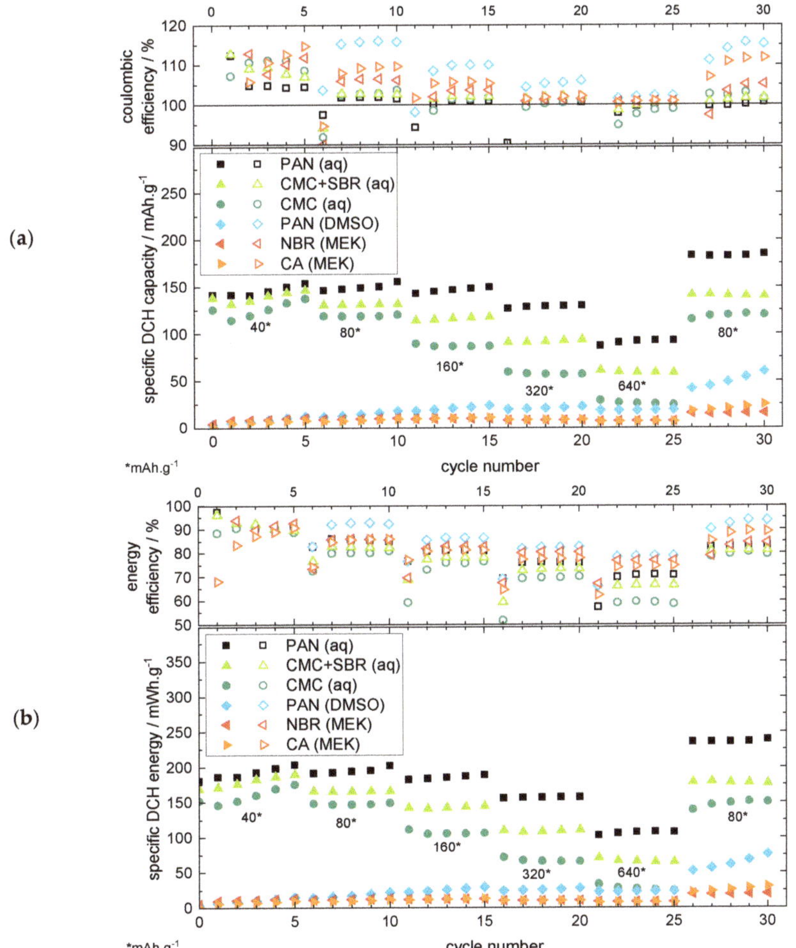

Figure 2. Results of the RCT plotting the specific discharge (DCH) capacity (**a**) and energy (**b**) under variation of the current rate over the cycle number for the non-aqueous electrode coatings.

The aqueous PAN-based electrode shows the highest specific capacity for all current rates with ~150 mAh·g^{-1} at 40 mA·g^{-1} and ~90 mAh·g^{-1} at 640 mA·g^{-1}. Both the CMC+SBR and the CMC binders show lower capacity values, with the lowest values for the pure CMC binder. The better performance of the CMC binder with SBR addition could be related to the better adhesion of the coating on the current collector, as well as a better flexibility of the coating through the addition of the SBR elastomer [51]. The PVP binder, which showed a flaking off the current collector in the MST, was not able to finish the RCT, which can be related to the low stability of the electrode coating (see RCT results in Supplementary Figure S6). Therefore, the PVP binder will not be discussed further in the following sections.

The non-aqueous binders show low-capacity values <20 mAh·g^{-1} in the first cycles, but interestingly it is with an increasing trend, even if the current rate increases. This is especially prevalent during the last cycles with 80 mA·g^{-1}, as the increasing discharge capacity trend becomes clearly visible. This behaviour can be related to the surface deposition/dissolution of MnO_2 as the predominant reaction mechanism for the non-aqueous electrode, whereby a growing MnO_2 surface layer could be formed on the positive electrode [19,33,37,40,41]. The lower overall discharge capacity of the non-aqueous binders compared to the aqueous binders could be related to the observations of the MST results of the previous section, where a swelling and wetting of the electrodes in the electrolyte was suggested for the aqueous but not (or to a lesser extent) for the non-aqueous electrode coatings. In consequence, the MnO_2 active material loading of the electrode coating could only become available for the aqueous binders by dissolution/deposition processes during charging/discharging, leading to additional capacity. This suggestion is confirmed by the latest literature for ARZIBs [19,33,37,40,41]. For the non-aqueous binders, the active material loading could mainly be isolated by the binder.

The coulombic efficiencies (CE, see Figure 2a) for all the binder types are located above or at ~100%, which again can be related to the MnO_2 deposition/dissolution process. The dissolution of the MnO_2 active material loading as well as the additionally deposited MnO_2 could lead to an additional discharge capacity, which results in CE values >100%. The CE values for the non-aqueous binders are generally located above the values for the aqueous binders, indicating a more dominant MnO_2 deposition/dissolution mechanism. This could be related to the inactivity of the isolated MnO_2 loading within the positive electrode coating, as no swelling of the binder is suggested (see MST section), which could enable a material deposition/dissolution only on the surface of the electrode with better efficiency values.

The RCT results showing the specific energy (see Figure 2b) additionally reveal the effects of overpotentials during charge/discharge by also considering the potential value (besides the capacity value, see the equation in Supplementary Figure S5) of the particular cycle. As the energy output of a battery system is of even higher importance for the application than only the capacity output, this way of plotting can be regarded as being a more application-oriented plot. For the specific energy plots, the comparison of the aqueous/non-aqueous binder types show the same characteristics as the specific capacity plots. Nevertheless, distinctions in terms of the energy efficiency (EE) values are noticeable: the EE values decrease for increasing current rates. This indicates an increasing inner resistance and overpotential of the cell [52]. For the non-aqueous binders, the energy efficiency values are generally located above the values of the aqueous binders, as seen before in the capacity plot, again indicating a more dominant MnO_2 deposition/dissolution mechanism.

In Figure 3, the potential curves for the 10th and 30th cycle, respectively, each with the current rate of 80 mA·g^{-1}, are shown for the non-aqueous binders. Generally, the shape of the discharge curves of all binder/solvent combinations show two potential plateaus divided by a potential bend (see exemplarily Figure 3, marker 1). The charge curve again shows two potential plateaus (see Supplementary Figure S11 for more details). The potential curves of the non-aqueous binders (see Figure 3b) each have a significant increase in the specific capacities compared to the aqueous binders (see Figure 3a). Nevertheless, the potential curves characteristics stay the same, as the characteristic potential bend for ARZIBs (with MnO_2 active material) indicates (see Figure 3, marker 1). The potential curves also show that the capacity increase from the 10th to the 30th cycle is the highest for both of the PAN-based electrodes (aqueous and non-aqueous), which could explain the good overall performance of the aqueous PAN-based electrode in the RCT (see Figure 2a).

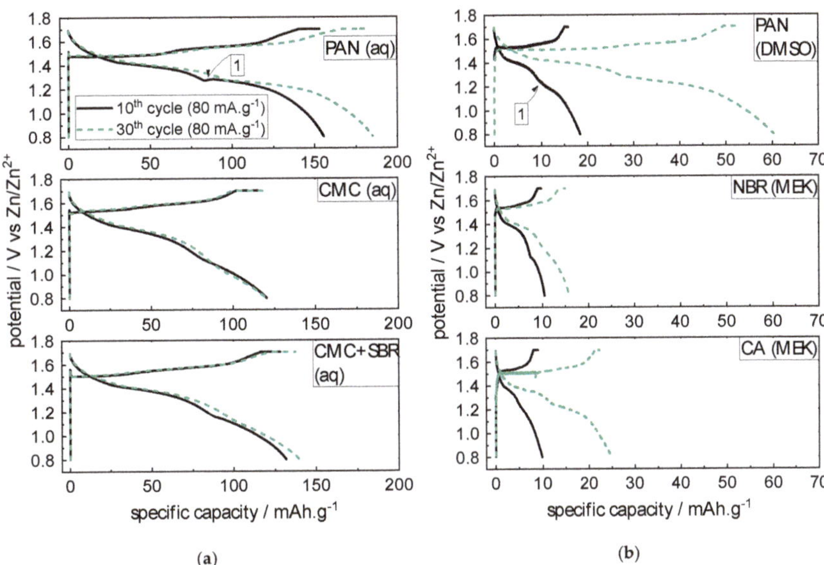

Figure 3. Overview over the potential curves of the aqueous (**a**) and non-aqueous (**b**) binder polymers for the 10th and 30th cycle, each with the current rate of 80 mA·g^{-1}.

By performing SEM+EDX characterizations (see Figure 4), the cross-sections of electrode coatings of PAN (aq) and PAN (DMSO), prepared by ion-polishing, were compared in a pristine state and in discharged state after the RCT procedure (post-mortem state). This way, the influence of an aqueous and non-aqueous BP could be investigated in more detail.

For both the pristine PAN (aq) and (DMSO) electrodes, the MnO$_2$ particles (coloured green) and the CB content (coloured orange) show comparable morphology characteristics such as the homogeneity, the porosity, and the material distribution (see Supplementary Figure S7 for EDX images of the single elements, respectively).

Interestingly, after the RCT procedure (post-mortem) in the discharged state, both electrodes show a surface layer deposition, which is thicker for the PAN (aq) electrode and could be attributed to a MnO$_2$ deposition (see Figure 4d,h, marker 1), as described in the previous literature [19,33,37,40,41]. Furthermore, both electrodes in a post-mortem state (discharged state) show a new flake structure inside the electrode coating (see Figure 4c,d,g,h, circle), which is not visible in the pristine state (see Figure 4a,b,e,f). This could be attributed to ZHS precipitations filling the pores of the electrode coating, as described in the previous literature [14,19,33,37,40–44]. However, this precipitation seems to be less distinctive for the PAN (DMSO) coating. This could be explained by the previously mentioned assumption of less wetting of the electrode coating by the electrolyte and thus a smaller active inner surface, as the CB-containing, electrically conductive coating could mainly act as an electrochemically active surface. Hence, this could result in a lower capacity by the dissolution of the (mostly unavailable) MnO$_2$ active material loading during discharge, compared to the PAN (aq) coating. The flake structure (see Figure 4c, marker 2) on the surface of the electrode can be referred to the separator with electrolyte salt residues, which was stuck on the electrode coating surface (possibly due to the aforementioned MnO$_2$ deposition).

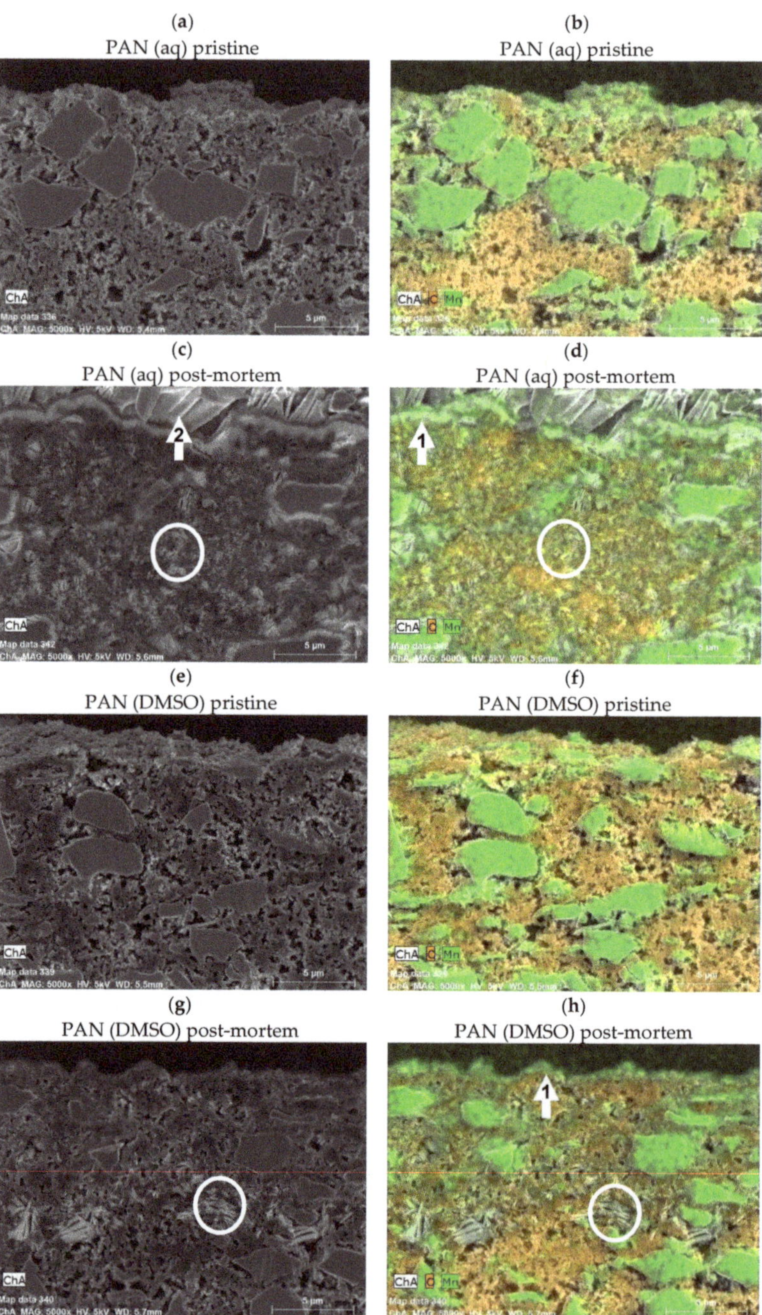

Figure 4. SEM and EDX images of the PAN (aq) electrode in a pristine (**a,b**) and post-mortem (**c,d**, discharged state) state, and the PAN (DMSO) electrode in a pristine (**e,f**) and post-mortem (**g,h**, discharged state) state. The MnO_2 particles are coloured green, and the carbon black is coloured orange, respectively. For the post-mortem state, the PAN (aq) and PAN (DMSO) both show a MnO_2 surface deposition, but with a thinner layer thickness for the PAN (DMSO) electrode.

Altogether, the comparison of the SEM+EDX images in a pristine and post-mortem (discharged) state could confirm the existence of a MnO_2 deposition during the charge steps, which takes place both in the pores of the electrode coating (if available, depending on the binder/solvent combination, as previously discussed) as well as on the surface of the electrode. In the following discharge steps, the previous MnO_2 deposition dissolves again, releasing capacity, which should be higher for those electrodes whose MnO_2 active material loading is available for the deposition/dissolution mechanism (here: PAN (aq) electrode coating, see RCT results). Due to efficiency reasons of the dissolution, MnO_2 deposition residues seem to stay, leading to an accumulating MnO_2 (surface) deposition layer (here: thicker surface layer for PAN (aq) than for PAN (DMSO) coating).

For a better understanding of the MnO_2 deposition/dissolution mechanism, further SEM+EDX characterizations of cross-sections of the positive electrode coatings in different states of charge should be carried out to further prove the latter assumptions of this study.

To underline the findings of the RCT tests for the aqueous and non-aqueous electrodes, the electrical impedance spectroscopy (EIS) measurements of the positive electrode (vs. Zn/Zn^{2+} reference) were analysed for the PAN (aq) and PAN (DMSO) binder in a charged (CH) and discharged (DCH) state to compare the effects of the different solvent categories as an example (for all the EIS results, see Supplementary Figure S8). The choice of these two binder combinations allowed for a direct comparison of the influence of the non-aqueous solvent (DI-water vs. DMSO) without the influence of the binder polymer (here, the *LA133* binder is considered as a PAN-based binder since the specification of PAN in the safety data sheet is >95 wt% of solid content). In Figure 5, the EIS results of the positive electrode are shown.

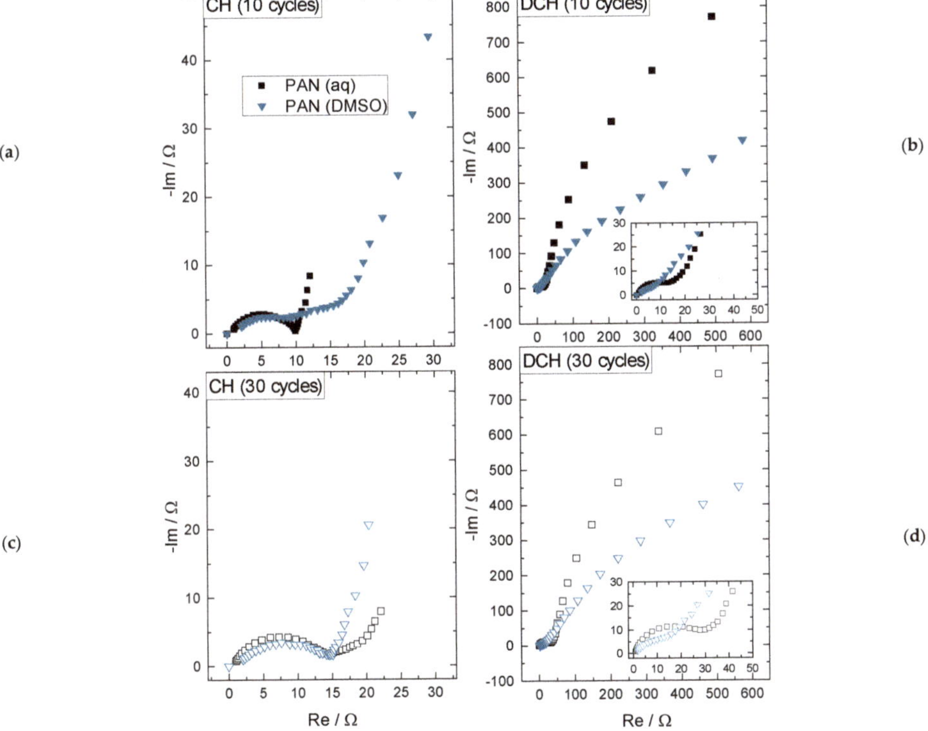

Figure 5. Results of the EIS measurements of the positive electrode for charged (CH, 1.7 V vs. Zn/Zn^{2+}) and discharged (DCH, 0.8 V vs. Zn/Zn^{2+}) state following both the 80 mA·g^{-1}-RCT-steps after 10 and 30 cycles, respectively.

For the different PAN binder solvents and state of charges, the following qualitative interpretations of the Nyquist plot can be made:

- In **charged state** after **10 cycles** (see Figure 5a), there is a lower impedance for the PAN (aq) than for the PAN (DMSO) electrode: The double layer (DL) capacity and the charge transfer (CT) resistance of the PAN (aq) binder are both clearly visible as a semi-circle, whereas the DL capacity and CT resistance of the electrode PAN (DMSO) binder is stretched, which indicates a higher CT resistance and could refer to the inactive and isolated MnO_2 active material particles in the electrode coating.
- In **charged state** after **30 cycles** (see Figure 5c), the impedance spectra show quite comparable semi-circles, again representing the DL capacity and the CT resistance. This could be explained by the growing deposition layer of MnO_2 on the positive electrode surface of the PAN (DMSO) coating. This assumption is supported by the capacity increase of the PAN (DMSO) electrode in Figure 2a.
- In **discharged state** after **10 and 30 cycles** (see Figure 5b,d), the overall impedances increase by about two orders of magnitude for both PAN (aq) and PAN (DMSO). This can be explained by the precipitation of ZHS as a consequence of the MnO_2 dissolution, as discussed in [19]. The precipitation results in a new DL capacity with a higher CT resistance, which is shown by the large low frequency half circle. Still, the impedance spectra for the lower impedance values are visible (see insets in Figure 5b,d), which were previously discussed for the charged state representing the MnO_2 active material layer. Relating to the PAN (DMSO) binder, the capacitive part of the impedance spectra is still below the PAN (aq) spectra, which can be explained by the thinner MnO_2 deposition layer and less ZHS precipitation (less wetting of the coating), resulting in lower specific capacity values (see Figure 2a).

Altogether, the comparison of the aqueous/non-aqueous electrodes generally shows higher discharge capacities/energies for the aqueous binders, which is related to the availability of the MnO_2 active material loading. The MnO_2 active material loading is not available in the non-aqueous electrodes, as the electrolyte should not lead to an electrode coating swelling as suggested, in contrast, for the aqueous electrodes. Nevertheless, the non-aqueous electrodes show increasing discharge capacities/energies with proceeding cycles, underlining the major role of MnO_2 deposition/dissolution processes (also on the surface of the electrode) for the ZIB with $ZnSO_4$/$MnSO_4$-based electrolytes being introduced in the recent literature.

As the aqueous PAN-based *LA133* binder shows the best cycling results, this binder was chosen for the mixing ratio variation shown in the following section.

2.2. Mixing Ratio Variation

Beside the variation of the BP itself (with fixed mixing ratios of AM/CB/BP 70/20/10 wt%), the influence of different mixing ratios was systematically investigated. The goal of the variation was to improve the energy density of the positive electrode. Therefore, this variation aimed at finding an advantageous mixing ratio with maximized active material share while still maintaining the electrochemical performance (i.e., the electrical conductivity driven by the CB content). The investigated mixing rations are graphically summarized using a ternary plot visualization in Figure 6.

The following criteria were considered for the positive electrode coatings:

- An active material proportion as high as possible to reduce the share of the passive materials (carbon black, binder polymer) in the electrode coating.
- A binder polymer proportion as low as possible to reduce the share of a passive and electrically non-conducting coating component.
- A carbon black proportion as low as possible, while still enabling a sufficient electrical conductivity of the coating.
- Good cycling performance in the RCT considering different application fields such as high-power (current rates >1 C, here: >160 $mA \cdot g^{-1}$ based on an experimen-

tally determined capacity of ~160 mAh·g^{-1} of MnO$_2$) and high-energy (<1C, here: <160 mA·g^{-1}) applications.

Table 3. Overview over the mixing ratios being tested using LA133 as the BP.

#	Name	Active Material (AM)/wt%	Carbon Black (CB)/wt%	Binder Polymer (BP)/wt%	CB/BP Ratio
1	70/20/10	70	20	10	2
2	75/20/05	75	20	5	4
3	75/15/10	75	15	10	1.5
4	80/15/05	80	15	5	3
5	80/10/10	80	10	10	1
6	82/15/03	82	15	3	5
7	85/10/05	85	10	5	2

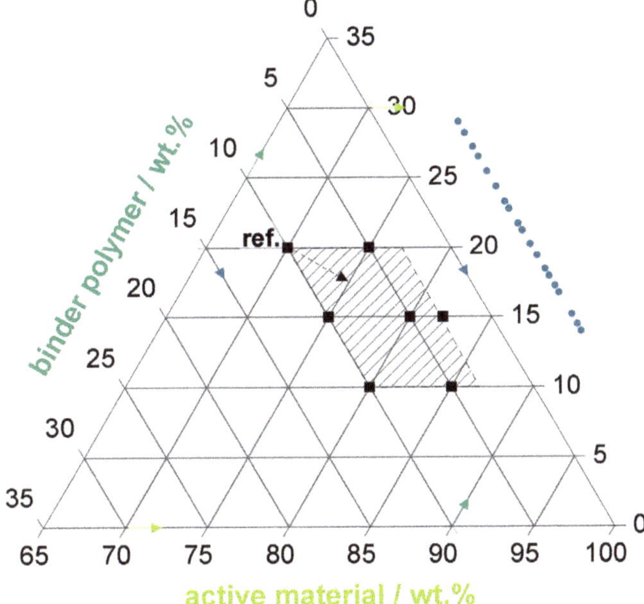

Figure 6. Graphical scheme with the mixing ratios being tested using the ternary plot visualization. The black squares represent the seven different mixing ratios of Table 3.

In Figure 7, the results of the RCT are shown by plotting the specific capacities and energies for different current rates over the cycle number. Here, the specific values were calculated by dividing the absolute capacity values by the total mass loading of the electrode (AM, CB, and BP, instead of only considering the AM mass loading as for the binder polymer variation in Section 2.1). This calculation method was chosen to be able to identify the influences of the different mixing ratios of the electrode coatings on the resulting capacity and energy values, respectively.

The results of the RCT with the specific discharge capacities (see Figure 7a) show that a low polymer binder percentage of 3 wt% is possible in the electrode coating, still enabling a cycling with a discharge capacity of ~140 mAh·g^{-1} @ 40 mA·g^{-1}. Nevertheless, the CB share, especially, and its ratio to the BP share has a high influence on the current rate stability:

- Lowering the CB share allows reducing the BP share.
- Reducing the CB share without reducing the BP share results in poor cycling behaviour due to insufficient conductivity within the electrode.

Therefore, a higher ratio of CB to BP improves the performance, as for the 75/20/05 mixture with a CB/BP ratio of 4 (20 wt% divided by 5 wt%) the performance for all current rates is significantly better compared to the 80/10/10 mixture with a CB/BP ratio of 1 (10 wt% divided by 10 wt%). For the other mixtures with ratio values between 1.5 to 3, the performances are located between the above-mentioned mixtures. The interpretation of the results by analysing the CB/BP ratio can be explained as the lower the CB/BP ratio, the lower the electrical conductivity of the electrode coating—resulting in a higher share of electrically isolating BP.

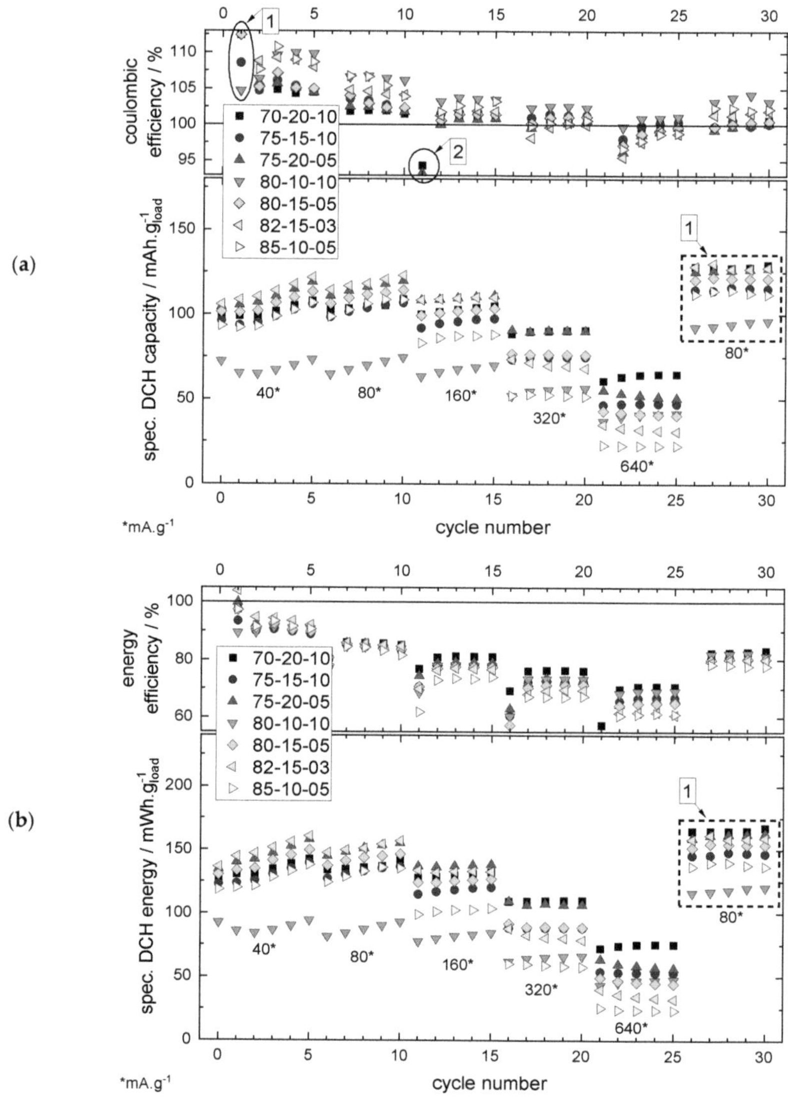

Figure 7. Specific discharge (DCH) capacity (**a**) and energy (**b**) with the efficiency values, respectively, for the investigated mixing ratios (the mixing ratios are given in the order AM/CB/BP and in wt%). The markers 1 and 2 show the efficiency values of the first cycle and after a change of the current rate, respectively, whereas they do not represent a typical efficiency value for the chosen current rate (see Supplementary Figure S9).

This observation leads to the differentiation of the interpretation of different CB shares depending on the application, i.e., high-power (HP) and high-energy (HE) applications.

- For HP applications, a CB share of >15 wt% is necessary to still enable a good RCT capacity also at higher current rates >320 $mA \cdot g^{-1}$. The CB/BP ratio should be >2 (see 70/20/10 or 75/20/05).
- For HE applications, an AM loading as high as possible is important for lower the specific costs of the battery, and therefore lower CB loadings are preferable. Considering this, a CB loading of 10 wt% and a CB/BP-ratio of ~2 (s. 85/10/05) still enables a good cycle performance at lower current rates <160 $mA \cdot g^{-1}$.

The RCT showing the specific discharge energies (see Figure 7b) also considers the potential, in addition to the capacity value, which provides a better insight into the inner resistance of the battery cell during cycling (see the previous section and Supplementary Figure S11). Although basically showing the same characteristics of the results of the different mixing ratios as the capacity representation (regarding the order of the different electrode material compositions, see marker 1), slight differences of the capacity vs. energy level order of the different mixture ratios can be observed, especially for the last RCT step with 80 $mA \cdot g^{-1}$, herein for the PAN (aq) binder as an example. The differences of the order of capacity/energy plots can refer to the different potential plateau levels during cycling, which is only considered by the energy representation (see the equation in Supplementary Figure S5).

Basically, this type of energy representation is introduced as a useful and more application-oriented alternative to the established capacity representation. This applies, in particular, to the aqueous battery chemistries such as the ARZIB technology with significant distinctions in the potential plateaus as a consequence of various cell constructions, as the literature shows.

Finally, it must be noted that the results were generated with the aqueous PAN-based *LA133* binder. For other binder polymers, the results could differ. Still, the characteristics of the variation of the mixing ratio of the electrode components should stay comparable.

3. Conclusions

3.1. Binder Variation

Mechanical stress tests of the positive electrode coatings revealed the maintenance of the structural integrity of the coatings based on aqueous binders in the electrolyte solution but not in DI-water, in comparison to the non-aqueous binders showing a stability in both DI-water and electrolyte. This could refer to the lower solvation free energy of the water molecules in the electrolyte in consequence of the $ZnSO_4$ and $MnSO_4$ addition, stabilizing the aqueous binders.

Rate capability tests (RCT) showed a lower capacity/energy of the positive electrodes for the binder polymers with non-aqueous solvents compared to those with aqueous solvents, with the best results achieved for the aqueous binder PAN. This could be explained by the availability of the MnO_2 active material loading of the coating: The aqueous binders allow a swelling of the coating by the electrolyte in contrast to the non-aqueous binders. The coating swelling makes the active material loading available for MnO_2 deposition/dissolution mechanisms in relation to the suggested reaction mechanism of the ARZIB, resulting in higher capacity/energy values. Still, the non-aqueous binders show an increasing capacity/energy with the proceeding cycle number. This could be explained by a MnO_2 deposition/dissolution mechanism on the electrode surface being confirmed by SEM+EDX characterizations of electrode cross sections. EIS measurements show that the characteristics of the spectra of the non-aqueous binder (here: PAN (DMSO)) in the charged state also changes by increasing the cycle number, which might indicate a growing MnO_2 deposition layer for proceeding cycling.

3.2. Mixing Ratio Variation

Considering the results, the optimized mixing ratios must be chosen following the favoured application of the battery. Here, the absolute carbon black (CB) share of the coating and the ratio of the CB and the binder polymer (CB/BP-ratio) shares were investigated as important parameters for the electrode coatings of ARZIBs. For a CB share $\leq$15 wt% and CB/BP-ratios $\leq$2, the cycling performance decreases, whereas for a CB share >15 wt% and CB/BP-ratios >2 and up to 5, the cycling performance increases, respectively, also for high current rates. Therefore, higher CB shares qualify the positive electrode for usage in high-power applications, while reduced CB shares (but high active material shares) are suitable for high-energy applications. Furthermore, in terms of an industrial production approach, the proportion of the passive materials (CB+BP) was successfully reduced to 15 wt% in total, and the BP proportion to 3 wt%, still enabling the cycling of these battery cells.

This study sheds light on the influence of the binder polymer/solvent combination on the cell performance of ARZIBs through comparing (non-)aqueous solvents, as well as on the influence of the mixing ratio by minimizing the passive material share of the positive electrode. The results of this study can be used for application-oriented cell fabrication of ARZIBs.

4. Materials and Methods

4.1. Materials

For the positive electrode preparation, manganese dioxide (AM, Manganese(IV) oxide, 99.9% (metals basis), 325 mesh powder, Alfa Aesar, Heysham, United Kingdom), Carbon black (CB, Carbon black, acetylene, 100% compressed, 99.9+%, Alfa Aesar, Heysham, United Kingdom), binder polymer (BP) and solvent were mixed in different combinations in a speed mixer (DAC 150.1 FVZ-K, Hauschild and Co.KG, Hamm, Germany) at 3000 rpm for 2 min. Subsequently, the positive electrode aqueous slurry was homogenized in a three-roll mill (EXAKT 80E, EXAKT Advanced Technologies GmbH, Norderstedt, Germany) with a final roll gap of 10 µm to control and unify the particle sizes. Due to the evaporation and security characteristics of the solvents, the non-aqueous slurries were homogenized for 30 min at least at 2000 rpm in a laboratory dissolver (DISPERMAT CN10, VMA-GETZMANN GmbH, Reichshof, Germany) instead of being processed in the three-roll mill. The slurry was coated with a doctor blade and 200 µm wet layer thickness on a stainless-steel foil (1.4301/AISI 304, thickness 25 µm, TBJ Industrieteile GmbH, Leipzig, Germany). The AM mass loading was ~2.5–4.6 mg/cm^{-2} and the electrode coating thickness ~40 µm (see Supplementary Figure S10). Afterwards, the coating was dried for 24 h at 40 °C. Then, 16 mm coins were used as positive electrodes. As the negative electrode, a 25 µm thick zinc foil with the diameter of 16 mm was used (zinc foil 99.95%, Goodfellow GmbH, Hamburg, Germany). The separator (18 mm diameter) was made of glass fibre filter (Whatman® GF/A, Cytiva Europe GmbH, Freiburg, Germany). For the electrolyte, zinc sulfate heptahydrate (AnalaR NORMAPUR® Reag. Ph.Eur., ACS for analysis, VWR International GmbH, Darmstadt, Germany) and manganese sulfate tetrahydrate (99%, metal basis, Alfa Aesar, Heysham, United Kingdom) were used. A zinc wire (99.9% pure zinc wire, Goodfellow GmbH, Hamburg, Germany) was used as the reference electrode.

For the BP study, the following binders have been investigated:

- LA133 (polyacrylonitrile (PAN)-based aqueous binder dispersion, GELON LIB Group, Linyi, China)
- SBR (styrene-butadiene rubber, SSBR 100, aqueous suspension, Targray, Kirkland, QC, Canada)
- CMC (carboxymethyl cellulose, average MW ~250,000, degree of substitution 0.9, sodium salt, powder, Sigma-Aldrich, Merck KgaA, Darmstadt, Germany)
- PVP (polyvinylpyrrolidone, MW 1,300,000, powder, Alfa Aesar, Heysham, United Kingdom)
- CA (cellulose acetate, MW ~100,000, powder, Acros Organics, Thermo Fisher Scientific, New Brunswick, NJ, USA)

- NBR (nitrile butadiene rubber, acrylonitrile 37–39 wt%, chunk, Sigma-Aldrich, Merck KgaA, Darmstadt, Germany)
- PAN (polyacrylonitrile, copolymer 99.5% AN/0.5% MA, MW 230,000, powder, Goodfellow GmbH, Hamburg, Germany)

For the slurry preparation, aqueous and non-aqueous binder solutions/suspensions were prepared: For the aqueous slurry preparation, DI-water and for the non-aqueous slurry preparation, methyl ethyl ketone (MEK, for NBR and CA) or dimethyl sulfoxide (DMSO, for PAN) were used. The binder preparations are listed in Table 2.

In order to evaluate the material shares of active material, conductive agent and binder polymer, different mixing ratios were tested. For this purpose, the PAN-based aqueous *LA133* was used exemplarily as the binder. The share combinations listed in Table 3 have been tested.

4.2. Cell Assembly

Cells were assembled using EL-Cell® ECC-Aqu experimental cells. A glass fibre separator with a diameter 18 mm was used. Two hundred µL of an aqueous electrolyte containing 2 M $ZnSO_4$ and 0.1 M $MnSO_4$ was used (pH~4). The zinc wire was used as the reference electrode in the EL-Cell® ECC-Aqu. As the current collector on the positive electrode, an 18 mm stainless steel coin was used. As the current collector on the negative electrode, an 18 mm copper coin was used. The cell setup is shown in Supplementary Figure S2.

4.3. Characterization Methods

The cell cycling was performed by making a rate capability test (RCT) with 5 cycles, each at 40, 80, 160, 320, 640, and 80 mA·g^{-1} (related to active material loading of the manganese dioxide electrode) from 0.8 to 1.7 V with a *VMP3* potentiostat from Biologic Science Instruments, Seyssinet-Pariset, France. The charging process included a constant current (CC) step followed by a constant voltage (CV) step until 10% of the current in the CC-step was reached. The discharge process was performed at CC. EIS measurements were performed between 10 kHz and 20 mHz at 1.7, 1.25 and 0.8 V cell voltage after the five 80 mA·g^{-1} charge/discharge cycles, respectively. The test plan procedure is shown in detail in Supplementary Figure S3.

For the mechanical stress test (MST), the electrode coins were put into a transparent PP beaker (20 mL, ø31 × 48 mm) filled with the same amount (~10 mL) of either DI-water or electrolyte solution together with a magnetic stirring bar on top of the electrode sheet. The stirring bar (PTFE, cylindrical, length 20 mm) was operated for 60 s at 200 rpm, respectively. The experimental setup is visualized in Supplementary Figure S4.

For analysing the distribution of the different components in the positive electrode, SEM-EDX images were realized with selected positive electrodes. The positive electrodes were dipped into deionized water and dried overnight before being ion polished. The cutting edges were prepared by ion milling (IM4000 Ion Milling System, Hitachi High-Tech Europe GmbH, Mannheim, Germany) and characterized by SEM (Zeiss Auriga 60, Carl Zeiss Microscopy Deutschland GmbH, Oberkochen, Germany) and EDX (Bruker Quantax XFlash 6160, Bruker Corporation, Billerica, USA).

Supplementary Materials: The following are available online at https://www.mdpi.com/article/10.3390/batteries7020040/s1, Figure S1: Overview of the experimental cell setup inside the EL-Cell® ECC aqu, Figure S2: Graphical overview of the test plan procedure for the rate capability test of the herein tested cells, Figure S3: Experimental setup of the mechanical stress test with a magnetic stirring bar in a PP beaker (20 mL, ø31 × 48 mm) filled with ~ 10 mL of liquid, Figure S4: RCT results including the PVP (aq) binder polymer, which is not finishing the RCT due to stability issues of the electrode coating (s. MST), Figure S5: Potential curves including the PVP binder, which was not able to finish the RCT, Figure S6: Comparison of the PAN (aq) and the PAN (DMSO) coating in pristine state. MnO_2 particles are coloured green, carbon black is coloured orange, Figure S7: Comparison of the PAN (aq) and the PAN (DMSO) coating in post mortem state. MnO_2 particles are coloured green, carbon black is coloured orange, Figure S8: Overview of the EIS spectra for all investigated binder polymers in charged (CH) and discharged (DCH) state after 10 and 30 cycles, respectively, Figure S9: Diagrams of the RCT showing the specific capacity and energy, respectively, together with the efficiency values, Figure S10: SEM+EDX images of the coating with aqueous LA133 (PAN-based) binder with the mixing ration 70/20/10 (AM/CB/BP, wt%) in pristine state. The electrode thickness is ~40 μm, Figure S11: SEM+EDX images of the coating with aqueous LA133 (PAN-based) binder with the mixing ration of 80/15/05 (AM/CB/BP, wt%) in pristine state. The electrode thickness is ~40 μm. (green coloured = MnO_2 particles, orange coloured = carbon black (CB)), Figure S12: Comparison of the 70/20/10 and 80/15/05 binders in pristine state. For the 80/15/05 coating, smaller distances of the MnO_2 active material particles (coloured green) and a more compact visual impression of the coating could be assumed, which could be referred to the lower CB content and the lower overall passive material (CB+BP) share (20 wt% instead of 30 wt% for the 70/20/10 coating), Figure S13: Overview over the charge/discharge curves of the 10th cycle (80 mA·g^{-1}) of the RCT shown in Figure S4 for the binder variation (a) and mixing variation (b) with insets showing the IR drops directly after the switchover from charge to discharge (marker 1) and the first potential plateau during discharge (marker 2), Table S1: Brief literature research on binder materials, solvents, mixing ratios and current collectors. This literature search was carried out as an example and does not claim to be complete, Table S2: Detailed Overview of the test plan procedure for the RCT.

Author Contributions: Conceptualization, O.F.; methodology, O.F., S.I. and C.B.; validation, O.F., S.I. and C.B.; formal analysis, O.F.; investigation, S.I., C.B. and O.F.; data curation, O.F. and S.I.; writing—original draft preparation, O.F.; writing—review and editing, H.G., S.I., C.B., K.P.B. and D.B.; visualization, O.F.; supervision, D.B., K.P.B. and D.S.; project administration, D.B. All authors have read and agreed to the published version of the manuscript.

Funding: This research received no external funding.

Data Availability Statement: The data presented in this study are available on request from the corresponding author.

Acknowledgments: Oliver Fitz acknowledges the German Federal Environmental Foundation (Deutsche Bundesstiftung Umwelt, DBU) and Christian Bischoff acknowledges the Heinrich Böll Foundation for the support. We thank Volker Kübler for carrying out the SEM+EDX characterizations.

Conflicts of Interest: The authors declare no conflict of interest.

References

1. Biswal, A.; Tripathy, B.C.; Subbaiah, T.; Meyrick, D.; Minakshi, M. Electrodeposition of manganese dioxide: Effect of quaternary amines. *J. Solid State Electrochem.* **2013**, *17*, 1349–1356. [CrossRef]
2. Biswal, A.; Tripathy, B.C.; Subbaiah, T.; Meyrick, D.; Minakshi, M. Dual effect of anionic surfactants in the electrodeposited MnO_2 trafficking redox ions for energy storage. *J. Electrochem. Soc.* **2015**, *162*, A30–A38. [CrossRef]
3. Song, M.; Tan, H.; Chao, D.; Fan, H.J. Recent advances in Zn-ion batteries. *Adv. Funct. Mater.* **2018**, *28*, 1802564. [CrossRef]
4. Chang, H.J.; Rodríguez-Pérez, I.A.; Fayette, M.; Canfield, N.L.; Pan, H.; Choi, D.; Li, X.; Reed, D. Effects of water-based binders on electrochemical performance of manganese dioxide cathode in mild aqueous zinc batteries. *Carbon Energy* **2020**. [CrossRef]
5. Olbasa, B.W.; Fenta, F.W.; Chiu, S.-F.; Tsai, M.-C.; Huang, C.-J.; Jote, B.A.; Beyene, T.T.; Liao, Y.-F.; Wang, C.-H.; Su, W.-N.; et al. High-rate and long-cycle stability with a dendrite-free zinc anode in an aqueous Zn-Ion battery using concentrated electrolytes. *ACS Appl. Energy Mater.* **2020**, *3*, 4499–4508. [CrossRef]
6. Pan, H.; Ellis, J.; Li, X.; Nie, Z.; Chang, H.J.; Reed, D. Electrolyte effect on the electrochemical performance of mild aqueous zinc-electrolytic manganese dioxide batteries. *ACS Appl. Mater. Interfaces* **2019**. [CrossRef]
7. Palaniyandy, N.; Kebede, M.A.; Raju, K.; Ozoemena, K.I.; Le Roux, L.; Mathe, M.K.; Jayaprakasam, R. α-MnO_2 nanorod/onion-like carbon composite cathode material for aqueous zinc-ion battery. *Mater. Chem. Phys.* **2019**, *230*, 258–266. [CrossRef]
8. Guo, X.; Li, J.; Jin, X.; Han, Y.; Lin, Y.; Lei, Z.; Wang, S.; Qin, L.; Jiao, S.; Cao, R. A hollow-structured manganese oxide cathode for stable Zn-MnO_2 batteries. *Nanomaterials* **2018**, *8*, 301. [CrossRef]
9. Xu, D.; Li, B.; Wei, C.; He, Y.-B.; Du, H.; Chu, X.; Qin, X.; Yang, Q.-H.; Kang, F. Preparation and characterization of MnO_2/acid-treated CNT nanocomposites for energy storage with zinc ions. *Electrochim. Acta* **2014**, *133*, 254–261. [CrossRef]
10. Bischoff, C.; Fitz, O.; Schiller, C.; Gentischer, H.; Biro, D.; Henning, H.-M. Investigating the impact of particle size on the performance and internal resistance of aqueous zinc ion batteries with a manganese sesquioxide cathode. *Batteries* **2018**, *4*, 44. [CrossRef]
11. Chamoun, M.; Brant, W.R.; Tai, C.-W.; Karlsson, G.; Noréus, D. Rechargeability of aqueous sulfate Zn/MnO_2 batteries enhanced by accessible Mn^{2+} ions. *Energy Storage Mater.* **2018**, *15*, 351–360. [CrossRef]

12. Zhang, N.; Cheng, F.; Liu, J.; Wang, L.; Long, X.; Liu, X.; Li, F.; Chen, J. Rechargeable aqueous zinc-manganese dioxide batteries with high energy and power densities. *Nat. Commun.* **2017**, *8*, 405. [CrossRef]
13. Jiang, B.; Xu, C.; Wu, C.; Dong, L.; Li, J.; Kang, F. Manganese sesquioxide as cathode material for multivalent zinc ion battery with high capacity and long cycle life. *Electrochim. Acta* **2017**, *229*, 422–428. [CrossRef]
14. Pan, H.; Shao, Y.; Yan, P.; Cheng, Y.; Han, K.S.; Nie, Z.; Wang, C.; Yang, J.; Li, X.; Bhattacharya, P.; et al. Reversible aqueous zinc/manganese oxide energy storage from conversion reactions. *Nat. Energy* **2016**, *1*, 16039. [CrossRef]
15. Islam, S.; Alfaruqi, M.H.; Mathew, V.; Song, J.; Kim, S.; Kim, S.; Jo, J.; Baboo, J.P.; Pham, D.T.; Putro, D.Y.; et al. Facile synthesis and the exploration of the zinc storage mechanism of β-MnO$_2$ nanorods with exposed (101) planes as a novel cathode material for high performance eco-friendly zinc-ion batteries. *J. Mater. Chem. A* **2017**, *5*, 23299–23309. [CrossRef]
16. Poyraz, A.S.; Laughlin, J.; Zec, Z. Improving the cycle life of cryptomelane type manganese dioxides in aqueous rechargeable zinc ion batteries: The effect of electrolyte concentration. *Electrochim. Acta* **2019**, *305*, 423–432. [CrossRef]
17. Huang, J.; Wang, Z.; Hou, M.; Dong, X.; Liu, Y.; Wang, Y.; Xia, Y. Polyaniline-intercalated manganese dioxide nanolayers as a high-performance cathode material for an aqueous zinc-ion battery. *Nat. Commun.* **2018**, *9*, 2906. [CrossRef] [PubMed]
18. Zeng, X.; Liu, J.; Mao, J.; Hao, J.; Wang, Z.; Zhou, S.; Ling, C.D.; Guo, Z. Toward a reversible Mn^{4+}/Mn^{2+} redox reaction and dendrite-free Zn anode in near-neutral aqueous Zn/MnO$_2$ batteries via salt anion chemistry. *Adv. Energy Mater.* **2020**, *10*, 1904163. [CrossRef]
19. Guo, X.; Zhou, J.; Bai, C.; Li, X.; Fang, G.; Liang, S. Zn/MnO$_2$ battery chemistry with dissolution-deposition mechanism. *Mater. Today Energy* **2020**, *16*, 100396. [CrossRef]
20. Sun, W.; Wang, F.; Hou, S.; Yang, C.; Fan, X.; Ma, Z.; Gao, T.; Han, F.; Hu, R.; Zhu, M.; et al. Zn/MnO$_2$ battery chemistry with H$^+$ and Zn^{2+} coinsertion. *J. Am. Chem. Soc.* **2017**, *139*, 9775–9778. [CrossRef]
21. Hou, Z.; Dong, M.; Xiong, Y.; Zhang, X.; Ao, H.; Liu, M.; Zhu, Y.; Qian, Y. A high-energy and long-life aqueous Zn/birnessite battery via reversible water and Zn^{2+} coinsertion. *Small* **2020**, *16*, e2001228. [CrossRef]
22. Jia, H.; Wang, Z.; Tawiah, B.; Wang, Y.; Chan, C.-Y.; Fei, B.; Pan, F. Recent advances in zinc anodes for high-performance aqueous Zn-ion batteries. *Nano Energy* **2020**, *70*, 104523. [CrossRef]
23. Ming, J.; Guo, J.; Xia, C.; Wang, W.; Alshareef, H.N. Zinc-ion batteries: Materials, mechanisms, and applications. *Mater. Sci. Eng. R Rep.* **2019**, *135*, 58–84. [CrossRef]
24. Biswal, A.; Chandra Tripathy, B.; Sanjay, K.; Subbaiah, T.; Minakshi, M. Electrolytic manganese dioxide (EMD): A perspective on worldwide production, reserves and its role in electrochemistry. *RSC Adv.* **2015**, *5*, 58255–58283. [CrossRef]
25. Xu, C.; Li, B.; Du, H.; Kang, F. Energetic zinc ion chemistry: The rechargeable zinc ion battery. *Angew. Chem. Int. Ed. Engl.* **2012**, *51*, 933–935. [CrossRef] [PubMed]
26. Lee, B.; Yoon, C.S.; Lee, H.R.; Chung, K.Y.; Cho, B.W.; Oh, S.H. Electrochemically-induced reversible transition from the tunneled to layered polymorphs of manganese dioxide. *Sci. Rep.* **2014**, *4*, 6066. [CrossRef]
27. Lee, B.; Lee, H.R.; Kim, H.; Chung, K.Y.; Cho, B.W.; Oh, S.H. Elucidating the intercalation mechanism of zinc ions into α-MnO$_2$ for rechargeable zinc batteries. *Chem. Commun.* **2015**, *51*, 9265–9268. [CrossRef]
28. Han, S.-D.; Kim, S.; Li, D.; Petkov, V.; Yoo, H.D.; Phillips, P.J.; Wang, H.; Kim, J.J.; More, K.L.; Key, B.; et al. Mechanism of Zn insertion into nanostructured δ-MnO$_2$: A nonaqueous rechargeable Zn metal battery. *Chem. Mater.* **2017**, *29*, 4874–4884. [CrossRef]
29. Alfaruqi, M.H.; Gim, J.; Kim, S.; Song, J.; Jo, J.; Kim, S.; Mathew, V.; Kim, J. Enhanced reversible divalent zinc storage in a structurally stable α-MnO$_2$ nanorod electrode. *J. Power Sources* **2015**, *288*, 320–327. [CrossRef]
30. Alfaruqi, M.H.; Islam, S.; Gim, J.; Song, J.; Kim, S.; Pham, D.T.; Jo, J.; Xiu, Z.; Mathew, V.; Kim, J. A high surface area tunnel-type α-MnO$_2$ nanorod cathode by a simple solvent-free synthesis for rechargeable aqueous zinc-ion batteries. *Chem. Phys. Lett.* **2016**, *650*, 64–68. [CrossRef]
31. Alfaruqi, M.H.; Mathew, V.; Gim, J.; Kim, S.; Song, J.; Baboo, J.P.; Choi, S.H.; Kim, J. Electrochemically induced structural transformation in a γ-MnO$_2$ cathode of a high capacity zinc-ion battery system. *Chem. Mater.* **2015**, *27*, 3609–3620. [CrossRef]
32. Xu, C.; Chiang, S.W.; Ma, J.; Kang, F. Investigation on zinc ion storage in alpha manganese dioxide for zinc ion battery by electrochemical impedance spectrum. *J. Electrochem. Soc.* **2012**, *160*, A93–A97. [CrossRef]
33. Qiu, C.; Zhu, X.; Xue, L.; Ni, M.; Zhao, Y.; Liu, B.; Xia, H. The function of Mn^{2+} additive in aqueous electrolyte for Zn/δ-MnO$_2$ battery. *Electrochim. Acta* **2020**, *351*, 136445. [CrossRef]
34. Khamsanga, S.; Pornprasertsuk, R.; Yonezawa, T.; Mohamad, A.A.; Kheawhom, S. δ-MnO$_2$ nanoflower/graphite cathode for rechargeable aqueous zinc ion batteries. *Sci. Rep.* **2019**, *9*, 8441. [CrossRef]
35. Ko, J.S.; Sassin, M.B.; Parker, J.F.; Rolison, D.R.; Long, J.W. Combining battery-like and pseudocapacitive charge storage in 3D MnO$_x$@carbon electrode architectures for zinc-ion cells. *Sustain. Energy Fuels* **2018**, *2*, 626–636. [CrossRef]
36. Alfaruqi, M.H.; Islam, S.; Putro, D.Y.; Mathew, V.; Kim, S.; Jo, J.; Kim, S.; Sun, Y.-K.; Kim, K.; Kim, J. Structural transformation and electrochemical study of layered MnO$_2$ in rechargeable aqueous zinc-ion battery. *Electrochim. Acta* **2018**, *276*, 1–11. [CrossRef]
37. Huang, Y.; Mou, J.; Liu, W.; Wang, X.; Dong, L.; Kang, F.; Xu, C. Novel insights into energy storage mechanism of aqueous rechargeable Zn/MnO$_2$ batteries with participation of Mn^{2+}. *Nano Micro Lett.* **2019**, *11*, 860. [CrossRef]
38. Zhao, Q.; Chen, X.; Wang, Z.; Yang, L.; Qin, R.; Yang, J.; Song, Y.; Ding, S.; Weng, M.; Huang, W.; et al. Unravelling H$^+$/Zn^{2+} synergistic intercalation in a novel phase of manganese oxide for high-performance aqueous rechargeable battery. *Small* **2019**, *15*, e1904545. [CrossRef] [PubMed]

39. Gao, X.; Wu, H.; Li, W.; Tian, Y.; Zhang, Y.; Wu, H.; Yang, L.; Zou, G.; Hou, H.; Ji, X. H$^+$-insertion boosted α-MnO$_2$ for an aqueous Zn-ion battery. *Small* **2020**, *16*, e1905842. [CrossRef] [PubMed]

40. Li, L.; Hoang, T.K.A.; Zhi, J.; Han, M.; Li, S.; Chen, P. Functioning mechanism of the secondary aqueous Zn-β-MnO$_2$ battery. *ACS Appl. Mater. Interfaces* **2020**, *12*, 12834–12846. [CrossRef]

41. Mateos, M.; Makivic, N.; Kim, Y.-S.; Limoges, B.; Balland, V. Accessing the two-electron charge storage capacity of MnO$_2$ in mild aqueous electrolytes. *Adv. Energy Mater.* **2020**, *10*, 2000332. [CrossRef]

42. Lee, B.; Seo, H.R.; Lee, H.R.; Yoon, C.S.; Kim, J.H.; Chung, K.Y.; Cho, B.W.; Oh, S.H. Critical role of pH evolution of electrolyte in the reaction mechanism for rechargeable zinc batteries. *ChemSusChem* **2016**, *9*, 2948–2956. [CrossRef] [PubMed]

43. Li, Y.; Wang, S.; Salvador, J.R.; Wu, J.; Liu, B.; Yang, W.; Yang, J.; Zhang, W.; Liu, J.; Yang, J. Reaction mechanisms for long-life rechargeable Zn/MnO$_2$ batteries. *Chem. Mater.* **2019**. [CrossRef]

44. Wang, J.; Wang, J.-G.; Liu, H.; Wei, C.; Kang, F. Zinc ion stabilized MnO$_2$ nanospheres for high capacity and long lifespan aqueous zinc-ion batteries. *J. Mater. Chem. A* **2019**, *7*, 13727–13735. [CrossRef]

45. Atkins, P.; de Paula, J. *Physical Chemistry*, 9th ed.; W. H. Freeman and Company: New York, NY, USA, 2010; ISBN 9781429218122.

46. Kim, S.H.; Oh, S.M. Degradation mechanism of layered MnO$_2$ cathodes in Zn/ZnSO$_4$/MnO$_2$ rechargeable cells. *J. Power Sources* **1998**, *72*, 150–158. [CrossRef]

47. Pourbaix, M. *Atlas of Electrochemical Equilibria in Aqueous Solutions*, 2nd ed.; National Association of Corrosion Engineers: Houston, TX, USA, 1974; ISBN 0915567989.

48. Minakshi, M.; Appadoo, D.; Martin, D.E. The anodic behavior of planar and porous zinc electrodes in alkaline electrolyte. *Electrochem. Solid State Lett.* **2010**, *13*, A77. [CrossRef]

49. Lim, M.B.; Lambert, T.N.; Chalamala, B.R. Rechargeable alkaline zinc–manganese oxide batteries for grid storage: Mechanisms, challenges and developments. *Mater. Sci. Eng. R Rep.* **2021**, *143*, 100593. [CrossRef]

50. Franks, F. *Aqueous Solutions of Simple Electrolytes*; Springer: Boston, MA, USA, 1973; ISBN 9781468429558.

51. Wang, R.; Feng, L.; Yang, W.; Zhang, Y.; Zhang, Y.; Bai, W.; Liu, B.; Zhang, W.; Chuan, Y.; Zheng, Z.; et al. Effect of different binders on the electrochemical performance of metal oxide anode for lithium-ion batteries. *Nanoscale Res. Lett.* **2017**, *12*, 575. [CrossRef] [PubMed]

52. Bard, A.J.; Faulkner, L.R. *Electrochemical Methods and Applications*, 2nd ed.; Wiley: Hoboken, NJ, USA, 2001; ISBN 0-471-04372-9.

Article

Methodology for Determining Time-Dependent Lead Battery Failure Rates from Field Data

Rafael Conradt [1],*, Frederic Heidinger [1] and Kai Peter Birke [2]

[1] Robert Bosch GmbH, Mittlerer Pfad 9, 70499 Stuttgart, Germany; frederic.heidinger@de.bosch.com
[2] Electrical Energy Storage Systems, Institute for Photovoltaics, University of Stuttgart, Pfaffenwaldring 47, 70569 Stuttgart, Germany; peter.birke@ipv.uni-stuttgart.de
* Correspondence: rafael.conradt@de.bosch.com

Abstract: The safety requirements in vehicles continuously increase due to more automated functions using electronic components. Besides the reliability of the components themselves, a reliable power supply is crucial for a safe overall system. Different architectures for a safe power supply consider the lead battery as a backup solution for safety-critical applications. Various ageing mechanisms influence the performance of the battery and have an impact on its reliability. In order to qualify the battery with its specific failure modes for use in safety-critical applications, it is necessary to prove this reliability by failure rates. Previous investigations determine the fixed failure rates of lead batteries using data from teardown analyses to identify the battery failure modes but did not include the lifetime of these batteries examined. Alternatively, lifetime values of battery replacements in workshops without knowing the reason for failure were used to determine the overall time-dependent failure rate. This study presents a method for determining reliability models of lead batteries by investigating individual failure modes. Since batteries are subject to ageing, the analysis of lifetime values of different failure modes results in time-dependent failure rates of different magnitudes. The failure rates of the individual failure modes develop with different shapes over time, which allows their ageing behaviour to be evaluated.

Keywords: lead batteries; safety concept; safety battery; battery monitoring; electronic battery sensor; failure modes; failure distribution; failure rates; field battery investigation; safe supply; power supply system

Citation: Conradt, R.; Heidinger, F.; Birke, K.P. Methodology for Determining Time-Dependent Lead Battery Failure Rates from Field Data. *Batteries* **2021**, *7*, 39. https://doi.org/10.3390/batteries7020039

Academic Editors: Carlos Ziebert and Joeri Van Mierlo

Received: 30 March 2021
Accepted: 2 June 2021
Published: 15 June 2021

Publisher's Note: MDPI stays neutral with regard to jurisdictional claims in published maps and institutional affiliations.

1. Introduction

The change in individual mobility is particularly evident in the electrification and automation of vehicles. The electrification of the powertrain leads to an increase in the weight of the vehicles. In addition to automation functions, this leads to higher safety requirements especially for electric braking and steering power assistance. Apart from these essential safety-relevant consumers steering (e.g., steer-by-wire) and braking (e.g., brake-by-wire), automated driving functions require sensors and actuators for position and environment detection as well as for control [1]. These components are supplied via the 12 V on-board power supply system (also known as powernet) of the vehicle and need a safe power supply to ensure a safe operation.

Therefore, knowing the failure behaviour and failure rates of the components and the power supply is crucial for the overall safety concept of the power supply system and the vehicle itself. Dominguez-Garcia et al. have already dealt with this increasing problem of safety-relevant consumers and their power supply in [2] by designing safe power supply system topologies. Subsequently, Kurita et al. also addressed the design of on-board power supply systems in [3] and placed the focus on the importance for future driving applications, which supports the actuality of the topic.

In such on-board power supply systems, the power supply is provided during normal operation by the alternator or DCDC converter in hybrid or electric vehicles. In addition, the 12 V battery is a main component of the power supply system and, besides buffering peak loads and supplying the vehicle with standby power, e.g., when parking, it also acts as a backup power supply. This backup function of the 12 V battery in the on-board power supply system is now becoming very important and the search for a safe power supply at the 12 V level with proof of reliability must be conducted.

Modern vehicle architectures consist to a large extent of electrical and/or electronic (E/E) systems. These systems are crucial for the safe operation of the vehicle and must therefore meet functional safety standards. The vehicle power supply system, as an obligatory component for safe operation, must therefore be developed according to the ISO 26262 standard [4,5]. Batteries are not in the scope of the ISO 26262 standard although the battery is a main component of the power supply system and its failure rates need to be included. In contrast to electrical components with constant failure rates, electrochemical components such as batteries are subject to ageing and thus do not have constant failure rates over time. Nevertheless, this failure behaviour and the resulting increasing failure rates must be included in the functional safety concept. Therefore, a method for determining battery failure rates must be established and checked for plausibility with the failure mechanisms investigated.

The lead battery can be part of a safe power supply system architecture if its ageing mechanisms and faults are manageable. In an early stage Albers et al. identified in [6] the necessity for further investigations of batteries in the context of functional safety requirements in the changing environment of vehicles with new functions. Proving reliability is difficult for electrochemical components such as the lead battery due to their ageing behaviour over time. Previous investigations on the reliability of the lead battery, such as those carried out by Albers et al. in [7], examined the batteries by tear down analysis to obtain a probability distribution of the failures that occurred. With the failure distribution, they were able to calculate fixed failure rates for the lead battery by using the corresponding ADAC failure statistics [8]. However, the ageing behaviour and thus the failure rate of electrochemical elements such as the lead battery does not follow the bathtub curve like electronic components, which are dominated by random failures in terms of time. The ageing process of electrochemical components starts with first use and batteries can be considered as wear out components. This means that the ageing behaviour behaves like the third area of the bathtub curve with increasing failure rate and therefore a fixed failure rate is less appropriate for wear out components.

In [9], Mürken et al. chose the approach of using workshop data on the battery exchange of the vehicle battery and its lifetime values. With the help of the individual lifetime values, it was possible to determine an ageing model based on a Weibull distribution for the failure of the battery. This made it possible to calculate the reliability of the overall battery over time and thus time-dependent failure rates. A consideration of the different failure modes and thus corresponding failure rates could not be investigated in this study due to lack of knowledge about the occured failure mechanisms. Furthermore, many batteries are replaced without faults and thus these functioning batteries are included in the failure rates in this approach.

Varying ageing mechanisms lead to different ageing behaviour; this was already recognised by Kumar et al. in [10]. It was shown that lifetime estimates of batteries are possible with Weibull analyses and that the Weibull parameters determined differ for varying failure mechanisms. Thus, an assignment of the failure mechanisms to the gradual faults or the sudden faults was possible. However, it remains open how the failure rate for certain failure modes develops over time and whether this is plausible with the knowledge of the ageing mechanism.

Since recent studies by the ADAC in [11] again show that the battery is responsible for a large number of vehicle failures as well as increasing safety requirements for future driving applications, it is important to include specific battery failure rates. Therefore, this article presents a method to calculate time-dependent failure rates of different investigated battery failures to get a better understanding of the progressing of ageing mechanisms.

For this purpose, the knowledge about the dominant failure mechanism is required in addition to the lifetime values of the failed batteries. With the knowledge gained about the development of the failure mechanisms over time, the diagnostic mechanisms of the electric battery sensor can be prioritised and further developments can be made to ensure a safe power supply from the battery.

The main focus of this study is on the methodology for determining time-dependent failure rates and not on absolute values of the failure rates. The reason for that is the data available so far from 12 V lead batteries come from conventional combustion vehicles and were partly operated without any energy management and therefore do not cover the future application of a safety battery in backup operation at all. The failure criterion in previous studies is the ability to start the vehicle by providing the high starting current, whereas the requirement for a safety battery is the safe supply in the event of a safe stop scenario. For future investigations, the current profile required in a safe stop scenario would therefore be more suitable as a failure criterion. Nevertheless, in order to present the methodology with real field data, the collected field data of the Battery Council International (BCI) are used, which, however, come from such conventional vehicles. The BCI publishes a *Report on Battery Failure Modes* every five years; the most recent one from 2020 [12] is only available for purchase, which is why the methodology in this paper is shown using the example of the study from 2015 [13].

This study begins with an introduction to the methodology for determining time-dependent failure rates for different failure modes based on field data. Subsequently, a database is presented to demonstrate the methodology. Finally, the results of time-dependent failure rates are presented in normalised format and then discussed.

2. Methodology

This section presents the methodology for determining time-dependent failure rates of different failure modes using field data. For this purpose, the lifetime distribution is presented, which can be used to represent the field data. In order to be able to examine different failure modes in relation to each other, censoring methods are introduced and bias correction is applied due to limited sample sizes in some cases. Finally, the derivation of the failure rate from the obtained parameters is presented, which can then be used to calculate the time-dependent failure rates.

2.1. Lifetime Distribution

The raw data of lifetime values are the basis for lifetime analyses. For statistical estimations, these empirical data need to be represented by functions. The failure probability $F(t)$ can represent such empirical lifetime values in a continuous matter. There are different types of lifetime distributions to properly represent various failure behaviours. A widely used distribution function that can be adapted to a large variety of shapes is the Weibull distribution. The Weibull distribution as the 2-parametric version has a shape parameter b and a scale parameter T to vary its behavior. The distribution also has the statistical variable t, which is in this case the lifetime value [14,15].

The failure probability $F(t)$ of the 2-parametric Weibull distribution is given by:

$$F(t) = 1 - e^{-\left(\frac{t}{T}\right)^b} \tag{1}$$

The derivative of the failure probability $F(t)$ of the 2-parametric Weibull is the probability density function $f(t)$:

$$f(t) = \frac{b}{T} \cdot \left(\frac{t}{T}\right)^{b-1} \cdot e^{-\left(\frac{t}{T}\right)^b} \tag{2}$$

Both the shape parameter b and the scale parameter T have to be estimated so that the Weibull distribution represents the collected lifetime values well. A widely used robust analysis method is the maximum likelihood estimation (MLE). The likelihood function is maximised to find the most likely values of the distribution parameters for a given data set. The logarithmic likelihood function, which is easier to maximize, is based on the probability density function $f(t)$ for a chosen lifetime distribution, in this case the 2-parametric Weibull distribution from Equation (2) with the parameters b and T to be estimated:

$$\Lambda = \ln L = \sum_{i=1}^{n} \ln f(t_i; b, T) \tag{3}$$

Subsequently, the values for the parameters must be found, which result in the highest value for this function. The partial derivatives of the likelihood function are set equal to zero for each parameter:

$$\frac{\partial \Lambda}{\partial b} = 0 \ \& \ \frac{\partial \Lambda}{\partial T} = 0 \tag{4}$$

These determined Weibull parameters represent the lifetime behaviour for the given data set [16].

In a previous work, Mürken et al. already investigated the failure behaviour of the entire battery based on workshop data on battery replacement. In this approach, however, there was no information about the reason for the failure and consequently aged batteries in a still functional state were also included [9].

The approach presented in this paper allows individual failure modes to be investigated separately but in relation to the entire sample. For this purpose, a methodology is presented to determine ageing models for individual ageing mechanisms from individual lifetime values of batteries, which are verified with the knowledge of the ageing mechanisms.

In order to perform separate Weibull analyses for different failure modes, the data must be censored with respect to the different failure modes. Furthermore, the approach is shown on a data set that shows the details of the individual failure modes only for a certain period of time, so additional censoring with respect to time must be applied. These adjustments are explained in the following section.

2.2. Censoring Data

This approach is shown with lifetime values of batteries with different failure modes; therefore, censoring according to the investigated failure modes is necessary. This censoring according to individual failure modes is shown in Table 1; the Failure mode A, B as well as Functional are listed next to the lifetime values in hours. The "1" indicates which lifetime values are to be assigned to the failure mode and accordingly the "0" indicates which lifetime values are censored for the specific reliability analysis. This approach is shown on lifetime values based on the BCI *Report on Battery Failure Modes* 2015 in which the breakdown into individual failure modes covers only a period up to 4 years, although the entire sample contains batteries that survived beyond these 4 years up to 12 years. Therefore, the BCI study, used to present this approach, is considered to be a study performed over a certain period of time, in this case for 4 years, and the batteries that survive this period are censored. Thus a censoring according to time has to be applied, which is known as a right censoring of type 1. The censoring according to time is applied for lifetime values from 4 years onwards, thus all remaining lifetime values of the batteries that survived longer than these 4 years are assigned a lifetime value of $t_{censored} = 4$ years $= 35,040$ h as shown in Table 1.

Table 1. Example of a table to censor data according to the failure mode and time.

Lifetime [h]	Failure Mode A	Failure Mode B	Functional
10,000	1	0	0
20,000	1	0	0
15,000	0	1	0
25,000	0	1	0
5000	0	0	1
15,000	0	0	1
35,040	0	0	0
35,040	0	0	0
35,040	0	0	0

With both censoring methods the Table 1 is an example of a censoring table as used for reliability analysis and ageing model building in this work.

The censoring of the data with regard to different failure modes may result in only a small sample number for rarely occurring failure mechanisms. For such small subsamples, the accuracy of the Weibull distribution fitting suffers and the Weibull parameters determined in the reliability analysis are overestimated. To avoid this, a bias correction is applied to the reliability analyses carried out in this work. The unbiased shape parameter b_u is obtained by multiplying the shape parameter b determined by MLE with the bias correction factor U:

$$b_u = b \cdot U \tag{5}$$

Since multiple failure modes are investigated, censored data are examined; the bias correction factor U according to Zhang et al. [17] is given by:

$$U = \frac{1}{1 + \frac{1.37}{r - 1.92}\sqrt{\frac{N}{r}}} \tag{6}$$

The calculated bias correction factor U varies with the number of failures considered r and the total sample size N. If a bias correction is carried out, first the biased shape parameter b determined in Equation (4) is unbiased by Equation (5). Then the corresponding scale parameter T is calculated using Equation (4). According to Tevetoglu et al. this provides valid results for reliability analysis even for small sample sizes and recommends the use of bias correction methods [18]. However, even with larger samples it is advisable to apply a bias correction to increase accuracy [19].

The techniques presented in this section can be used to determine characteristic Weibull parameters for individual failure modes using the Weibull distribution from Section 2.1. In Section 2.3, time-dependent failure rates can be calculated with these determined shape parameters b and scale parameters T.

2.3. Derivation of Time-Dependet Failure Rates

Since Weibull analyses have been carried out and the specific shape parameters b and scale parameter T have been determined, time-dependent failure rates can now be calculated. With the probability density function $f(t)$ in Equation (2) and the failure probability $F(t)$ in Equation (1) the failure rate $\lambda(t)$ is obtained with Equation (7). By using the determined shape parameters b and scale parameter T, the failure rate λ can be calculated for different times t. This also applies to Weibull parameters, which were determined using bias correction as described in Section 2.2.

$$\lambda(t) = \frac{f(t)}{1 - F(t)} = \frac{b}{T} \cdot \left(\frac{t}{T}\right)^{b-1} \tag{7}$$

In addition to the shape parameter b, the development of the failure rate over time gives an indication of the ageing and failure behaviour due to different failure modes of a component. In order to be able to show the presented methodology for batteries with different failure modes, it is applied to lifetime values derived from the field investigations of the BCI with the *Report on Battery Failure Modes* 2015. Therefore, in the following Section 3, this database is presented and the individual failure modes are explained for a better understanding in relation to the development of the failure rate over time.

3. Database of Lifetime Values

The presented methodology in Section 2 requires a suitable database to prove its validity. The *Battery Council International* presented a field investigation of lead batteries in *Report on Battery Failure Modes* in 2015 [13]. This study is used to prove the presented approach from Section 2 with real battery data and thus the applicability to electrochemical elements and their different ageing mechanisms. Therefore, the database and the major battery failure modes are presented in this section.

3.1. Field Data of Batteries

The BCI collected a total sample of 1454 batteries from different locations in the USA and presented the results in [13]. The batteries were collected within five months between August and December 2014 and analysed using a teardown analysis to find the dominant failure mode that led to the battery failure. Although several ageing mechanisms can occur in parallel and in total can affect the performance of the battery, there is one failure mode that is most pronounced and has thus been identified as the dominant failure mode for this battery failure. If this dominant ageing mechanism did not occur to the same extent, the battery would possibly still be functional at this time.

The sample reflects the different influences of temperature on batteries because of the different climatic conditions of the sample locations. Due to the large climatic differences in the USA between the North and the South, the sample can be divided into North and South accordingly. In the North with lower mean temperatures 857 batteries were collected; in the South with higher mean temperatures 597 batteries were collected.

In addition to climatic differences due to the locations where the batteries were collected, the sample contains batteries of varying quality and from a wide variety of cars. With an age of up to 12 years, the oldest batteries date back to 2002 and thus come from vehicles that rarely had energy management or start-stop function. Furthermore, besides the manufacturing quality of the batteries, nothing is known about the type, so the sample contains few or no advanced lead batteries such as absorbent glass mat or enhanced flood batteries but mainly older flood batteries.

Furthermore, the design of the batteries differs. For example, the sample includes battery housings according to the European standard with a protective terminal niche [20]. However, this type of housing represents a maximum of about 20% of the sample, most batteries are built according to the US standard or the Japanese Industrial Standards (JIS) [21]. This means that the majority of the batteries in the sample have a battery housing without terminal niche and the battery terminals are exposed. This makes mechanical impacts on the battery housing, e.g., during installation or removal or even during transportation when stacked, much more critical. Due to improper handling, the battery terminals are prone to mechanical damage. This can cause fractures between the battery terminal and the cell, which partially or directly destroys the battery.

All in all the lead batteries investigated in this sample are no longer relevant for today's driving applications with high safety requirements. Since the operational demands on the battery were different back then, energy management was rarely implemented and a wide variety of qualities and types of lead batteries can be found in the sample. For future driving applications, however, only high-quality lead batteries of the latest generation will be used in combination with a powerful battery sensor and energy management.

In addition to the failure probability $F(t)$ for the total sample, those of the sample parts North and South over a time period of 12 years are also published. In this study, the data of the total sample is used for the evaluation, because it covers locations from all over the USA. In combination with batteries surviving up to 12 years, this sample can be recognised as a representative sample of lead batteries in the field in 2014. These batteries represent the various stress levels, qualities and types used at that time. To perform the statistical approach the graphs are digitised again and converted into individual lifetime values of the batteries. The total sample used for this analysis is shown in Figure 1 as probability of failure $F(t)$ for all failure modes including exchanged but functional batteries in one curve. The failure probability $F(t)$ always starts for $t = 0$ with $F(t=0) = 0\%$ failed parts and always results after $t = n$ until all parts of the sample failed in $F(t=n) = 100\%$.

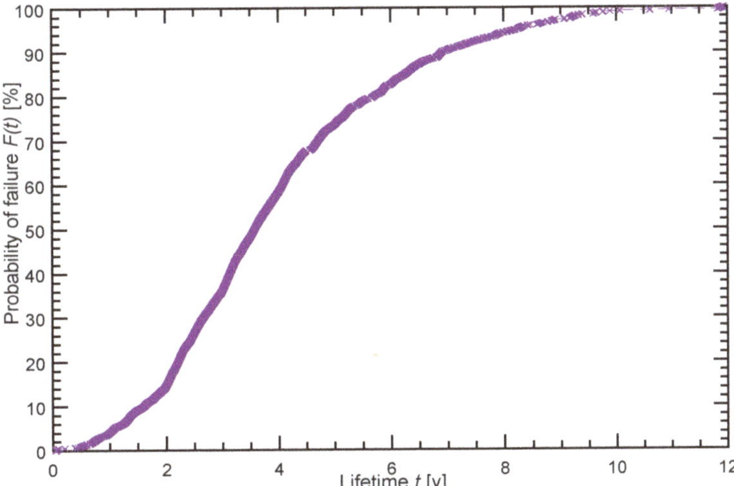

Figure 1. The total sample as a sum of all failures modes is shown in this graph for a time period of 12 years. On the left side the empirical probability of failure $F(t)$ is shown [13].

Besides the failure probability $F(t)$ of all failures modes as a sum, the failure probabilities $F(t)$ of the major five failure modes found by BCI while examining the batteries are published as well. The probability of failures $F(t)$ of each failure mode is separately presented for a time period of 4 years for the sample parts North and South by BCI. The failure probabilities $F(t)$ of these five failure modes are shown for the total sample in Figure 2.

Figure 2 shows the major failure modes. For a better differentiation of the individual curves, they are mentioned in relation to the time of 4 years. Starting from the top, *Serviceable*, *Short Circuit*, *Plates and Grids*, *Worn out and Abused* and *Open Circuit* are displayed.

Both Figures 1 and 2 are based on single lifetime values derived from the *Report on Battery Failure Modes* by BCI in 2015 [13]. In particular, the data of the five major failure modes in Figure 2 are the basis to present the methodology for determining time-dependent failure rates $\lambda(t)$ in Section 2.

In order to discuss the time-dependent failure rates of the individual failure modes in relation to their ageing behaviour, the five major failure modes are presented in the following sections.

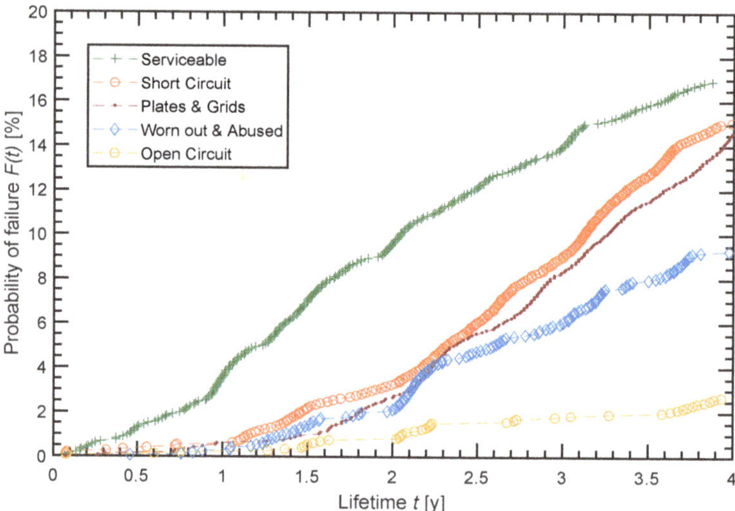

Figure 2. The major five failure modes found by BCI are shown for the total sample. The individual probability of failures *F(t)* is displayed for a time period of 4 years [13].

3.2. Failure Modes of Lead Batteries

Various ageing effects influence the performance of batteries and contribute to battery failure. Ruetschi investigates the ageing mechanisms of lead batteries in [22], where more detailed explanations of the failure modes mentioned in this section can also be found. Brik et al. presented their causal relation structured in a fault tree in [23]. However, the focus of this work is on the statistical consideration and the principle development of the different failure modes. The field data with the failure modes presented by BCI in [13] are only used to show the methodology. For safety applications, higher requirements apply for the design of a safety concept. This also includes the 12 V battery as a safe power supply. Therefore, suitable monitoring with energy management is a premise for safe operation of the battery. However, this cannot be assumed for the BCI sample, and hence this sample only serves to demonstrate the methodology. In addition to very old batteries, this sample includes various technologies that were operated with different load scenarios, that cannot be derived from the published data. However, the operation was rarely done with a suitable energy management. Moreover, the failure criterion used for this sample is the ability to start the vehicle. Nevertheless, this database is used to demonstrate the methodology and for this purpose the failure modes are described again in this section. These failure modes summarise different battery ageing mechanisms and can thus be categorised as sudden faults and gradual faults. The sudden faults occur abruptly and without notice and thus cannot be detected or predicted by diagnostic functions with a battery sensor. The sudden faults lead to a complete loss of power. The gradual faults refer to battery failures caused by ageing mechanisms and wearing out that occurs slowly during use and can thus be detected by appropriate diagnostic functions [10,24].

3.3. Open Circuit

The failure mode Open Circuit can be caused by broken connections at various parts of the battery. The electrical connection between the battery terminal and the connected cell can break due to massive mechanical impact from outside the battery housing and especially the battery terminals. As a result, the battery directly loses its full performance capability. Since many battery housings do not have terminal niches and the battery terminals are exposed, they are probably very prone to unintentional mechanical influences from the outside. This can lead to more frequent occurrences of *Open Circuit*, whereas terminal niche may protect the battery terminal from careless handling and prevent frequent *Open Circuit*. The individual cells of the lead battery are electrically contacted internally via solid inter-cell connectors to connect the six individual cells to one battery. If one of these inter-cell connectors between the cells breaks, there is also an *Open Circuit*. Finally, within each of the six individual cells, there are connections from the positive and negative electrodes to the respective inter-cell connector. These partial elements of the electrodes, known as electrode lugs, can break as well and thus make this electrode inactive. However, this means that only one part of a cell and thus one part of its capacity is no longer available; the remaining electrodes and cells work as usual. Due to the failure of *Open Circuit* the battery can lose its entire performance, as its circuit is interrupted. The *Open Circuit* happens suddenly without prior indication and thus is not detectable. Consequently it is allocated to the sudden faults and clearly defined by BCI with *Open Circuit*.

3.4. Plates and Grids

The failure mode *Plates and Grids* summarises ageing mechanisms that contribute to the increase of the internal resistance. Besides the main ageing mechanism corrosion in this category, this is to some extent also sulphation. Corrosion is favoured by a low acid concentration in the electrolyte, i.e., at low states of charge. However, high charging voltage and especially operation at high temperatures have a major influence. The general cycling of the battery also leads to corrosion over time. Corrosion mainly affects the positive electrode. In this case, corrosion leads to a reduction of the grid cross-section and thus to less active area for electron transport between the active mass of the electrode and the grid itself, and thus the grid resistance increases. With the decreasing grid cross-section, the mechanical stability also decreases. If the battery is very corroded, even parts of the active mass can lose contact with the grid and thereby no longer be reached and contribute to the reaction. In extreme cases, this can also reduce the available capacity. Primarily, however, corrosion leads to an increase in internal resistance. In addition to corrosion, sulphation can contribute to an increase in internal resistance. Sulphation occurs when the battery is frequently operated in states of partial charge and when the battery is rarely fully charged. This operation sometimes also leads to the occurrence of acid stratification. In addition to the operating conditions, high temperatures favour the formation of sulphation on the one hand but also lead to good dissolution of fine sulphate crystals if the charging voltage is sufficiently high. Low temperatures, on the other hand, have negative effects on the charging ability and can lead to the already mentioned rare full charges and frequent partial states of charge. In particular, the low states of charge lead to a higher solubility of $Pb2^+$ ions due to a lower acid concentration in the electrolyte, which form a sulphate layer by recrystallisation to $PbSO_4$. The sulphate layer develops on both electrodes during discharge; as the pores of the negative electrode are larger, the sulphate crystals can become much larger than on the positive electrode. The larger the sulphate crystals are, the more difficult it is to dissolve them again by extensive full charging; during normal charging processes in the vehicle, reversible dissolution of the fine sulphate crystals hardly takes place. Due to the coarse sulphate crystal structure on the active electrode surface, this is reduced and loses conductivity. The lower conductivity increases the internal resistance. The binding of active material in the sulphate crystals reduces the available active mass and leads to a lower acid density of the electrolyte. The lower acid density means that very high maximum open-circuit voltages can no longer be achieved, thus limiting the capacity from above. This influence is assigned to the next category *Worn out and Abused* as well. Overall, the failure mode *Plates and Grids* with its ageing mechanisms corrosion and sulphation is to be assigned to the gradual faults, as they can only arise through extensive use over time.

3.5. Worn out and Abused

This failure mode covers the ageing mechanisms that lead to loss of capacity. The capacity of the battery determines to a large extent how much electrical energy it can store.

If it loses capacity, it may no longer be able to supply enough energy for an application. This process thus has a decisive influence on the lifetime of a battery. The loss of capacity is primarily influenced by loss of active mass and due to cycling and sulphation. The sulphation has already been shortly explained in the previous Section 3.4, but decrease is mainly due to the processes described in the following. During the normal use of the battery, it is discharged and charged, that means cycled. This leads gradually to the wearing out of the active mass, mainly at the positive electrode, which is decisive for the capacity. This cycling stress is intensified by deep discharging or discharging with high currents, because the mechanical stress for the active material is correspondingly higher. When charging or discharging, parts of the electrode material are transformed and the volume changes. If large rapid volume changes take place, parts of the active material can become internally decontacted and thus can no longer contribute in the reaction. The porosity and conductivity decreases, which mainly reduces the available capacity but also slightly increases the internal resistance. In summary, cycling in particular contributes significantly to the loss of capacity, as does sulphation to some extent. These processes only become apparent when the battery is used or remains in a low state of charge and therefore *Worn out and Abused* is allocated to gradual faults.

3.6. Short Circuit

The failure mode *Short Circuit* implies an additional electrical path between positive and negative electrodes within a cell of the battery [25]. This additional connection has a resistance that varies depending on the characteristics of the short circuit and can discharge the cell at different rates. In addition to a changed open-circuit voltage, a short circuit can be recognised by an increased overall resistance of the battery. In principle, there are four possibilities for internal short circuits in lead batteries. Due to strong corrosion, the grid can change considerably and grow together above the active electrode surface in the area of the busbar. In addition, due to severe corrosion and accompanying change in volume of the electrodes, the separator can be damaged, causing a connection between the positive and negative electrodes. The mechanism of corrosion has already been explained in the previous Section 3.4. On the other hand, deep discharge of the battery and operation out of the specification can cause dendrites to grow, which can create a connection between the electrodes. Similarly to sulphation, at low states of charge and accompanying low acid density, the solubility of the lead increases and fine nonconductive dendrites of lead sulphate can form. Frequent use of the battery outside the specification at very low states of charge can cause the dendrites to grow and form fine filaments which can grow through the separator. At higher states of charge, the nonconductive lead sulphate is reduced to electrically conductive metallic lead, which can create an electrical connection. The fine short-circuit thus develops slowly during use, as do the other ageing mechanisms, and it can be avoided by appropriate operational management and monitoring. Only the hard short-circuit, by penetration with, e.g., a nail, could produce a sudden hard short-circuit, which, however, can only be achieved by external action and is not possible by normal operation.

3.7. Serviceable

The last failure mode listed is labelled *Serviceable*, which defines actually still good, functional batteries whose ageing progress has not yet led to failure. These batteries were replaced prematurely, presumably due to customer request. This failure mode is therefore not a failure mode of the battery but only indicates how many batteries have been replaced before the end of their lifetime.

These five major failure modes presented in the previous sections can be found in the field data in Figure 2 and in Section 3.1. The following Section 4 presents the results of the Weibull analysis and the calculated time-dependet failure rates $\lambda(t)$.

4. Results

Since this work is about the methodology for determining time-dependent failure rates for individual failure modes and no field data from batteries are available that represent future load scenarios and failure criteria, the results are presented in normalised form. This section shows the results of the Weibull analysis performed with the field data derived from the *Report on Battery Failure Modes* by BCI 2015 as described in Section 3.

Nevertheless, the older database is used to demonstrate the methodology from Section 2, but the absolute failure rates $\lambda(t)$ are not relevant for the reasons mentioned above. The Weibull parameters are presented, but the scale parameter T is normalised with respect to a typical vehicle lifetime. Subsequently, the calculated time-dependent failure rates $\lambda(t)$ are also presented normalised. The time is normalised with respect to a typical vehicle lifetime $\tau_{vehicle}$. The failure rates $\lambda_{FailureMode}(t)$ of the individual failure modes from Section 3 are normalised to the failure rate $\lambda_S(t)$ of the still functional batteries of the failure mode *Serviceable* after $\tau_{vehicle,0.6} = 0.6 \cdot \tau_{vehicle}$ vehicle lifetimes according to the following Equation (8):

$$\lambda_{FailureMode,normalised}(t) = \frac{\lambda_{FailureMode}(t)}{\lambda_S(\tau_{vehicle,0.6})} = \frac{\lambda_{FailureMode}(t)}{\lambda_{S,0.6}} \tag{8}$$

The parameters determined by Weibull analyses are summarised in a table in the following section and the determined time-dependent failure rates $\lambda(t)$ are listed individually.

4.1. Estimated Weibull Parameters

The results of the Weibull analyses are obtained by parameter estimation with maximum likelihood estimation (MLE) and the application of bias correction. Table 2 shows the results of the estimated Weibull parameters. For the available five failure modes, the determined shape parameters b and the normalised scale parameter T are listed.

Table 2. Determined Weibull parameters for the failure modes of the total sample. The table shows the estimated shape parameter b for each failure mode and the normalised scale parameter T by a typical vehicle lifetime $\tau_{vehicle}$. (Analysis method: MLE; Rank method: Median ranks; Confidence bounds method: Fisher matrix; unbiasing parameters)

Failure Mode	Shape Parameter b	Normalised Scale Parameter T
Serviceable	1.239	1.134
Open Circuit	1.819	2.127
Plates & Grids	2.812	0.582
Worn out & Abused	2.255	0.826
Short Circuit	2.637	0.596

The parameters listed in Table 2 give an indication for the failure behaviour of the obtained component. In general, components can fail due to different reasons. The classic bathtub curve is divided into three areas that show different stages of failure behaviour and thus are linked to different magnitudes of the shape parameter b. Early failures with a decreasing failure rate λ_{early} and a shape parameter $b \ll 1$ are often due to manufacturing defects, incorrect storage and long storage times. Random failures with a constant failure rate λ_{random} and a shape parameter $b \approx 1$ are often encountered in electronic components. The third area of the bathtub curve is indicated as wear out or ageing failures with an increasing failure rate λ_{ageing} and a shape parameter $b \gg 1$. As batteries age with first use, their failure rate is expected to be continuously or exponentially increasing during use like λ_{ageing} with a shape parameter $b \gg 1$. The second parameter determined is the scale parameter T. This parameter indicates the time at which the probability of failure $F(T) = 63.2\%$ is reached and a corresponding amount of parts of the sample have failed.

In the following Section 4.2, the results of the Weibull analyses are presented as time-dependent failure rates for the different failure modes listed in Table 2 and its parameters and failure behaviour are discussed.

4.2. Failure Rates of Lead Battery Failure Modes

The Weibull parameters listed in Table 2 are obtained by Weibull adaptions according to the methodology presented in Section 2 to lifetime values of the different failures modes as presented in Section 3. According to Section 2.3 the failure rate $\lambda(t)$ can be calculated with Equation (7). By using the determined shape parameters b and scale parameter T, the failure rate λ can be calculated for different times t. The five failure modes and their Weibull parameters from Table 2 are used to calculate these failure rates $\lambda(t)$. In the following, this is shown individually with the Weibull parameters from Table 2 in normalised graphs by a typical vehicle lifetime $\tau_{vehicle}$. All failure rates are displayed with their upper and lower confidence interval for a confidence level of 95% (on time, type 1).

The failure rates are normalised by failure rate $\lambda_{S,0.6}$ of the functional batteries of the failure mode *Serviceable* according to Equation (8). Consequently the still good batteries and their failure rate $\lambda_S(t)$ is displayed at first in Figure 3.

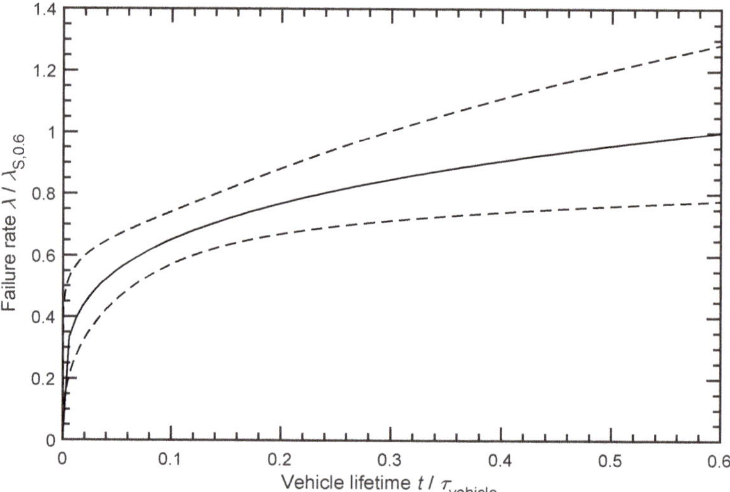

Figure 3. The normalised failure rate $\lambda_S(t)$ of the failure mode *Serviceable* with a shape parameter $b = 1.239$ and a normalised scale parameter $T = 1.134$ is shown. In addition, the upper and lower confidence limits for a confidence level of 95% are displayed.

The pattern of the failure rate $\lambda_S(t)$ of the failure mode *Serviceable* in Figure 3 clearly belongs to the category of random failures with a shape parameter very close to $b = 1$. This makes sense with the information about this failure mode according to Section 3.7. Since the batteries are still functional and there is no fault due to ageing, the battery changes in car workshops due to *Serviceable* are highly random and logically there is no progression, as expected with an ageing mechanism.

The failure rate $\lambda_{OC}(t)$ of the first real failure mode *Open Circuit* is shown in Figure 4.

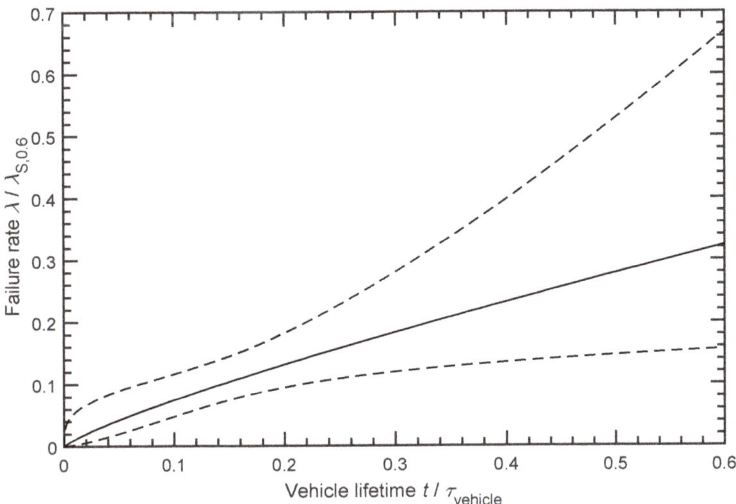

Figure 4. The normalised failure rate $\lambda_{OC}(t)$ of the failure mode *Open Circuit* with the estimated shape parameter $b = 1.819$ and a normalised scale parameter $T = 2.127$ is shown. In addition, the upper and lower confidence limits for a confidence level of 95% are displayed.

The failure rate $\lambda_{OC}(t)$ in Figure 4 shows a pattern that cannot be clearly assigned to either random or ageing mechanisms. The failure rate $\lambda_{OC}(t)$ increases with time for the shape parameter $b = 1.819$ determined by Weibull analysis. However, no clear assignment to ageing effects with linear or exponential increase and shape parameter $b \geq 2$ or to completely random failures with shape parameter $b \approx 1$ is possible. This uncertainty is also shown by the wide confidence bounds. Only a few batteries failed with this failure mode, as can be seen in Figure 2. The small sample size of this failure mode results in very wide confidence bounds. Thus, on the one hand, the progression of the failure rates could be rather exponential, as shown by the upper confidence limit. On the other hand, it could be almost randomly distributed, as the lower confidence limit shows. This could be explained by the knowledge of the failure mode *Open Circuit* from Section 3.3. It can be assumed that the *Open Circuit* is mainly caused by a mechanical impact from the outside and only in rare cases, e.g., corrosion inside leads to such strong instabilities that a rupture can occur due to normal vibrations in the vehicle. Before an *Open Circuit* occurs, a failure of the battery is more likely due to its greatly increased internal resistance. This Weibull statistical analysis shows that for the *Open Circuit* failure mode, different ageing mechanisms overlap and the failure rate does not show a consistent failure behaviour. According to the failure description in Section 3.3, an assignment of the failure mode *Open Circuit* to the sudden failures would fit but cannot be done definitively with this database. For this purpose, a larger sample and a longer observation period would make sense in order to obtain more meaningful results.

The next failure mode according to Table 2 is *Plates and Grids* and the corresponding failure rate $\lambda_{PG}(t)$ is shown in Figure 5.

The development of the failure rate $\lambda_{PG}(t)$ in Figure 5 shows a moderate exponential increase. The shape parameter $b = 2.812$ clearly belongs to the ageing effects, which are characterised by shape parameter $b \gg 1$. With the knowledge of the ageing mechanism *Plates and Grids* from Section 3.4, the curve in Figure 5 is understandable. When the battery is not in use, only minor ageing takes place. Only with increasing time and use of the battery does the ageing effect become apparent and eventually lead to a failure of the battery due to increased internal resistance. The allocation of the failure mode *Plates and Grids* to the gradual faults is suitable for the curve according to Figure 5, resulting from the shape parameter b.

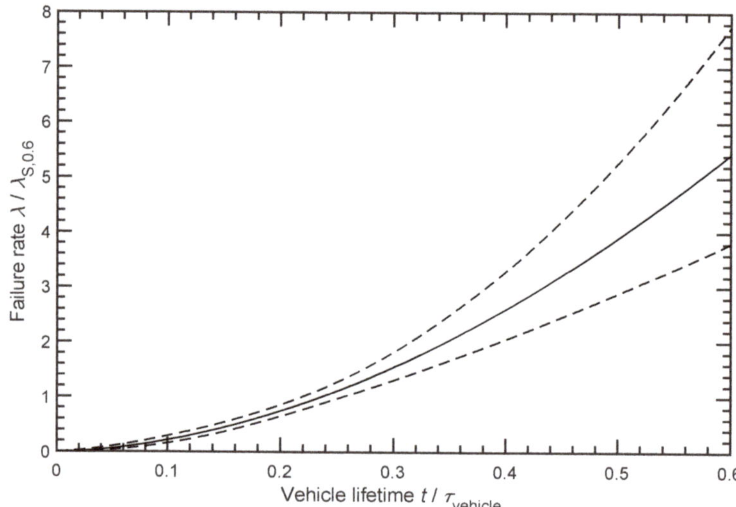

Figure 5. The normalised failure rate $\lambda_{PG}(t)$ of the failure mode *Plates and Grids* with a shape parameter $b = 2.812$ and a normalised scale parameter $T = 0.582$ is shown. In addition, the upper and lower confidence limits for a confidence level of 95% are displayed.

The following Figure 6 shows the normalised failure rate $\lambda_{WOA}(t)$ of the failure mode *Worn out and Abused*.

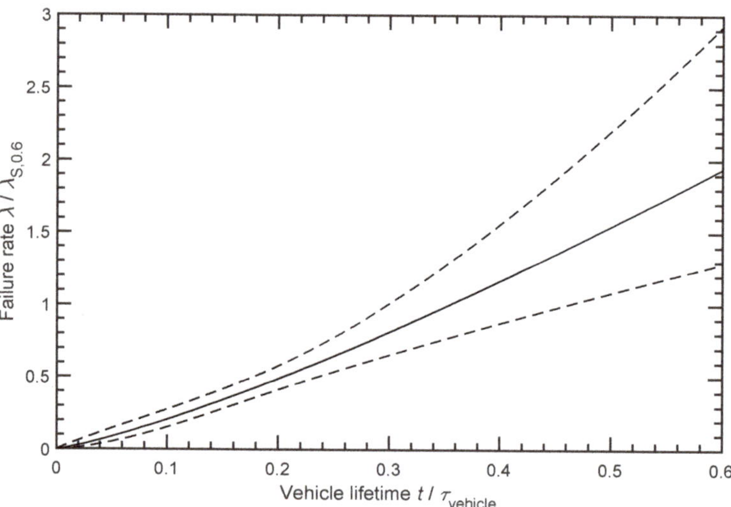

Figure 6. The normalised failure rate $\lambda_{WOA}(t)$ of the failure mode *Worn out and Abused* with a shape parameter $b = 2.255$ and a normalised scale parameter $T = 0.826$ is shown. In addition, the upper and lower confidence limits for a confidence level of 95% are displayed.

The shape of the failure mode *Worn out and Abused* in Figure 6 is almost linear with a shape parameter close to $b \approx 2$. This clearly indicates a wear failure or ageing mechanism. In addition, the scale parameter T shows that many batteries survive for a very long time of a typical vehicle lifetime until this failure mode becomes sufficiently dominant to cause the battery to fail. This can be well explained by Section 3.5, since capacity loss is decisively caused by cyclisation of the battery, which requires the battery to be used for a long time. The clear assignment of the failure mode to the ageing effects fits well with the gradual faults.

The failure mode *Short Circuit* is shown in Figure 7.

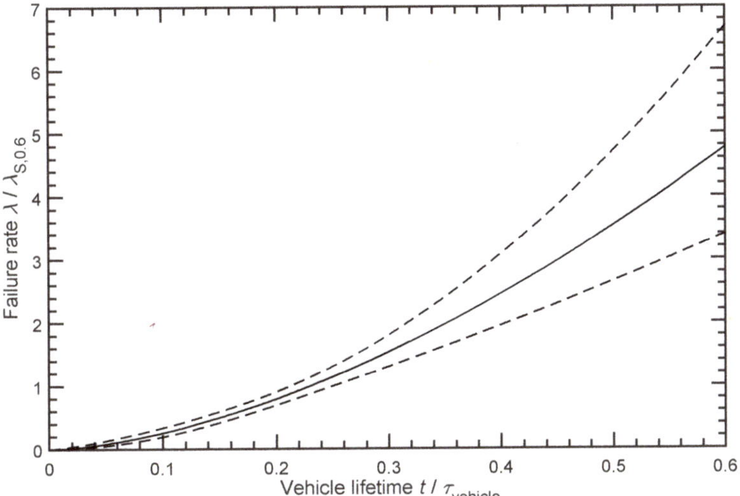

Figure 7. The normalised failure rate $\lambda_{SC}(t)$ of the failure mode *Short Circuit* with a shape parameter $b = 2.637$ and a normalised scale parameter $T = 0.596$ is shown. In addition, the upper and lower confidence limits for a confidence level of 95% are displayed.

The development of the failure rate $\lambda_{SC}(t)$ in Figure 7 shows a slightly exponential increase with the shape parameter $b = 2.637$. This indicates a strongly use-dependent ageing mechanism. According to Section 3.6, the failure mode *Short Circuit* can occur for different reasons. It can only occur completely randomly due to misuse from the outside or extreme operation out of specifications with frequent severe deep discharges, so dendrites can grow at an early stage. In normal use, energy management should protect the battery from frequent deep discharge conditions so that dendrite growth, if it occurs at all, takes a long time to develop. The comparatively small scale parameter T indicates vehicles in which the energy management is poor from the start and the battery is thus exposed to frequent deep discharges early on.

The results of the Weibull adaptation and the Weibull parameters determined on the basis of the BCI 2015 *Report on Battery Failure Modes* field data and the visualised failure rates $\lambda(t)$ for the individual failure modes are summarised and discussed in the following Section 5.

5. Conclusions

Reliability analyses in vehicles are becoming increasingly important due to the complexity of vehicles. To ensure safe operation, proof of reliability must be provided for more components. In this study, an approach to determine failure rates for lead batteries from field data is presented. It has been shown that the statistical reliability analysis can also be applied to electrochemical components to derive appropriate failure rates. In particular, by knowing the reason for the failure of the lead battery, the developments of the failure modes over time as shown can be understood but also justified.

Batteries with the failure mode *Serviceable* show a progression of the failure rate over time, which clearly indicates random failures, as the failure description indicated. The failure mode *Open Circuit*, on the other hand, shows a failure behaviour over time that cannot be clearly assigned to random or ageing failures. The small sample size and the short observation period can have a negative influence on the accuracy of the Weibull analysis. For a longer observation period, a clearer result could emerge for the failure mode *Open Circuit* and a clearer assignment to sudden faults or gradual faults could be possible. The two failure modes, *Plates and Grids* as well as *Worn out and Abused*, show progressions of the failure rates over time, which can clearly be assigned to the ageing mechanisms. The last investigated failure mode *Short Circuit* can, according to the description in Section 3.6, arise due to dendrites slowly developing through deep discharges, also called soft short, or due to hard short circuits caused by external rather random influences. The determined failure rate for the failure mode *Short Circuit* shows an exponential progression over time and is rather to be assigned to the wear or ageing mechanisms for these data.

With the presented approach, ageing models for the investigated failure modes can be determined from field data, which are useful for the safety analysis in the power supply system and the entire vehicle. The field data examined are still based on conventional vehicles in which the lead battery is used as a typical starter battery. In future driving applications, however, a 12 V battery will be needed in the power supply system, which will not be used for starting the engine anymore but much more for the safe supply of energy in critical cases. This means that the 12 V battery must provide the energy and power for a so-called safe stop scenario, which brings the vehicle to a safe stop at the emergency lane by a double lane change. These requirements are of course different from those of a conventional, mainly manually driven vehicle. Thus, the battery is continuously monitored and operated in a favourable operating range for the battery. By monitoring the battery's condition and ageing progress, it is possible to replace the battery before it fails, so that critical failures can be avoided.

Nevertheless, this study used these old field data from outdated batteries just to show the approach for future field investigations on batteries to determine relevant failure rates for future driving applications. Therefore, in further investigations, the collected lead batteries could be tested with the new failure criterion of a defined safe stop scenario when collecting field data. Furthermore, the determination of the failure mode by tear down analysis is prone to subjective evaluations. One possibility would be to electrically test the collected batteries and evaluate their ageing progress with objective criteria. This would be useful for assessing the reliability of 12 V batteries such as lead batteries for future driving applications.

Author Contributions: Conceptualization, R.C., F.H. and K.P.B.; methodology, R.C. and F.H.; software, R.C.; validation, R.C. and F.H.; formal analysis, R.C. and F.H.; investigation, R.C. and F.H.; resources, R.C.; data curation, R.C.; writing—original draft preparation, R.C.; writing—review and editing, R.C.; visualization, R.C.; supervision, F.H. and K.P.B.; project administration, R.C. and F.H.; All authors have read and agreed to the published version of the manuscript.

Funding: This research received no external funding.

Institutional Review Board Statement: Not applicable.

Informed Consent Statement: Not applicable.

Data Availability Statement: Not applicable.

Conflicts of Interest: The authors declare no conflict of interest.

References

1.	Stolte, T.; Bagschik, G.; Maurer, M. Safety goals and functional safety requirements for actuation systems of automated vehicles. In Proceedings of the IEEE 19th International Conference on Intelligent Transportation Systems (ITSC), Rio de Janeiro, Brazil, 1–4 November 2016.
2.	Dominguez-Garcia, A.D.; Kassakian, J.G.; Schindall, J.E. Reliability evaluation of the power supply of an electrical power net for safety-relevant applications. *Reliab. Eng. Syst. Saf.* **2006**, *5*, 505–514. [CrossRef]
3.	Kurita, Y.; Münzing, P.; Koller, O. Future Powernet Topology for Automated Driving. In Proceedings of the 2017 JSAE Annual Congress, Yokohama, Japan, 24–26 May 2017; pp. 109–114.

4. Koehler, A.; Bertsche, B. An Approach of Fail Operational Power Supply for Next Generation Vehicle Powernet Architectures. In Proceedings of the 30th European Safety and Reliability Conference and 15th Probabilistic Safety Assessment and Management Conference (ESREL2020-PSAM15), Venice, Italy, 1–5 November 2020; pp. 32–58. [CrossRef]
5. Koehler, A.; Bertsche, B. Cyclisation of Safety Diagnoses: Influence on the Evaluation of Fault Metrics. In Proceedings of the 67th Annual Reliability & Maintainability Symposium (RAMS2021), Orlando, FL, USA, 24–27 May 2021; to be published.
6. Albers, J.; Koch, I. Functional Safety of Lead-acid Batteries in New Vehicle Applications. In Proceedings of the 14th European Lead Battery Conference, Edinburgh, Scotland, 9–12 September 2014.
7. Albers, J.; Koch, I. Reliability of lead-acid batteries. In the printed proceedings of EEHE 2016 *Elektrik/Elektronik in Hybrid- und Elektrofahrzeugen und elektrisches Energiemanagement VII*; Expert Publishers: Tübingen, Germany, 2016, ISBN: 978-3-8169-3346-5.
8. ADAC e.V. Pannenstatistik 2014. Allgemeiner Deutscher Automobil-Club e.V. 2014. Available online: https://bit.ly/3eqyYyf (accessed on 9 June 2021).
9. Mürken, M.; Kübel, D.; Thanheiser, A.; Gratzfeld, P. Analysis of automotive lead-acid batteries exchange rate on the base of field data acquisition. In Proceedings of the IEEE International Conference on Electrical Systems for Aircraft, Railway, Ship Propulsion and Road Vehicles International Transportation Electrification Conference (ESARS-ITEC), Nottingham, UK, 7–9 November 2018; pp. 1–6.
10. Kumar, E.S.; Sarkar, B. Improvement of life time and reliability of battery. *Int. J. Eng. Sci. Adv. Technol.* **2012**, *2*, 1210–1217.
11. ADAC e.V. Pannenstatistik 2020. Allgemeiner Deutscher Automobil-Club e.V. 2020. Available online: https://bit.ly/3srT4gn (accessed on 9 June 2021).
12. Knauer, D. *Report on Battery Failure Modes*; Battery Council International: Chicago, IL, USA, 2020.
13. Knauer, D. *Report on Battery Failure Modes*; Battery Council International: Chicago, IL, USA, 2015.
14. Pasha, G.R.; Khan, M.S.; Pasha, A.H. Empirical analysis of the Weibull distribution for failure data. *J. Stat.* **2006**, *10*, 33–45.
15. Bertsche, B.; Lechner, G. *Zuverlässigkeit im Fahrzeug-und Maschinenbau: Ermittlung von Bauteil-und System-Zuverlässigkeiten*; Springer: Berlin, Germany, 2006.
16. Cohen, A.C. Maximum likelihood estimation in the Weibull distribution based on complete and on censored samples. *Technometrics* **1965**, *7*, 579–588. [CrossRef]
17. Zhang, L.F.; Xie, M.; Tang, L.C. Bias correction for the least squares estimator of Weibull shape parameter with complete and censored data. *Reliab. Eng. Syst. Saf.* **2006**, *8*, 930–939. [CrossRef]
18. Tevetoglu, T.; Bertsche, B. On the Coverage Probability of Bias-Corrected Confidence Bounds. In Proceedings of the Asia-Pacific International Symposium on Advanced Reliability and Maintenance Modeling (APARM), Vancouver, BC, Canada, 20–23 August 2020; pp. 1–6.
19. Ross, R. Bias and standard deviation due to Weibull parameter estimation for small data sets. *IEEE Trans. Dielectr. Electr. Insul.* **1996**, *3*, 28–42. [CrossRef]
20. *Blei-Akkumulatoren-Starterbatterien Teil 2: Maße von Batterien und Kennzeichnung von Anschlüssen—DIN EN 50342-2*; Verband der Elektrotechnik, Elektronik und Informationstechnik: Frankfurt am Main, Germany, 2021.
21. *Lead-Acid Starter Batteries—JSA JIS D 5301*; Japanese Standards Association: Tokyo, Japan, 2019.
22. Ruetschi, P. Aging mechanisms and service life of lead–acid batteries. *J. Power Source* **2004**, *127*, 33–44. [CrossRef]
23. Brik, K.; Ammar, F. Causal tree analysis of depth degradation of the lead acid battery. *J. Power Source* **2013**, *228*, 39–46. [CrossRef]
24. Culpin, B.; Rand, D.A.J. Failure modes of lead/acid batteries. *J. Power Source* **1991**, *4*, 415–438. [CrossRef]
25. Zeng, Y.; Hu, J.; Ye, W.; Zhao, W.; Zhou, G.; Yonglang, G. Investigation of lead dendrite growth in the formation of valveregulated lead-acid batteries for electric bicycle applications. *J. Power Source* **2015**, *286*, 182–192. [CrossRef]

Article

Determination of the Distribution of Relaxation Times by Means of Pulse Evaluation for Offline and Online Diagnosis of Lithium-Ion Batteries

Erik Goldammer * and Julia Kowal

Electrical Energy Storage Technology, Departement of Energy and Automation Technology, Faculty IV, Secr. EMH 2, Technische Universität Berlin, Einsteinufer 11, D-10587 Berlin, Germany; julia.kowal@tu-berlin.de
* Correspondence: goldammer@tu-berlin.de; Tel.: +49-30-314-73851

Abstract: The distribution of relaxation times (DRT) analysis of impedance spectra is a proven method to determine the number of occurring polarization processes in lithium-ion batteries (LIBs), their polarization contributions and characteristic time constants. Direct measurement of a spectrum by means of electrochemical impedance spectroscopy (EIS), however, suffers from a high expenditure of time for low-frequency impedances and a lack of general availability in most online applications. In this study, a method is presented to derive the DRT by evaluating the relaxation voltage after a current pulse. The method was experimentally validated using both EIS and the proposed pulse evaluation to determine the DRT of automotive pouch-cells and an aging study was carried out. The DRT derived from time domain data provided improved resolution of processes with large time constants and therefore enabled changes in low-frequency impedance and the correlated degradation mechanisms to be identified. One of the polarization contributions identified could be determined as an indicator for the potential risk of plating. The novel, general approach for batteries was tested with a sampling rate of 10 Hz and only requires relaxation periods. Therefore, the method is applicable in battery management systems and contributes to improving the reliability and safety of LIBs.

Keywords: lithium-ion battery; DRT by time domain data; pulse evaluation; relaxation voltage; online diagnosis; degradation mechanisms; EIS

Citation: Goldammer, E.; Kowal, J. Determination of the Distribution of Relaxation Times by Means of Pulse Evaluation for Offline and Online Diagnosis of Lithium-Ion Batteries. *Batteries* **2021**, *7*, 36. https://doi.org/10.3390/batteries7020036

Academic Editor: Kai Peter Birke

Received: 26 February 2021
Accepted: 2 April 2021
Published: 1 June 2021

Publisher's Note: MDPI stays neutral with regard to jurisdictional claims in published maps and institutional affiliations.

1. Introduction

The use of energy storage systems is essential for the transition to renewable energies. Due to the unsteady energy supply from renewable energies such as wind power and photovoltaics, energy storage devices are required to compensate for fluctuations in output and to reliably provide energy at all times. For the electrification of the transport sector, energy storage systems are required in order to move vehicles independently of an external energy supply and without burning fossil fuels. In battery electric vehicles in particular, LIBs are by far the most frequently used battery technology due to their high energy and power density.

Large batteries are required in electric vehicles (EVs) and stationary energy storage devices, which are therefore the main cost factor in these applications. Accordingly, the requirements for the service life of the LIBs are high and a high degree of reliability is a prerequisite.

Therefore, non-invasive investigation methods are used to achieve a comprehensive understanding of the underlying degradation mechanisms of LIBs to enable system optimizations with regard to the service life. In addition, the non-invasive methods are required for the online diagnosis of LIBs and to identify potential risks during operation in order to ensure safe operation.

In current studies, the continous determination of the changes in kinetic parameters of LIBs made it possible to draw conclusions about progressive aging and degradation mechanisms [1–5]. Compared to recording of the open-circuit-voltage (OCV) or to directly determining the capacity, measuring the dynamic behavior of LIBs is usually less time-consuming [1].

EIS as a non-invasive technique is widely used for aging studies in order to measure the impedance directly and thus to determine the kinetic parameters of the LIBs [6–9]. Aging processes such as lithium plating [3] and the formation of the solid electrolyte interface (SEI) [4] could be investigated using EIS.

The derivation of the DRT on the basis of the EIS measurement data as well as the further DRT analysis is a common method for the investigation of LIBs and other batteries. In [10–15] methods are investigated and described to obtain the DRT from impedance data. In several publications, the number of dominant electrochemical processes of LIB as well as their time constants and polarization contributions could be quantified by analyzing the DRT of the measured spectra [14–17]. Changes in the time constants and polarization contributions during the aging of the LIBs under investigation can also be determined using the DRT, as recently shown by Sabet et al. [5].

The main disadvantage of EIS is the long measurement time, especially for low frequencies. The measurement time ranges from several hours for frequencies in the millihertz range to several days for frequencies in the microhertz range. Therefore the bandwidth and resolution of the impedance measurement data available for the DRT is limited [15]. Additionally, only small signal amplitudes are permitted to ensure that the steady state condition is not violated during the EIS. The long measuring times and low signal amplitudes cannot be implemented under operating conditions and with commercially available measuring electronics [18], which is why DRT analyzes have so far been limited to laboratory tests. But even under laboratory conditions, when measuring lower frequencies, maintaining the steady state and current drift are becoming increasingly critical [19].

Due to the general availability in online applications, the characterization of cell impedance based on pulse data is of great interest and the subject of some current studies (e.g., [20,21]). The time domain data of current and voltage curves can be evaluated in the event of pulse excitations in order to determine the impedance of the LIB and other battery types. The calculation on the basis of time domain data is generally faster than the direct measurement of the impedance, since several frequencies are excited simultaneously with excitations in the time domain [19]. Thus, on the basis of time domain data, a higher frequency resolution and wider bandwidth of the impedance measurements can be achieved, which is of particular interest with regard to the DRT.

An overview of the options for calculating the impedance using time domain data is given in [22]. In [19,23], the time domain data resulting from pulse measurements were brought into the frequency domain using a fast Fourier transform (FFT) in order to calculate the low frequency impedance. However, the excitation of a pulse is non-periodic. A window function must therefore be used for the transformations, which in turn generates an offset of the signal [18]. To avoid additional sources of error, as an alternative to the FFT or Laplace transformation, a pre-selected equivalent circuit model (ECM) can be parameterized using the time domain data [19]. The impedance is obtained by transforming the ECM into the frequency domain. The ECM is chosen in a manner that the impedance behavior of the battery can be reproduced. Therefore, the model must be selected individually for each battery type, cell chemistry and the electrochemical processes to be expected [24]. In addition, the impedance of the batteries can change significantly with aging, state of charge (SOC) as well as temperature and thus also the dynamic behavior in the time domain. For this reason, prior knowledge of the electrochemical processes involved and of the aging behavior and further dependencies is required in order to select a suitable model.

The DRT can be derived from the calculated impedance. If the selected ECM consists only of RC elements (parallel connection of resistance and capacitor) and an internal resistance (Thévenin model), the DRT is given directly by the model in the time domain and a transformation to frequency domain is not necessary. Analytical methods have already been proposed for the online parameterization of a Thévenin model [20,25] but the number of RC elements is limited and too small to be able to derive a DRT with adequate resolution. The sufficient number of RC elements can be characterized by iterative numerical optimization methods such as genetic algorithms and method of least squares. Compared to the analytical methods, however, the computationally intensive iterative calculations are disadvantageous for online applications [25] and the number of RC elements is limited again due to the computing power available [26]. The initial values to be determined also require prior knowledge of the electrochemical processes and their approximate time constants and polarization contributions.

In summary, it can be stated that based on the analysis of the DRT, aging mechanisms have already been examined in numerous investigations. The DRT is determined according to the state of the art on the basis of the impedance. The direct measurement of the impedance by means of EIS cannot be implemented in current online applications without adaptations and at low frequencies, even under laboratory conditions, it is time-consuming and subject to inaccuracies. The indirect determination of the impedance based on the evaluation of time domain data leads to additional inaccuracies or requires an individual definition of a model for different cell chemistries and types. A direct derivation of the DRT based on time domain data, which can also be used online, would therefore be desirable.

In this study a method is proposed and described to derive the DRT directly from the measured voltage course during cell relaxation. The method is validated experimentally by comparing the results with a DRT obtained from EIS measurements. In an aging study, the introduced method and a subsequent DRT analysis are used to identify the time constants and polarization contributions of the electrochemical processes of LIBs. The method is shown to be sufficiently sensitive to quantify the changes in these parameters during aging without the need for EIS equipment or other devices not available in online applications. The study demonstrates that the DRT derived from time domain data offers significantly more insights into processes with large time constants respectively low-frequency impedance with less expenditure of time than a DRT derived from EIS measurements. In addition, a correlation between a polarization contribution of one of the processes with a large time constant and a degradation mechanism could be established. Thus, the proposed method offers a time-effective estimate of the kinetic parameters of a LIB, provides insights into battery degradation and is offline and online applicable.

2. Methods

Multiple electrochemical processes with a wide range of time constants occur in LIBs. In addition to the the charge transfer processes of the individual electrodes, diffusion processes such as solid state diffusion and diffusion of lithium ions through the surface films of the electrodes also influence the dynamic characteristics of LIBs [9]. Some processes have similar time constants and are therefore not easy to distiguish from each other only considering the spectra obtained by EIS. In Figure 1a, a spectrum of the cells examined in this research is shown.

The diffusion branch at low frequencies can be clearly assigned to the solid state diffusion as the time constant of these processes is usually a few decades larger compared to other dynamic processes [9]. The semicircle in the spectra is caused by charge transfer processes and the diffusion trough the SEI [27]. Since the time constants are of a comparable order of magnitude, these processes show overlapping effects in the impedance spectra and only a single semicircle is visible. Changes of the polarization contribution and time constants of the various processes during aging can therefore not be reliably separated and assigned to the specific processes. An approach to identify the single processes and their parameters is an analysis of the DRT [14].

2.1. DRT of Frequency Domain Data

Since the DRT analysis of frequency domain data is well-established and the focus of this study is the DRT of time domain data the method is only briefly described. We refer to [15] for a more detailed explanation of the procedure used.

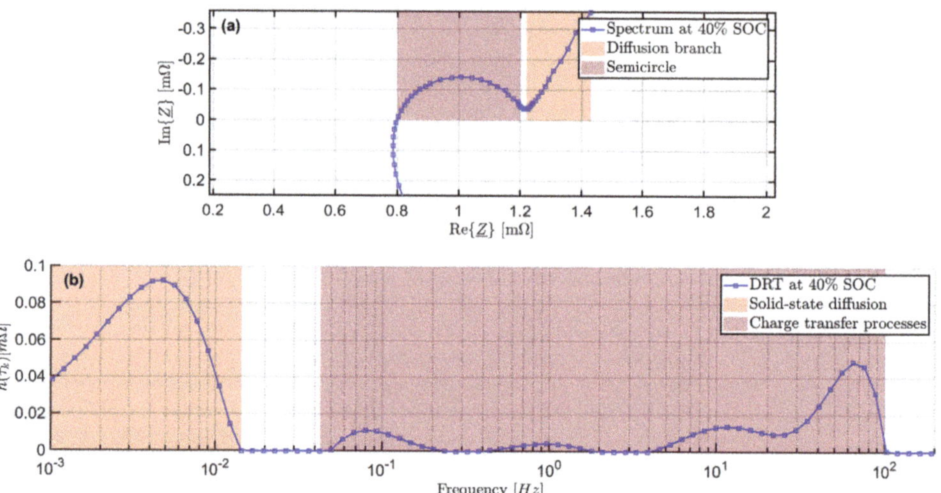

Figure 1. (**a**) Impedance spectra of pristine cell at 40 % SOC. (**b**) Corresponding DRT.

In order to calculate the DRT using frequency domain data, the resistive–capacitive part of the given spectra is reconstructed with a series of m RC-elements, a parallel circuit of an ohmic resistor and a capacitor.

$$Z_{RC,single} = \frac{R}{1 + j\omega RC} = \frac{R}{1 + j\omega\tau} \tag{1}$$

$$Z_{RC,series} = \sum_{k=1}^{m} \frac{h(\tau_k)}{1 + j\omega\tau_k} \tag{2}$$

where $h(\tau_k)$ are the RC-elements unknown polarization contributions. By pre-defining the time constants $\tau = R \cdot C$, the equation of the RC elements and thus the optimization problem become linear. In view of the nature of electrochemical processes, an equidistant distribution on a logarithmic scale is assumed for the time constants. The range of time constants is expanded by two decades and the number of RC-elements per decade is increased by a factor of three compared to the measured frequencies in order to enhance the accuracy and to obtain an adequate resolution of the DRT (see e.g., [15]).

Since the number of RC-elements per decade is higher than the resolution of the impedance data, the optimization problem becomes ill-posed. Therefore, a regularization technique which was introduced by Tikhonov [28] is used to enable an analytical solution. The Tikhonov regularization for determining the DRT has already been shown in some publications [11,13–15]. The regularization parameter was elected according to [15]. The value is kept constant for the evaluation of all spectra, since the results of the DRT analysis are sensitive to the parameter. Thus, the DRTs of the different spectra are comparable and changes in the dispersion due to the aging process which can be tracked.

The parts of the measured spectra with positive imaginary values cannot be modelled by a series of RC-elements and have to be discarded. Additionally, the internal resistance R_i is subtracted from impedance values of the measured spectra prior to the fitting procedure. It is assumed that R_i is equal to the zero crossing of the imaginary axis [9]. After preprocessing of the measurement data, the spectra mainly consist of resistive–capacitive contributions, which can be modelled by RC-elements.

The optimization problem is solved with a non-negative least squares fit because only positive resistances are physically meaningful. Figure 1b depicts the DRT of the spectrum shown in Figure 1a. The position of the peaks indicate the time constants of the processes involved and the area under the peaks correlates to the polarization contribution of the process [14–16]. In contrast to the Nyquist diagram, which shows only a single semicircle at moderate frequencies, various processes with different polarization contributions and time constants are visible in the same frequency range.

2.2. DRT of Time Domain Data

2.2.1. General Approach

Figure 2a shows the simulation of the cell voltage of an LIB during a charge pulse with constant current and the subsequent relaxation period. The impedance of the battery was modelled by three RC-elements and internal resistance. White noise was added to the voltage signal to take measurement inaccuracies of real-world applications into account.

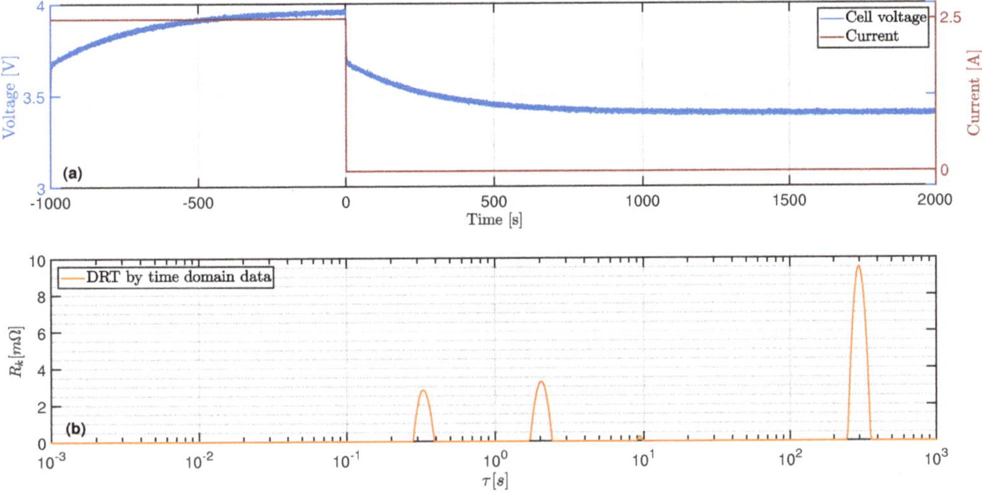

Figure 2. (**a**) Simulated cell voltage of a LIB during a pulse test. (**b**) DRT derived from the voltage relaxation.

For the proposed method of an DRT analysis of time domain data, the voltage relaxation is evaluated. Similar to the DRT analysis of frequency domain data, an attempt is made to reconstruct the voltage behaviour with a series of RC-elements. The voltage course of a single RC-element during a relaxation period can be described as follows:

$$u_{RC}(t) = U_0 \cdot e^{-\dfrac{t}{\tau_{RC}}} \tag{3}$$

where $\tau_{RC} = R \cdot C$, the characteristic time constant of a RC-element. The voltage of the RC-element at the beginning of the relaxation period U_0 at $t = 0\,\mathrm{s}$ can be calculated if the duration t_{cp} and current I_{cp} of the previous pulse are known:

$$U_0 = R \cdot I_{cp} \left(1 - e^{-\dfrac{t_{cp}}{\tau_{RC}}} \right) \tag{4}$$

Equation (4) is only valid if the cell has been sufficiently relaxed before the current pulse and any overvoltages can be neglected. Assuming a finite number m of RC-elements, the overall voltage response can be computed numerically:

$$U_{RC,series}(t) = \sum_{k=1}^{m} R_k \cdot I_{cp} \left(1 - e^{-\frac{t_{cp}}{\tau_k}} \right) \cdot e^{-\frac{t}{\tau_k}} \tag{5}$$

By defining fixed time constants τ_k, the optimization problem becomes linear. The values of the resistances R_k must be set in such a way that the error between the voltage prediction by the series of RC-elements and the measured voltage data (or in this case the simulated voltage data) is minimized. Using the sum of squared errors as a cost function leads to the following optimization problem $min\{J\}$:

$$J = \| A \cdot R_{vec} - U_{vec} \|^2 \tag{6}$$

where the vector R_{vec} corresponds to the unknown polarization contributions of the RC-elements

$$R_{vec} = [R_1 \ldots R_k \ldots R_m]^T \tag{7}$$

and U_{vec} is the vector of the measured (or simulated) voltage course:

$$U_{vec} = [U_1 \ldots U_n] \tag{8}$$

The number n corresponds to the number of measurement points. The measurement data should be preprocessed in advance. In Section 2.2.3, a detailed description of the process is given. The dimension of the matrix A is $n \cdot m$ and the matrix is calculated for the different predefined constants and the time after the end of the pulse t which corresponds to the respective voltage measurement in U_{vec}.

$$A = I_{cp} \cdot \begin{bmatrix} \left(1 - e^{-\frac{t_{cp}}{\tau_1}}\right) \cdot e^{-\frac{t_1}{\tau_1}} & \cdots & \left(1 - e^{-\frac{t_{cp}}{\tau_m}}\right) \cdot e^{-\frac{t_1}{\tau_m}} \\ \vdots & \ddots & \vdots \\ \left(1 - e^{-\frac{t_{cp}}{\tau_1}}\right) \cdot e^{-\frac{t_n}{\tau_1}} & \cdots & \left(1 - e^{-\frac{t_{cp}}{\tau_m}}\right) \cdot e^{-\frac{t_n}{\tau_m}} \end{bmatrix} \tag{9}$$

2.2.2. Predefinition of Time Constants τ

A logarithmically uniform distribution is defined for the time constants $[\tau_1 \cdots \tau_m]$. The minimum and the maximum time constants τ_{min} and τ_{max} are determined based on the maximum sampling rate $f_{s,max} = \frac{1}{\Delta t_{min}}$ and the duration of the relaxation process t_{max}. The Shannon theorem states that frequencies that are higher than half of the sampling rate cannot be evaluated without further processing of the measured signal.

$$f_{eval,max} \leq \frac{1}{2 \cdot \Delta t_{min}} \tag{10}$$

Taking into account the transient frequency $f_t = \frac{1}{2 \cdot \pi \cdot \tau}$ of a RC-element, the following relationship result for the smallest observable time constant $\tau_{eval,min}$ [29]:

$$\frac{1}{2 \cdot \pi \cdot \tau_{eval,min}} \leq \frac{1}{2 \cdot \Delta t_{min}} \tag{11}$$

$$\Rightarrow \tau_{eval,min} = \frac{\Delta t_{min}}{\pi} \tag{12}$$

The largest evaluable time constant $\tau_{eval,max}$ is limited by the duration of the relaxation phase according to [19].

$$f_{eval,min} \geq \frac{4}{t_{max}} \tag{13}$$

$$\Rightarrow \tau_{eval,max} = \frac{t_{max}}{8\pi} \tag{14}$$

The minimum and maximum time constants of the distribution τ_{min} and τ_{max} are chosen to be two decades smaller, respectively, andlarger than the evaluable time constants in order to increase the accuracy at both ends of the frequency dispersion:

$$\tau_{min} = \frac{\tau_{eval,min}}{100} = \tau_1 \tag{15}$$

$$\tau_{max} = \tau_{eval,max} \cdot 100 = \tau_m \tag{16}$$

This procedure was similarly proposed by [15] for the DRT analysis of frequency domain data. In order to obtain a smooth DRT curve within the evaluated frequency range, the number of RC elements or time constants m must be selected to be high enough. A number of one hundred τ per decade was used for the analysis of the DRT in this study.

2.2.3. Pre-Processing of Measurement Data

Before calculating the DRT, the existing measurement data should be checked for usability and preprocessed. According to the Equation (12), the sampling rate must be high enough (or the timestep between two successive voltage measurements Δt_{min} small enough) to enable a meaningful evaluation of the DRT down to the time constant $\tau_{eval,min}$. Therefore, when selecting the the sampling rate beforehand, the minimum time constant relevant for the investigation should be taken into account. If the maximum sampling rate of the measurement device is lower, the DRT can only be evaluated for time constants that satisfy Equation (12). This consideration is particularly important for BMS, as they often have sample rates of 10 Hz or less.

The processes with low time constants only have measurable contributions at the beginning of the relaxation phase and quickly subside. Therefore, the sampling rate does not necessarily have to be constant over the entire measurement period and can be reduced in order to avoid large amounts of measurement data. With very high sampling frequencies and the evaluation of very small time constants, the speed of the current control needs to be considered. As long as the current has not dropped to zero, Equation (5) is not valid. These voltage measurements should therefore be discarded and only larger time constants evaluated. Apart from that, only the voltage values should be evaluated that were recorded at least one time step after the load drop. Since pure ohmic resistances cannot be modelled with Equation (5), the influence of internal resistance on the voltage response should not be taken into account.

The OCV U_{OCV} must be subtracted from the voltage measurements $u_{meas}(t)$ before determining the DRT. Assuming a completely relaxed cell at the end of the relaxation phase, this can be achieved by subtracting the last measured voltage value $u_{meas}(t_{max})$ from all measurements. A considerably long relaxation phase is therefore imperative.

$$u_{processed}(t) = u_{meas}(t) - U_{OCV} = u_{meas}(t) - u_{meas}(t_{max}) \tag{17}$$

At the beginning of the relaxation process, all processes contribute to changes in cell voltage. As relaxation progresses, only processes with larger time constants are relevant. Following the measurement data should be interpolated on a logarithmic scale in order to weigh the single measurement points accordingly. Due to the low impedance of automotive cells, the signal-to-noise ratio is usually rather low. As the cell becomes increasingly relaxed, the signal-to-noise ratio deteriorates further, as can be seen in Figure 2. Forming the moving average before interpolation smooths the measured values and can therefore improve the ratio.

2.2.4. Calculation of the DRT

Due to the long measurement time for low-frequency impedances, the number of time constants m usually exceeds the number of frequencies measured. In contrast to frequency domain data, the resolution in the time domain is comparatively high. The number of interpolation points can also be selected to be higher than m in order to use the additional information provided by the high-resolution pulse measurement. To solve the over-determined system, the Tikhonov regularization is applied again. Equation (6) is extented by the regularization term:

$$J = \|A \cdot R_{vec} - U\|^2 + \|I \cdot \lambda U\|^2 \tag{18}$$

where λ is the regularization parameter and I is the $m \cdot m$ indentity matrix. The value of the regularization parameter is optimized with respect to the sum of the square errors of the reconstructed voltage signal. A non-negative least square algorithm is proposed to solve the optimization problem. Due to the restriction to positive results, only physically possible polarizations can be calculated. Using the determined polarizations R_{vec} and the predefined time constants τ, the spectrum can be reconstructed for different angular frequencies ω:

$$Z_{DRT} = \sum_{k=1}^{m} \frac{R_{vec}(k)}{1 + j\omega\tau_k} \tag{19}$$

If a measured or modelled spectrum is available, the sum of the squares of errors of the capacity-resistive part can be calculated and used as an additional selection criterion for the regularization parameter. This method for the evaluation of the parameter was already proposed by [15] for the DRT calculation of frequency domain data.

In Figure 2b, the DRT is given; this was derived from the voltage course during the relaxation period shown in Figure 2a. The DRT reveals three processes vizualized by the peaks. The area under the peaks corresponds to the polarization contribution of the process and the position of the peak to the time constant. The time constants as well as the polarizations of the identified processes are consistent with the parameters of the RC-elements used in the battery model. Table 1 shows the calculated values and the original parameter set of the battery model as well as the relative deviation.

Table 1. Comparison of the parameter set of the battery model and the values derived by the DRT.

Parameter	Calculated Value	Model Value	Relative Error
R_1	31.5 mΩ	30 mΩ	5 %
τ_1	0.328 s	0.3 s	9.2 %
R_2	37.5 mΩ	39 mΩ	3.8 %
τ_2	2.046 s	1.95 s	4.9 %
R_3	116.9 mΩ	117 mΩ	<0.1 %
τ_3	296.899 s	292.5 s	1.5 %

The relative error of the parameters with larger time constants decreases due to the diminishing influence of the superimposed noise signal. The processes with short time constants only contribute to voltage changes for a short time and their voltage contribution quickly drops to zero. The voltage drop is also most pronounced in processes with the long time constants at the beginning of the relaxation, but it lasts longer. Therefore, the effect of these processes is visible at more measuring points, which enables an increasing averaging of the noise. In Figure 3a, the reconstructed voltage signal is plottet together with the simulated voltage course in order to prove the accuracy of the DRT. Figure 3b shows the reconstructed impedance, derived by using the calculated distribution function and Equation (19), alongside the impedance spetrum of the battery model.

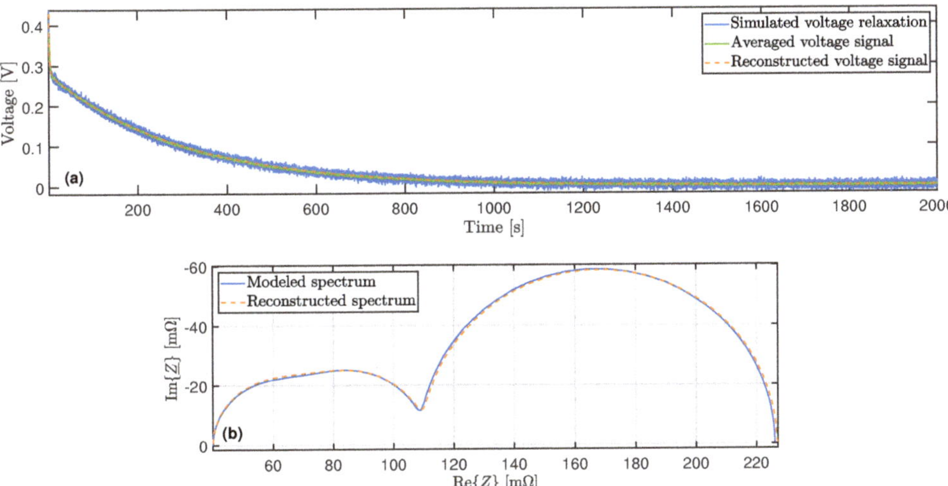

Figure 3. (**a**) Comparison of simulated and reconstructed voltage during the relaxation period. (**b**) Comparison of modeled and reconstructed impedance spectrum.

Regardless of whether the real part or the imaginary part is considered, the relative deviation between the two spectra of Figure 3b is less than 1% for the frequency range $[f_{eval,min} \cdots f_{eval,max}]$. In general, the accuracy is higher for lower frequencies within the specified frequency range. Because of the added noise in the simulation, the relative error of the reconstructed voltage signal (Figure 3a) cannot be used as an indicator for the quality of the distribution function. However, the reconstructed voltage follows the mean value of the simulated signal, even with dynamic behavior in the first seconds. Despite the good reconstruction of the dynamics, the impact of the noise is substantially minimized, which proves that no overfitting has taken place.

Based on the sampling rate of commercially available BMS, the resolution of the simulated values was only 100 ms. Since the impedance of automotive sized cells is usually very low and hence the voltage response, a strong noise signal has been added compared to the excitation. Nevertheless, processes with time constants that are only slightly larger than the resolution can be identified and their contributions determined. The results therefore show that the method introduced is theoretically suitable for determining the DRT and the cell impedance and examining the dynamics and processes parameters of an LIB.

3. Experimental

In order to further investigate the applicability and accuracy of the introduced method under real conditions, an experimental study was carried out. To validate the introduced method, automotive LIBs were characterized with both EIS and pulse tests and the resulting DRT and derived process parameters were compared.

The automotive cells were then cyclically aged under various conditions. The changes in processes and their parameters were determined using both methods. Thus, the results obtained through time and frequency domain data can be directly compared and evaluated with regard to the traceability of degradation mechanisms.

For experimental investigations, large format pouch-type LIBs with a nominal capacity of 50 Ah were used. The anode of the examined cells are made of graphite and the cathode consists mainly of nickel–manganese–cobalt-oxide (NMC). All cells were taken from the same batch to minimize the impact of manufacturing tolerances.

3.1. Experimental Validation

For the experimental validation of the method introduced one of the cell was fully charged and subsequently discharged to determine the available cell capacity. In order to minimize the influence of the cell impedance, a constant-current–constant-voltage (CCCV) protocol was used for both charging and discharging. The CV phases were held until the current dropped below 50 mA. In the CC phase, 25 A were applied during charging and 50 A during discharging. According to the determined capacity the cell was then charged with a constant current of 25 A to certain SOC values (20/40/60/80%). After reaching the desired SOC level, the current was set to zero. The relaxation of the voltage was recorded in high resolution with a sampling rate of 2 MHz in the first six seconds after the load jump. After six seconds, the sampling rate was reduced down to 1 Hz and the voltage relaxation recorded for 4 h.

The pulse tests were followed by EIS measurements. Galvanostatic excitation with an amplitude of 3 A was applied. The frequency was varied in the range from 50 kHz to 0.5 mHz. For frequencies above 100 Hz, 16 measurements per decade were carried out, otherwise only eight were performed. According to Barai et al. [30] the relaxation phase of the previous pulse test is long enough to meet stationary conditions.

3.2. Aging Study

As part of the aging study, the LIBs were cycled under certain operating conditions that are within the manufacturer's cell specifications. A total of three aging scenarios were carried out, with the average SOC value being varied between the individual scenarios. With the exception of the charging current and the state of charge, the other operating conditions like the depth of discharge (DOD) (30%) and the discharge current (1.6 C) were set to be the same for all scenarios. Table 2 gives an overview about the deviating operating conditions for each scenario.

Table 2. Different operating conditions.

Cycling Scenario	$\overline{SOC}$ [%]	I_{ch} [A]
Sc_1	60	37.5
Sc_2	20	37.5
Sc_3	80	18.75

The SOC range of each scenario was set in such a way that, according to [31], different degrees of capacity depletion and deterioration of the individual electrodes could be expected. The aging study thus enables an assessment of whether the sensitivity of the introduced method is sufficient to determine aging mechanisms qualitatively or even quantitatively.

Before starting the cycling, a checkup was carried out for each cell in order to determine the initial cell parameters. During the cycling of the cells, the long-term tests were occasionally interrupted and further checkups were carried out. The checkup procedure contains a CCCV capacity measurement. In addition, at 20/40/60 and 80% a pulse test and an EIS measurement are performed, following the procedure described in Section 3.1. However, the maximum sampling rate during the pulse test is limited to 10 Hz, which is more realistic for online applications such as battery management systems.

4. Results

4.1. Experimental Validation

In this section, the results of the pulse tests and the EIS measurements of the experimental validation are shown and evaluated.

Figure 4 depicts the recorded cell voltage during the relaxation at 20%. In addition to the originally measured signal, the moving average of the signal is given in order to be able to assess the accuracy of the reconstructed signal. It can be stated that the deviation between the reconstructed and the averaged signal over the entire range under consideration is less than 0.5 mV.

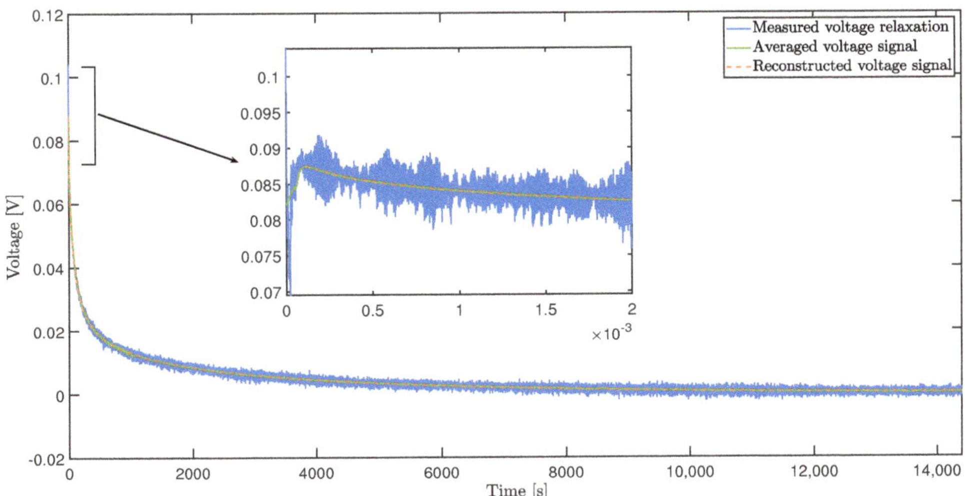

Figure 4. Pulse test: Voltage relaxation at 20 % SOC.

The high measurement resolution reveals the inductive behavior of the cell voltage. Due to the inductance, the voltage over the cell connectors and current collectors drop to negative values immediately after the load jump. For less than 0.1 ms, the voltage rises again as the voltage across the inductive components drops quickly. Therefore, the voltage measurements for $t < 0.1$ ms have been omitted, as electrochemical processes (time constants of usually larger than 1 ms) are of interest for this study. However, it is pointed out that the inductive behavior can also be simulated by expanding the matrix A (Equation (9)) by the mathematical description of several parallel circuits of ohmic resistance and inductance (RL-elements).

Equation (19) is used to calculate the impedance of the cell for the pulse test at 20% percent. For better comparability with the spectrum measured directly after the pulse test, Figure 5a,b show the imaginary and real parts of both spectra over frequency.

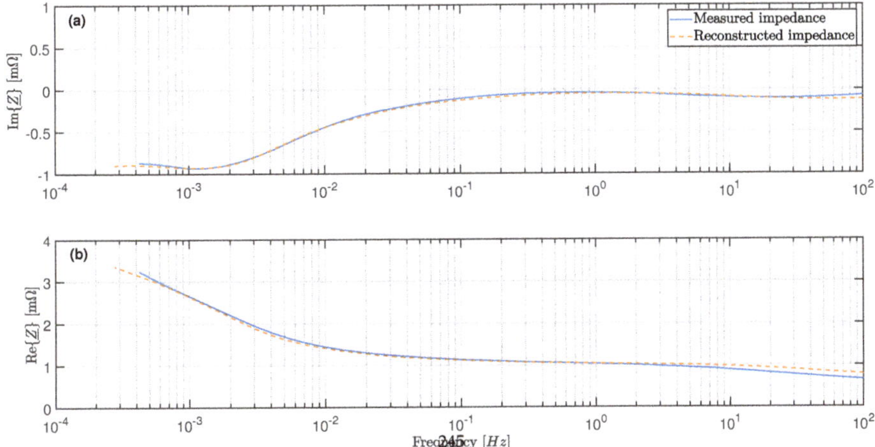

Figure 5. Comparison of measured and reconstructed impedance at 20 % SOC (**a**) Imaginary part of impedance. (**b**) Real part of impedance.

Figure 5a shows that the imaginary part coincides with frequencies below 50 Hz. Due to the limited measuring range, only impedances at frequencies above 0.5 mHz can be specified for the measured spectrum.

The plot of the real part of the impedance in Figure 5b provides similar insights. The course of the reconstructed spectrum deviates at high frequencies, but follows the measured spectrum at frequencies of ≤5 Hz.

The increasing deviations at >5 Hz can be explained by the Butler–Volmer kinetics. The current excitation at relaxation period was set to zero, while the excitation during the direct measurement of the impedance was set between −3 A and 3 A. According to the Butler–Volmer equation, the kinetics of charge transfer processes are not linear. Thus, due to increasing charge transfer resistances, the potential response does not decrease linearly with the current excitation [32]. In addition, it has been experimentally proven in some publications that the relaxation time, which differs between the pulse test and the EIS, influences the charge transfer resistances [30,32]. It is known that charge transfer processes at the anode [4,15] and at the cathode [5,33] occur at moderate frequencies of 1–100 Hz. The deviations at frequencies of 5–100 Hz between pulse test and EIS were therefore to be expected.

The ranges of the spectrum at frequencies above 100 Hz could not be adequately reconstructed because the optimization function is limited to resistive–capacitive elements. The impedance values measured at over 100 Hz were therefore discarded and not plotted in Figure 5.

Figure 6 shows the DRT obtained from time domain data and the DRT determined from frequency domain data. The resolution of the DRT from time domain data is twice as high. Therefore, the heights of peaks in one DRT cannot be compared directly with those in the other DRT. In order to compare the polarization contributions of the identified processes, the sum of all polarization contributions within a peak must be determined instead. When calculating the polarization contributions for the first (peak at smallest time constant) to fifth peak, the same orders of magnitude result for the respective peak for both DRTs. In addition, both DRTs show that the contribution of the first and fifth peaks are greatest, followed by the fourth and second peaks. The third peak has a comparatively small contribution in both DRTs.

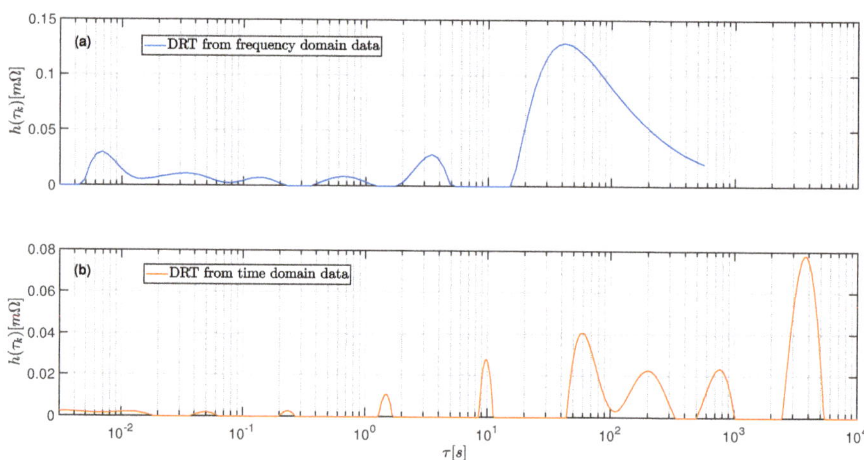

Figure 6. DRT derived from frequency domain data (**a**) and from time domain data (**b**).

Both distributions reveal four relevant processes with time constants in the range of $2 \cdot 10^{-3}$ s $< \tau < 2 \cdot 10^{0}$ s. According to $f_T = \frac{1}{2 \cdot \pi \tau}$, these time constants approximately correspond to frequencies between 0.1 Hz and 100 Hz. Processes in this frequency range were assigned to charge transfer [4,5,15,33] and surface films of the electrodes [4,34]. Due to the Butler–Volmer kinetics, time constants and polarization contributions differ between the two distributions, which explains the discrepancy that occurs in the first four peaks.

For time constants >10 s, the DRT obtained from time domain data shows a comparatively better resolution of the processes. The DRT based on frequency domain data only indicates a single process for time constants between 10 and 1000 s. In contrast, the DRT obtained through time domain data reveals the existence of three different processes. In addition, time constants that are two orders of magnitude higher can be resolved. If the spectrum is recorded via EIS, the measurement takes more than a day to resolve large time constants. The evaluation of pulse data with the method introduced can therefore be advantageous for examining processes with large time constants such as solid state diffusion.

4.2. Aging Study

4.2.1. EIS Measurement Evaluation

In Figure 7 shows the measured spectra and the corresponding DRT against frequency of an examined cell as well as the development during aging. The measurements from one cell from aging scenario Sc_1 were selected for the exemplary representation. The upper diagrams show the values determined at 20% and the lower part at 80% SOC.

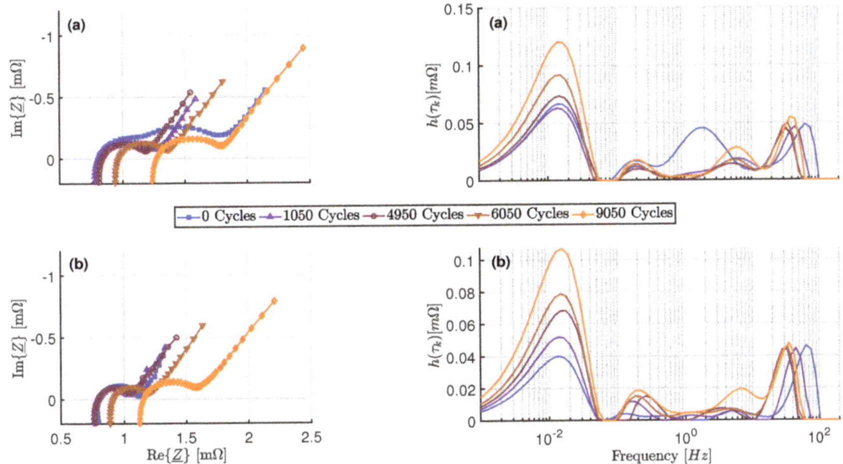

Figure 7. Measured impedance spectra and corresponding DRTs over aging (**a**) At 20 % SOC; (**b**) At 80 % SOC.

In the Nyquist diagram, two major changes can be seen as the cell ages:

- An increase of the internal cell resistance R_i during aging (Shifting of the zero crossing of the imaginary axis to higher real parts)
- Disappearance of the second semicircle at 20% SOC during the first 1050 cycles of the long-term test

According to the second bullet point, the DRT at 20% reveals distinct changes in the polarization contribution and the time constant for a process at moderate frequencies. This peak can be assigned to the second semicircle in the spectrum because it occurs particularly when the SOC is low and the cell is new. As already mentioned, this peak can be assigned to a charge transfer process due to the range of the time constant.

In contrast to the observation of the spectra, the analysis of the DRT shows a slight increase in the contribution of this process during the last cycles. Furthermore, slight increases in the polarization of the charge transfer processes are visible at moderate frequencies and 80%. In addition, small changes in the parameters of the process with the smallest time constant are visible, but they do not follow a clear trend. Due to the frequency range, this process can be assigned to the transfer through the SEI of the anode [4,34].

For frequencies <0.1 Hz, a single process is visible. At low SOC, the polarization of the process remains almost constant during the first 4950 cycles. After the first 4950 cycles, a continuous increase can be observed. At 80% SOC, this continuous increase begins with the start of the long-term test.

The mentioned changes in the process parameters were recognizable for all cells, except for the increase in polarization contribution at low frequencies. This behavior was not pronounced in the cells of scenario Sc_2. Changes of the process parameters were made at a similar rate within a scenario. However, there were deviations in the rate of change between the scenarios, especially with regard to the increase in polarization contribution at frequencies below 0.1 Hz.

4.2.2. Pulse Test Evaluation

Figure 8 shows the cell voltage course during the relaxation period and the DRT derived from the time domain data. Again, the values at 20% (upper graphs) and 80% SOC (lower graphs) are given and the same cell as in Section 4.2.1 is evaluated.

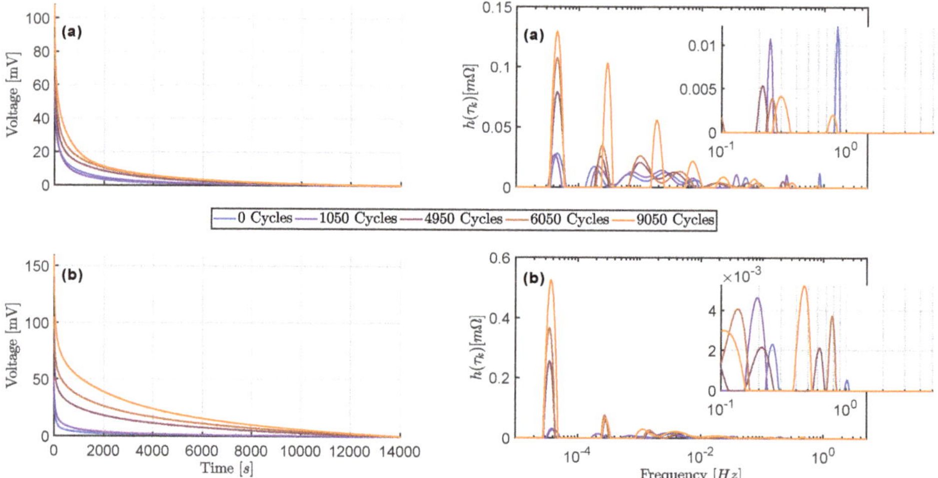

Figure 8. Measured voltage relaxation and derived DRTs over aging (**a**) At 20 % SOC; (**b**) At 80% SOC.

The voltage courses show a slowdown of the relaxation process as aging progresses. The DRT derived by the time domain data reveals changing parameters of processes with large time constants as the reason. A strong increase of polarization contributions can be observed for low frequencies. Compared to the DRT obtained from frequency domain data, not only a more pronounced increase can be observed. In addition, more processes with large time constants can be identified, since the DRT can also be specified for frequencies that are lower than 1 mHz.

According to Equation (10), the DRT only provides representative values for the frequencies ≤ 5 Hz. The DRT is cut off accordingly for frequencies above 5 Hz. The process with the smallest time constant shown in Figure 7 (assigned to the SEI) is therefore not visible.

The disappearance of a polarization contribution of a process at moderate frequencies and 20% SOC can be confirmed. The number of processes in the range of moderate frequencies between 0.1 and 5 Hz corresponds to the number determined on the basis of frequency domain data. However, the greater spread of the time constants and the polarization with no discernible trend shows that due to the lower sampling frequency the accuracy is too low to reliably quantize the parameters of the identified processes at moderate frequencies.

4.2.3. Comparison of DRT by Time and Frequency Domain Data

The results of the experimental studies can be summarized as follows:

- Processes with characteristic time constants that do not meet Equation (12) cannot be identified with the set sampling rate using the time domain data alone.
- Already with a maximum sampling rate of 10 Hz, which is realistic for online applications, the DRT by time domain data can be used to identify charge transfer processes and to provide a qualitative description. However, the change in process parameters during aging can only be traced to a limited extent.
- Quantitative statements for charge transfer processes, even if higher sampling rates are used, differ between the two methods due to the Butler–Volmer kinetics. In fact, due to the strongly non-linear excitations in real applications, the better transferability of the results obtained from frequency domain data is questionable.
- The DRT based on time domain data is more sensitive for processes with large characteristic time constants such as solid state diffusion.
- When using frequency domain data, either longer measuring periods are required or the measuring range is limited to higher frequencies, since different frequencies have to be excited successively during EIS.

4.2.4. Correlation of the Capacity and the Identified Process Parameters

Figure 9a shows the capacity course during the aging of a cell and the corresponding polarization contribution of the process with the largest time constant, which is determined by time domain data at 60% SOC. The polarization contribution is referred to below as $R_{tau_{max}}$. The results belong to the cell that has already been evaluated in the previous Sections 4.2.1 and 4.2.2.

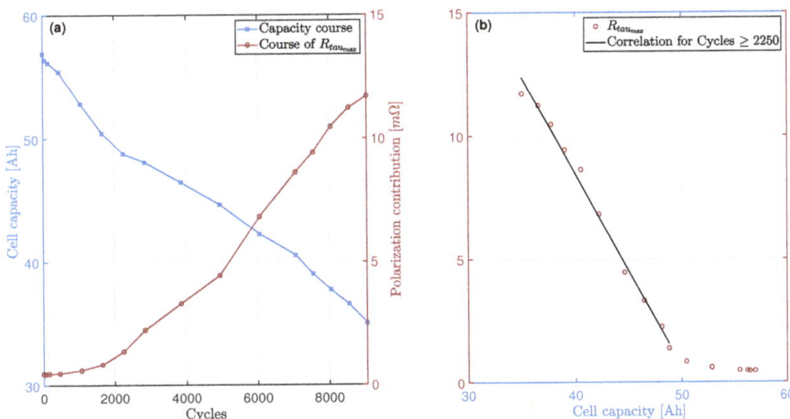

Figure 9. (**a**) Capacity course and polarization contribution $R_{tau_{max}}$ vs. cycles. (**b**) Correlation between capacity and $R_{tau_{max}}$.

For a better overview of the correlation between the two parameters, the development of $R_{tau_{max}}$ is plotted over the capacity. A strong negative correlation can be found for $\geq$2250 cycles. During the first cycles, the increase in polarization in relation to the capacity fade is less pronounced. Regardless of which scenario, a significantly weaker increase in $R_{tau_{max}}$ could be confirmed for all cells at the beginning of the aging tests. Thus, the average correlation coefficients C_{coeff} for each scenario and for the cycles $\geq$2250 are given in Table 3.

Table 3. Different operating conditions.

Cycling scenario	$\overline{SOC}$ [%]	C_{Coeff} [%]
Sc_1	60	99.1
Sc_2	20	92.8
Sc_3	80	98.0

Taking into account the measurement data only for cycles $\geq$2250, a Pearson correlation coefficient of $\geq$98% can be specified for scenarios $Sc1$ and $Sc3$, which indicates a high negative correlation between the two parameters. For scenario $Sc2$, the correlation is weaker, but still noticeable.

Figure 10 shows the capacity courses over the cycles for all cells examined.

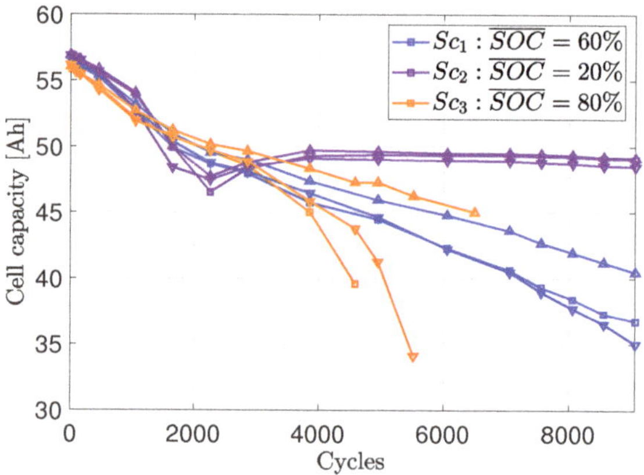

Figure 10. Capacity course vs. cycles during the long-term cycling.

For scenario $Sc2$, a recovery of cell capacity can be seen. After 2250 cycles, the capacity remains relatively constant at around 48 A h. Because of the flat curve, the accurate correlation is more difficult to determine because of the corresponding small changes in polarization $R_{tau_{max}}$ are difficult to detect. This could be one reason why the correlation coefficient is lower in scenario $Sc2$.

To further investigate the reasons for the lower correlation coefficient, the correlation between the capacity of the anode and the polarization $R_{tau_{max}}$ was determined. The determination of the loss of active material at the anode (LAM$_{An}$) on the basis of the non-invasive measurements during the checkups was carried out in [27]. The correlation between the LAM$_{An}$ and $R_{tau_{max}}$ was $\geq$ 96% for each cell of scenario $Sc1$ from cycle 0 to 9050. Taking into account the uncertainties in the estimation of the LAM$_{An}$, 96% is a high value and it can be stated that the increase in $R_{tau_{max}}$ is associated with the LAM$_{An}$. For all cells of scenario $Sc1$ the decrease in capacity at the anode was more pronounced compared to the cells of $Sc2$. Gantenbein et al. has already experimentally confirmed a higher LAM$_{An}$ for cells that were cycled at SOC levels above 65%, while the LAM$_{An}$ for cells that were cycled below 45% was neglible [31]. The lower LAM$_{An}$ provides a further reason for the lower correlation in $Sc2$ and the lower capacity fade during aging. For the cells of $Sc3$, the LAM$_{An}$ after 3850 cycles could not be determined due to a change in the electrode balance [27]. According to the results of [31] a comparatively high LAM$_{An}$ is to be expected. This statement is supported by the correlation factor determined for scenario $Sc3$ (see Table 3).

In contrast to the correlation to the cell capacity (see Figure 9), the linear relationship between LAM$_{An}$ and increase in $R_{tau_{max}}$ is visible from the first few cycles of the long-term test. It is known from the literature that the cell capacity can be limited by different degradation mechanisms in different aging stages of a LIB (e.g., [35,36]). In a previous publication, the loss of lithium inventory (LLI) was identified as the main aging mechanism during the first few cycles of the aging study described [27]. The LAM_{An} becomes the limiting factor for the capacity of the cells of scenarios $Sc1$ and $Sc3$ after 2250 cycles. Therefore, a correlation between the cell capacity and the $R_{tau_{max}}$ only becomes visible after 2250 cycles. It is thus assumed that the correlation between the cell capacity and $R_{tau_{max}}$ originates from the correlation of LAM_{An} and $R_{tau_{max}}$.

5. Discussion

In the context of the aging study, the method introduced has proven to be advantageous for investigating processes with time constants in the range of minutes to hours. In addition to the more precise resolution of the low-frequency processes compared to the DRT, which is derived from frequency domain data, the time required for the measurements is significantly lower. Besides higher resolution and lower time efforts, the steady state criterion with regard to the SOC is not violated during the relaxation period. In order to avoid a change in the SOC during the EIS at low frequencies, the excitation current must be limited, which leads to a comparatively poor signal-to-noise ratio.

However, high sampling rates are required in order to investigate frequency-dependent processes with characteristic time constants in the range of a few seconds or even milliseconds such as charge transfer and SEI. According to the Shannon theorem, time constants in the range of milliseconds should theoretically be able to be evaluated with a sampling rate of 10 Hz, which is realistic for an online application. In contrast to this, the aging study has shown that only the parameters of processes with time constants of almost two decades higher ($\tau \geq 1.6\,s$) can be quantified with sufficient accuracy using the aforementioned sampling rate. In many applications, and especially in most online applications, the method presented is therefore limited to the aging monitoring of processes with time constants in the range of seconds to several hours, such as solid state diffusion.

Therefore, for applications in which both pulse tests and EIS are possible, a combined use of both techniques is proposed. By merging the impedance data obtained through time and frequency domain data as suggested by [19], the impedance spectrum and DRT can be determined for a wide frequency range.

Using the DRT derived from the time domain data, a correlation between the polarization contribution and LAM_{An} could be determined at low frequencies around 35 µHz. Due to the limited frequency range of the EIS, this connection could not be established using the frequency domain data. A correlation between a degradation mechanisms and the impedance of LIBs could already be established in earlier studies. In a comprehensive experimental study, the correlations between kinetic parameters of an LIB and degradation mechanisms could be derived [1]. Recently, a correlation between low frequency impedance and the LLI was found in [24] and Schindler et al. [37] confirmed a correlation between the low frequency impedance and LAM. Using cell kinetics to gain insights into the degradation mechanisms is beneficial because measuring impedances at high and moderate frequencies is fast compared to deriving the mechanisms by recording open circuit voltage curves [1]. However, EIS measurements suffer from long measurement durations for low-frequency impedances and a lack of availability in most state-of-the-art online applications. In contrast, relaxation periods occur in every online application. As a result, the introduced method enables a time-effective investigation that can be used online in order to obtain the relevant parameters of the processes with large time constants and thus enable an online estimate of the degradation mechanisms.

It is known that LIBs can suffer from accelerated capacity loss towards the end of life due to lithium plating originally caused by a lack of anodic material [35,36]. For online applications, an early detection of an increasing capacity reduction and lithium plating is of great importance for the reliability and safety of LIBs. Many publications like [38] have therefore dealt with the prediction of the sudden reduction in capacity. As a further outcome of this study, the correlation found could be used to identify an advanced LAM_{An}

online and thus the risk of lithium plating and the resulting rapid loss of capacity at an early stage.

In summary, this work showed that the evaluation of the voltage relaxation after a current pulse is suitable for investigating the kinetic parameters of a LIB even without prior knowledge of the polarization processes involved. The DRT obtained from the time domain data offers a time-effective determination of processes with large characteristic time constants and enables the parameters to be estimated online. The correlations found provide insights into the degradation mechanisms both offline and online and thus support the future battery design and enable optimization of the operating strategies. The findings about the early detection of a potential risk of lithium plating are available for subsequent studies and contribute to improving the reliability and safety of LIBs.

In a future step, Post-Mortem analyses will be carried out to directly quantify the LAM_{An} and validate the assumptions made. In addition, the transferability of the results to other cell chemistries and formats must be checked.

Author Contributions: Conceptualization, E.G.; methodology, E.G.; software, E.G.; validation, E.G.; formal analysis, E.G.; investigation, E.G.; resources, J.K.; data curation, E.G.; writing—original draft preparation, E.G.; writing—review and editing, J.K. and E.G.; visualization, E.G.; supervision, J.K.; project administration, E.G.; funding acquisition, J.K. All authors have read and agreed to the published version of the manuscript.

Funding: This research was funded by Bundesministerium für Bildung und Forschung grant number 16EMO0262 (SiCWell).

Institutional Review Board Statement: Not applicable.

Informed Consent Statement: Not applicable.

Data Availability Statement: Not applicable.

Conflicts of Interest: The authors declare no conflict of interest. The funders had no role in the design of the study; in the collection, analyses, or interpretation of data; in the writing of the manuscript, or in the decision to publish the results.

Abbreviations

The following abbreviations are used in this manuscript:

DRT	Distribution of relaxation times
LIB	Lithium-ion battery
EIS	Electrochemical impedance spectroscopy
EV	Electric vehicle
OCV	Open circuit voltage
SEI	Solid electrolyte interface
FFT	Fast Fourier transform
ECM	Equivalent circuit model
SOC	State of charge
NMC	Nickel-manganese-cobalt-oxide
CCCV	constant-current-constant-voltage
DOD	Depth of discharge
LAM_{An}	Loss of active material at the anode
LLI	Loss of lithium inventory

References

1. Pastor-Fernández, C.; Uddin, K.; Chouchelamane, G.H.; Widanage, W.D.; Marco, J. A Comparison between Electrochemical Impedance Spectroscopy and Incremental Capacity-Differential Voltage as Li-ion Diagnostic Techniques to Identify and Quantify the Effects of Degradation Modes within Battery Management Systems. *J. Power Sources* **2017**, *360*, 301–318. [CrossRef]
2. Pastor-Fernández, C.; Yu, T.F.; Widanage, W.D.; Marco, J. Critical review of non-invasive diagnosis techniques for quantification of degradation modes in lithium-ion batteries. *Renew. Sustain. Energy Rev.* **2019**, *109*, 138–159. [CrossRef]
3. Schindler, S.; Bauer, M.; Petzl, M.; Danzer, M.A. Voltage relaxation and impedance spectroscopy as in-operando methods for the detection of lithium plating on graphitic anodes in commercial lithium-ion cells. *J. Power Sources* **2016**, *304*, 170–180. [CrossRef]

4. Steinhauer, M.; Risse, S.; Wagner, N.; Friedrich, K.A. Investigation of the Solid Electrolyte Interphase Formation at Graphite Anodes in Lithium-Ion Batteries with Electrochemical Impedance Spectroscopy. *Electrochim. Acta* **2017**, *228*, 652–658. [CrossRef]
5. Shafiei Sabet, P.; Warnecke, A.J.; Meier, F.; Witzenhausen, H.; Martinez-Laserna, E.; Sauer, D.U. Non-invasive yet separate investigation of anode/cathode degradation of lithium-ion batteries (nickel–cobalt–manganese vs. graphite) due to accelerated aging. *J. Power Sources* **2020**, *449*, 227369. [CrossRef]
6. Schmitt, J.; Maheshwari, A.; Heck, M.; Lux, S.; Vetter, M. Impedance change and capacity fade of lithium nickel manganese cobalt oxide-based batteries during calendar aging. *J. Power Sources* **2017**, *353*, 183–194. [CrossRef]
7. Andre, D.; Meiler, M.; Steiner, K.; Wimmer, C.; Soczka-Guth, T.; Sauer, D.U. Characterization of high-power lithium-ion batteries by electrochemical impedance spectroscopy. I. Experimental investigation. *J. Power Sources* **2011**, *196*, 5334–5341. [CrossRef]
8. Tröltzsch, U.; Kanoun, O.; Tränkler, H.R. Characterizing aging effects of lithium ion batteries by impedance spectroscopy. *Electrochim. Acta* **2006**, *51*, 1664–1672. [CrossRef]
9. Jossen, A. Fundamentals of battery dynamics. *J. Power Sources* **2006**, *154*, 530–538. [CrossRef]
10. Tuncer, E.; Macdonald, J.R. Comparison of methods for estimating continuous distributions of relaxation times. *J. Appl. Phys.* **2006**, *99*, 074106. [CrossRef]
11. Saccoccio, M.; Wan, T.H.; Chen, C.; Ciucci, F. Optimal Regularization in Distribution of Relaxation Times applied to Electrochemical Impedance Spectroscopy: Ridge and Lasso Regression Methods - A Theoretical and Experimental Study. *Electrochim. Acta* **2014**, *147*, 470–482. [CrossRef]
12. Wan, T.H.; Saccoccio, M.; Chen, C.; Ciucci, F. Influence of the Discretization Methods on the Distribution of Relaxation Times Deconvolution: Implementing Radial Basis Functions with DRTtools. *Electrochim. Acta* **2015**, *184*, 483–499. [CrossRef]
13. Boukamp, B.A.; Rolle, A. Analysis and Application of Distribution of Relaxation Times in Solid State Ionics. *Solid State Ion.* **2017**, *302*, 12–18. [CrossRef]
14. Danzer, M.A. Generalized Distribution of Relaxation Times Analysis for the Characterization of Impedance Spectra. *Batteries* **2019**, *5*, 53. [CrossRef]
15. Hahn, M.; Schindler, S.; Triebs, L.C.; Danzer, M.A. Optimized Process Parameters for a Reproducible Distribution of Relaxation Times Analysis of Electrochemical Systems. *Batteries* **2019**, *5*, 43. [CrossRef]
16. Ivers-Tiffée, E.; Weber, A. Evaluation of electrochemical impedance spectra by the distribution of relaxation times. *J. Ceram. Soc. Jpn.* **2017**, *125*, 193–201. [CrossRef]
17. Manikandan, B.; Ramar, V.; Yap, C.; Balaya, P. Investigation of physico-chemical processes in lithium-ion batteries by deconvolution of electrochemical impedance spectra. *J. Power Sources* **2017**, *361*, 300–309. [CrossRef]
18. Alavi, S.; Birkl, C.R.; Howey, D.A. Time-domain fitting of battery electrochemical impedance models. *J. Power Sources* **2015**, *288*, 345–352. [CrossRef]
19. Klotz, D.; Schönleber, M.; Schmidt, J.P.; Ivers-Tiffée, E. New approach for the calculation of impedance spectra out of time domain data. *Electrochim. Acta* **2011**, *56*, 8763–8769. [CrossRef]
20. Karger, A.; Wildfeuer, L.; Maheshwari, A.; Wassiliadis, N.; Lienkamp, M. Novel method for the on-line estimation of low-frequency impedance of lithium-ion batteries. *J. Energy Storage* **2020**, *32*, 101818. [CrossRef]
21. Schröer, P.; Khoshbakht, E.; Nemeth, T.; Kuipers, M.; Zappen, H.; Sauer, D.U. Adaptive modeling in the frequency and time domain of high-power lithium titanate oxide cells in battery management systems. *J. Energy Storage* **2020**, *32*, 101966. [CrossRef]
22. Zou, C.; Zhang, L.; Hu, X.; Wang, Z.; Wik, T.; Pecht, M. A review of fractional-order techniques applied to lithium-ion batteries, lead-acid batteries, and supercapacitors. *J. Power Sources* **2018**, *390*, 286–296. [CrossRef]
23. Schmidt, J.P.; Berg, P.; Schönleber, M.; Weber, A.; Ivers-Tiffée, E. The distribution of relaxation times as basis for generalized time-domain models for Li-ion batteries. *J. Power Sources* **2013**, *221*, 70–77. [CrossRef]
24. Zhu, J.; Dewi Darma, M.S.; Knapp, M.; Sørensen, D.R.; Heere, M.; Fang, Q.; Wang, X.; Dai, H.; Mereacre, L.; Senyshyn, A.; et al. Investigation of lithium-ion battery degradation mechanisms by combining differential voltage analysis and alternating current impedance. *J. Power Sources* **2020**, *448*, 227575. [CrossRef]
25. Hentunen, A.; Lehmuspelto, T.; Suomela, J. Time-Domain Parameter Extraction Method for Thévenin-Equivalent Circuit Battery Models. *IEEE Trans. Energy Convers.* **2014**, *29*, 558–566. [CrossRef]
26. Xiaosong, H.; Fengchun, S.; Yuan, Z.; Huei, P. Online estimation of an electric vehicle Lithium-Ion battery using recursive least squares with forgetting. In Proceedings of the 2011 American Control Conference, San Francisco, CA, USA, 29 June–1 July 2011; pp. 935–940. [CrossRef]
27. Goldammer, M.S.E.; Kowal, J. Investigation of degradation mechanisms in lithium-ion batteries by incremental open-circuit-voltage characterization and impedance spectra. In Proceedings of the 2020 IEEE Vehicle Power and Propulsion Conference (VPPC), Gijon, Spain, 18 November–16 December 2020; pp. 1–8. [CrossRef]
28. Tikhonov, A.N.; Goncharsky, A.V.; Hazewinkel, M. *Numerical Methods for the Solution of Ill-Posed Problems*; Springer: Dordrecht, The Netherlands, 2010.
29. Schindler, S. *Diskrete Elektrochemische Modellierung und Experimentelle Identifikation von Lithium-Ionen-Zellen Basierend auf Halbzellpotentialen*; Universität Bayreuth: Bayreuth, Germany, 2018.
30. Barai, A.; Widanage, W.D.; Marco, J.; McGordon, A.; Jennings, P. A study of the open circuit voltage characterization technique and hysteresis assessment of lithium-ion cells. *J. Power Sources* **2015**, *295*, 99–107. [CrossRef]

31. Gantenbein, S.; Schönleber, M.; Weiss, M.; Ivers-Tiffée, E. Capacity Fade in Lithium-Ion Batteries and Cyclic Aging over Various State-of-Charge Ranges. *Sustainability* **2019**, *11*, 6697. [CrossRef]
32. Kindermann, F.M.; Noel, A.; Erhard, S.V.; Jossen, A. Long-term equalization effects in Li-ion batteries due to local state of charge inhomogeneities and their impact on impedance measurements. *Electrochim. Acta* **2015**, *185*, 107–116. [CrossRef]
33. Shafiei Sabet, P.; Sauer, D.U. Separation of predominant processes in electrochemical impedance spectra of lithium-ion batteries with nickel-manganese-cobalt cathodes. *J. Power Sources* **2019**, *425*, 121–129. [CrossRef]
34. Illig, J.; Ender, M.; Weber, A.; Ivers-Tiffée, E. Modeling graphite anodes with serial and transmission line models. *J. Power Sources* **2015**, *282*, 335–347. [CrossRef]
35. Dubarry, M.; Baure, G.; Devie, A. Durability and Reliability of EV Batteries under Electric Utility Grid Operations: Path Dependence of Battery Degradation. *J. Electrochem. Soc.* **2018**, *165*, A773–A783. [CrossRef]
36. Anseán, D.; Dubarry, M.; Devie, A.; Liaw, B.Y.; García, V.M.; Viera, J.C.; González, M. Operando lithium plating quantification and early detection of a commercial LiFePO4 cell cycled under dynamic driving schedule. *J. Power Sources* **2017**, *356*, 36–46. [CrossRef]
37. Schindler, S.; Danzer, M.A. A novel mechanistic modeling framework for analysis of electrode balancing and degradation modes in commercial lithium-ion cells. *J. Power Sources* **2017**, *343*, 226–236. [CrossRef]
38. Fermín-Cueto, P.; McTurk, E.; Allerhand, M.; Medina-Lopez, E.; Anjos, M.F.; Sylvester, J.; dos Reis, G. Identification and machine learning prediction of knee-point and knee-onset in capacity degradation curves of lithium-ion cells. *Energy AI* **2020**, *1*, 100006. [CrossRef]

batteries

Article

Advanced Monitoring and Prediction of the Thermal State of Intelligent Battery Cells in Electric Vehicles by Physics-Based and Data-Driven Modeling

Jan Kleiner *, Magdalena Stuckenberger, Lidiya Komsiyska and Christian Endisch

Technische Hochschule Ingolstadt, Institute of Innovative Mobility, Esplanade 10, 85049 Ingolstadt, Germany; Magdalena.Stuckenberger@thi.de (M.S.); Lidiya.Komsiyska@thi.de (L.K.); Christian.Endisch@thi.de (C.E.)
* Correspondence: Jan.Kleiner@thi.de

Abstract: Novel intelligent battery systems are gaining importance with functional hardware on the cell level. Cell-level hardware allows for advanced battery state monitoring and thermal management, but also leads to additional thermal interactions. In this work, an electro-thermal framework for the modeling of these novel intelligent battery cells is provided. Thereby, a lumped thermal model, as well as a novel neural network, are implemented in the framework as thermal submodels. For the first time, a direct comparison of a physics-based and a data-driven thermal battery model is performed in the same framework. The models are compared in terms of temperature estimation with regard to accuracy. Both models are very well suited to represent the thermal behavior in novel intelligent battery cells. In terms of accuracy and computation time, however, the data-driven neural network approach with a Nonlinear AutoregRessive network with eXogeneous input (NARX) shows slight advantages. Finally, novel applications of temperature prediction in battery electric vehicles are presented and the applicability of the models is illustrated. Thereby, the conventional prediction of the state of power is extended by simultaneous temperature prediction. Additionally, temperature forecasting is used for pre-conditioning by advanced cooling system regulation to enable energy efficiency and fast charging.

Keywords: lithium-ion battery; electro-thermal model; smart cell; intelligent battery; neural network; temperature prediction

Citation: Kleiner, J.; Stuckenberger, M.; Komsiyska, L.; Endisch, C. Advanced Monitoring and Prediction of the Thermal State of Intelligent Battery Cells in Electric Vehicles by Physics-Based and Data-Driven Modeling. *Batteries* **2021**, 7, 31. https://doi.org/10.3390/batteries7020031

Academic Editor: Kai Peter Birke

Received: 5 March 2021
Accepted: 7 May 2021
Published: 11 May 2021

Publisher's Note: MDPI stays neutral with regard to jurisdictional clai-ms in published maps and institutio-nal affiliations.

1. Introduction

In the past few decades, the transport mobility sector, and especially the automotive industry, has experienced considerable changes. In order to address the climate change problems and particularly, to reduce CO_2 emissions, among other technologies, Electric Vehicles (EVs) are becoming increasingly important [1]. Due to their high energy and power density, lithium-ion (li-ion) batteries are the most preferred battery type for EV applications [2]. Nevertheless, the challenges and limitations related to costs, safety and aging need to be addressed and are part of current research [3]. One important factor influencing performance, safety and life time of the battery pack, is temperature [2,4,5]. Low temperatures lead to less available power and capacity and can result in irreversible battery degradation when reaching subzero values. High temperatures outside the optimal range of 15–35 °C [4] lead to accelerated aging and may result in a thermal runaway when exceeding the safety limit of 60 °C [4,5]. Monitoring battery cell temperatures is, therefore, necessary, which is a task of the Battery Thermal Management System (BTMS) as part of the Battery Management System (BMS) [4]. The basic approach in commercial vehicles is to measure the temperature at a few discrete points at the surface or tab of li-ion cells in the system [6]. In large format battery cells this temperature may greatly differ from the temperature reached in the battery cell core [7–9], which is the critical temperature in terms of performance and safety. Therefore, the temperature estimation by using thermal models is a first step and a necessary part to enable precise thermal state monitoring. In addition, predictive thermal management with several advantages, such as the core temperature information, can be utilized [10–13].

The development of next generation intelligent battery systems takes this into account by advanced monitoring of the individual cell states [14,15]. Sensors are integrated in cell-level electronics and intelligent algorithms determine the cell state, detect faults or perform bad-block management [16–18]. In our previous work [19], a prismatic cell was equipped with electronics for single cell data acquisition and system reconfiguration. The use of sensors integrated in mass-produced electronics in combination with thermal models enables the core temperature of the battery cells to be monitored. This avoids the need to integrate additional temperature sensors into the cell, as proposed by [5,7], which would lead to increased production costs and safety issues [8]. However, for thermal state monitoring, the internal cell behavior needs to be modeled and the developed models have to be integrated in the vehicle architecture. Cell-level modeling is necessary, since cooling system gradients and cell-to-cell variations lead to different thermal conditions for cells in a system [20]. A substantial thermal gradient also arises due to the thermal coupling between cells. Therefore, the thermal states of multiple cells differ and need to be monitored individually.

Many different approaches for thermal models for conventional battery cells can be found in the literature. The modeling approaches can generally be separated into physics-based and data-driven models. Another common designation is white-box and black-box modeling depending on the way the results are derived from the inputs. In general, the prediction of a white-box model is physically and geometrically motivated and can be understood more intuitively compared to black-box models which are solely data-driven. There are detailed physical-based electrochemical-thermal models that are coupled, e.g., with geometries modeled by Computational Fluid Dynamics (CFD) or Finite Element Method (FEM) approaches [21,22] or even in combination with mechanical models [23]. Those models are useful for battery design but their main drawback is their low computational speed, which does not allow using them for online temperature estimation in a BTMS [24,25]. Lumped thermal models consider the relevant physical phenomena and simplify the differential equation system by concentrating the important cell characteristics on a few points [24,25]. The model parameters can be derived analytically for known material parameters [24,26,27] or fitted to experimental measurements [25,28]. The latter are a first step towards data-driven models and, therefore, one representative of the so called gray-box approaches. Completely data-driven models do not represent the cell internal physical and geometrical properties and model the output behavior implicitly. Thereby, mostly with the help of machine learning methods mathematical relations between inputs and outputs are trained [29]. Examples are models using Support Vector Machines [30] and Artificial Neural Networks (ANNs) [9,29,31]. The latter are a novel topic of research in the field of thermal battery modeling and have proven to have advantages in modeling non-linear dynamic relations as they can be found in batteries [9]. Nevertheless, the ANN approaches found in the literature are mainly simple network architectures, such as Feedforward (FF) [29,31], and none of them considers integrating the ANN as thermal model in a total model system. Data-driven time-series prediction, implemented, for example, as Nonlinear AutoregRessive with eXogeneous input (NARX) architecture, is particularly reported in the literature to be adequate as thermal parameter forecasting in energy systems [32,33] or surface temperature predictions of cylindrical cells [34]. However, there are currently no models that use NARX networks for core temperature modeling of large format cells, neither for conventional, nor for intelligent batteries.

Since both approaches, physical-based and data-driven modeling, seem to be adequate as thermal battery models, it is important to compare the modeling approaches related to the application in Battery Electric Vehicles (BEVs). Thereby, the latest developments in terms of hardware and structure of intelligent cells are taken into account in this work.

Currently, there are neither physical-based nor data-driven models that consider the hardware influences within intelligent cells on the battery's temperature.

The application of cell individual temperature monitoring by using a thermal model in a real BEV system creates an additional opportunity to estimate the core temperature, and also the new possibility to predict the temperature for future events and to utilize corresponding operation strategies. Only a few approaches exist for using battery temperature prediction in BEV scenarios [10,12,35,36]. The results are promising with thermal models that can be used to improve the prediction of the driving range [35], control the regulation of an air- and liquid-dual cooling system [10], or predict the future available battery power [36]. However, there are certainly more applications in which real-time thermal modeling can lead to advances in the specific prediction scenario. In this work, the cases of State-Of-Power (SOP) prediction and advanced cooling system regulation are proposed.

Prediction functionalities, such as SOP prediction, are important for the BEV application as a means of knowing the available power in situations such as user acceleration requirement or long-term performance availability without fear of overcharging or over-discharging [37]. Current publications focus on the calculation of the available SOP based on current, voltage and State-Of-Charge (SOC) limitations of the cells [37–43]. For low SOCs, the discharge current is the limiting factor, while for high SOCs, the charging current is taken accordingly. At the same time, the maximum permissible load and the voltage limits of the cell restrict the maximum power output. However, in reality, the SOP is also limited by the rising temperatures during maximum performance. For that purpose, temperature estimation for an adequate SOP prediction is necessary, as performed in this work.

Another investigated use case of temperature prediction is related to the cooling system regulation. Amini et al. [44] argue that temperature prediction achieves a planning horizon to activate or deactivate the cooling system which is necessary as the cooling system contains high thermal masses. By reducing the initial phase for a battery thermal management system, the power necessary for cooling can be reduced in general [45]. Additionally, the goal of the cooling system is always to keep all cells in the optimal thermal range. However, there are applications discussed in the literature that prefer increased temperatures. For example, heating the battery to 40 °C instead of cooling it to 25 °C is known to reduce the stress and degradation caused by the intercalation and deintercalation processes and, therefore, aging [46]. Another application is the thermally challenging fast charging of EVs. Yang et al. [47–49] reveal slow preheating as favorable for fast charging procedures, leading to much less heat generation during the fast charging and, therefore, to less energy demand for the cooling. Collin et al. [50] mention, in their work on advanced fast charging technologies, the first approaches in commercial BEVs that perform defined pre-conditioning to improve fast charging.

In this work, the focus is on developing a thermal model for an intelligent prismatic 25 Ah cell prototype, including electronics for a BEV application. The aim is to represent the thermal interactions within the large format cell and the actual influence of the electronics. For the first time, the existing approaches for conventional battery cells, e.g., a physics-based and a data-driven model using the example of a Thermal Equivalent Circuit Model (TECM) and an ANN respectively, are implemented and compared. The two thermal modeling methods are parametrized and implemented for an existing intelligent cell hardware published in our former work [19] and integrated in a total framework for a BEV application. The models are used for the cell-level temperature estimation of different local temperatures and are compared in terms of parametrization, accuracy and computation time. Using intelligent cells and the thermal models for advanced thermal management, novel temperature prediction applications in a realistic BEV scenario are presented. Improvements for SOP prediction are then presented by considering the thermal cell state. In a second application, an predictive cooling system regulation is presented that enables pre-conditioning for fast charging.

2. General Modeling Approach

The general electro-thermal model framework is shown in Figure 1. It consist of three submodels: an electrical model, a heat generation model and a thermal model. The goal of the total model is to estimate the core temperature of the battery cell. Therefore, a commonly used electro-thermal co-simulation is performed to capture the temperature dependency of several model parameters [21,24,25,27]. The core temperature as and output of the thermal submodel is fed back to the other two submodels, which are described in the following sections. For the integration of the model in BEV applications, an observer is implemented that is described in detail in Section 3.4. The current I, which is determined via measurement in a realistic BEV system, is the input to the total model structure, as well as the observer. For the investigation in this work, two thermal models are implemented and compared that both fit in the same framework. The thermal models are described in detail in Section 3.

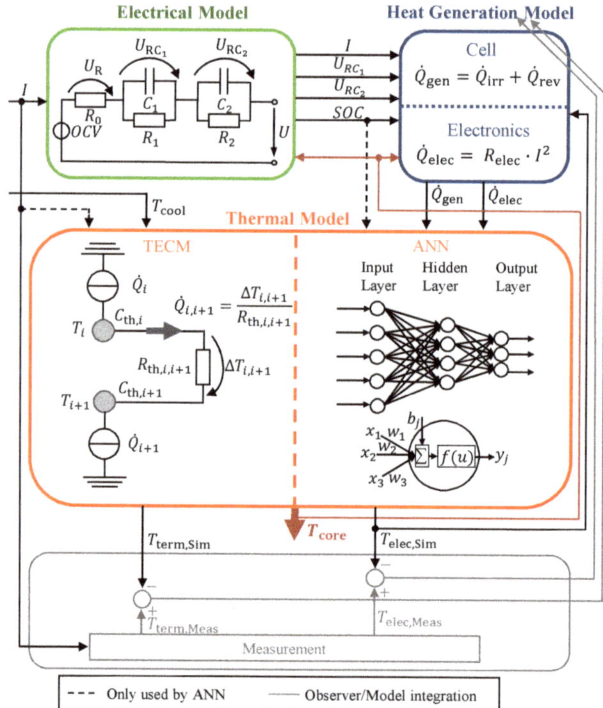

Figure 1. Schematic representation of the modeling framework of an electro-thermal battery cell model, consisting of an electrical, a heat generation, a thermal submodel, and an observer. Two different thermal models are implemented for comparison.

2.1. Electrical Model

An Equivalent Circuit Model (ECM) is used as an electrical model, as frequently seen in thermal modeling of battery cells [27,51]. In this work, the model consists of a voltage source, representing the Open Circuit Voltage (*OCV*) of the cell, a series resistance R_0 corresponding to the cell's ohmic resistance and two RC-elements, which stand, for the voltage drop due to overvoltages, e.g., charge transfer, diffusion, phase change overvoltages and other losses. The cell terminal voltage U can be calculated from Equation (1) below.

$$U = OCV + R_0 \cdot I + U_{RC_1} + U_{RC_2} \tag{1}$$

All the components of the ECM, e.g., resistances, capacitances and the OCV are dependent on the cell core temperature and the SOC. The temperature is fed back from the thermal submodel and the SOC is calculated inside the electrical submodel using coulomb counting. The resistances and capacitances used for the model parametrization were determined analytically for the 25 Ah cell and published in a previous publication [52].

2.2. Heat Generation Model

The heat generation model comprises two parts. It separately calculates the heat generation in the cell and the electronics components. For the heat generation in the jelly roll, Equation (2) is used based on the simplified energy balance in electrochemical systems by Bernardi et al. [53]:

$$\dot{Q}_{\mathrm{gen}} = \dot{Q}_{\mathrm{irr}} + \dot{Q}_{\mathrm{rev}} = I^2 \cdot R_0 + \frac{U_{\mathrm{RC_1}}^2}{R_1} + \frac{U_{\mathrm{RC_2}}^2}{R_2} + I \cdot T \cdot \frac{\mathrm{d}OCV}{\mathrm{d}T} \tag{2}$$

In Equation (2), the first three terms stand for the irreversible heat generation resulting from the voltage drop at the electrical resistances shown in the electrical submodel. It is always positive and therefore, leads to the heating of the cell. The last term in Equation (2) is reversible heat generation, which results from the entropy change during intercalation and the deintercalation of the lithium ions. It may be positive or negative, depending on the SOC and the direction of the current and, therefore, may heat or cool the cell, respectively. The entropy coefficient $\mathrm{d}OCV/\mathrm{d}T$ is dependent on the temperature and was also determined experimentally in [52].

The second part of the heat generation model calculates the heat generation in the electronics based on joule heating by

$$\dot{Q}_{\mathrm{elec}} = \sum_{i=1}^{n} I_i^2 \cdot R_{\mathrm{elec},i} \tag{3}$$

$R_{\mathrm{elec},i}$ is the ohmic resistance of the electronics' pieces, respectively, and is strongly dependent on each components' temperature. Thereby, n is the number of all different electronics segments. A detailed description of the electronics and the corresponding model is given in Section 3.1 .

3. Thermal Models of an Intelligent Cell

The purpose of the thermal models is to estimate the jelly roll temperature for the given heat input by the heat generation model. In order to compare a physics-based to a data-driven model, a TECM and an ANN are implemented for the example of a prismatic cell prototype for intelligent batteries published in [19]. Thereby, the geometrical as well as boundary conditions for this prototype are defined by the reference system and taken into account for the modeling. Since the focus of this work is on the differences of the two approaches, they are implemented in the same electro-thermal framework to compare accuracy, computational effort and time and applicability to an in-use temperature estimation in a BTMS.

3.1. Reference System

The cell under investigation is a prismatic 25 Ah cell from SANYO/Panasonic with a Nickel-Mangan-Cobalt-Oxide (NMC) cathode and graphite anode material. The dimensions of the cell are 14.8 cm, 2.65 cm and 9.1 cm.

In our previous work [19], the cell was combined with electronics to form an intelligent cell prototype. A schematic representation of this setup is shown in Figure 2a. For the prismatic cell equipped with electronics, a detailed electro-thermal 3D CFD model was implemented to investigate the specific influences of the electronics to the cell. The shown cell with electronics is utilized for parametrization of the models of the cell and the electronics in this work. The dataset of the previous investigation is used as target data for both simplified real-time modeling approaches. Considering that no measurement data of the jelly roll core temperature are available, the neural network is trained using the highly resolved and already experimentally validated CFD model. This teacher approach, using a detailed model for the training of a simplified real-time model, is similarly done for other applications, e.g., by Fang et al. [54]. In addition, the neural network uses independent datasets for training and validation. For this purpose, datasets for the temperature range of 15–45 °C are available, which contain many different static and dynamic load profiles, as well as the information about local heat generation and temperatures. Figure 2b shows, as an example, a part of the data for a reference temperature of 25 °C.

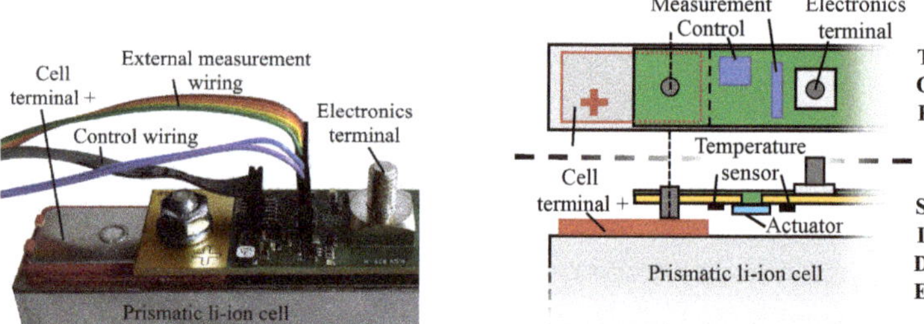

(**a**) Cell prototype and schematic geometry for detailed 3D CFD model in [19].

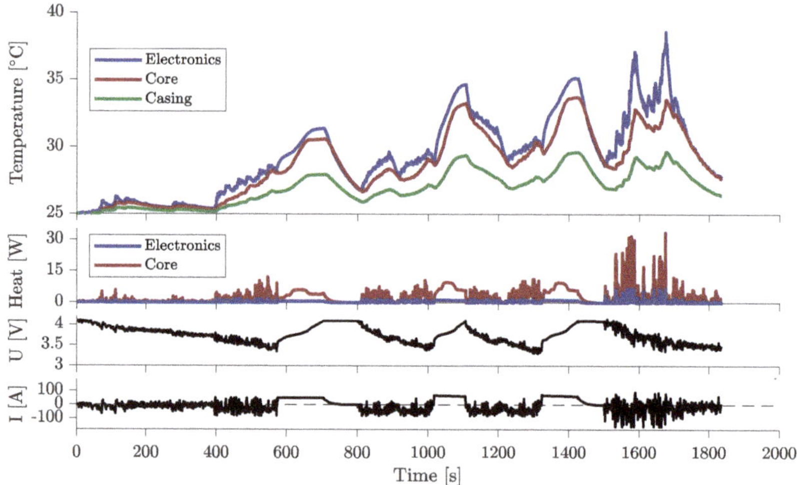

(**b**) Example investigation results used as training or target data.

Figure 2. Reference system based on our previous investigation [19]: (**a**) Cell prototype with electronics and corresponding schematic geometry for the detailed 3D spatially dependent electro-thermal CFD model. (**b**) Example results for the temperature and heat generation behavior of the prototype cell for 25 °C starting temperature. Various comparable datasets in the range of 15–45 °C are used as a training or target profile.

For a meaningful scenario, BEV boundary conditions are considered [21]. The cell is assumed to be adiabatic in all directions with the exception of the connection to the cooling system at the bottom with the temperature T_{cool}. For the investigation on a system level, cells are additionally thermally coupled via the casing and the busbars, which are described in Section 3.4. For a realistic cooling system behavior, a simplified model is introduced that is comparable to [55–57]. Thereby, the cooling system is regulated stepwise and rule-based which is an extension of the thermostat controller [58,59]. The maximum heat flow per cell $\dot{Q}_{max}$, that can be dissipated by the cooling system, is controlled in three steps as shown in Figure 3. Starting with a deactivated cooling system for an optimal cell temperature below 30 °C in step 0, the cooling system performance rises stepwise to a maximum of 9 W at 38 °C for step 3. A hysteresis of 0.3 K , visualized by the marked line, is used for the regulation of the cooling power [60]. Excessive heat production of the cell, exceeding the cooling power limit, leads to an increase in the cooling system temperature.

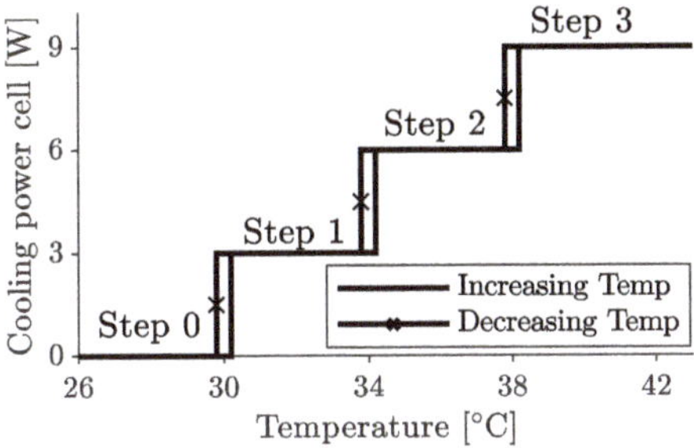

Figure 3. Overview of stepwise cooling system regulation with a hysteresis of 0.3 K for decreasing and increasing temperatures.

3.2. Physics-Based Thermal Equivalent Circuit Model

A commonly used physics-based modeling approach for conventional cells is a lumped thermal model, also called TECM. The basic structure of TECMs is symbolized in the left-hand section of the thermal submodel in Figure 1. In general, in a TECM, the component to be observed is discretized by small volumes. Each volume is represented by a thermal node, which contains the thermal capacity and parameters of the volume respectively. The thermal capacity of one volume i is calculated using Equation (4).

$$C_{\mathrm{th},i} = \rho_i \cdot V_i \cdot c_{\mathrm{p},i} \tag{4}$$

where ρ_i and $c_{\mathrm{p},i}$ are the averaged density and specific heat capacity of the volume i, and V_i is its geometric volume. Thermal capacities, therefore, describe the heat accumulation analogously to electrical capacitances [24]. In each thermal node, the volumetric fraction of the total heat generation is used as a heat source, which is analogous to a current source in electrical models. The thermal nodes are connected by thermal resistances defining the heat transfer between them [24]:

$$R_{\mathrm{th,cond}} = \frac{L}{\lambda \cdot A} \tag{5}$$

L is the distance between two thermal nodes, λ is the heat conduction coefficient of the respective material and A is the cross-sectional area of the heat transfer path between two nodes.

The total TECM for the cell and the detailed electronics model is shown in the left-hand section of Figure 4. Significant temperature gradients can result in large format prismatic battery cells [21]. Therefore, the cell-internal structure is included in the TECM to model a more realistic temperature distribution through a physics-based model in comparison to other implemented lumped thermal models [28]. For that purpose, $3 \times 3 \times 3$ nodes standing for volumes of the same size are arranged in the mid of the jelly roll. In order to model the complex curved geometry of the jelly roll, three nodes are added below and above the $3 \times 3 \times 3$ cuboid, respectively, resulting to a total number of 33 nodes inside the jelly roll. In the same manner, the case is discretized by each of the nine nodes on the x-y-side, three nodes on the y-z-side and two nodes for cell top and one for the bottom. Additional nodes are added for the current collectors and the cell terminals. Computer tomography scans have shown that electrolyte is remaining at the bottom of the case [21].

Therefore, the connection between the jelly roll and the bottom case is considered via heat conduction through the electrolyte.

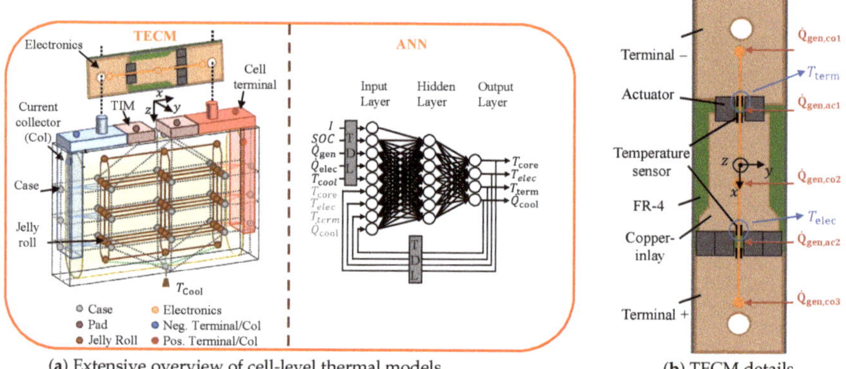

(**a**) Extensive overview of cell-level thermal models   (**b**) TECM details

Figure 4. Real-time thermal models for a intelligent cell in a BEV battery system: (**a**) Overview of the TECM and the ANN modeling details. (**b**) Bottom view of the electronics hardware geometry and the related TECM model.

The electronics can have significant influence on the jelly roll temperature [19]. Therefore, the thermal influences of the specific electronics are considered by a lumped electronics model of the real hardware. In Figure 4a the general positioning and thermal connection of the electronics is revealed. Additionally, Figure 4b depicts a detailed view of the electronics components and the modeling scheme. An one-dimensional heat conduction and temperature distribution along the current path in x-direction is considered. Three thermal nodes are integrated representing the thermal masses of the current conducting copper inlays. Temperature and current dependent heat generation is located in all current carrying parts. The information of the current and the SOC are provided by the framework emulating a BMS, but they are not further used in the present implementation. The TECM calculates the temperature at the thermal nodes and also at the positions where thermistors are integrated on the hardware prototype. Using this presented structure of the TECM makes it possible to estimate the temperature in multiple positions and dimensions of the cell which is advantageous for the later system level implementation described in Section 3.4. The TECM is implemented in MATLAB/Simulink Version 2020a using Simscape.

Thermal masses and thermal resistances are determined analytically on the basis of material parameters, dimensions and manufacturers data of the investigated prismatic cell and electronics. The initial material parameters used are listed in Table 1.

Table 1. General physical and thermal parameters of the used components for the TECM model.

Component	Material	ρ [kgm^{-3}]	c_p [Jkg^{-1}K^{-1}]	λ [Wm^{-1}K^{-1}]
Case [1]	aluminum	2700	900	220
Pos. Term./Col. [1]	aluminum	2700	900	220
Neg. Term./Col. [1]	copper	8700	385	400
Jelly Roll [2]	mixed	2043	1371	33($\parallel$)/0.7($\perp$)
Insulation [1]	plastic foil	1190	1470	0.18
PCB inlay [1]	copper	8700	385	400
Electrolyte [3]	solvent	1130	2055	0.6
Therm. Interface Material (TIM) [4]	silicone	2300	1000	3.5

[1] [61], [2] Cell manufacturer data sheet (25 °C), [3] [21], [4] manufacturer data sheet.

3.3. Data-Driven ANN-Based Thermal Model

In comparison to the physics-based approach, a data-driven model is implemented by means of an ANN shown in the right part of the thermal model in Figures 1 and 4a. ANNs mimic the information processing in human brains by modeling of the nonlinear input-output-relations, which explicitly occur in battery cells [29,62]. Their main advantage is that complex correlations do not have to be physically modeled as networks can be trained to represent the correlations. It is, therefore, necessary to provide the data basis that includes the data correlations to be trained.

Regarding the model architecture and implementation, a network consists of a number of neurons that are arranged in layers. The neurons are connected via edges that contain weights that indicate the strength of the connection. In every neuron, its weighted inputs and a bias are summed, and the sum is mapped, in a nonlinear manner, to the output by the activation function [29,54,62] in Equation (6).

$$y_j = f\left(\sum_{i}^{n} w_i \cdot x_i + b_j\right) \qquad (6)$$

In Equation (6), x_i are the inputs to one neuron, w_i are the respective edge weights of the inputs, b_j is the bias in the neuron, f is the activation function and y_j is the neuron's output. In the training process the weights and biases are updated using known output results. The goal is to minimize the difference between the network output and the target data. The difference is mostly described by the Mean Squared Error (MSE) and is termed the loss function. Several optimizations exist in the literature for minimizing the loss function.

The total ANN thermal model developed in this work is depicted in detail in the right-hand section of Figure 4a. As already mentioned, the implementation of an ANN a dataset containing the input–output correlations to be learned is necessary. The present network is trained based on a dataset generated by the reference system described in Section 3.1. The underlying load profile contains different static (constant charge/discharge) and dynamic load profiles (e.g., the ECE, EUDC, US06, RTS95, FTP72, Artemis Motorway cycles), as well as fast charging and cooling sections without load. In order to simulate varying ambient and cooling conditions, the dataset is generated for different starting temperatures of 15–45 °C in steps of 5 °C leading to a total of over 273 k timesteps. A dataset provides information of current, SOC, heat generation in cell and electronics, the local temperature of electronics, terminal and core as well as the removed heatflux by cooling for every second. The network inputs are scaled in a range between 0 and 1 using the minimum and maximum values of each feature in the training dataset. This makes the later training more robust and efficient [9,33] and prevents it from putting more emphasis on signals with higher absolute value [45].

The network architecture consists of one input layer, one hidden layer and one output layer. The NARX architecture is implemented for the situation of an cell equipped with electronics. As characteristic of the NARX as recurrent network, the outputs are fed back to the input layer. In contrast to conventional FF networks, this enables the representation of temporal dynamics, which improves the modeling of the thermal cell behavior. Thus, the outputs of different timesteps $y(t-1), y(t-2) \ldots, y(t-d_{\text{Output}})$ are stored in a Tapped Delay Line (TDL) and used as inputs for the current timestep $y(t)$. In the same manner, the inputs form different timesteps $x(t), x(t-1), x(t-2), \ldots, x(t-d_{\text{Input}})$, which are stored in a TDL and used as inputs to calculate the output $y(t)$ at the current timestep. d_{Input} and d_{Output} are the maximum backward timesteps stored in the TDL of the inputs and outputs, respectively [33]. The activation function used for the hidden layer is the Rectified Linear Unit (ReLU) function (Equation (7)):

$$f(x) = ReLU(x) = max(0, x) = \begin{cases} x & \text{if } x \geq 0 \\ 0 & \text{if } x < 0 \end{cases} \qquad (7)$$

The ReLU function is advantageous because of its lower computing time requirements, compared to other activation functions, such as logsig or tansig [62], which makes it more suitable for the current application in a vehicle BMS.

The main advantage of the NARX approach compared to other data-driven regression networks is the network architecture. A NARX network is implemented with neurons and a structure as simple as a FF network. Therefore, it can be trained with the fast and accurate FF training procedures and algorithms with good convergence. The training algorithm used in this work is the Levenberg-Marquardt (LM) algorithm, which offers a fast convergence for small networks [29]. The network is trained open-loop and independent of the other two submodels, meaning that the loop from the outputs to the inputs is opened by utilizing the known target temperatures. The open-loop configuration allows to efficiently train the NARX net, analogous to FF nets and stable inputs are available [33]. This is the main advantage compared to other recurrent network approaches.

Important features influencing the training time and especially the accuracy of the results are the NARX hyperparameters, such as number of hidden neurons or input delays. In order to determine a suitable architecture for the network, the authors applied the grid-search method [63], meaning that the number of input and output delays and the number of neurons on the hidden layer are varied between 0 to 10, 1 to 10 and 3 to 10 respectively. The span of the parameters was found with prior investigations on testing many parameters on a large scale. For measuring the quality of the current network topology, another independent validation dataset containing different profiles to the training dataset is used. The network with the lowest MSE on the validation set is used as the current best configuration at the initialization by chance. For this application, a net is chosen, which consists of seven neurons on the hidden layer, and TDLs lengths of 7 and 3 for the inputs and outputs respectively. The net and training parameters are summed up in Table 2. In order to avoid extrapolation scenarios, the training profiles are chosen and combined in a way that the physically possible limits of SOC, cell voltage and load are reached. The total dataset generated is split into 70%, 15% and 15% for the training, validation and testing, respectively. In order to avoid overfitting, the early-stopping method is used by means of stopping the training if the MSE on the validation dataset increases in six consecutive epochs [33]. The maximum number of epochs is limited to 150, whereby an epoch is one representation of the total training dataset with adaptation of the weights afterwards [45]. The whole setup and training of the network is performed utilizing the MATLAB Deep Learning Toolbox.

Table 2. Parameters for the neural network applied as thermal model.

Parameter	Variation	Result
Training dataset	39 k timesteps per temperature	
Starting temperatures	15–45 °C in 5 °C steps	
Training, validation, test ratio	70 %, 15 %, 15 %	
Training algorithm	LM	
Activation function	ReLU (hidden layer) identity (output layer)	
Max. number of epochs	150	
Max. input delays	0 to 10	7
Max. output delays	1 to 10	3
Number of hidden neurons	3 to 10	7

3.4. BEV Integration and System Level Simulation

In order to compare the two modeling approaches, the two thermal submodels are integrated into the presented model framework, including electrical and heat generation submodels. There is a need to avoid drifting of the total model, as the three submodels are strongly intercoupled. An observer is, therefore, included, as shown by the components in gray color in Figure 1. Generally, the observer compares estimated and measured values and the difference is fed back to adopt the estimation [64]. A few approaches can be found in the literature which make use of an observer to estimate the battery cell core temperature by measuring the surface temperature of a battery cell [64–66]. In this work, a different observer structure is implemented that compares the temperatures measured and estimated at the electronics T_{elec} and a terminal T_{term}. The difference in the electronics' temperatures $T_{elec,Sim} - T_{elec,Meas}$ is used to update the heat generation at the electronics $\dot{Q}_{elec}$ and the difference in the terminal temperatures $T_{term,Sim} - T_{term,Meas}$ for adaption of heat generation in the jelly roll $\dot{Q}_{gen}$.

Obviously, there are multiple cells in a battery system that require multiple single cell models. The cells in the system thermally interact with each other [67], and this needs to be considered in system-level modeling. To achieve a fast computing system level simulation, a possible coupling scheme is presented in Figure 5. Temperatures at the boundary of the cell model are used as a boundary condition between the individual cell models connected in series. The coupling can be performed by temperature signal routing instead of full physical coupling to reduce the computational effort. In the system simulation approach, based on single cell models, either the TECM, or the ANN, or another compatible cell-level model, can be used at any time. Possible system representations are that every cell within a module receives data from the case and terminal temperatures of its neighboring cells in order to represent the thermal interactions within a module. The cells at the border of the modules can be assumed to be adiabatic or are also coupled. The modeled boundary conditions depend on the specific application. Overall, the approach is able to represent the thermal coupling or cooling conditions in state-of-the-art battery systems. Nevertheless, system simulation is an ongoing field of research and part of future work.

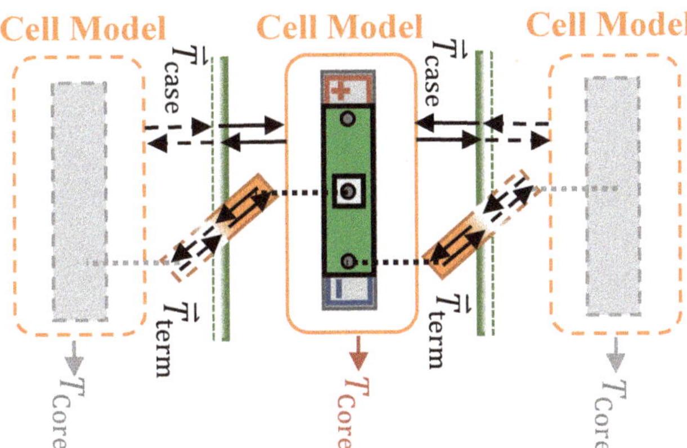

Figure 5. System simulation approach based on thermal coupling of multiple cell models via busbar and casing connections.

4. Results and Discussion

4.1. Spatial Temperature Estimation of TECM and ANN

In order to compare the temperature estimation of the real-time thermal models to the target results of the reference system in Section 3.1, an independent dataset of the reference system based on the ADAC electric vehicle cycle is chosen. The corresponding 2000 s current profile is repeated until the lower cut-off voltage of 3 V is reached. For adequate comparison, a homogeneous constant temperature of 25 °C is defined for the starting and cooling condition. To emulate realistic application conditions, the target data for the observer contain a normally distributed noise with a mean value of 0 K and a standard deviation of 0.2 K. Figure 6 displays the estimated transient temperatures of the core, the electronics and a terminal, as well as the related temperature deviation ΔT of the real-time models. The need for spatial resolution modeling is clearly illustrated by the differences between the local measurement points. At the point of the maximum temperature, differences of up to 4 K are detected. However, there are differences for the modeling approaches and the positions. At high thermal loads, the TECM slightly overestimates the core temperature and underestimates the terminal temperature. For the temperature estimation at the electronics, the dynamic changes in the electronics heat generation in combination with the three representative thermal masses lead to a maximum deviation of 2.7 K for the TECM. The deviation of the electronics results from the lumped representation. Thereby, thermal masses and resistances are based on the initial geometries and material properties. Thus, parameter deviations and contact resistances lead to differences. Corresponding model deviations can be reduced by simple optimization procedures, which is a step towards data-driven models. Overall, the TECM model reveals a good temperature estimation with the highest mean deviation of 0.15 K at the electronics.

The ANN shows in total very small deviations with a maximum difference of -0.67 K at the electronics temperature. The low noise on the deviation also shows that the dynamics of the specific location is represented accordingly. The advantage of the ANN compared to the TECM is due to the training with comparable datasets, which contain the dynamic behavior. For all locations, the mean deviation is below 0.05 K without any static offset or time-dependent drift of the estimation results. Other comparison differences are summarized in Table 3. The overall estimation accuracy of both models is very good within a RMSE of 0.23 K for the TECM and 0.08 K for the ANN. Regarding the computation of the results, the total solving time with the same commercial laptop is 30 s for the ANN model and 60 s for a maximum timestep of 1 s. Thus, both approaches are adequate for real-time estimation with calculation times of 20 ms and 10 ms per timestep. However, the computation time will differ for realistic BEV hardware, which needs to be the focus of future work.

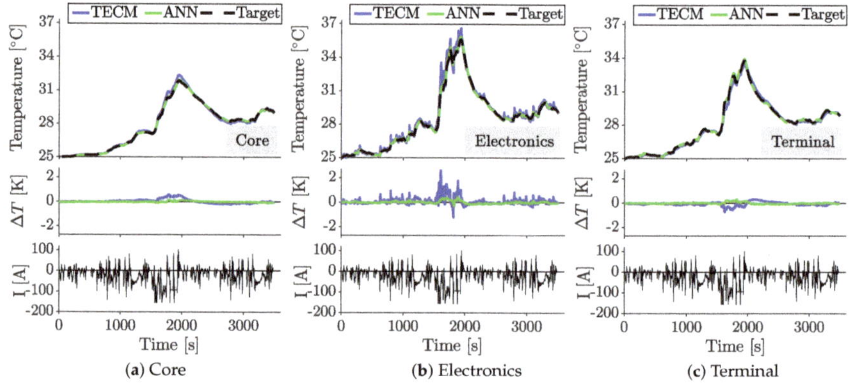

Figure 6. Transient temperature profiles of the TECM and the ANN and the related estimation deviation in comparison to the target data of the reference system. The locations of (**a**) the cell core, (**b**) the electronics and (**c**) the terminal.

Table 3. Comparison of the TECM and the ANN model for the estimation scenario of Figure 6.

	TECM			**ANN**		
Parameter	**Core**	**Elec.**	**Term**	**Core**	**Elec.**	**Term**
Max. local deviation	0.6 K	2.7 K	0.06 K	0.15 K	-0.67 K	0.3 K
Mean local deviation	0.06 K	0.15 K	-0.04 K	0.01 K	0.04 K	0.03 K
Overall RMSE		0.23 K			0.08 K	
Computation effort per timestep [1]		20 ms			10 ms	
Modeling approach		physical-based			data-driven	

[1] On a commercial laptop with Intel i5 CPU.

In general, physics-based or data-driven approaches can be used for estimation and prediction applications, as outlined in the introduction. Both have proven high accuracy for the challenging task of a large format prismatic cell with electronics. Nevertheless, the general concept of the models offers certain advantages and disadvantages for use. The present ANN is more accurate and provides at least small benefits related to the computation time. However, adequate training datasets must exist for training and parametrization. Additionally, changes in the hardware and materials that have not been part of the training data would generate discrepancies in the estimated results. With the physics-based TECM model hardware changes can be implemented easier or the area of application can be extended. The parametrization does not need multiple datasets for different behavior, but a detailed analysis of corresponding geometries and materials. For example, using the model for thermal runaway investigations can be provided by implementing the corresponding Arrhenius-based heat generation equations above the normal temperature range of 60 °C, as proposed, e.g., in [68]. Since the lumped approach does not represent all physical phenomena, estimation deviations arise with the present model. However, these can be reduced by suitable optimization processes. The use of both models is discussed in the second part of this paper, which describes the implementation of new BTMS functionalities.

4.2. Prediction Applications

One important task of the BMS is to estimate different battery states [3,8]. A statement about the current state, e.g., the SOP, often requires a prediction of future situations. More information about the present and future behavior can improve the state prediction and regulation procedures. In the current work, precise thermal models are available and, based on their temperature prediction capability novel BTMS, functionalities are developed. Thereby, an optimization of existing SOP prediction approaches is demonstrated by taking into account the thermal cell state. On the other hand, a novel battery pre-conditioning approach for fast charging optimization is presented.

4.2.1. Improvement of the SOP Prediction

Present SOP prediction strategies are based primarily on current, voltage, and SOC limits [37,40]. The thermal battery state is largely neglected [69]. Taking the example of a common scenario of highway driving followed by fast charging, the importance of taking the temperature into account is presented. Figure 7 depicts the current profile and the related temperature increase in the target data of the reference system. While driving, both thermal models are used to estimate the core temperature of a cell. Both models conform very well to the target data with a small deviation of less than 0.1 K for the ANN and a maximum of 0.8 K for the TECM. The maximum deviation appears in both cases during fast charging at the highest thermal stress of the cell. Subsequently, a conventional power prediction is performed without the observer for the upcoming continuation of the driving and the expected thermal behavior is predicted for different permissible current scenarios. A detailed view of the mid-term prediction scenario is displayed on the right side of Figure 7. As revealed by the results of both models, the only electrical-based conventional calculation of the SOP will exceed the safety temperature limit of 60 °C. In addition, predicting the point in time for reaching the voltage or SOC limit is not performed correctly without taking into account the temperature dependency of the corresponding electrical parameters. Bearing in mind the temperature limit, one may reduce the total permissible output current so that the maximum allowable temperature is not reached during the total discharging (see dashed lines in both cases). This, in turn, restricts the available power in the early stage.

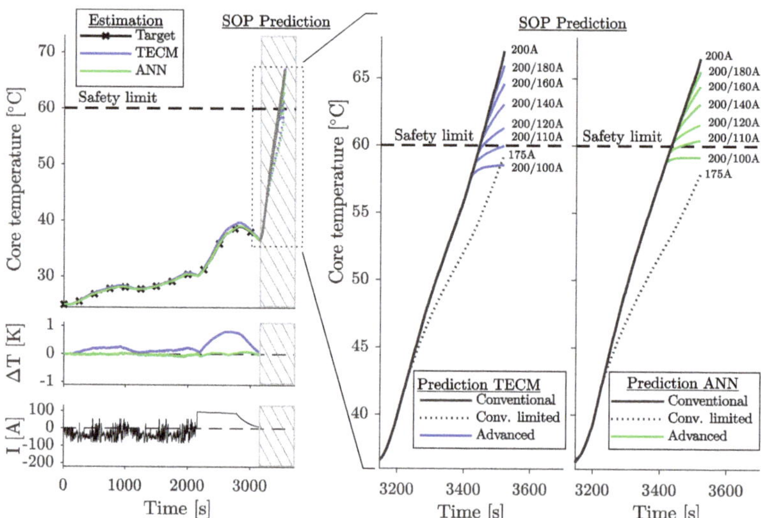

Figure 7. Highway driving and fast charging scenario with in-use SOP prediction in various ways: ANN/TECM approaches are used for the temperature estimation while driving, as well as the prediction of the temperature during SOP determination.

An optimized SOP prediction by an individual output current limitation based on temperature prediction is, therefore, desirable. Multiple output current scenarios can be calculated in advance and the resulting temperature is predicted, which enables maximized performance, while staying within the battery safety limits. Both modeling approaches reveal comparable predicted temperature results. The differences arise from the model's specific absolute temperature predictions.

On the basis of the detailed models, multiple different reference temperatures can be combined with adaptive and intelligent algorithms for safe prediction optimization. Nevertheless, concrete algorithms are beyond the scope of this work and form part of future investigations. In the present investigation, a simple optimization of the point in time for limiting the output current and the current value after limitation is performed (Figure 7). Once a thermally safe scenario has been found, the associated current profile can be used to calculate the available power. Thus, the investigation clearly shows that considering the temperature development is necessary for an accurate, safe and maximized performance SOP prediction.

4.2.2. Predictive Thermal Management

In addition to advanced state monitoring and prediction, detailed thermal models also enable the specific control of the cooling system. Conventional cooling system approaches for BEVs normally react to measured temperatures, e.g., at the cell terminals, and do not act predictively to achieve target temperatures. There are already approaches in the literature for novel BTMS that use model-based temperature prediction in order to react to varying conditions [10,11]. At this point, we add another approach to the existing ones, which relates to the thermal pre-conditioning of batteries. In their work, Collin et al. [50] propose to elevate the battery system temperature in order to enable an optimized fast charging. To meet this goal, the future cooling regulation does not only depend on static temperature rules but also needs to be adjusted.

For the same BEV driving scenario as for the SOP prediction, the step-wise cooling system behavior (see Figure 3) and the resulting terminal temperature are shown in Figure 8. The current profile, the related temperature increase and the cooling system behavior of the target data are depicted in black until the end of fast charging at 3200 s.

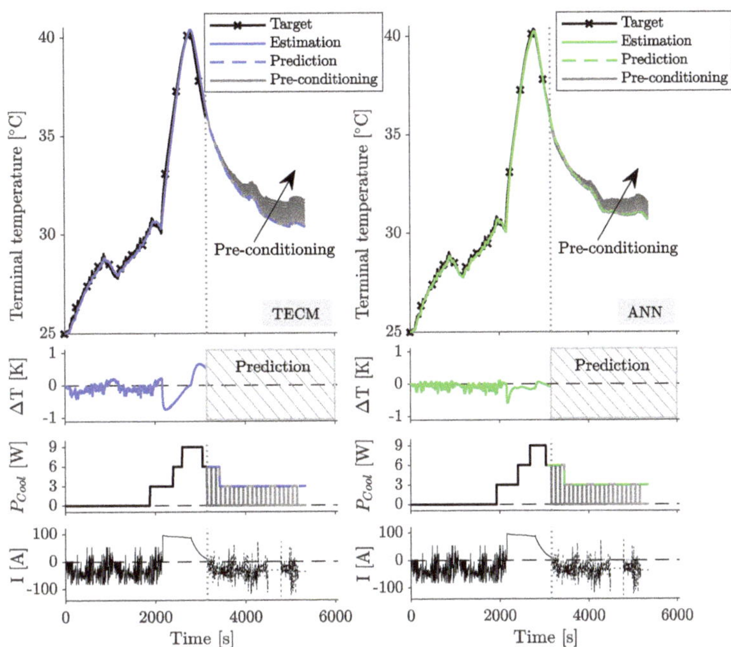

Figure 8. Temperature estimation and predictive cooling system regulation based on ANN/TECM thermal model integrated with an observer in a BEV environment.

The models show comparable estimation behavior with good representation of the transient temperature development. In both cases, the largest deviation arises at the beginning of fast charging. The terminal temperature is underestimated for the initial fast charging section with a maximum deviation of 0.7 K for the TECM and 0.5 K for the NARX. Afterwards, the NARX converges to the target temperature more quickly, while the TECM requires the period towards the end of charging. The increased deviation for the terminal temperature estimation compared to the core temperature estimation in Figure 7 corresponds with the findings from the validation. Subsequent to the fast charging, the models are used to predict the temperature for the basic cooling system behavior of Figure 3 for continued highway driving after fast charging stops.

Here, it should be mentioned that a key aspect of thermal prediction is load prediction. However, the prediction of the future current is a stand-alone field of current research and, for example, performed based on navigation profiles and geographical data [70]. In this work, we assume that the future vehicle velocity profile on a motorway contains comparable elements in the same order of magnitude as the previous highway driving. Based on this velocity profile and a simplified vehicle model, a load profile is calculated. In a real BEV application, this simple prediction can be optimized, e.g., by navigation information or driver behavior models.

It can be seen in the case of both models, that the present terminal temperatures lead to the activation of the cooling system and a transient temperature decrease towards 30 °C. However, if the destination is the next fast charging stop, raising the temperature towards the end could be more suitable [48]. Using the thermal models, in a simple optimization procedure, temperature development during dynamic driving towards the next charging stop can be predicted while testing different pre-conditioning timelines. The resulting differences in the transient temperature development of the core temperature are visualized in gray color. Using the simple battery cooling model in combination with model-based temperature prediction, the resulting temperature at the next charging station can be increased. In such a case, the range of the variation depends on concrete application, including the thermal mass of the battery, the ambient temperature and the distance to the next charging stop. It is shown that, depending on the respective cooling step and the switching time, the resulting cell temperature can be varied. Using thermal models for predictive thermal management in applications of this type has several advantages. The favored target temperatures can be reached precisely. In addition, unnecessary cooling power to temporarily reduce the cell temperature is available as electrical energy [44] and, therefore, extends, e.g., the vehicle's range. Furthermore, the thermal conditions for fast charging procedures can be optimized to improve the performance and extend the lifetime [50].

A possible enhanced BTMS needs to evaluate and weight the different influencing factors, e.g., battery temperature, future driving state and effect on driving range, to choose an optimal strategy for cooling system power regulation. In contrast to the already described approaches in the introduction, monitoring the cells' individual temperatures also enables a new degree of freedom for reacting to maximum temperatures inside the battery system and specific cooling regulation.

5. Conclusions

In this work, a physics-based TECM and a data-driven ANN model have been implemented as thermal models for a prismatic li-ion cell equipped with electronics for the application of intelligent battery systems. For the TECM, a lumped thermal model of the cell has been combined with a lumped electronics model. The novel ANN approach with a NARX network is implemented with the same model interfaces as the TECM to fit into an electro-thermal model framework for a BEV application. Both models are parametrized/trained with datasets based on a reference hardware setup of our previous work [19]. Subsequently, the model behavior is validated and several applications of temperature estimation and prediction are investigated. The following conclusions can be made:

- The temperature estimation of both modeling approaches is in good accordance with the reference temperatures for multiple locations. Thereby, the RMSE is 0.23 K for the TECM and only 0.08 K for the ANN for a dynamic BEV driving cycle. Thus, for the first time, thermal models are presented, that are able to represent the thermal interactions in novel cell assemblies for intelligent batteries. The detected cell internal temperature differences of 4 K confirm the need of cell level thermal modeling.
- Comparing physical-based and data-driven modeling, the advantages of the data-driven ANN lie in its high accuracy and fast computation time. The TECM offers advantages in parametrization and the flexibility of the configuration, since it does not need to be trained with target data. Since both models reveal adequate estimation results, the selection depends on the application of intelligent batteries.
- In addition to thermal state estimation, thermal models in combination with the present model framework enable prediction functionalities for a BTMS. The information base for SOP prediction can be enlarged through the consideration of the thermal state. As a result, the thermal safety limits are respected and the available short-term power is maximized.
- In a second prediction application, predictive cooling system regulation is presented. Thereby, pre-conditioning for special BEV applications, e.g., fast charging, for maximizing the BEV energy efficiency and aging conditions is possible.

The focus of future work is on the development and optimization of the corresponding algorithms for SOP prediction and advanced thermal management functionalities and related verification on an intelligent battery system hardware.

Author Contributions: Conceptualization, J.K., M.S. and L.K.; methodology, J.K.; software, J.K. and M.S.; validation, J.K. and M.S.; investigation, J.K., M.S. and L.K.; writing—original draft preparation, J.K. and M.S.; writing—review and editing, J.K., M.S. and L.K.; visualization, J.K.; supervision, L.K. and C.E.; project administration, C.E. All authors have read and agreed to the published version of the manuscript.

Funding: This research received no external funding.

Acknowledgments: The authors would like to express their appreciation to Michael Hinterberger (AUDI AG, Ingolstadt) for the valuable expert advice and acknowledge F. Haselbeck and L. Lechermann for the extensive discussions related to the implementation and training of learning systems. This work was supported by the AUDI AG, Ingolstadt, Germany.

Conflicts of Interest: The authors declare no conflict of interest.

Abbreviations

The following abbreviations are used in this manuscript:

ANN	Artificial Neural Network
BEV	Battery Electric Vehicle
BMS	Battery Management System
BTMS	Battery Thermal Management System
CFD	Computational Fluid Dynamics
EV	Electric Vehicle
ECM	Equivalent Circuit Model
FF	Feedforward
FEM	Finite Element Method
li-ion	lithium-ion
LM	Levenberg-Marquardt
MSE	Mean Squared Error
NARX	Nonlinear AutoregRessive with eXogeneous input
NMC	Nickel-Mangan-Cobalt-Oxide
OCV	Open Circuit Voltage
ReLU	Rectified Linear Unit
SOC	State-Of-Charge
SOP	State-Of-Power
TDL	Tapped Delay Line
TECM	Thermal Equivalent Circuit Model

References

1. Canals Casals, L.; Martinez-Laserna, E.; Amante García, B.; Nieto, N. Sustainability analysis of the electric vehicle use in Europe for CO$_2$ emissions reduction. *J. Clean. Prod.* **2016**, *127*, 425–437. [CrossRef]
2. Karimi, G.; Li, X. Thermal management of lithium-ion batteries for electric vehicles. *Int. J. Energy Res.* **2013**, *37*, 13–24. [CrossRef]
3. Lu, L.; Han, X.; Li, J.; Hua, J.; Ouyang, M. A review on the key issues for lithium-ion battery management in electric vehicles. *J. Power Sources* **2013**, *226*, 272–288. [CrossRef]
4. Xia, G.; Cao, L.; Bi, G. A review on battery thermal management in electric vehicle application. *J. Power Sources* **2017**, *367*, 90–105. [CrossRef]
5. Zhang, G.; Cao, L.; Ge, S.; Wang, C.Y.; Shaffer, C.E.; Rahn, C.D. In Situ Measurement of Radial Temperature Distributions in Cylindrical Li-Ion Cells. *J. Electrochem. Soc.* **2014**, *161*, A1499–A1507. [CrossRef]
6. Wang, Y.; Chen, Z.; Zhang, C. On-line remaining energy prediction: A case study in embedded battery management system. *Appl. Energy* **2017**, *194*, 688–695. [CrossRef]
7. Lee, C.Y.; Lee, S.J.; Hung, Y.M.; Hsieh, C.T.; Chang, Y.M.; Huang, Y.T.; Lin, J.T. Integrated microsensor for real-time microscopic monitoring of local temperature, voltage and current inside lithium ion battery. *Sens. Actuators A Phys.* **2017**, *253*, 59–68. [CrossRef]
8. Wang, Y.; Tian, J.; Sun, Z.; Wang, L.; Xu, R.; Li, M.; Chen, Z. A comprehensive review of battery modeling and state estimation approaches for advanced battery management systems. *Renew. Sustain. Energy Rev.* **2020**, *131*, 110015. [CrossRef]
9. Liu, K.; Li, K.; Peng, Q.; Guo, Y.; Zhang, L. Data-Driven Hybrid Internal Temperature Estimation Approach for Battery Thermal Management. *Complexity* **2018**, *2018*, 1–15. [CrossRef]

10. Zhu, C.; Lu, F.; Zhang, H.; Mi, C.C. Robust Predictive Battery Thermal Management Strategy for Connected and Automated Hybrid Electric Vehicles Based on Thermoelectric Parameter Uncertainty. *IEEE J. Emerg. Sel. Top. Power Electron.* **2018**, *6*, 1796–1805. [CrossRef]

11. Zhu, C.; Lu, F.; Zhang, H.; Sun, J.; Mi, C.C. A Real-Time Battery Thermal Management Strategy for Connected and Automated Hybrid Electric Vehicles (CAHEVs) Based on Iterative Dynamic Programming. *IEEE Trans. Veh. Technol.* **2018**, *67*, 8077–8084. [CrossRef]

12. Wan, J.; Wu, C. The Effects of Driver Speed Prediction-Based Battery Management System on Li-ion Battery Performance for Electric Vehicles. *Proc. Hum. Factors Ergon. Soc. Annu. Meet.* **2014**, *58*, 515–519. [CrossRef]

13. Kim, E.; Shin, K.G.; Lee, J. Real-time battery thermal management for electric vehicles. In Proceedings of the 2014 ACM/IEEE International Conference on Cyber-Physical Systems (ICCPS), Berlin, Germany, 14–17 April 2014; IEEE: Piscataway, NJ, USA, 2014; pp. 72–83. [CrossRef]

14. Steinhorst, S.; Lukasiewycz, M.; Narayanaswamy, S.; Kauer, M.; Chakraborty, S. Smart Cells for Embedded Battery Management. In Proceedings of the Cyber-Physical Systems, Networks, and Applications (CPSNA), Hong Kong, China, 25–26 August 2014; IEEE: Piscataway, NJ, USA, 2014; pp. 59–64. [CrossRef]

15. Otto, A.; Rzepka, S.; Mager, T.; Michel, B.; Lanciotti, C.; Günther, T.; Kanoun, O. Battery Management Network for Fully Electrical Vehicles Featuring Smart Systems at Cell and Pack Level. In *Advanced Microsystems for Automotive Applications 2012*; Meyer, G., Ed.; Springer: Berlin/Heidelberg, Germany, 2012; pp. 3–14.

16. Schneider, D.; Vögele, U.; Endisch, C. Model-based sensor data fusion of quasi-redundant voltage and current measurements in a lithium-ion battery module. *J. Power Sources* **2019**, *440*, 227156. [CrossRef]

17. Schmid, M.; Kneidinger, H.; Endisch, C. Data-Driven Fault Diagnosis in Battery Systems through Cross-Cell Monitoring. *IEEE Sens. J.* **2020**, *1*. [CrossRef]

18. Schmid, M.; Gebauer, E.; Hanzl, C.; Endisch, C. Active Model-Based Fault Diagnosis in Reconfigurable Battery Systems. *IEEE Trans. Power Electron.* **2021**, *36*, 2584–2597. [CrossRef]

19. Kleiner, J.; Heider, A.; Hanzl, C.; Komsiyska, L.; Elger, G.; Endisch, C. Thermal Behavior of an Intelligent Li-Ion Cell under Vehicle Conditions. In Proceedings of the IECON 2020 The 46th Annual Conference of the IEEE Industrial Electronics Society, Singapore, 18–21 October 2020; pp. 2081–2086. [CrossRef]

20. Feng, F.; Hu, X.; Hu, L.; Hu, F.; Li, Y.; Zhang, L. Propagation mechanisms and diagnosis of parameter inconsistency within Li-Ion battery packs. *Renew. Sustain. Energy Rev.* **2019**, *112*, 102–113. [CrossRef]

21. Kleiner, J.; Komsiyska, L.; Elger, G.; Endisch, C. Thermal Modelling of a Prismatic Lithium-Ion Cell in a Battery Electric Vehicle Environment: Influences of the Experimental Validation Setup. *Energies* **2020**, *13*, 62. [CrossRef]

22. Lundgren, H.; Svens, P.; Ekström, H.; Tengstedt, C.; Lindström, J.; Behm, M.; Lindbergh, G. Thermal Management of Large-Format Prismatic Lithium-Ion Battery in PHEV Application. *J. Electrochem. Soc.* **2016**, *163*, A309–A317. [CrossRef]

23. Rieger, B.; Erhard, S.V.; Kosch, S.; Venator, M.; Rheinfeld, A.; Jossen, A. Multi-Dimensional Modeling of the Influence of Cell Design on Temperature, Displacement and Stress Inhomogeneity in Large-Format Lithium-Ion Cells. *J. Electrochem. Soc.* **2016**, *163*, A3099–A3110. [CrossRef]

24. Damay, N.; Forgez, C.; Bichat, M.P.; Friedrich, G. Thermal modeling of large prismatic LiFePO$_4$/graphite battery. Coupled thermal and heat generation models for characterization and simulation. *J. Power Sources* **2015**, *283*, 37–45. [CrossRef]

25. Farag, M.; Sweity, H.; Fleckenstein, M.; Habibi, S. Combined electrochemical, heat generation, and thermal model for large prismatic lithium-ion batteries in real-time applications. *J. Power Sources* **2017**, *360*, 618–633. [CrossRef]

26. Zhao, Y.; Patel, Y.; Zhang, T.; Offer, G.J. Modeling the Effects of Thermal Gradients Induced by Tab and Surface Cooling on Lithium Ion Cell Performance. *J. Electrochem. Soc.* **2018**, *165*, A3169–A3178. [CrossRef]

27. Stocker, R.; Lophitis, N.; Mumtaz, A. Development and Verification of a Distributed Electro-Thermal Li-Ion Cell Model. In Proceedings of the IECON 2018-44th Annual Conference of the IEEE Industrial Electronics Society, Washington, DC, USA, 21–23 October 2018.

28. Li, D.; Yang, L. Identification of spatial temperature gradient in large format lithium battery using a multilayer thermal model. *Int. J. Energy Res.* **2020**, *44*, 282–297. [CrossRef]

29. Fang, K.; Mu, D.; Chen, S.; Wu, B.; Wu, F. A prediction model based on artificial neural network for surface temperature simulation of nickel–metal hydride battery during charging. *J. Power Sources* **2012**, *208*, 378–382. [CrossRef]

30. Chen, Z.; Xiong, R.; Lu, J.; Li, X. Temperature rise prediction of lithium-ion battery suffering external short circuit for all-climate electric vehicles application. *Appl. Energy* **2018**, *213*, 375–383. [CrossRef]

31. Panchal, S.; Dincer, I.; Agelin-Chaab, M.; Fraser, R.; Fowler, M. Experimental and theoretical investigation of temperature distributions in a prismatic lithium-ion battery. *Int. J. Therm. Sci.* **2016**, *99*, 204–212. [CrossRef]

32. Kim, J.H.; Seong, G.H.; Choi, W. Cooling Load Forecasting via Predictive Optimization of a Nonlinear Autoregressive Exogenous (NARX) Neural Network Model. *Sustainability* **2019**, *11*, 6535. [CrossRef]

33. Afroz, Z.; Urmee, T.; Shafiullah, G.M.; Higgins, G. Real-time prediction model for indoor temperature in a commercial building. *Appl. Energy* **2018**, *231*, 29–53. [CrossRef]

34. Hussein, A.A.; Chehade, A.A. Robust Artificial Neural Network-Based Models for Accurate Surface Temperature Estimation of Batteries. *IEEE Trans. Ind. Appl.* **2020**, *56*, 5269–5278.

35. German, R.; Shili, S.; Desreveaux, A.; Sari, A.; Venet, P.; Bouscayrol, A. Dynamical Coupling of a Battery Electro-Thermal Model and the Traction Model of an EV for Driving Range Simulation. *IEEE Trans. Veh. Technol.* **2020**, *69*, 328–337. [CrossRef]

36. Tang, X.; Yao, K.; Liu, B.; Hu, W.; Gao, F. Long-Term Battery Voltage, Power, and Surface Temperature Prediction Using a Model-Based Extreme Learning Machine. *Energies* **2018**, *11*, 86. [CrossRef]

37. Sun, F.; Xiong, R.; He, H. Estimation of state-of-charge and state-of-power capability of lithium-ion battery considering varying health conditions. *J. Power Sources* **2014**, *259*, 166–176. [CrossRef]

38. Fleischer, C.; Waag, W.; Bai, Z.; Sauer, D.U. Adaptive On-line State-of-available-power Prediction of Lithium-ion Batteries. *J. Power Electron.* **2013**, *13*, 516–527. [CrossRef]

39. Xiong, R.; Sun, F.; He, H.; Nguyen, T.D. A data-driven adaptive state of charge and power capability joint estimator of lithium-ion polymer battery used in electric vehicles. *Energy* **2013**, *63*, 295–308. [CrossRef]

40. Pei, L.; Zhu, C.; Wang, T.; Lu, R.; Chan, C.C. Online peak power prediction based on a parameter and state estimator for lithium-ion batteries in electric vehicles. *Energy* **2014**, *66*, 766–778. [CrossRef]

41. Balagopal, B.; Chow, M.Y. The state of the art approaches to estimate the state of health (SOH) and state of function (SOF) of lithium Ion batteries. In Proceedings of the 2015 IEEE 13th International Conference on Industrial Informatics (INDIN), Cambridge, UK, 22–24 July 2015; IEEE: Piscataway, NJ, USA, 2015; pp. 1302–1307. [CrossRef]

42. Xavier, M.A.; Trimboli, M.S. Lithium-ion battery cell-level control using constrained model predictive control and equivalent circuit models. *J. Power Sources* **2015**, *285*, 374–384. [CrossRef]

43. Burgos-Mellado, C.; Orchard, M.E.; Kazerani, M.; Cárdenas, R.; Sáez, D. Particle-filtering-based estimation of maximum available power state in Lithium-Ion batteries. *Appl. Energy* **2016**, *161*, 349–363. [CrossRef]

44. Amini, M.R.; Sun, J.; Kolmanovsky, I. Two-Layer Model Predictive Battery Thermal and Energy Management Optimization for Connected and Automated Electric Vehicles. In Proceedings of the 2018 IEEE Conference on Decision and Control (CDC), Miami, FL, USA, 17–19 December 2018; pp. 6976–6981.

45. Korthals, F.; Stöcker, M.; Rinderknecht, S. Artificial Intelligence in predictive thermal management for passenger cars. In *20. Internationales Stuttgarter Symposium*; Bargende, M., Reuss, H.C., Wagner, A., Eds.; Springer Fachmedien Wiesbaden GmbH and Springer Vieweg: Wiesbaden, Germany, 2020; pp. 529–543.

46. Keil, P. Aging of Lithium-Ion Batteries in Electric Vehicles. Ph.D. Thesis, Technical University Munich, Munich, Germany, 2017. Available online: http://nbn-resolving.de/urn/resolver.pl?urn:nbn:de:bvb:91-diss-20170711-1355829-1-5 (accessed on 19 February 2021).

47. Yang, X.G.; Liu, T.; Wang, C.Y. Innovative heating of large-size automotive Li-ion cells. *J. Power Sources* **2017**, *342*, 598–604. [CrossRef]

48. Yang, X.G.; Zhang, G.; Ge, S.; Wang, C.Y. Fast charging of lithium-ion batteries at all temperatures. *Proc. Natl. Acad. Sci. USA* **2018**, *115*, 7266–7271. [CrossRef]

49. Yang, X.G.; Liu, T.; Gao, Y.; Ge, S.; Leng, Y.; Wang, D.; Wang, C.Y. Asymmetric Temperature Modulation for Extreme Fast Charging of Lithium-Ion Batteries. *Joule* **2019**, *3*, 3002–3019. [CrossRef]

50. Collin, R.; Miao, Y.; Yokochi, A.; Enjeti, P.; von Jouanne, A. Advanced Electric Vehicle Fast-Charging Technologies. *Energies* **2019**, *12*, 1839. [CrossRef]

51. Jiang, J.; Ruan, H.; Sun, B.; Zhang, W.; Gao, W.; Le Wang, Y.; Zhang, L. A reduced low-temperature electro-thermal coupled model for lithium-ion batteries. *Appl. Energy* **2016**, *177*, 804–816. [CrossRef]

52. Schmid, M.; Vögele, U.; Endisch, C. A novel matrix-vector-based framework for modeling and simulation of electric vehicle battery packs. *J. Energy Storage* **2020**, *32*, 101736. [CrossRef]

53. Bernardi, D.; Pawlikowski, E.; Newman, J. A General Energy Balance for Battery Systems. *J. Electrochem. Soc.* **1985**, *132*, 5–12. [CrossRef]

54. Fang, Q.; Li, Z.; Wang, Y.; Song, M.; Wang, J. A neural-network enhanced modeling method for real-time evaluation of the temperature distribution in a data center. *Neural Comput. Appl.* **2019**, *31*, 8379–8391. [CrossRef]

55. Kuang, X.; Li, K.; Xie, Y.; Wu, C.; Wang, P.; Wang, X.; Fu, C. Research on Control Strategy for a Battery Thermal Management System for Electric Vehicles Based on Secondary Loop Cooling. *IEEE Access* **2020**, *8*, 73475–73493. [CrossRef]

56. Angermeier, S.; Ketterer, J.; Karcher, C. Liquid-Based Battery Temperature Control of Electric Buses. *Energies* **2020**, *13*, 4990. [CrossRef]

57. Caramia, G.; Cavina, N.; Capancioni, A.; Caggiano, M.; Patassa, S. *Combined Optimization of Energy and Battery Thermal Management Control for a Plug-in HEV*; SAE Technical Paper Series; SAE International400 Commonwealth Drive: Warrendale, PA, USA, 2019. [CrossRef]

58. Park, S.; Ahn, C. Stochastic Model Predictive Controller for Battery Thermal Management of Electric Vehicles. In Proceedings of the 2019 IEEE Vehicle Power and Propulsion Conference (VPPC), Hanoi, Vietnam, 14–17 October 2019. [CrossRef]

59. Hung, Y.H.; Lue, Y.F.; Gu, H.J. Development of a Thermal Management System for Energy Sources of an Electric Vehicle. *IEEE/ASME Trans. Mechatron.* **2015**, *21*, 402–411. [CrossRef]

60. He, F.; Ma, L. Thermal management of batteries employing active temperature control and reciprocating cooling flow. *Int. J. Heat Mass Transf.* **2015**, *83*, 164–172. [CrossRef]

61. Neubronner, M; Properties of Solids and Solid Materials. In *VDI Heat Atlas*, 2nd ed.; VDI-Buch; Springer: Berlin, Germany, 2010; pp. 551–614. [CrossRef]

62. Ding, B.; Qian, H.; Zhou, J. Activation functions and their characteristics in deep neural networks. In Proceedings of the 2018 Chinese Control And Decision Conference (CCDC), Shenyang, China, 9–11 June 2018; [CrossRef]
63. Bergstra, J.; Bengio, Y. Random Search for Hyper-Parameter Optimization. *J. Mach. Learn. Res.* **2012**, *13*, 281–305.
64. Lin, X.; Fu, H.; Perez, H.E.; Siege, J.B.; Stefanopoulou, A.G.; Ding, Y.; Castanier, M.P. Parameterization and Observability Analysis of Scalable Battery Clusters for Onboard Thermal Management. *Oil Gas Sci. Technol. Rev. D'IFP Energies Nouv.* **2013**, *68*, 165–178. [CrossRef]
65. Richardson, R.R.; Ireland, P.T.; Howey, D.A. Battery internal temperature estimation by combined impedance and surface temperature measurement. *J. Power Sources* **2014**, *265*, 254–261. [CrossRef]
66. Liu, Z.; Du, J.; Stimming, U.; Wang, Y. Adaptive observer design for the cell temperature estimation in battery packs in electric vehicles. In Proceedings of the 2014 9th IEEE Conference on Industrial Electronics and Applications, Hangzhou, China, 9–11 June 2014; IEEE: Piscataway, NJ, USA, 2014; pp. 348–353. [CrossRef]
67. Liu, X.; Ai, W.; Naylor Marlow, M.; Patel, Y.; Wu, B. The effect of cell-to-cell variations and thermal gradients on the performance and degradation of lithium-ion battery packs. *Appl. Energy* **2019**, *248*, 489–499. [CrossRef]
68. Liang, G.; Zhang, Y.; Han, Q.; Liu, Z.; Jiang, Z.; Tian, S. A novel 3D-layered electrochemical-thermal coupled model strategy for the nail-penetration process simulation. *J. Power Sources* **2017**, *342*, 836–845. [CrossRef]
69. Zou, C.; Klintberg, A.; Wei, Z.; Fridholm, B.; Wik, T.; Egardt, B. Power capability prediction for lithium-ion batteries using economic nonlinear model predictive control. *J. Power Sources* **2018**, *396*, 580–589. [CrossRef]
70. Amini, M.R.; Wang, H.; Gong, X.; Liao-McPherson, D.; Kolmanovsky, I.; Sun, J. Cabin and Battery Thermal Management of Connected and Automated HEVs for Improved Energy Efficiency Using Hierarchical Model Predictive Control. *IEEE Trans. Control. Syst. Technol.* **2020**, *28*, 1711–1726. [CrossRef]

 batteries

MDPI

Article

Global Warming Potential of a New Waterjet-Based Recycling Process for Cathode Materials of Lithium-Ion Batteries

Leonard Kurz [1,*], Mojtaba Faryadras [1], Ines Klugius [1], Frederik Reichert [1,*], Andreas Scheibe [1], Matthias Schmidt [2] and Ralf Wörner [1]

[1] Institute for Sustainable Energy Technology and Mobility (INEM), Hochschule Esslingen—University of Applied Sciences, Kanalstraße 33, 73728 Esslingen, Germany; mofags@hs-esslingen.de (M.F.); Ines.Klugius@hs-esslingen.de (I.K.); Andreas.Scheibe@hs-esslingen.de (A.S.); Ralf.Woerner@hs-esslingen.de (R.W.)

[2] Erlos GmbH, Reichenbacher Straße 67, 08056 Zwickau, Germany; Matthias.Schmidt@weckpluspoller.de

* Correspondence: Leonard.Kurz@hs-esslingen.de (L.K.); Frederik.Reichert@hs-esslingen.de (F.R.)

Abstract: Due to the increasing demand for battery electric vehicles (BEVs), the need for vehicle battery raw materials is increasing. The traction battery (TB) of an electric vehicle, usually a lithium-ion battery (LIB), represents the largest share of a BEV's CO_2 footprint. To reduce this carbon footprint sustainably and to keep the raw materials within a closed loop economy, suitable and efficient recycling processes are essential. In this life cycle assessment (LCA), the ecological performance of a waterjet-based direct recycling process with minimal use of resources and energy is evaluated; only the recycling process is considered, waste treatment and credits for by-products are not part of the analysis. Primary data from a performing recycling company were mainly used for the modelling. The study concludes that the recycling of 1 kg of TB is associated with a global warming potential (GWP) of 158 g CO_2 equivalents (CO_2e). Mechanical removal using a water jet was identified as the main driver of the recycling process, followed by an air purification system. Compared to conventional hydro- or pyrometallurgical processes, this waterjet-based recycling process could be attributed an 8 to 26 times lower GWP. With 10% and 20% reuse of recyclate in new cells, the GWP of TBs could be reduced by 4% and 8%, respectively. It has been shown that this recycling approach can be classified as environmentally friendly.

Keywords: lithium-ion battery; traction battery; waterjet-based recycling; direct recycling; life cycle assessment; global warming potential

Citation: Kurz, L.; Faryadras, M.; Klugius, I.; Reichert, F.; Scheibe, A.; Schmidt, M.; Wörner, R. Global Warming Potential of a New Waterjet-Based Recycling Process for Cathode Materials of Lithium-Ion Batteries. *Batteries* **2021**, *7*, 29. https://doi.org/10.3390/batteries7020029

Academic Editor: Kai Peter Birke

Received: 1 April 2021
Accepted: 20 April 2021
Published: 1 May 2021

Publisher's Note: MDPI stays neutral with regard to jurisdictional claims in published maps and institutional affiliations.

1. Introduction

Since the development of lithium-ion batteries (LIBs) in the 1970s, and due to the exclusive attributes of a long lifespan and high energy capacity, LIBs have grown to be important in the field of portable electronic devices [1,2]. In 2015, at least 5.6 billion LIB cells were traded globally [3,4], and the market size of LIBs has been predicted to grow by another 12% from 2016 to 2024, achieving a market value of USD 56 billion by 2024 [3]. Due to their significantly higher energy density, economic, and ecological advantages over other cell technologies, LIBs are particularly suitable as traction batteries, and therefore are predestined to be used in electric vehicles. The shift from internal combustion engines (ICEs) to e-mobility is expected to lead to even greater increases in registrations of electrically powered vehicles in the coming years. Kosai et al. claimed that, by 2050, up to 56 million battery electric vehicles (BEVs) are expected to be registered, 28 times as many as in 2016, which increases the demand for LIBs significantly [5].

After reaching their end of life (EOL), because they become hazardous materials in the waste stream of the environment, EOL LIBs need to be managed responsibly [6,7]. In Europe, between the years 2013 and 2014, the LIB market reported a total consumption of 65,500 tons of LIBs [8], while only about 1900 tons was recycled in the same period [3].

In another report, Wang et al. [9] claimed that the number of waste LIBs in vehicles that would need to be recycled, in an optimistic case, would be 6.76 million in 2035. Through a combination of innovations in recycling technologies and proactive regulations regarding collection and disposal of spent batteries, a significant fraction of the materials required for production of new LIBs could be supplied by EOL LIBs [10]. With the recently published proposal concerning batteries and waste batteries, the European Union (EU) aims to boost the circular economy of the battery value chain to reduce the environmental impact of batteries. To achieve this, the EU is striving for directives that set minimum rates for the recycling of EOL batteries and usage of recyclate within the production of new batteries for the EU market [11]. Therefore, the development of applicable technologies for LIB recycling, to recover particularly the elements that are low in the earth crust, are necessary [9]. Another aim of recycling LIBs is to keep hazardous materials from entering landfills [12] and to close the loop of raw materials utilized in LIBs for long-term sustainability and resource conservation [13]. In other words, since recycling processes enable internal material flows, it is considered to be a fundamental aspect of the circular economy [3].

Apart from reusing the batteries, recycling of LIBs can be categorized into the following three different technologies [9], used alone or in combination—namely, hydrometallurgy (e.g., retrieve technologies); pyrometallurgy (e.g., Umicore); and direct recycling (example—On to tech) [10,14,15], as shown in Figure 1. The pretreatment process includes discharging by immersing in a salt solution (e.g., NaCl), dismantling, which must be done in a sealed environment for safety reasons, and separating of damaged spent LIBs [9]. The hydrometallurgy route involves the leaching of valuable elements from a solid matrix into ions formed in a solution of inorganic or organic acid [16] and, subsequently, recovery of these metals by selective separation, electrochemical deposition, solvent extraction [16] into mono-metal material, or a precursor for the fabrication of new electrode materials [9,17]. Before hydrometallurgical treatment, the spent LIBs usually get shredded for an easier leaching procedure. The pyrometallurgical route is a smelting process which concentrates metal species such as Co, Cu, and Ni in the molten phase, while other constituents such as Li, Al, Si, Ca, and some Fe in the slag phase [9,14]. In order to obtain pure metals or metal salts, the slags are processed using the hydrometallurgical technique. In direct recycling, battery material is recovered without breaking it down into its components [6,18]. The aim is to reuse this material in new cells with minimal effort and without additional processes [6]. The most important battery material for direct recycling is considered to be the cathode material, as it is usually the most valuable component of an LIB cell with 33% of the total cost [19]. The most important question, however, is whether the recovered material can have comparable quality properties to virgin material [6]. While recovered material from direct recycling methods might not perform the same as primary material, however, compared to the hydrometallurgy and pyrometallurgy, direct recycling methods appear to be more environmentally friendly with lower amounts of emissions and energy consumption [10].

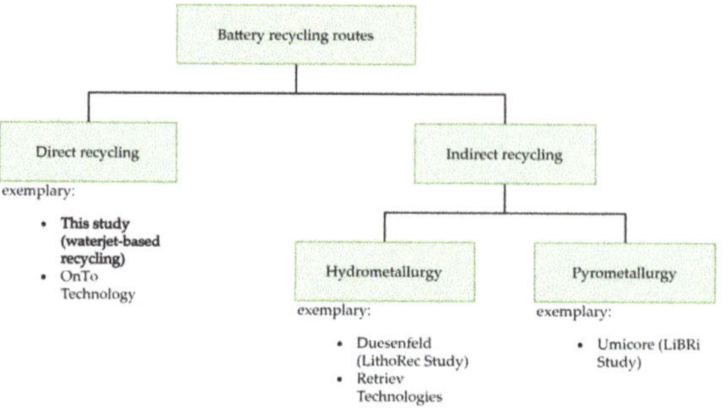

Figure 1. Battery recycling routes [20].

Beside these, recently, there have been other methods for battery recycling reported. These methods, include cathode healing, which combines two simple steps of hydrothermal processing of the spent cathode material followed by heat treatment in a cost-effective way [21], deagglomeration of polyvinylidene fluoride binder from cathode composite, which is based on a degree of surface hydrophobicity using the froth flotation method [14], and an oxalic coprecipitation method, which is a closed-loop regeneration method for $LiNi_5Co_2Mn_3O_2$ material of spent LIBs using mixed acid leaching, oxalate co-precipitation and solid-phase reaction [16].

Sieber et al. [17] claimed that, as we move into the future, resource- and energy-inefficient hydrometallurgy and pyrometallurgy recycling methods must be avoided. Since there are no reports on waterjet-based battery recycling processes to the best of our knowledge, the here-presented procedure is new in the field of battery recycling. This new approach involves a waterjet-based direct recycling method (or functional recycling) with minimal use of chemicals and energy for separation of the cathode-coating layer, which contains $LiNi_xMn_yCo_zO_2$ (NMC), carbon black, and organic binder from the Al substrate foil, while preserving their chemical, physical, and morphological characteristics [17]. This recycling process does not use energy-intensive shredding or melting down of the cells, as is otherwise used in conventional recycling processes for LIBs. Instead, the process focuses on dismantling the battery and the cell itself as far as possible. The separated electrode then goes on a conveyor belt through a water jetting process, whereby the electrode coating is separated from the collector foil by a high-pressure water jet. The intention of this direct (or functional) recycling approach is to add recovered battery-grade cathode material to new NMC active material up to a certain proportion. Doose et al. [22] claimed that recycling processes must be flexible and adaptable to future cell chemistry and production technologies. Whether or not this waterjet-based recycling can meet these criteria, and can be used for production of new cells, must be examined.

In this study, this new waterjet-based direct recycling approach was ecologically evaluated with the help of a life cycle assessment (LCA) to evaluate its potential eco-logic performance. The structure of this LCA is divided into four sections according to the standards DIN EN ISO 14040:2006 [23] and ISO 14044:2006 [24]—the objectives and methodology of the study, life cycle inventory (LCI), life cycle impact assessment (LCIA), and life cycle evaluation.

2. Objectives and Methodology of the Study

2.1. Objective of the Study

In November 2019, the joint project "Industrial Disassembly of Battery Modules and E-Motors to Secure Economically Strategic Raw Materials for E-Mobility (DeMoBat)" [25] was approved by the government of Baden-Wuerttemberg in Germany. The aim is, inter alia, to investigate the feasibility of a waterjet-based recycling process for traction batteries (TBs), to explore the limits of possible admixtures and to identify the potentials for reuse in the TB manufacturing process. Part of this project is to assess the environmental impacts of such an approach. Therefore, we modelled a waterjet-based recycling method for TBs to create an LCA, as well as to be able to draw conclusions about the environmental impacts associated with the recovered cathode coating. Furthermore, comparisons with indirect recycling processes should be made where possible, as well as to investigate possible impacts on new cathode coatings through different admixtures. The focus of this study is primarily based on the impact category of the global warming potential (GWP), which represents an excerpt from a complete LCA.

2.2. Methodology of the Study

2.2.1. Product Sustainability Software: GaBi

For modelling the LCI and calculating the LCIA, the GaBi product sustainability software tool from Sphera Solutions, Inc. (Chicago, IL, USA) with version 10.0.0.71 and the professional database was used. With the help of GaBi, the preparation of the LCI and the determination of the LCIA was significantly simplified. After all processes were modelled, the desired method of impact assessment was selected from various available methods. GaBi assigned the impact indicator values taken from the LCI for the selected categories, calculated the corresponding values, and displayed the results [26].

2.2.2. Function and Functional Unit

The function of the system is the recovery of the cathode material of TBs according to the described waterjet-based recycling process. Therefore, the functional unit is defined as the recycling of one kilogram of TBs and the associated reference flow is defined as the path of the cathode material from one kilogram of TBs. The amount of cathode material depends on the battery composition. In this study, the amount corresponds to 171 g of NMC cathode material.

2.2.3. System Boundaries and Cut-Off Criteria

Figure 2 shows the system boundaries which define the valid assessment space for the LCA of the waterjet-based TB recycling. This system represents a gate-to-gate constellation, because only the production of the recyclate cathode material is considered [27]. Impacts on an environmental balance can be distributed as a result of by-products recovered through a recycling process, such as metals from the battery housing and others. Those impacts are resulting from primary extractions, which could be avoided for the materials of those by-products. Therefore, credits can be distributed to this extent, which usually leads to a positive impact on the environmental balance [27]. In this LCA, by- products of the battery are not further treated or evaluated. Since no credits are to be collected for the LCIA for the resulting by-products, their further treatments are outside the system boundary. The same applies to transport and waste treatments. The remaining electrical energy from the TBs is not considered, and therefore leaves the system boundary. Thus, the system boundaries only include the environment of the battery recycling plant. The geographical boundaries of this system are defined as Germany. The time frame of the study is set to the publication date of this study. For this study, in terms of mass, energy, or environmental relevance, no cut-off criteria are set.

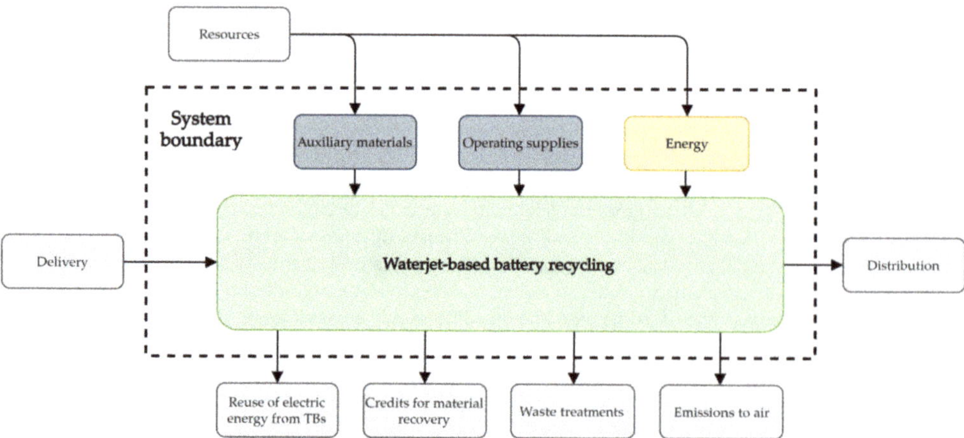

Figure 2. System boundaries of the life cycle assessment (LCA) for the waterjet-based direct traction batteries (TBs) recycling process. Processes outside of the system boundary are not included in this study.

2.2.4. Method of Impact Assessment

The LCI of this ecological study is evaluated according to the CML2001 method [28]. The CML method is an impact-oriented approach that assigns the data of the LCI to the environmental impacts. This multidimensional approach aims at the direct material and energetic allocation of all impact flows between the environment and the production system [29]. In this study, only the impact category of the GWP, measured in kg CO_2 equivalents (CO_2e), is discussed, and evaluated.

2.2.5. Types and Sources of Data

For modelling, various LCA data sources such as primary and secondary data were used. For the energy demand and the amount of auxiliary and operating materials required, primary data were available from the pilot plant in operation at the industrial partner's site. Generic data sets from the GaBi balancing software were used for modelling the auxiliary and operating materials as well as the energy supply. For material quantities, for which no primary data were available, reasonable assumptions were made.

To represent the material flows of the individual battery components, data on the battery composition were required; the material composition of the battery was taken from the LithoRec II study [30], a generic battery composition, which was developed by a consortium of various battery manufacturers and other experts within the framework of that project. This battery composition represented the average battery composition commonly used at that time and is shown in Table 1. As this battery composition was also used in other projects, the comparability of the different recycling approaches was simplified [30].

Table 1. Generic battery composition created by expert panels within the LithoRec II study. The battery composition is intended to represent an average traction battery at this time. The battery contains $LiNi_xMn_yCo_zO_2$ (NMC) cells and $LiPF_6$ as the electrolyte [30].

Complete Battery		100%
	Cathode	**25.2%**
	Manganese	7.4%
	Lithium	1.2%
	Cobalt	1.0%
	Nickel	2.0%
	Aluminium	6.2%
	Oxygen	5.5%
	Others	1.9%
	Anode	**20.4%**
	Separator	**4.4%**
	Electrolyte	**9.6%**
	Cell housing	**5.1%**
	Pack and Module	**34.7%**
	Others	**0.6%**

3. Life Cycle Inventory (LCI)

3.1. Waterjet-Based Recycling: Process Description

This process for recycling TBs is a direct recycling process that aims to recover the electrode coatings of a lithium-ion cell and is currently being practiced in a pilot plant of the industrial DeMoBat project partner Erlos GmbH. In addition to $LiNi_xMn_yCo_zO_2$ (NMC) cells, other cell chemistries with liquid electrolytes are processable with this recycling approach. The recycling process is described below using an NMC cell with a liquid electrolyte. The stoichiometry of the cathode coating is irrelevant and the NMC material recovered by this approach should subsequently be able to be used directly to produce new TBs. To clarify whether this is possible or not is currently being investigated as part of the DeMoBat project. The graphite of the anode can also be recovered, but in this case is not. Due to the usual cathode material recovery rate of at least 90%, or usually higher, this value has been used for this LCA. The other battery components can be recovered almost completely, except for the electrolyte, as this is not treated and therefore not recovered. In Figure 3, the waterjet-based recycling process flow is illustrated with its material input flows. Material flows leaving the process are not shown, due to the defined system boundary. The waterjet-based recycling process can be divided into six process steps as shown in Figure 3, which shows the energy, compressed air, and cathode material flows. Most of the process steps are automated and some are carried out manually. The entire recycling process requires 0.486 MJ of electrical energy for processing 1 kg of TBs.

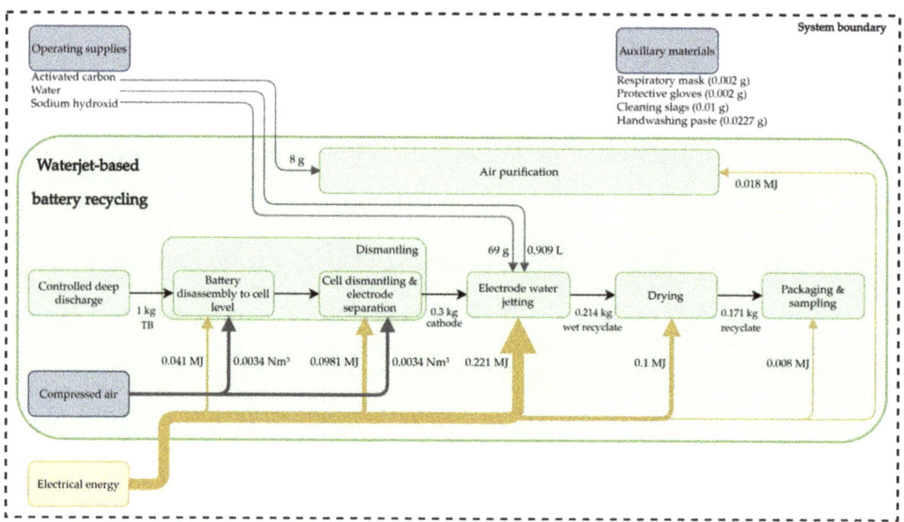

Figure 3. Process flow chart with material input flows and route of the recyclate, with 1 kg of traction batteries as the initial value.

When delivered, the TBs are opened manually and completely deep discharged. In this step, electrical energy recovered by the deep discharge can be fed into the production grid. To minimize the dangers posed by TBs, they remain short-circuited for 24 h.

After the step of complete deep discharge, the TBs are manually dismantled to the cell level and the separated components, such as the battery housing, the cooling system, the electric and electronics are fed into the usual recycling channels. The energy required for dismantling is low, as it is largely carried out manually at the present time. After removal of the modules, the cells can be separated individually from the battery modules, and, due to the manual disassembly, this step is not energy-intensive. The required energy is allocated to pneumatic, electric power tools and the production environment. Since the compressed air requirement cannot be divided more concretely, it is divided equally between the two sub-process steps of dismantling, as shown in Figure 3. Then, these cells are automatically opened and further disassembled using different methods, depending on the cell typ. In pouch cells, the electrodes and the separator are usually stacked, while in prismatic cells, a stack of cathode, separator, and anode is rolled up. This step is mainly automated. During the cell disassembly, there is no direct treatment of the electrolyte and therefore, some of the electrolyte is trapped in the activated carbon filter.

Since the saturated activated carbon filter is not desorbed in the pilot plant, these materials are replaced regularly. The gases obtained through the opening of the cell, the water jetting, and the drying are captured by a permanent suction system, filtered by activated carbon, and released into the air.

After separation of these components, the anode and cathode are each fed separately into the electrode decoating process. Separating the anode and the cathode significantly increases the purity of the recyclate and eliminates the need of inerting the process, and therefore leads to energy and cost savings. The decoating of the electrodes is primarily a mechanical removal of the electrode coatings with the aid of a water jet, where the anode and cathode are passed through this process step separately, and this is described as water jetting.

High-pressure water jetting processes are being used more and more [31]. The possible applications include cleaning processes, decoating processes, cutting processes and others [32]. When the water flows through the nozzle, the pressurised flowing water is accelerated. The radius of the water jet becomes larger the further the nozzle is positioned from the target. In addition, the water jet speed decreases with increasing distance from the nozzle. This enables using waterjet processes for various applications. For cutting processes using a waterjet, the distance between the nozzle and the target workpiece is kept small, while decoating processes are carried out at a greater distance. This is because as the distance to the workpiece increases, the difference in jet speed between the center of the water jet and the outer edge of the water jet decreases. In this way, a more consistent removal result can be achieved over a larger area [32].

For this process step, the entire electrode is placed on a conveyor belt, which moves at a constant speed through a washing system. In this washing system, a water jet removes the coating from the cathode and anode collector foil. This process water circulates within the washing system and is rarely exchanged, which is why the total water requirement can be classified as low. During this process, the coating material sediments in the catch basin of the washing system can be recovered by filtration. No special temperature is required for water jetting of the electrodes, therefore, this process step is carried out at room temperature. The remaining electrolytes in the coatings of anode and cathode are washed out during the water jetting, thus, they mix with the process water. Those remaining electrolytes cause the process water to become acidic, which is neutralized by adding sodium hydroxide (NaOH). Since the pH value of process water is kept in a slightly basic range, the amount of NaOH required depends on the pH value of the process water. This process step is highly automated, and it is the most energy-intensive process step in this recycling process. In addition to the electrode coating, the water jetting process also recovers the carrier foil, which can then also be specifically recycled.

Then, the recovered electrode material is dried in an oven for 3 h. Dust produced in the process step is also treated by an air purification system. Packaging and sampling are the final steps in the process chain which require the least amount of energy. In this step, the recyclate is ground to a desired particle size and examined according to its composition. The amount of recovered cathode or anode coating depends mainly on the processed TBs, as there are differences in coating thickness and resistance to water jetting.

3.2. Greenhouse Gases and Cumulative Energy Consumption

Figure 4 shows the LCI of the described process flow with a selection of greenhouse gases, particulate matter (PM10), the cumulative energy consumption, and the water consumption, each subdivided into the different process steps. A cumulative energy input (primary energy demand from regenerative and non-regenerative resources) of 3.3 MJ is calculated. As shown in Figure 4, water jetting and air purification are the most energy-intensive process steps, and they are also the biggest emitters of NMVOCs, methane, carbon dioxide, and nitrogen oxides. A closer look at the water jetting step reveals that both NaOH (50%) and the necessary electrical energy are the biggest contributors to gas emissions. For the entire recycling process, a water consumption of 4.6 L could be determined, and water jetting has the largest share of water consumption among all process steps, with approximately 3 L. Although this recycling approach is a waterjet-based process, the water consumption required for water jetting plays only a minor role, with 0.909 L. The remaining water consumption is mainly attributable to the provision of electric energy and the NaOH. The relatively high impacts of the air purification process step for many emissions are mainly attributable to the filter material (activated carbon), while the necessary energy consumption is less decisive. For the nitrous oxide and fluorinated and chlorinated hydrocarbons, more than 50% of the emitted gas belong to the water jetting step, because of its high energy consumption. The second largest contributor is dismantling, which is due to the high energy requirements and the necessary protective equipment, such as gloves and respiratory protection, as shown in Figure 4. Since sampling and packaging only require a very small amount of electrical energy, this step produces up to 1% of the emitted gases, which is the lowest share as compared with other steps.

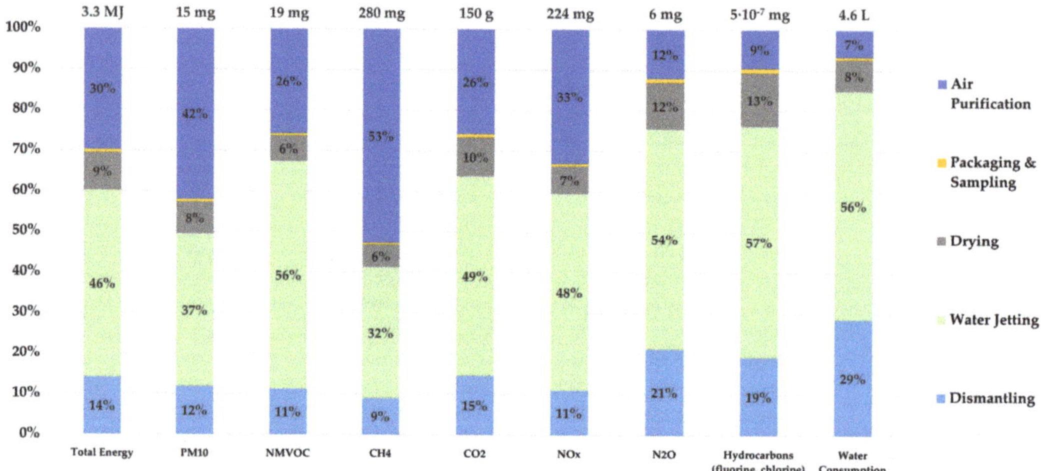

Figure 4. Life cycle inventory (LCI) with selected emissions and cumulative energy consumption of the waterjet-based recycling with the functional unit of 1 kg traction batteries. Amounts smaller than 1% are not shown.

4. Life Cycle Impact Assessment: Global Warming Potential (LCIA-GWP)

A GWP of 158 g CO_2e is calculated for the entire recycling process of 1 kg of TBs. This implies that the recovery of 171 g of NMC cathode material with the recycling process described here results in the abovementioned GWP (158 g CO_2e) or 924 g CO_2e for recycling of 1 kg NMC active material with the battery composition shown in Table 1. Figure 5 shows the LCIA, and thus presents the GWP of the entire process for 1 kg of the TBs. The largest share of the GWP is attributable to the water jetting step, which emits 76 g CO_2e for the defined functional unit and thus causes 48% of the total process and the lowest share of the GWP, which is 1 g CO_2e (<1%), belonging to the packaging and sampling step. The water used for water jetting contributes only a small share (approximately 1.5 g CO_2e or 2% of the water jetting step). Two other factors, the electrical energy required for water jetting the electrodes and NaOH (50%), contribute significantly to the GWP with shares of 30 and 41 g CO_2e, respectively. In the case of air purification, which is the second largest emitter of greenhouse gases with 27% of the whole process, the required electrical energy causes approximately 3 g CO_2e (7%) of the GWP, while the remaining 40 g CO_2e (93%) is due to the filter medium, which is the activated carbon. The drying oven is powered by electricity, which is why this is fully responsible for the GWP in this process step, with a share of 9% of the whole recycling process. In the packaging and sampling step, there are no emitters of greenhouse gases, except the electrical energy, which leads to a share of 1% of the total process.

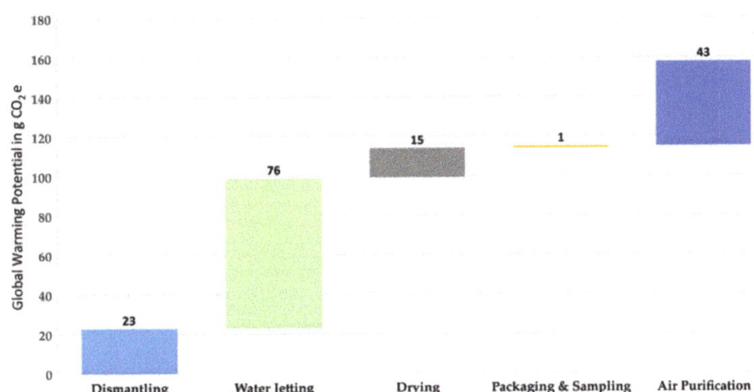

Figure 5. Life cycle impact assessment (LCIA) as the breakdown of the global warming potential according to the process steps of the waterjet-based recycling method for 1 kg of traction batteries.

5. Life Cycle Evaluation

5.1. Comparison between Systems

It is reasonable to compare this waterjet-based recycling approach with the recycling processes presented by Buchert et al., namely LithoRec II, EcoBatRec and LiBRi [30,33,34], however, the comparability of the systems should be ensured beforehand or the differences should be considered. By considering the scope definitions, because of differences in the system boundaries, it is difficult to compare the different systems. One of these differences is based on the studies by Buchert et al. [30,34] that evaluated material flows which leave the system boundary via credits, which was deliberately omitted in this study. Table 2 shows the pure direct debits from the cited studies. It is worthwhile mentioning that, in this study, transportation was considered, and therefore, debits for transportation of TBs are not included in the values presented here. More important than the comparison of the abovementioned methodological framework conditions, is the function of these recycling approaches. This is because the recyclates from the LithoRec II, EcoBatRec and LiBRi processes are not comparable with the recyclate from the recycling process presented here. By comparing the LithoRec II process and this waterjet-based recycling process, the differences are clearer. As described in Section 3.1, this waterjet-based recycling process separates cathode coating from the rest of the cell and aims to proportionately add the obtained recyclate directly to a new cathode coating. In contrast, the products of the LithoRec II process are nickel, manganese and cobalt in a sulphate solution, and lithium as lithium hydroxide [30], which can be used again for cathode coating production.

Table 2. A comparison of the global warming potential (GWP) and the cumulative energy consumption of different recycling approaches using the functional unit of 1 kg of traction batteries. The system boundaries are adjusted to each other.

Studies	GWP	Cumulative Energy Consumption
LiBRi (indirect recycling) [33]	4.248 kg CO_2e	61.3 MJ
EcoBatRec (indirect recycling) [34]	1.282 kg CO_2e	14.5 MJ
LithoRec II (indirect recycling) [30]	2.133 kg CO_2e	35.5 MJ
DeMoBat (direct recycling)	0.158 kg CO_2e	3.3 MJ

Therefore, the comparison in Table 2 must be considered carefully and is not intended to be a benchmark, but rather to highlight the possible advantages of a direct recycling process as compared with the indirect recycling route. Buchert et al. [30,33,34] focused on indirect battery recycling that used pyrometallurgical or hydrometallurgical processes to recover the raw materials of the cathodes, which contributed the most to the GWP of these recycling processes. In the LithoRec II process, hydrometallurgical processing caused about 70% of the total GWP of this process, while in the LiBRi process, the pyrometallurgical process step and the slag preparation caused 77% of the total GWP [33]. The biggest driver in the EcoBatRec study was mechanical treatment with 33%, however, the product of the EcoBatRec recycling approach required further treatment, for example by the hydrometallurgical process as used in the LiBRi approach [34].

5.2. Comparison between Products

The benefits of this method of adding recovered battery-grade cathode material to the new NMC active material up to a certain proportion are that almost the entire production chain of new cathode coating can be avoided for this proportion, and furthermore, the mining of primary raw materials is not necessary to this extent. While the DeMoBat project determines whether such an approach is possible and, if so, up to what proportion admixtures can be realised without having to record too great a loss of quality in the product (new traction batteries), this study quantifies the possible savings for the GWP of cathode coatings through this procedure.

Since significant production steps to produce cathode coating from primary materials could be neglected, a comparison of the active material recycled in the way described here with cathode coating from primary production is reasonable. Accardo et al. [35] claimed that the stoichiometric composition of the NMC materials played a minor role for the GWP, which was why, in this study, only the addition of recyclate to NMC111 material is illustrated. Dai et al. [36] reported a GWP of 16.1 kg CO_2e for 1 kg of NMC111 cathode material [36], which is 100 times higher GWP potential than the cathode material (recyclate) obtained through the waterjet-based battery recycling. The savings that can be achieved by blending different mass fractions of recyclate to the primary cathode material are described in Figure 6, which shows that blending 10% and 20% recyclate can save 9.3% and 18.6% CO_2e, respectively. Due to the very low GWP of the recyclate, the GWP of the cathode material can be reduced by approximately the same percentage as the recyclate content. The NMC111 cathode material studied by Dai et al. [36] has a share of about 40% in the GWP of the entire TB, which makes it possible to estimate the impact of the admixture for the GWP of the entire battery. Consequently, the GWP of a traction battery with NMC111 cells can be reduced by 4–8% by blending the abovementioned proportions. While Accardo et al. [35] illustrated that there are no significant differences in the GWP of different NMC compositions, at the same time, their study also showed that NMC111 material has the highest GWP, which suggests that the savings shown in Figure 6 represent the highest possible savings among the cathode coatings NMC111, NMC622, and NMC811.

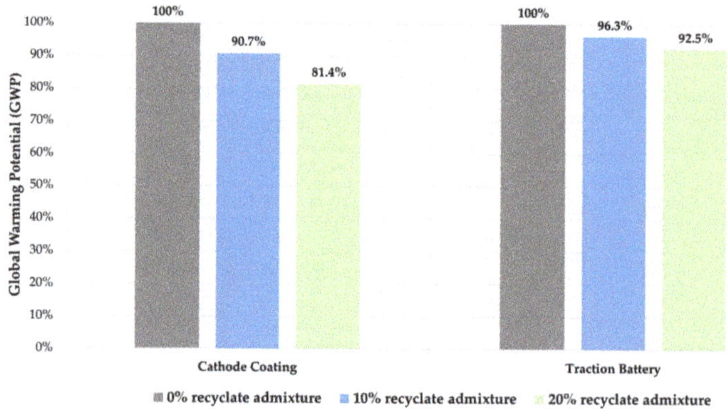

Figure 6. Impact of recyclate admixture in different proportions on the GWP of NMC111 cathode material and the production stage of a traction battery.

6. Conclusions

In this study, a new waterjet-based recycling process for traction battery cells is presented. This direct recycling process detaches the cathode material of a battery cell from the carrier foil with the aid of a water jet. An advantage of this procedure over indirect approaches is that the recovered recyclate can be mixed with new cathode material directly, and therefore energy-intensive manufacturing steps for the admixed fraction can be saved. In addition, the cells or modules are not shredded or melted down, as in conventional recycling processes, but are dismantled to the cellular level, and therefore the pure battery components can be separated by type.

Furthermore, the waterjet-based recycling approach was ecologically evaluated with the help of a life cycle assessment (LCA). The aim is to evaluate its potential ecological performance as compared with indirect recycling approaches and virgin cathode material with the focus on global warming potential (GWP). The results from this study provide an initial assessment of the ecological performance of this recycling approach. A comparison with indirect recycling processes shows that the GWP associated with indirect recycling of traction batteries (TBs) can be reduced by a factor up to 26 through substitution with this direct recycling approach. Compared to cathode coating from production with primary raw materials, the recyclate obtained through the waterjet-based recycling process shows 100 times less GWP. Consequently, the GWP of new battery cells can be significantly reduced through admixture of recyclate obtained through the waterjet-based recycling process. The data basis of the water-based recycling process for the LCA is consistent, as the data are provided first-hand from the recycling company. However, a critical review of the life cycle analysis in this study shows that with the focus on recycling the cathode material, the further battery components are not included. They should be added to the assessment scope in upcoming LCAs to gain a holistic picture of the ecological performance of this process.

From an ecological point of view, the waterjet-based recycling approach can be classified as more environmentally friendly than conventional hydro- or pyrometallurgical recycling processes. Due to the effective ecological performance of this recycling approach, the technical limits should be explored and tested. It also needs to be clarified whether the process can be used for future cell chemistry to fully exploit the given potential of this recycling approach in the long term.

Author Contributions: Conceptualization, F.R. and R.W.; methodology, F.R., I.K.; software, L.K.; validation, A.S., F.R.; formal analysis, I.K. and F.R.; investigation, M.S., M.F. and F.R.; resources, M.S. and R.W.; data curation, M.S. and L.K.; writing—original draft preparation, L.K. and M.F.; writing—review and editing, F.R. and I.K; visualization, F.R. and L.K.; supervision, R.W., F.R. and A.S.; project administration, R.W. and I.K.; funding acquisition, R.W. and L.K. All authors have read and agreed to the published version of the manuscript.

Funding: The authors wish to thank the Ministry of the Environment, Climate Protection and the Energy Sector Baden-Wuerttemberg for funding this work under the funding code L7520105 as part of the joint project "DeMoBat". The article processing charge was funded by the Baden Wuerttemberg Ministry of Science, Research and the Arts and the Hochschule Esslingen—University of Applied Sciences in the funding program Open Access Publishing.

Institutional Review Board Statement: Not applicable.

Informed Consent Statement: Not applicable.

Data Availability Statement: The data presented in this study are available on reasonable request from the corresponding author.

Conflicts of Interest: The authors declare no conflict of interest.

References

1. Reddy, T.B. *Linden's Handbook of Batteries*, 4th ed.; Linden, D., Ed.; McGraw-Hill: New York, NY, USA, 2011; ISBN 9780071624213.
2. Tarascon, J.M.; Armand, M. Issues and challenges facing rechargeable lithium batteries. *Nature* **2001**, *414*, 359–367. [CrossRef] [PubMed]

3. Velázquez-Martínez, O.; Valio, J.; Santasalo-Aarnio, A.; Reuter, M.; Serna-Guerrero, R. A Critical Review of Lithium-Ion Battery Recycling Processes from a Circular Economy Perspective. *Batteries* **2019**, *5*, 68. [CrossRef]

4. Pillot, C. The Rechargeable Battery Market and Main Trends 2014–2025. In Proceedings of the 31st International Battery Seminar & Exhibit, Nica, France, 6 October 2015.

5. Kosai, S.; Takata, U.; Yamasue, E. Natural resource use of a traction lithium-ion battery production based on land disturbances through mining activities. *J. Clean. Prod.* **2021**, *280*, 124871. [CrossRef]

6. Gaines, L. The future of automotive lithium-ion battery recycling: Charting a sustainable course. *Sustain. Mater. Technol.* **2014**, *1–2*, 2–7. [CrossRef]

7. Bernardes, A.M.; Espinosa, D.C.R.; Tenório, J.A.S. Recycling of batteries: A review of current processes and technologies. *J. Power Source* **2004**, *130*, 291–298. [CrossRef]

8. EBRA. *Noticeable Growth of the Quantity of Batteries Recycled*; European Battery Recycling Association: Brussels, Belgium, 2012.

9. Wang, Y.; An, N.; Wen, L.; Wang, L.; Jiang, X.; Hou, F.; Yin, Y.; Liang, J. Recent progress on the recycling technology of Li-ion batteries. *J. Energy Chem.* **2020**, *55*, 391–419. [CrossRef]

10. Steward, D.; Mayyas, A.; Mann, M. Economics and Challenges of Li-Ion Battery Recycling from End-of-Life Vehicles. *Proc. Manuf.* **2019**, *33*, 272–279. [CrossRef]

11. European Commission. Proposal for a Regulation of the European Parliament and of the Council Concerning Batteries and Waste Batteries, Repealing Directive 2006/66/EC and Amending Regulation (EU) No 2019/1020. Available online: https://ec.europa.eu/environment/waste/batteries/pdf/Proposal_for_a_Regulation_on_batteries_and_waste_batteries.pdf (accessed on 17 February 2021).

12. Boyden, A.; Soo, V.K.; Doolan, M. The Environmental Impacts of Recycling Portable Lithium-Ion Batteries. *Proc. CIRP* **2016**, *48*, 188–193. [CrossRef]

13. Gao, W.; Zhang, X.; Zheng, X.; Lin, X.; Cao, H.; Zhang, Y.; Sun, Z. Lithium Carbonate Recovery from Cathode Scrap of Spent Lithium-Ion Battery: A Closed-Loop Process. *Environ. Sci. Technol.* **2017**, *51*, 1662–1669. [CrossRef]

14. Zhan, R.; Payne, T.; Leftwich, T.; Perrine, K.; Pan, L. De-agglomeration of cathode composites for direct recycling of Li-ion batteries. *Waste Manag.* **2020**, *105*, 39–48. [CrossRef]

15. Chen, M.; Ma, X.; Chen, B.; Arsenault, R.; Karlson, P.; Simon, N.; Wang, Y. Recycling End-of-Life Electric Vehicle Lithium-Ion Batteries. *Joule* **2019**, *3*, 2622–2646. [CrossRef]

16. Gao, R.; Sun, C.; Xu, L.; Zhou, T.; Zhuang, L.; Xie, H. Recycling LiNi0.5Co0.2Mn0.3O2 material from spent lithium-ion batteries by oxalate co-precipitation. *Vacuum* **2020**, *173*, 109181. [CrossRef]

17. Sieber, T.; Ducke, J.; Rietig, A.; Langner, T.; Acker, J. Recovery of Li(Ni0.33Mn0.33Co0.33)O$_2$ from Lithium-Ion Battery Cathodes: Aspects of Degradation. *Nanomaterials* **2019**, *9*, 246. [CrossRef]

18. Larouche, F.; Tedjar, F.; Amouzegar, K.; Houlachi, G.; Bouchard, P.; Demopoulos, G.P.; Zaghib, K. Progress and Status of Hydrometallurgical and Direct Recycling of Li-Ion Batteries and Beyond. *Materials* **2020**, *13*, 801. [CrossRef]

19. Zang, G.; Zhang, J.; Xu, S.; Xing, Y. Techno-economic analysis of cathode material production using flame-assisted spray pyrolysis. *Energy* **2021**, *218*, 119504. [CrossRef]

20. Gaines, L. Lithium-ion battery recycling processes: Research towards a sustainable course. *Sustain. Mater. Technol.* **2018**, *17*, e00068. [CrossRef]

21. Sloop, S.E.; Crandon, L.; Allen, M.; Lerner, M.M.; Zhang, H.; Sirisaksoontorn, W.; Gaines, L.; Kim, J.; Lee, M. Cathode healing methods for recycling of lithium-ion batteries. *Sustain. Mater. Technol.* **2019**, *22*, e00113. [CrossRef]

22. Doose, S.; Mayer, J.K.; Michalowski, P.; Kwade, A. Challenges in Ecofriendly Battery Recycling and Closed Material Cycles: A Perspective on Future Lithium Battery Generations. *Metals* **2021**, *11*, 291. [CrossRef]

23. Deutsches Insitut für Normung e.V. *Umweltmanagement—Ökobilanz—Grundsätze und Rahmenbedingungen*; Beuth Verlag GmbH: Berlin, Germany, 2009.

24. Deutsches Insitut für Normung e.V. *Umweltmanagement—Ökobilanz—Anforderungen und Anleitungen*; Beuth Verlag GmbH: Berlin, Germany, 2006.

25. Ministerium für Umwelt, Klima und Energiewirtschaft Baden-Württemberg. *Umweltminister Franz Untersteller Überreicht Förderbescheide für Demontagefabrik "DeMoBat" in Höhe von 13 Millionen Euro*; Ministerium für Umwelt, Klima und Energiewirtschaft Baden-Württemberg: Stuttgart, Germany, 2020.

26. Sphera Solutions, Inc. GaBi Manual. Available online: http://www.gabi-software.com/fileadmin/GaBi_Manual/GaBi_6_manual.pdf (accessed on 27 January 2021).

27. European Commission. *International Reference Life Cycle Data System (ILCD) Handbook: General guide for Life Cycle Assessment—Detailed Guidance*; European Commission: Luxemburg, 2010.

28. Bruijn, H.; Duin, R.; Huijbregts, M.A.J.; Guinee, J.B.; Gorree, M.; Heijungs, R.; Huppes, G.; Kleijn, R.; Koning, A.; Oers, L.; et al. *Handbook on Life Cycle Assessment: Operational Guide to the ISO Standards*; Kluwer Academic Publishers: Dordrecht, The Netherlands, 2004; ISBN 978-0-306-48055-3.

29. Kaltschmitt, M.; Schebek, L. *Umweltbewertung für Ingenieure: Methoden und Verfahren*; Springer Vieweg: Berlin, Germany, 2015; ISBN 978-3-642-36989-6.

30. Buchert, M.; Sutter, J. *Aktualisierte Ökobilanzen zum Recyclingverfahren LithoRec II für Lithium-Ionen-Batterien*; Oeko-Institut e.V.: Berlin, Germany; Darmstadt, Germany, 2016.

31.	Momber, A.W. Image processing as a tool for high-pressure water jet coating removal assessment. *Int. J. Adv. Manuf. Technol.* **2016**, *87*, 571–578. [CrossRef]
32.	Ning, D.; Wang, Q.; Tian, J.; Gong, Y.; Du, H.; Chen, S.; Hou, J. Experimental Study on the Coating Removing Characteristics of High-Pressure Water Jet by Micro Jet Flow. *Micromachines* **2021**, *12*, 173. [CrossRef]
33.	Buchert, M.; Jenseit, W.; Merz, C.; Schüler, D. *Verbundprojekt: Entwicklung Eines Realisierbaren Recyclingkonzepts für die Hochleistungsbatterien Zukünftiger Elektrofahrzeuge—LiBRi: Teilprojekt: LCA der Recyclingverfahren*; Oeko-Institut e.V.: Berlin, Germany; Darmstadt, Germany, 2011.
34.	Buchert, M.; Sutter, J. *Aktualisierte Ökobilanz zum Recyclingverfahren EcoBatRec für Lithium-Ionen-Batterien*; Oeko-Institut e.V.: Berlin, Germany; Darmstadt, Germany, 2016.
35.	Accardo, A.; Dotelli, G.; Musa, M.L.; Spessa, E. Life Cycle Assessment of an NMC Battery for Application to Electric Light-Duty Commercial Vehicles and Comparison with a Sodium-Nickel-Chloride Battery. *Appl. Sci.* **2021**, *11*, 1160. [CrossRef]
36.	Dai, Q.; Kelly, J.C.; Gaines, L.; Wang, M. Life Cycle Analysis of Lithium-Ion Batteries for Automotive Applications. *Batteries* **2019**, *5*, 48. [CrossRef]

Article

Effect of Vinylene Carbonate Electrolyte Additive on the Surface Chemistry and Pseudocapacitive Sodium-Ion Storage of TiO₂ Nanosheet Anodes

Rudi Ruben Maça [1,2] and Vinodkumar Etacheri [1,*]

1   Electrochemistry Division, IMDEA Materials Institute, 28906 Madrid, Spain; rudi.maca@imdea.org
2   Faculty of Science, Universidad Autónoma de Madrid, 28049 Madrid, Spain
*   Correspondence: vinodkumar.etacheri@imdea.org

Abstract: Although titanium dioxide has gained much attention as a sodium-ion battery anode material, obtaining high specific capacity and cycling stability remains a challenge. Herein, we report significantly improved surface chemistry and pseudocapacitive Na-ion storage performance of TiO₂ nanosheet anode in vinylene carbonate (VC)-containing electrolyte solution. In addition to the excellent pseudocapacitance (~87%), the TiO₂ anodes also exhibited increased high-specific capacity (219 mAh/g), rate performance (40 mAh/g @ 1 A/g), coulombic efficiency (~100%), and cycling stability (~90% after 750 cycles). Spectroscopic and microscopic studies confirmed polycarbonate based solid electrolyte interface (SEI) formation in VC-containing electrolyte solution. The superior electrochemical performance of the TiO₂ nanosheet anode in VC-containing electrolyte solution is credited to the improved pseudocapacitive Na-ion diffusion through the polycarbonate based SEI (coefficients of 1.65×10^{-14} for PC-VC vs. 6.42×10^{-16} for PC). This study emphasizes the crucial role of the electrolyte solution and electrode–electrolyte interfaces in the improved pseudocapacitive Na-ion storage performance of TiO₂ anodes.

Keywords: electrolyte; additive; interface; pseudocapacitance; intercalation; energy storage; secondary battery; sodium-ion

Citation: Maça, R.R.; Etacheri, V. Effect of Vinylene Carbonate Electrolyte Additive on the Surface Chemistry and Pseudocapacitive Sodium-Ion Storage of TiO₂ Nanosheet Anodes. *Batteries* **2020**, 7, 1. https://dx.doi.org/10.3390/batteries7010001

Received: 20 November 2020
Accepted: 21 December 2020
Published: 24 December 2020

Publisher's Note: MDPI stays neutral with regard to jurisdictional claims in published maps and institutional affiliations.

1. Introduction

Lithium-ion batteries (LIBs) are widely used in the current generation of portable electronics, electric vehicles, and smart grids coupled with renewable energy sources [1–4]. Despite their high energy density, moderate power density, and good cycle life, there are concerns regarding the future large-scale implementation of LIBs due to the limited availability of expensive ($17,000 per metric ton) lithium resources, where market demand will be up to two to six times extraction capacity in the next two decades [5,6]. Rechargeable sodium-ion batteries (SIBs) are very attractive in this regard, due to the abundance of inexpensive sodium resources [3,5,7,8]. Moreover, the similar electrochemistry and redox potentials of lithium and sodium (−3.02 and −2.71 V vs. SHE, respectively), make it a suitable candidate for efficient electrochemical energy storage [9–11]. However, the larger size of Na-ions compared to Li-ions (1.02 and 0.76 Å radius, respectively) hinders their intercalation in the most commonly used Li-ion battery anode material, graphite (interlayer d-spacing of 3.4 Å) [1,12,13]. Shuttling of solvated Na-ions between individual electrodes is also sluggish, leading to the poor rate performance and cycling stability of Na-ion batteries [14]. Hence, numerous studies have focused on the development of alternative high-performance anode materials [15–20].

Carbonaceous materials with large interlayer spacing, such as hard carbon, graphene, and amorphous carbon have been widely investigated as anode materials [16,17]. However, these anodes exhibited low specific capacities, poor rate performances, and mediocre cycling stability. Despite the high specific capacity of conversion (Co₃O₄, SnO₂, Fe₃O₄,

NiO, CuO, MnO$_2$, etc.) and alloying type (Sn, Ge, Sb, etc.) anodes, rapid capacity fading associated with huge volume changes make them less attractive for Na-ion battery applications [18–20]. Insertion type metal oxides, such as Nb$_2$O$_5$, Na$_2$Ti$_3$O$_7$, TiO$_2$, and Li$_4$Ti$_5$O$_{12}$ are another class of Na-ion battery anodes [21,22]. Titanium dioxide (TiO$_2$) has received much attention among anode materials due to its high chemical stability, non-toxicity, abundance, low volume change during Na-ion intercalation, and inexpensive nature [23–26]. Amorphous and different crystalline (anatase, bronze, rutile, etc.) polymorphs of TiO$_2$ have been investigated as promising Na-ion battery anodes [7,27–30]. The existence of 2D intercalation channels in crystalline polymorphs make them superior to amorphous TiO$_2$ for Na-ion storage [4,9,31]. Hence, anatase TiO$_2$ composed of TiO$_6$ octahedra and zigzag edges of the 3-D network have become the most studied polymorph [4,32]. Bronze polymorph of TiO$_2$ has recently attracted a lot of attention as a Na-ion battery anode due to the presence of more open channels and layered crystal structure [29,33]. The Na-ion intercalation kinetics' dependence on crystal structure and orientation has also been proved in recent studies [22,33]. Nevertheless, TiO$_2$ anodes suffer from low electronic conductivity, mediocre specific capacity, poor rate-performance, and low cycling stability [25,34].

Numerous strategies such as carbon coating, doping with other transition metal ions, and the synthesis of different morphologies in the nanoscale have been established for improving the Na-ion storage performance of TiO$_2$ anodes [23,26,34]. Nevertheless, these methods have only resulted in a trivial improvement of specific capacities and cycling stabilities. Another advanced approach to improve the electrochemical performance is to increase the diffusion independent pseudocapacitive-type Na-ion storage [14,21]. This surface/near-surface ion storage is independent of the electronic and ionic conductivity of the electrode material [21]. In addition to the excellent rate performance and cycling stabilities, due to negligible structural changes during the charge–discharge process, high specific capacities can also be achieved due to the synergy with diffusion limited Na-ion storage [30,35]. Improved pseudocapacitance can usually be achieved by either precise nanostructuring, or the formation of hybrids with carbonaceous materials. These approaches were recently demonstrated for enhancing the intrinsic (~4%) pseudocapacitive Na-ion storage of TiO$_2$ anodes [22]. For instance Chen et al. demonstrated the excellent electrochemical performance of TiO$_2$-nanosheets grown on RGO [36]. Our recent study demonstrated the significantly improved pseudocapacitance of nanointerface engineered anatase-bronze hybrid TiO$_2$ nanosheets [9]. The formation of additional Na-ion diffusion pathways, and efficient charge separation were responsible for the improved pseudocapacitance of these electrodes [37].

Solid electrolyte interface (SEI) characteristics are crucial in deciding the electrochemical performance, in addition to the physiochemical properties of an electrode material [38,39]. Carefully engineered electrode–electrolyte interfaces can enable producing batteries with superior energy/power densities and cycling stabilities [38,40]. Since SEIs are formed as a result of electrolyte decomposition, the most appropriate approach to engineer SEI properties is to tune the electrolyte composition [39,40]. Carbonate based electrolyte compositions usually contain solvents such as ethylene carbonate (EC), propylene carbonate (PC), ethyl methyl carbonate (EMC), etc., and a suitable sodium salt (NaPF$_6$, NaTFSI, NaClO$_4$, etc.). Electrochemical decomposition of these electrolytes results in the formation of an electrode–electrolyte interface composed of sodium alkyl carbonates, sodium alkoxides, polyenes, polycarbonates, and inorganic salts [5,38,41,42]. Excess electrolyte decomposition, due to the extreme reactivity of high surface area nanostructured electrodes, can often lead to increased SEI formation, which deteriorates the Na-ion storage performance [24,39]. Only a thin SEI layer capable of protecting the electrode from continuous reaction with the electrolyte solution is necessary to maintain optimal Na-ion diffusion, and a high degree of reversibility [43]. The chemical stability of SEI components is also critical to prevent dissolution in the electrolyte solution, which can lead to continuous SEI formation at the expense of electrolyte decomposition, and cell failure [38,41,44].

Using electrolyte additives is one widely employed strategy to endow the SEIs with superior electrochemical properties [40,44]. Vinylene carbonate (VC) is one of the most commonly used electrolyte additives in the case of Li-ion batteries [41,44,45]. This has been mainly aimed at flexible SEI formation in large volume expansion conversion/alloying type anodes, and at preventing metal-ion leaching from high voltage cathodes [38,42,46]. The high electrochemical reactivity of VC results in the preferential formation of polymeric SEIs, which prevent further reaction of the electrode material with the electrolyte solution during extreme operating environments, including high temperature, and large volume change [40,44,47]. Use of VC in the case of Na-ion batteries is also mainly aimed at improving the cycling stability of high capacity conversion/alloying type anodes [46]. However, none of these studies investigated SEI composition tuning to enhance the pseudocapacitive Na-ion storage of TiO_2 anodes.

Herein, we report significantly improved pseudocapacitive Na-ion storage and surface chemistry of TiO_2 nanosheet anodes in VC-containing electrolyte solution. Vinylene carbonate, a widely used Li-ion battery additive was chosen in this case due to its ability to form SEIs with superior passivation and Na-ion diffusion properties. The presence of a double bond is mainly responsible for the preferential decomposition/polymerization of VC associated with the generation of polycarbonate based SEIs. Ultrathin, mesoporous, and high surface area TiO_2 nanosheets were selected as the preferred anode material in this case, to improve the contact with the electrolyte solution and thereby amplify the effect of the electrolyte additive on the surface chemistry and electrochemical performance. The excellent electrochemical performance of TiO_2 nanosheet anode in VC-containing electrolyte solution is credited to the superior pseudocapacitance, resulting from the faster Na-ion diffusion through the polycarbonate based surface film.

Figure 1 shows the different solid electrolyte interface (SEI) composition on TiO_2 nanosheet anodes formed in vinylene carbonate (VC)-containing and VC-free electrolyte solutions.

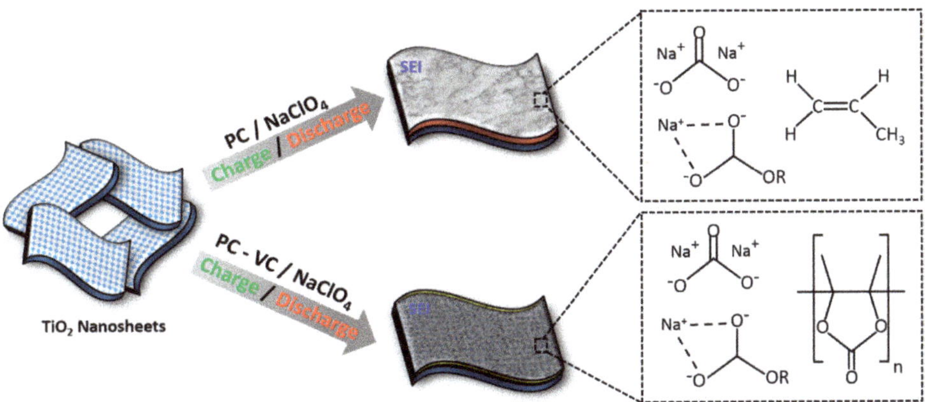

Figure 1. Schematic of the different solid electrolyte interface (SEI) composition on TiO_2 nanosheet anodes formed in vinylene carbonate (VC)-containing and VC-free electrolyte solutions.

2. Experimental

2.1. Materials Synthesis

TiO_2 nanosheets were synthesized through a solvothermal method [48]. In a typical synthesis, 4 mL of deionized water and 20% aqueous $TiCl_3$ (99.99%, Acros Organics) each were added dropwise in 50 mL of ethylene glycol (99.99%, Fisher Scientific) under continuous stirring. The dark brown solution was then heated to 150 °C for 6 h in a PTFE-lined stainless steel autoclave. The subsequent slurry was washed several times with deionized water and ethanol, and dried at 80 °C for 24 h to obtain bronze TiO_2 (TiO_2-B) nanosheets. Hierarchical anatase-bronze hybrid TiO_2 nanosheets were obtained by heat-treating TiO_2-B nanosheets at 400 °C for 2 h under air flow (heating and cooling rates of 10 °C/min).

2.2. Materials Characterization

X-ray diffraction (XRD) patterns of the samples were collected using a PANalytical Empyrean high-resolution diffractometer equipped with a Cu-Kα X-ray source (λ = 1.5406 Å). The average crystallite size of the TiO_2 nanosheets was calculated using the Debye–Scherrer equation ($D = K \lambda/\beta \cos\theta$), where D is particle size, K is the shape factor 0.9, λ is the wavelength of the X-ray radiation of Cu-Kα, and β is the full width at half maximum (FWHM) of the highest intensity peak. Room temperature Raman analysis was performed with a Renishaw PLC Raman spectrometer equipped with a 532 nm Nd: YAG laser. Sample damage was avoided by limiting the laser power to 5 mW. FTIR spectra of pristine and cycled electrodes after washing with anhydrous acetonitrile (Sigma-Aldrich, 99.9%) were recorded using a Thermo Scientific Nicolet iS50 in attenuated total reflectance (ATR) mode, in the range of 4000–400 cm^{-1}. FTIR spectra of pristine, cycled and non-washed electrodes were also collected for comparison. Microstructural characterization of TiO_2 nanosheets was performed using a scanning electron microscope (FEI Helios NanoLab 600i) and high-resolution transmission electron microscope (FEI Talos F200X FEG). X-ray photoelectron spectroscopy (XPS) measurements were performed using a SPECS PHOIBOS 150 9MCD instrument equipped with a Multi-Channeltron detector and a monochromatic X-ray source of twin Mg anodes. The binding energies of elements in the XPS spectra were set accordingly to the CC/CH component of the C 1s peak at 284.8 eV. For the quantitative analysis of SEI, high-resolution core-level spectra were used after removing the nonlinear Shirley background.

2.3. Electrochemical Measurements

A Vigor glovebox filled with high-purity ($\geq$99.999%) Ar-gas (H_2O and O_2 < 1.0 ppm) was used to fabricate 2032 type Na-ion half-cells. Metallic sodium foil was used for the counter and reference electrodes. The separator consisted of Whatman glass fiber (GF/B type) soaked in 1 M $NaClO_4$ ($\geq$98.0%, Sigma-Aldrich,), in propylene carbonate (PC, 99.7%, Sigma-Aldrich), and 3% vinylene carbonate (VC, 97%, Sigma-Aldrich). Working electrodes contained 70wt.% TiO_2 nanosheets, 20wt.% acetylene black (MTI Chemicals), and 10wt.% of polyvinylidene fluoride (PVDF, MW: 600.000, MTI Chemicals) binder. N-methylpyrrolidone (NMP, 99.9%, Aladdin Chemicals) solvent was used for preparing the slurry containing the active material, conductive additive, and binder, followed by coating on a 10 μm thick copper foil. Working electrodes consisted of an active material loading of 2–3 mg/cmA Neware BTS4000 model multichannel battery tester was employed to galvanostatically cycle the assembled half-cells between 0.01–3.0 V (vs. Na$^{+/0}$) at 25 °C. Specific capacities were expressed, calculated according to the active material weight, within a 3% error limit. A ZIVE SP1 electrochemical workstation was used for electrochemical impedance spectroscopy (EIS) measurements (1 MHz to 10 mHz) and for cyclic voltammetry (CV) measurements (0.1 to 100 mV/s). EIS measurements were performed using a 3-electrode coin-cell configuration. Diffusion-controlled and diffusion-independent processes were calculated according to Equation (1).

$$i(V) = k_1 v + k_2 v^{\frac{1}{2}} \tag{1}$$

$k_1 v$ and $k_2 v$ represent current contributions for diffusion-independent and diffusion-controlled processes, respectively. k_1 and k_2 coefficients are the slope and intercept from the linear fit of the plot $i(V)$ vs. v, respectively. Na-ion diffusion coefficients of TiO_2 nanosheets were determined using Equation (2).

$$D = \frac{1}{2}\left(\frac{RT}{AF^2\sigma_w C}\right)^2 \tag{2}$$

where D is Na-ion diffusion coefficient at absolute temperature T, and R denotes the gas constant. Faraday's constant, Na-ion concentration in electrolyte, and electrode area are represented as F, C and A, respectively. Warburg impedance, σ_w, was calculated from the slope of the linear plot of Z′ vs. $\omega^{-1/2}$.

3. Results and Discussion

3.1. Synthesis and Characterization of TiO_2 Nanosheets

Hierarchical TiO_2 nanosheets composed of anatase and bronze nanocrystallites were synthesized through a solvothermal method, followed by controlled heat treatment. Ethylene glycol was used as solvent to facilitate nanosheet formation through the interaction of OH groups of glycol with Ti-OH groups [48,49]. This resulted in the free energy change of the TiO_2 crystallographic planes and an associated anisotropic crystal growth [29,49]. Selection of a less reactive precursor $TiCl_3$ was aimed at controlled hydrolysis and condensation, which is necessary for the nanosheet formation [50]. The uniform size/shape distributions and ultrathin nature of the sheets were also assisted by the high viscosity of ethylene glycol [9]. Reaction between $TiCl_3$ and ethylene glycol resulted in the formation of Ti-glycolate complex during the hydrothermal reaction. Further hydrolysis and condensation formed TiO_2 bronze nanosheets, which was transformed to anatase-bronze hybrid nanosheets upon controlled heat treatment. Scanning electron microscopy (SEM) images of the TiO_2 nanosheets demonstrated their 2D morphology and (Figure 2a) hierarchical flower like microstructure, composed of numerous petals. High-resolution transmission electron microscopy (TEM) images (Figure 2b,c) of individual nanosheets verified the existence of nanograins of 7 ± 2 nm size. Lattice spacings of 0.34 and 0.62 nm represent the (101) and (110) planes of anatase and bronze phases, respectively. The presence of well-defined anatase-bronze nanograin boundaries is also evident from these images [9]. This was anticipated, due to the mismatch between anatase and bronze crystal structures. Nanograin boundaries present in metal oxides often result in unique physical/ chemical properties, including excellent catalytic and electrochemical performance [51]. Nanointerfaces present in these hierarchical TiO_2 nanosheets can also act as additional sites for Na-ion storage. Our previous study confirmed the advantages of such nanointerfaces for enhancing pseudocapacitive type Na-ion storage [9].

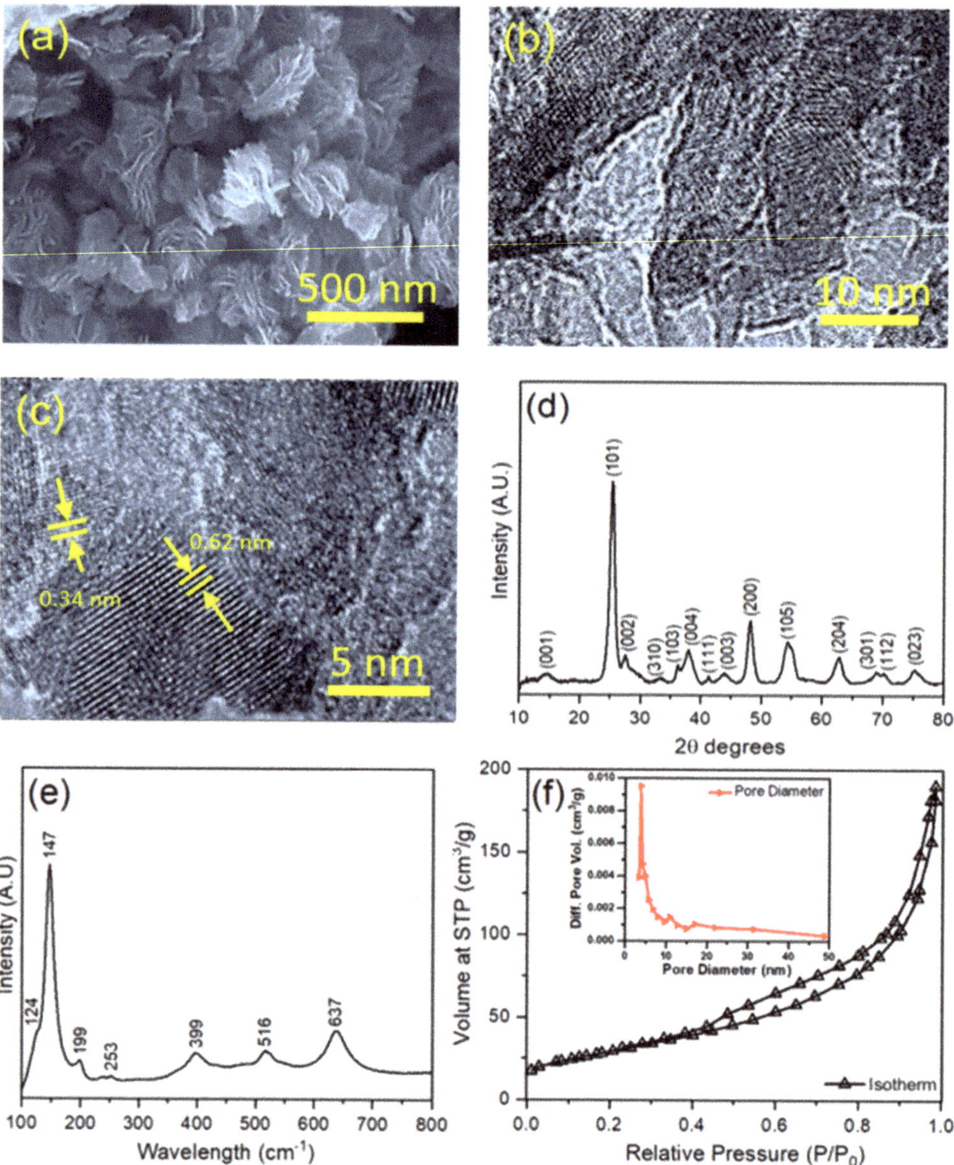

Figure 2. (**a**) SEM image, (**b**,**c**) TEM images, (**d**) X-ray diffraction pattern, (**e**) Raman spectrum, (**f**) N_2 adsorption–desorption isotherm, and pore size distribution (inset) of TiO_2 nanosheets.

The X-ray diffraction pattern of the TiO_2 nanosheets (Figure 2d) also confirmed the coexistence of anatase and bronze polymorphs in the sample. Peaks designated as (101), (004), (200), (105), (211), (204) corresponded to anatase phase (JCPDS 21–1272). Whereas, (001), (110), (002), (310), (103), (003), and (204) signals represented bronze phase (JCPDS 46-1237) [9]. The anatase and bronze content, calculated from the relative intensity of the (310) bronze peak and (004) anatase peak, was found to be 83% and 17%, respectively. Particle size calculation, using a Debye–Scherrer equation also revealed the existence of

7 ± 2 nm sized crystallites, which was in good agreement with the TEM results. A more surface sensitive technique, Raman spectroscopy was used to further confirm the phase purity and uniform nanoscale distribution of the anatase and bronze polymorphs (Figure 2e). Raman active modes were 144, 197, 399, 514, and 639 cm^{-1} for anatase, and 123, 145, 161, 172, 196, 201, and 259 cm^{-1} for bronze polymorphs [52]. These results and the quantification of anatase/bronze content from high-resolution Raman spectra (Figure S1) were also in line with the XRD results, confirming the coexistence and nanoscale distribution of anatase and bronze nanocrystallites [52]. The textural property investigation of TiO$_2$ nanosheets was performed through N$_2$ adsorption–desorption analysis (Figure 2f). This active material exhibited a high surface area of 106 m^2/g. Type IV isotherms and H3 type hysteresis, characteristic of a mesoporous structure were identified in this case [13]. Mesoporosity was also evidenced by the higher steepness of the isotherm at high relative pressure (P/P0 = 0.4–1.0) [13]. Barrett-Joyner-Halenda (BJH) method pore size distribution measurements (Figure 2f inset) further confirmed the mesoporosity of the TiO$_2$ nanosheets. Such high surface area and mesoporosity of the active material are advantageous for improved Na-ion storage due to the superior contact with the electrolyte solution, and the possibility of pore/defect assisted pseudocapacitive type ion storage [49]. Thus, it can be summarized that controlled hydrolysis and condensation of TiCl$_3$ in ethanol water mixture followed by calcination resulted in the formation of mesoporous hierarchical anatase–bronze hybrid TiO$_2$ nanosheets.

3.2. Electrochemical Performance of the TiO$_2$ Nanosheets

The sodium-ion storage electrochemical performance of the TiO$_2$ nanosheets was investigated in VC-containing (PC-VC/NaClO$_4$) and VC-free (PC/NaClO$_4$) electrolyte solutions. Second galvanostatic voltage profiles at different current densities (Figure 3a,b) indicated the marginally superior specific capacities of the TiO$_2$ nanosheet anodes in PC-VC/NaClO$_4$ electrolyte at various charge–discharge rates. First cycle coulombic efficiencies of 36% and 44.5% were observed in the PC/NaClO$_4$ and PC-VC/NaClO$_4$ electrolyte solutions, respectively (Figure S2). Irreversible capacity loss in the first cycle was attributed to unavoidable electrolyte decomposition, and irreversible Na$_2$O formation associated with sodiation. Coulombic efficiency reached ~100% after the first charge–discharge in PC-VC/NaClO$_4$ electrolyte, demonstrating the complete formation of high quality SEI during the first cycle. However, coulombic efficiency never reached ~100% on extended cycling in PC/NaClO$_4$ electrolyte, which indicated the formation of an SEI with inferior passivation properties. This is also apparent from the consecutive cyclic voltammograms presented (Figure 3c,d). The TiO$_2$ nanosheet anodes exhibited a similar electrochemical response, irrespective of the electrolyte composition. Three different electrochemical processes constitute the first cathodic response in both systems. The high-voltage region between 2.25 and 0.7 V represents diffusion-independent pseudocapacitive Na-ion intercalation [53]. SEI formation resulting from the electrolyte decomposition could be identified from the cathodic response between 0.7 and 0.5 V [9]. Lack of this signal in the second cathodic scan confirmed complete SEI formation in the first charge–discharge cycle [32,54]. A low voltage region between 0.2 and 0 V corresponded to the diffusion controlled Na-ion intercalation into the anatase and bronze crystal structure [36]. These electrochemical responses were in line with the plateau at low potential and sloping voltage profiles. Anodic response in the voltage range of 0.2 V, and a broad signal between 0.5 and 2.25 V represented Na-ion deintercalation through pseudocapacitive and diffusion dependent process, respectively.

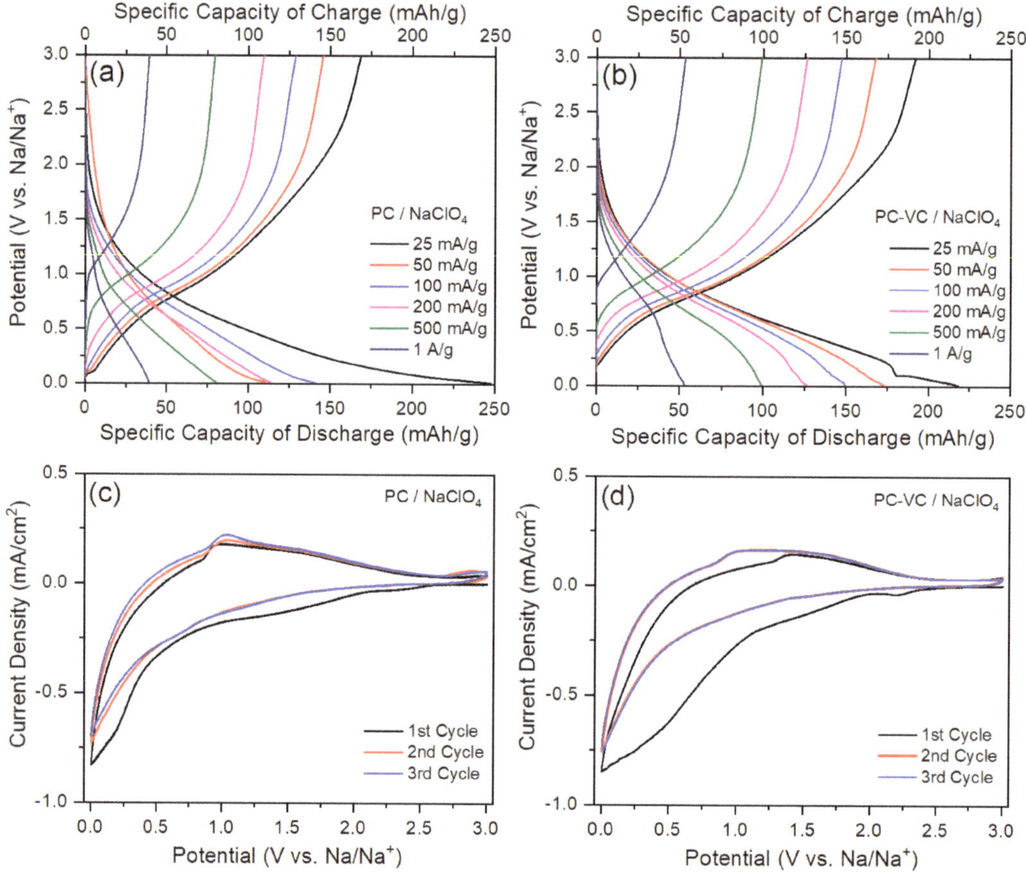

Figure 3. Galvanostatic voltage profiles of TiO$_2$ nanosheet anodes in Na-ion half-cells containing: (**a**) PC/NaClO$_4$, and (**b**) PC-VC/NaClO$_4$ electrolyte solutions at various current densities. Cyclic voltammograms of TiO$_2$ nanosheet anodes at a scan rate of 1 mV/s in (**c**) PC/NaClO$_4$, and (**d**) PC-VC/NaClO$_4$ electrolyte solution.

Sodiation over a wide voltage range was observed in this case, which was noticeably different from the intercalation reaction at specific potentials for previously reported TiO$_2$ anodes [27]. Distinct Na-ion storage was also evidenced by the marginally different shape of the voltammograms compared to earlier reports [37]. Although cathodic signals corresponding to SEI formation disappeared in the first cycle, there was a clear indication of irreversibility and incomplete SEI formation during the consecutive cycles of the VC-free battery, which was not observed in the case of the VC-containing Na-ion half-cells. This was a clear indication of efficient SEI formation on the TiO$_2$ nanosheet anodes in the VC-containing electrolyte solution. Thus, disappearance of the SEI formation peak in the first cycle and stable consecutive cycles confirmed the highly reversible Na-ion intercalation reactions of the TiO$_2$ nanosheet anodes in VC-containing electrolyte solution. The galvanostatic rate performance of the TiO$_2$ nanosheet anodes in both VC-free and VC-containing electrolyte solutions is presented in Figure 4a. Second cycle specific capacities were 247 and 219 mAh/g, respectively. Slightly higher capacities were identified in the case of VC-containing electrolyte solution during the following cycles at 25 mA/g. Capacity differences became more prominent at higher current densities, and also increased on reducing the current density to 25 mA/g. This was a clear indication of the capacity fading of TiO$_2$ anode in VC-free electrolyte, which was further verified during the extended galvanostatic cycling (Figure 4b).

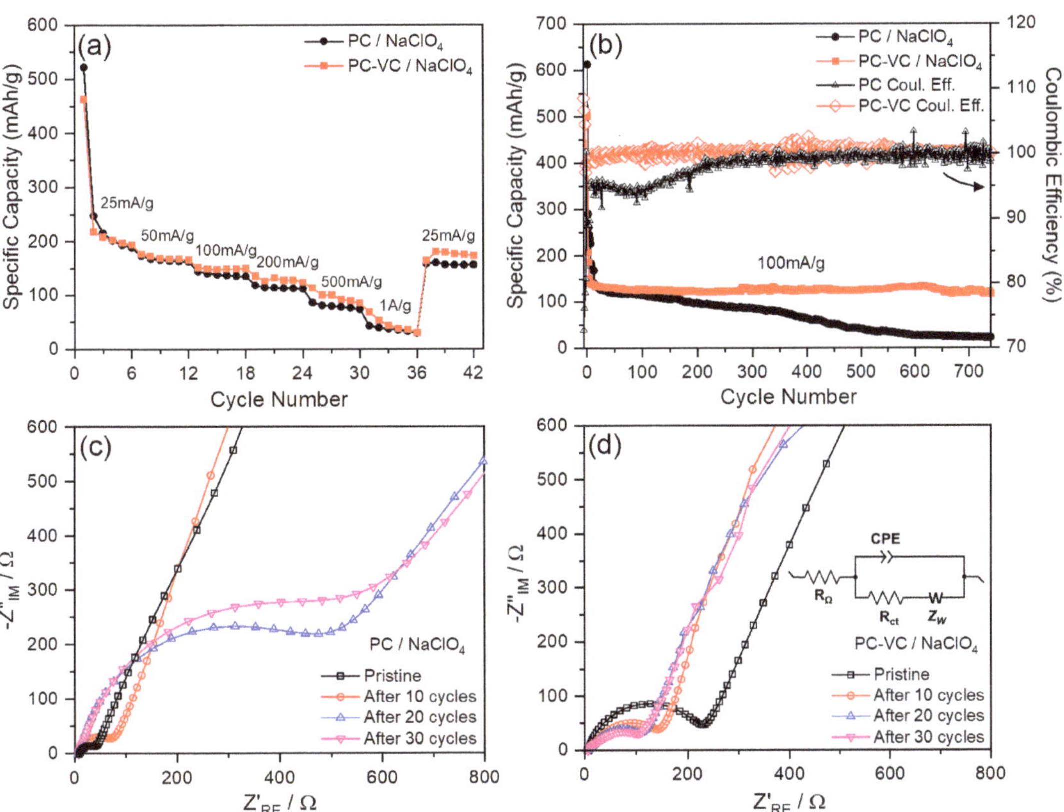

Figure 4. (**a**) Galvanostatic rate performance, and (**b**) cycling performance of TiO$_2$ nanosheet anodes in PC/NaClO$_4$ and PC-VC/NaClO$_4$ electrolyte solutions. Nyquist plots and equivalent circuit (inset) of TiO$_2$ nanosheets in (**c**) PC/NaClO$_4$ and (**d**) PC-VC/NaClO$_4$ electrolyte solutions.

Both cells demonstrated ~100% coulombic efficiency after 750 cycles at a current density of 100 mA/g. Galvanostatic long-cycling performance of electrolyte solutions containing different amounts of VC confirmed 3% as the optimum concentration (Figure S3). For instance, at a current density of 100 mA/g, TiO$_2$ nanosheet anode in Na-ion half-cell configuration retained 90% of the initial capacity, with ~100% coulombic efficiency after 750 galvanostatic charge–discharge cycles. On the other hand, Na-ion half-cells containing VC-free electrolyte solution retained only 16% of the initial capacity, with ~98% coulombic efficiency.

Electrochemical impedance spectroscopy (EIS) of pristine and cycled electrodes (Figure 4c,d) is performed to gain further insights into the effect of SEI films on the charge transfer, and Na-ion diffusion characteristics. High-to-medium frequency part of the EIS pattern can be assigned to the sum of SEI/contact resistance and charge-transfer resistance and represented as R_Ω and R_{ct} respectively in the Randles-like equivalent circuit (Figure 4d inset). Low frequency sloping line corresponds to solid-state diffusion kinetics of Na-ions shown as Warburg impedance (Z_w) while constant phase element (CPE) is used to model the surface storage of Na-ions [7,49,55]. Charge transfer resistances of pristine TiO$_2$ nanosheet anodes in VC-free and VC-containing electrolyte solutions, obtained by fitting the Nyquist plot to the equivalent circuit, are 39 Ω and 171 Ω respectively. It is interesting to observe such impedance difference in pristine batteries as VC is supposed to influence SEI formation only during the first charge-discharge cycle. This could be related to the spontaneous SEI formation resulting from the increased reactivity of high surface area TiO_2 nanosheets with the electrolyte solution even in the absence of applied potential. Such surface film formations were also observed previously in the case of pyrolytic graphite, carbon and silicon nanowire electrodes [41,56,57]. Charge transfer resistance in VC-free electrolyte solution increased to 517 Ω after 30 galvanostatic cycles. In contrast, resistance decreased to 105 Ω in the case of VC-containing half-cells. Na-ion diffusion coefficients of TiO_2 nanosheet anodes after 30 galvanostatic cycles in VC-free and VC containing electrolyte solutions calculated from the Nyquist plot are 6.42×10^{-16} and 1.65×10^{-14} correspondingly. Hence it is clear that VC-addition significantly reduces the charge transfer resistance and enhances the Na-ion diffusion kinetics, facilitating improved Na-ion diffusion into the TiO_2 nanosheets. These results are also in good agreement with the superior cycling performance of TiO_2 nanosheet electrodes in VC-containing electrolyte solution.

The sodium-ion storage mechanism of TiO_2 nanosheets in VC-free and VC-containing electrolyte solutions was further investigated by collecting cyclic voltammograms at various scan rates (Figure 5a,b). Considerable differences existed in the voltammograms in the two different electrolyte solutions. High peak current and area were identified for the TiO_2 nanosheets in the VC-containing electrolyte, which is in line with their improved Na-ion storage performance observed from the galvanostatic rate performance and cycling studies. The shapes of the voltammograms were also different suggesting different Na-ion storage mechanisms. Anodic and cathodic current increased with an increase of scan rate in both electrolyte solutions, which is indicative of pseudocapacitive type Na-ion storage [21].

Sodium ion intercalation of TiO_2 nanosheets, especially at higher scan rates, occurs over a wide voltage range in VC-containing electrolyte solution. Such broad signals indicate the increased diffusion independent behavior of Na-ion intercalation, which is typical in capacitive type electrode materials [58].

The sodium-ion storage of TiO_2 nanosheet anodes consists of both diffusion limited and surface controlled process [59]. These components can be differentiated from the relationship between peak current density and scan rates, as presented in Equation (3) [21]

$$i = av^b \tag{3}$$

where current shown as i, the scan rate as v, and a and b are adjustable variables. If b = 0.5, storage process is identified as diffusion-controlled, and surface controlled if b = 1.0 [60]. Moreover, as the Randles–Ševcik equation dictates, the increase of current is proportional to the square root of scan rate [55,61,62]. The slope of peak current density vs. scan rate plot (Figure 5c,d) represented dominant pseudocapacitive type Na-ion storage (increased slope) in TiO_2 nanosheets, with a high degree of pseudocapacitance in the PC-VC/$NaClO_4$ electrolyte solution.

The diffusion limited and pseudocapacitive contributions of specific capacities were quantified to further investigate the effect of electrolyte solutions on the Na-ion storage mechanism. Capacitive and diffusion controlled contributions of Na-ion half-cells containing PC/$NaClO_4$ and PC-VC/$NaClO_4$ electrolyte solutions at different scan rates are presented in Figure 6a,b. TiO_2 nanosheet anode exhibited superior pseudocapacitive Na-ion storage in PC-VC/$NaClO_4$ electrolyte solution (83% @ 1 mV/s) compared to PC/$NaClO_4$ (63% @ 1 mV/s). Capacitance calculations were also performed after 30 galvanostatic cycles (Figure 6c,d) in order to investigate the capacity fading on extended cycling. The pseudocapacitance contribution of the TiO_2 nanosheets reduced considerably (37% @ 1 mV/s), and slightly increased (87% @ 1 mV/s) in the PC/$NaClO_4$ and PC-VC/$NaClO_4$ electrolyte solutions, respectively (Figure 6e,f). The entirely different shape of the cyclic voltammograms also signified different Na-ion storage mechanisms in these TiO_2 nanosheet anodes during extended cycling. It is worth noting that these results are in agreement with the EIS results presented earlier, where the impedance of the TiO_2 anodes continually increased during galvanostatic cycling in PC/NaClO$_4$, and decreased in the PC-VC/NaClO$_4$ electrolyte solutions. These observations led to the conclusion that the SEI formed in the VC-containing electrolyte solutions facilitated improved pseudocapacitive Na-ion diffusion into the TiO$_2$ nanosheet anode. Hence, it is clear that the electrolyte composition played a crucial role in the cycling stability, impedance, and Na-ion storage mechanism of the TiO$_2$ nanosheet anodes.

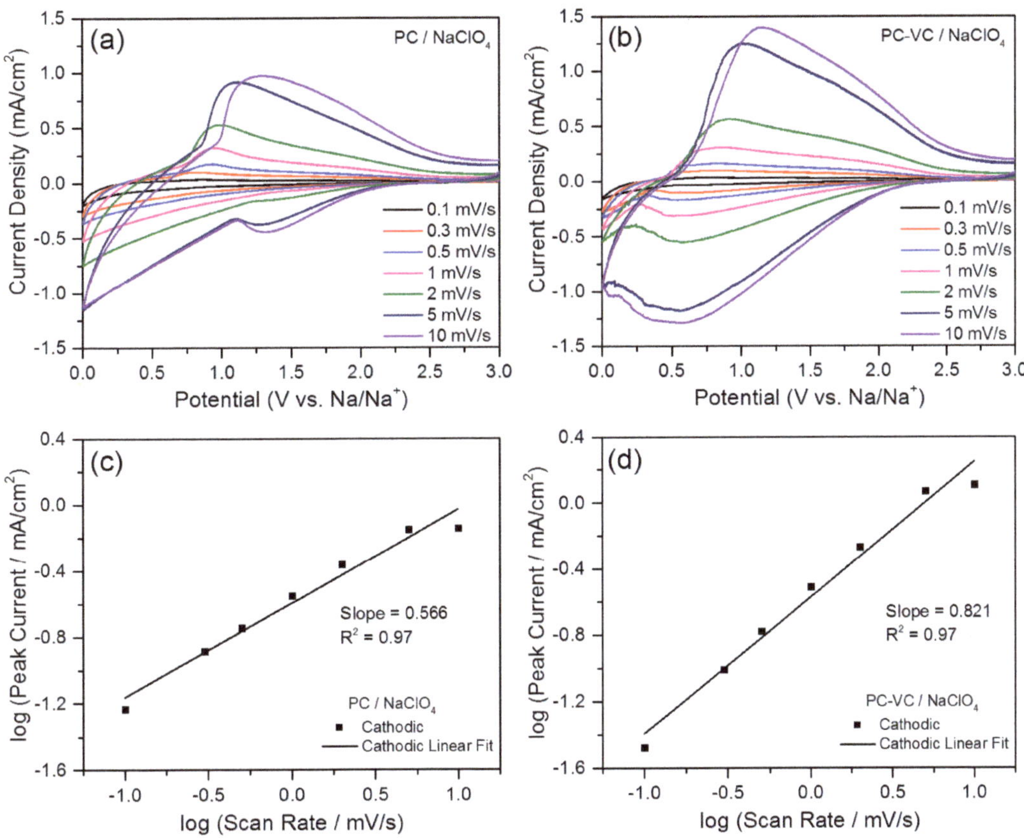

Figure 5. Cyclic voltammograms of TiO$_2$ nanosheet anodes at various scan rates in (**a**) PC/NaClO$_4$ and, (**b**) PC-VC/NaClO$_4$ electrolyte solutions. Scan rate dependence of peak current density for TiO$_2$ nanosheet anodes in (**c**) PC/NaClO$_4$, and (**d**) PC-VC/NaClO$_4$ electrolyte solutions.

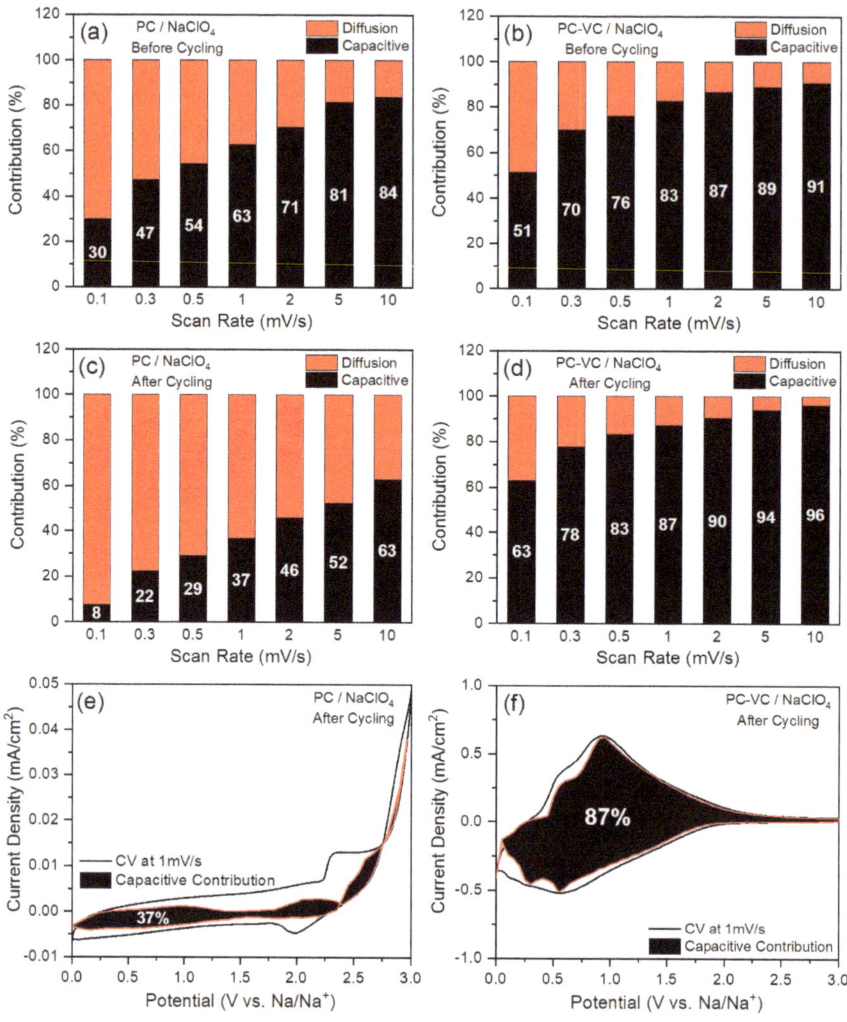

Figure 6. Pseudocapacitance contributions of the TiO_2 nanosheet anodes (**a**,**b**) before cycling, and (**c**,**d**) after cycling in PC/NaClO$_4$ and PC-VC/NaClO$_4$ electrolyte solutions. Pseudocapacitance contributions at a scan rate of 1 mV/s after 30 cycles in (**e**) PC/NaClO$_4$, and (**f**) PC-VC/NaClO$_4$ electrolyte solutions.

3.3. Surface Chemical Characterization of TiO$_2$ Nanosheets

It is crucial to understand the surface chemistry of TiO_2 nanosheet electrodes cycled in VC-free and VC-containing electrolyte solutions due to the significant role of surface films on electrochemical performance. Surface chemical characterizations of these electrodes cycled in PC/NaClO$_4$ and PC-VC/NaClO$_4$ electrolyte solutions were performed using ex-situ XPS and FTIR techniques. Electrodes were thoroughly washed with acetonitrile and dried under vacuum to make sure that there was no residual NaClO$_4$ salt and solvent left to interfere with these spectroscopic measurements. The quantification of elements present in the surface films formed in VC-free and VC-containing electrolyte solutions (Table 1) indicated the contribution of both solvent and Na-salt towards SEI formation. Higher oxygen and carbon concentrations found in the surface film confirmed solvent reduction as the major reaction responsible for SEI formation. Surface films formed in VC-containing electrolyte had a higher carbon and oxygen content, signifying the increased formation of numerous oxygen containing species, such as Na-carbonate, Na-alkyl carbonate, Na-alkoxides, and Na-polycarbonates, due to PC and VC reduction [63]. Lower Na and Cl content in this case was a clear indication of reduced Na-salt decomposition, which is vital to maintain good ionic conductivity during extended cycling. Increased Ti-content signifies the formation of surface films that allow superior X-ray/ electron penetration and detection of Ti-atoms underneath.

Table 1. Quantification of elements present in the TiO_2 nanosheet electrode surface film after 30 cycles in VC-containing and VC-free electrolyte solutions (based on XPS measurements).

Composition	C Content (at.%)	O Content (at.%)	Cl Content (at.%)	Na Content (at.%)	Ti Content (at.%)
$PC/NaClO_4$	21.101	33.412	0.706	44.590	0.191
$PC\text{-}VC/NaClO_4$	27.711	37.928	0.496	33.614	0.251

XPS spectra of the elements of interest present in the TiO_2 anode surface films formed in both electrolyte solutions ($PC/NaClO_4$ and $PC\text{-}VC/NaClO_4$) are presented in Figures 7 and 8. These spectra demonstrated a significant difference in the chemical composition of surface films. Alkyl carbonate solvents (PC and VC in this case) can be easily reduced to organic and inorganic Na-salts under an applied potential [64].

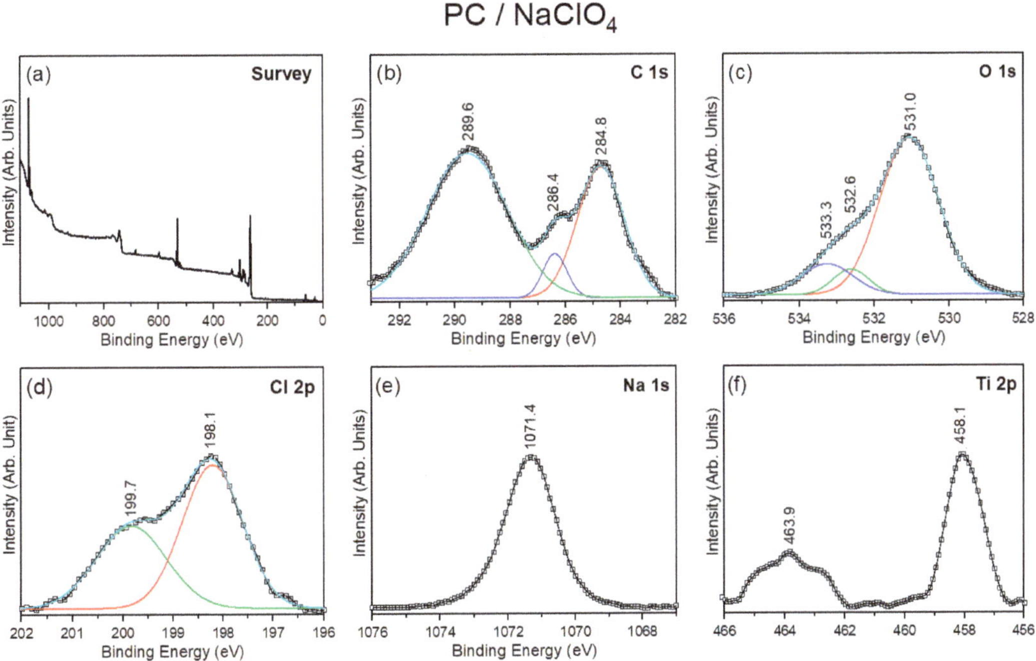

Figure 7. High-resolution XPS spectra of TiO_2 nanosheet electrodes after 30 cycles in $PC/NaClO_4$ electrolyte solution for (**a**) survey, (**b**) C 1s, (**c**) O 1s, (**d**) Cl 2p, (**e**) Na 1s, (**f**) Ti 2p.

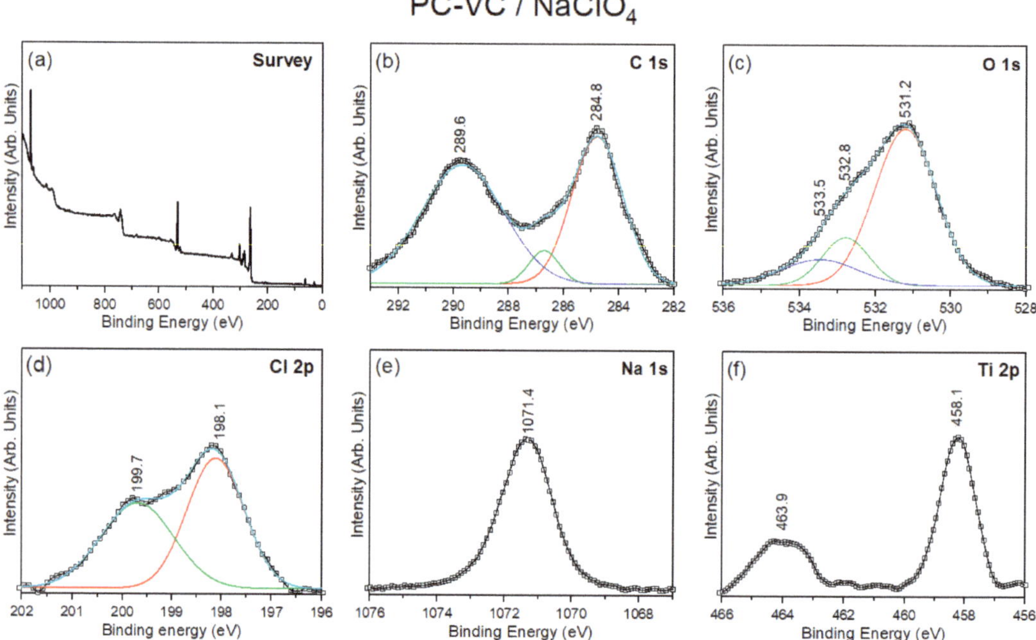

Figure 8. High-resolution XPS spectra of TiO$_2$ nanosheet electrodes after 30 cycles in PC-VC/NaClO$_4$ electrolyte solution for (**a**) survey, (**b**) C 1s, (**c**) O 1s, (**d**) Cl 2p, (**e**) Na 1s, (**f**) Ti 2p.

As presented in Scheme 1, two electron PC reduction in the presence of Na-ions leads to sodium carbonate (Na$_2$CO$_3$) and propylene (CH$_2$=CH-CH$_3$), whereas one electron reduction results in sodium alkyl carbonate (CH$_2$OCO$_2$Na)$_2$ and ethylene (CH$_2$=CH$_2$) formation. On the other hand, PC reaction with perchlorate anion (ClO$_4^-$) forms hydrogen perchlorate (HClO$_4$), carbondioxide, and acetaldehyde radical [65]. The sodium alkyl carbonate produced interacting with the trace amounts of water present in the electrolyte could produce more sodium carbonate (where R = -CH$_3$, or CH$_3$-CH$_2$), carbondioxide, sodium hydroxide, and/or ROH species [66]. Protons generated by the dissociation of HClO$_4$ could react with NaClO$_4$ to form HCl, and sodium hydroxide (NaOH). Previously formed NaOH species reacted with the PC solvent to produce sodium alkyl carbonates [67]. Initially formed sodium alkyl carbonates react with the sodium carbonate and hydrochloric acid (HCl), and in the following steps generating sodium chloride (NaCl), desodiated alkyl carbonate (ROCO$_2$H), and carbonic acid (H$_2$CO$_3$). This step can also form sodium bicarbonate (NaHCO$_3$) instead of carbonic acid [66]. Most importantly, the increased reactivity of VC due to the presence of a double bond could form higher complexity sodiated polymers, such as sodium ethylene dicarbonate (NaO$_2$CO-C$_2$H$_4$-OCO$_2$Na), sodium butylene carbonate (CH$_2$CH$_2$OCO$_2$Na)$_2$, and poly-vinylene carbonate [41,64,66].

$$PC + 2\ e^- + 2\ Na^+ \longrightarrow Na_2CO_3 + CH_2{=}CH\text{-}CH_3$$

$$2\ PC + 2\ e^- + 2\ Na^+ \longrightarrow (CH_2OCO_2Na)_2 + 2\ CH_2{=}CH_2$$

$$NaClO_4 \rightleftharpoons Na^+ + ClO_4^-$$

$$PC + ClO_4^- + 2\ e^- + 2\ Na^+ \longrightarrow Na_2CO_3 + (CH_3COCH_2)^- + CO_2 + HClO_4$$

$$ROCO_2Na + H_2O\ (trace) \longrightarrow Na_2CO_3 + ROH\ /\ NaOH + CO_2$$

$$NaClO_4 + 7\ H^+ \longrightarrow HCl + 2\ H_2O + NaOH + OH^-$$

$$NaOH + PC \longrightarrow CH_3CH(OCO_2Na)CH_2OH$$

$$CH_2OCO_2Na + Na_2CO_3 + 3\ HCl \longrightarrow 3\ NaCl + CH_2OCO_2H + H_2CO_3$$

$$CH_2OCO_2Na + Na_2CO_3 + 2\ HCl \longrightarrow 2\ NaCl + CH_2OCO_2H + NaHCO_3$$

$$NaClO_4 + HCl \longrightarrow NaCl + HClO_4$$

$$RONa + HCl \longrightarrow NaCl + ROH$$

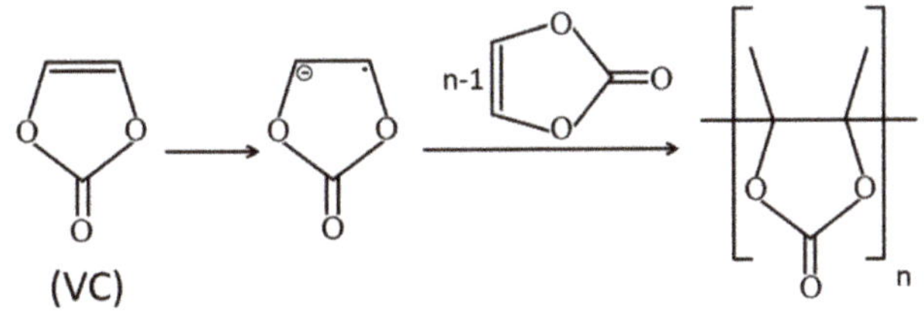

$$nR\cdot \longrightarrow R_n\ (polymer)$$

$$R\cdot + Na^+ + e^- \longrightarrow RNa$$

Scheme 1. Reduction reactions of PC-VC/NaClO$_4$ electrolyte solution.

Hence, the effect of solvents on the salt reduction can be investigated by comparing the Cl-concentration in the surface film. Among the carbonate solvents used in this case, VC is more reactive than PC towards electrochemical reduction, which leads to the difference in the SEI composition and properties [47]. In the case of PC/NaClO$_4$ electrolyte solution, high-resolution C 1s spectrum (Figure 7b) consisted of three distinct peaks located at 284.8 eV, 286.4 eV, and 289.6 eV, and characteristic of elemental carbon (C-C bonds), C-O-C/ C-O groups of RONa, and C=O groups of ROCO$_2$Na/ Na$_2$CO$_3$ species, respectively [28,68]. Signals corresponding to alkoxy groups were absent in the case of the VC-containing electrolyte solution (Figure 8b). The relative intensity of the ROCO$_2$Na peaks was also less, which clearly indicated difference in the surface film composition. The O 1s spectra of the TiO$_2$ anodes cycled in VC-free electrolyte solution (Figure 7c) could be deconvoluted into individual peaks located at 531.0 eV, 532.6 eV, and 533.3 eV, corresponding to alkoxides, ether, and carbonate groups present in the surface film [49,68,69].

Due to the presence of similar functional groups, it was difficult to distinguish ROCO$_2$Na present in VC-free and polycarbonate groups formed in VC-containing electrolyte solutions. However, slightly increased O 1s binding energies (531.2, 532.8, 533.5 eV, respectively) and increased width in the case of PC-VC/NaClO$_4$ indicated (Figure 8c) the presence of higher carbonate concentration in the SEI, possibly due to polycarbonate formation resulting from VC-decomposition. Increased oxygen and carbon content quantified from the high-resolution spectra (Table 1) also indicated the presence of higher carbonate content in the case of the VC-containing electrolyte solution. Hence, it can be concluded that the surface films formed in both VC-free electrolyte solutions consisted of alkoxides, carbonates, and ether species. Whereas alkoxides were absent in the polycarbonate rich surface film formed in the VC-containing electrolyte solution.

High-resolution Cl 2p spectra related to both the VC-free and VC-containing electrolyte solutions exhibited similar features (Figures 7d and 8d). Individual peaks at 198.1 and 199.7 eV correspond to the Cl $2p_{3/2}$ and Cl $2p_{1/2}$ components, respectively, of inorganic chlorides (mainly NaCl) present in the surface film [70]. Signals corresponding to organic chlorides were not identified in the case of both electrolyte solutions. Increased salt (NaClO$_4$) decomposition in the case of the VC-free electrolyte solution is also evidenced by the lower relative amount (Table 1) of inorganic chlorides in the surface film. This can be credited to the more pronounced ClO$_4^-$ reduction in the absence of highly reactive VC electrolyte additive. High-resolution Na 1s spectra in the case of both electrolyte solutions (Figures 7e and 8e) were identical, with a prominent peak at 1071.4 eV indicating the presence of NaCl in the surface film [70]. High-resolution Ti 2p spectra in the case of the surface films formed in both electrolyte solutions (Figures 7f and 8f) exhibited signals at 458.1 eV and 463.9 eV, which are characteristic of the Ti $2p_{3/2}$ and Ti $2p_{1/2}$ components of Ti^{4+} ions [53].

This is in good agreement with the intercalation-type pseudocapacitive Na-ion storage mechanism we reported earlier for these anatase–bronze hybrid nanosheets [9]. Identical Na 1s and Ti 2p signals also verified a similar Na-ion storage mechanism of TiO$_2$ nanosheets in both VC-free and VC-containing electrolyte solutions. Increased Na and Ti concentration in the case of PC/NaClO$_4$ and PC-VC/NaClO$_4$ can be attributed to the enhanced NaClO$_4$ decomposition in the absence of VC, and superior transparency of polycarbonate based surface film for X-rays/electron beam, correspondingly.

Further surface chemical studies of the TiO$_2$ were performed by ATR-FTIR spectral measurements. FTIR spectra of non-cycled TiO$_2$ nanosheet electrodes, after cycling in VC-free and VC-containing electrolyte solutions (Figure 9a), exhibited clear differences in the surface chemical composition. Presence of ROCO$_2$Na and Na$_2$CO$_3$ as the major components resulting from PC reduction was evident from these spectra [64,67,70]. Peaks indicative of the symmetric and asymmetric stretch mode of C-O bond (v C-O) present in ROCO$_2$Na appeared in the 1450–1360 cm^{-1} and 1650–1540 cm^{-1} regions, respectively [67,71]. Scissoring vibrations for OCO$_2^-$ (δ OCO$_2^-$) from the same species appeared at 871/877 cm^{-1} for both electrolyte compositions [67,72]. C-H stretching (v C-H) bands appear at higher frequencies between 2999 and 2930, which were identical to the previous reports for electrodes cycled in VC-free electrolytes [57,67]. Signals in the 2340–2350 cm^{-1} range indicated the presence of atmospheric CO$_2$ [41]. Peaks at 1295 cm^{-1}, 1075 cm^{-1}, and 1067 cm^{-1} belong to CO stretching/ CH$_3$ deformation coming from the organic species containing –ONa and/or –OCO$_2$Na functional groups and double bonds [57,71]. Peaks originating from the carbonate group of Na$_2$CO$_3$ (v CO$_3^{-2}$) were clearly visible at 1398 cm^{-1}/1400 cm^{-1} in the spectra [72]. These signals were absent in the spectra of the pristine TiO$_2$ electrode, confirming surface film formation only during the electrochemical charge–discharge process. Interestingly, additional peaks corresponding to PC/VC solvents and NaClO$_4$ salts are not visible in the case of the non-washed electrodes (Figure S4). This could be due to the dominant FTIR intensities of the SEI components compared to trace amounts of solvent and Na-salt. Similar FTIR spectra before and after washing confirmed that the SEI components were not damaged by acetonitrile solvent.

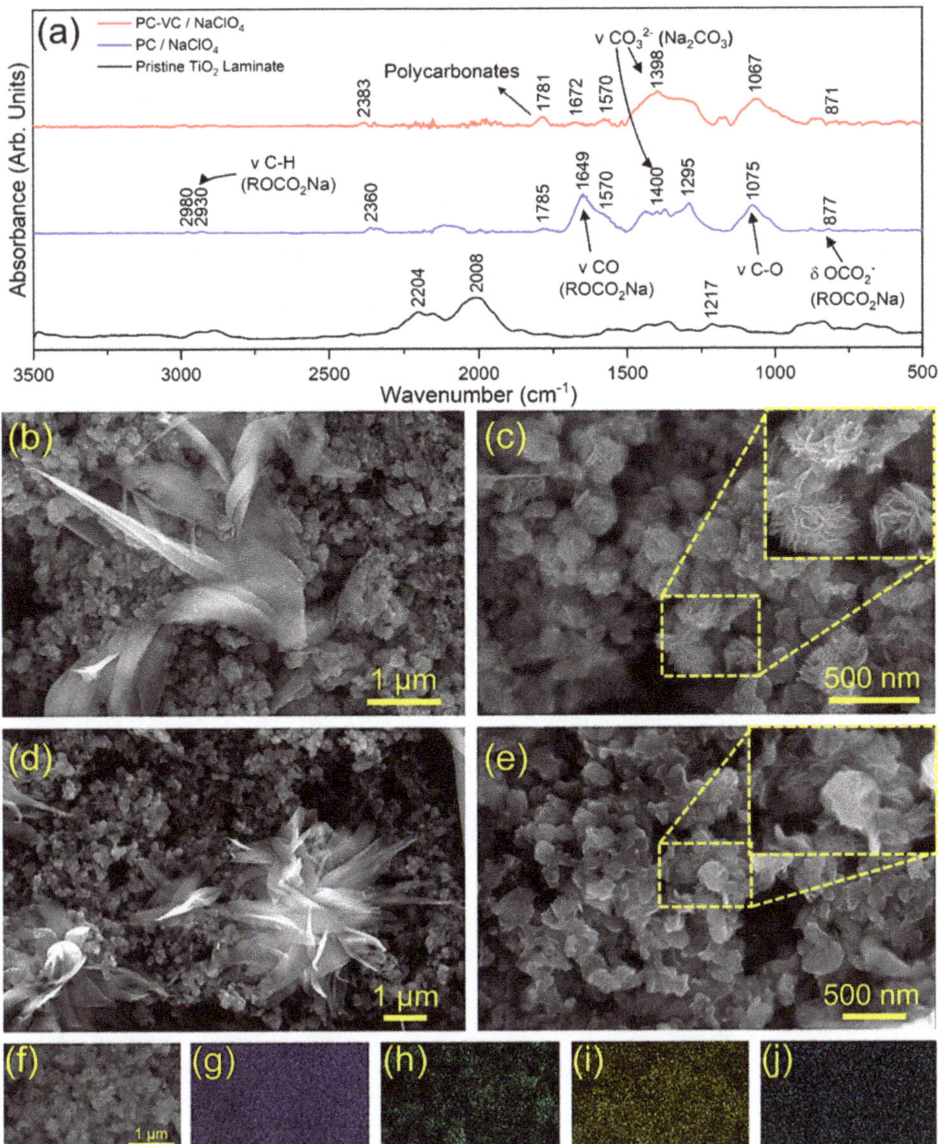

Figure 9. (**a**) ATR-FTIR spectra of non-cycled TiO_2 nanosheet electrodes, and after cycling in PC/$NaClO_4$ and PC-VC/$NaClO_4$ electrolyte solutions. Post-cycling SEM images of TiO_2 nanosheet electrodes in (**b,c**) PC/$NaClO_4$, and (**d,e**) PC-VC/$NaClO_4$ electrolyte solution. (**f**) SEM image and (**g–j**) corresponding SEM-EDX elemental mapping of TiO_2 nanosheet electrodes after 30 cycles in PC-VC/$NaClO_4$ electrolyte solution.

Thus, the FTIR signals clearly demonstrated the formation of alkoxides and carbonates formed as a result of electrolyte decomposition (Scheme 1). The spectra related to PC/$NaClO_4$ and PC-VC/$NaClO_4$ electrolyte solutions are somewhat similar. The major difference between these spectra is the presence of polycarbonate peaks at around 1780 cm^{-1} in the case of the VC-containing electrolyte [57]. It is indeed clear that VC-addition to PC/$NaClO_4$ electrolyte solution resulted in the formation of a polycarbonate rich surface film, along with Na_2CO_3 and $ROCO_2Na$ species. Post-cycling SEM images (Figure 9b–e) of the TiO_2 nanosheet anode confirmed their structural stability during the sodiation–desodiation process. Individual petals of the agglomerated TiO_2 nanosheets were clearly visible through the surface films formed in both electrolyte solutions. EDX mapping of the TiO_2 nanosheet anodes cycled in PC-VC/$NaClO_4$ electrolyte solution (Figure 9f–j) also verified homogeneous SEI formation. Hence, it is clear that the difference in pseudocapacitance and cycling stabilities was a consequence of the different chemical compositions of the surface films rather than their thickness. This should be expected, considering the minimal volume changes of the TiO_2 anode during the charge–discharge process, which is unlikely to form thicker SEI due to limited electrolyte consumption only in the initial cycles.

Surface chemical analysis results obtained from the ATR-FTIR measurements were in line with the XPS results discussed above. These spectral studies enabled us to differentiate the surface chemical aspects responsible for the increased pseudocapacitive Na-ion storage in VC-containing electrolyte solution. Surface film formation in the two different electrolyte solutions can be described as follows: In PC/$NaClO_4$ electrolyte, PC reduction into Na_2CO_3 and $ROCO_2Na$ are the primary processes contributing to the surface film. There exists a competition between the PC and VC reduction in the case of PC-VC/$NaClO_4$ electrolyte solution [47]. VC is the most reactive component, and surface films formed in this case constitute both VC polymerization and PC reduction products. Polycarbonates formed during the initial charge–discharge cycles, as a result of VC-decomposition, protect the highly reactive TiO_2 anode from further reaction with the electrolyte solution. Improved passivation also resulted in the reduced decomposition of the $NaClO_4$ salt, which is crucial for maintaining good ionic conductivity during extended charge–discharge cycles. Significantly improved pseudocapacitance resulted from the ultrafast Na-ion diffusion through the polycarbonate based surface film. It should be noted that improved electrochemical performance of graphite and silicon nanowire anodes in VC/ FEC containing electrolyte solutions has been observed previously [41]. Nevertheless, such drastic difference in the pseudocapacitive Na-ion storage mechanism has never been identified. In conclusion VC electrolyte additive resulted in excellent long term cycling stability and pseudocapacitive Na-ion storage, a consequence of the superior passivation and Na-ion transport properties of the polycarbonate-rich surface film.

4. Conclusions

The effects of vinylene carbonate (VC) as an electrolyte additive on the surface chemistry and Na-ion storage mechanism was thoroughly investigated. TiO_2 nanosheet anodes achieved enhanced long-term cycling stability and pseudocapacitance in VC-containing electrolyte solution compared to VC-free composition. FTIR and XPS surface chemical characterizations of TiO_2 nanosheets cycled in VC-containing electrolyte solution confirmed the formation of polycarbonate rich surface film. Whereas, the VC-free electrolyte solution resulted in the formation of a sodium alkyl carbonate and sodium carbonate based surface film. The superior electrochemical performance of the TiO_2 nanosheets in VC-containing electrolyte solution is credited to the faster pseudocapacitive Na-ion diffusion through the polycarbonate-based surface film. These results demonstrated that TiO_2 nanosheet anodes in VC-containing alkyl carbonate electrolyte solutions can be utilized for advanced high-performance Na-ion batteries.

Supplementary Materials: The following are available online at https://www.mdpi.com/2313-0105/7/1/1/s1. Figure S1: High-resolution Raman spectra, Figure S2: Galvanostatic long cycling performance with different amount of VC additive, Figure S3: First cycle charge-discharge profiles of TiO_2 nanosheet anodes, Figure S4: FTIR spectra of cycled TiO_2 laminates without washing.

Author Contributions: R.R.M. and V.E., conceptualize the idea; R.R.M., performed materials synthesis and characterization; R.R.M. and V.E., together wrote the manuscript; V.E., also performed supervision of the work. All authors have read and agreed to the published version of the manuscript.

Funding: We thank IMDEA Materials Institute for generous start-up funding. Vinodkumar Etacheri acknowledges Spanish Ministry of Economy, Industry and Competitiveness (MINECO), Spanish Ministry of Science and Innovation, and Comunidad de Madrid for Juan de la Cierva fellowship (IJCI-2015-25488), Retos investigacion project (MAT2017-84002-C2-2-R)/Ramon y Cajal fellowship (RYC-2018-025893-I), and Talent attraction fellowship (2016-T1/IND-1300) respectively.

Data Availability Statement: The data presented in this study are available within the article and/or supplementary material.

Conflicts of Interest: The authors declare no conflict of interest.

References

1. Hwang, J.-Y.; Myung, S.-T.; Sun, Y.-K. Sodium-ion batteries: Present and future. *Chem. Soc. Rev.* **2017**, *46*, 3529–3614. [CrossRef] [PubMed]
2. Palomares, V.; Serras, P.; Villaluenga, I.; Hueso, K.B.; Carretero-González, J.; Rojo, T. Na-ion batteries, recent advances and present challenges to become low cost energy storage systems. *Energy Environ. Sci.* **2012**, *5*, 5884–5901. [CrossRef]
3. Etacheri, V.; Marom, R.; Elazari, R.; Salitra, G.; Aurbach, D. Challenges in the development of advanced Li-ion batteries: A review. *Energy Environ. Sci.* **2011**, *4*, 3243–3262. [CrossRef]
4. Madian, M.; Eychmüller, A.; Giebeler, L. Current advances in TiO_2-based nanostructure electrodes for high performance lithium ion batteries. *Batteries* **2018**, *4*, 7. [CrossRef]
5. Dahbi, M.; Yabuuchi, N.; Fukunishi, M.; Kubota, K.; Chihara, K.; Tokiwa, K.; Yu, X.F.; Ushiyama, H.; Yamashita, K.; Son, J.Y.; et al. Black Phosphorus as a High-Capacity, High-Capability Negative Electrode for Sodium-Ion Batteries: Investigation of the Electrode/Electrolyte Interface. *Chem. Mater.* **2016**, *28*, 1625–1635. [CrossRef]
6. Battistel, A.; Palagonia, M.S.; Brogioli, D.; La Mantia, F.; Trócoli, R. Electrochemical Methods for Lithium Recovery: A Comprehensive and Critical Review. *Adv. Mater.* **2020**, *32*, 1905440. [CrossRef] [PubMed]
7. Rubio, S.; Maça, R.R.; Aragón, M.J.; Cabello, M.; Castillo-Rodríguez, M.; Lavela, P.; Tirado, J.L.; Etacheri, V.; Ortiz, G.F. Superior electrochemical performance of TiO_2 sodium-ion battery anodes in diglyme-based electrolyte solution. *J. Power Sources* **2019**, *432*, 82–91. [CrossRef]
8. Muñoz-Márquez, M.Á.; Saurel, D.; Gómez-Cámer, J.L.; Casas-Cabanas, M.; Castillo-Martínez, E.; Rojo, T. Na-Ion Batteries for Large Scale Applications: A Review on Anode Materials and Solid Electrolyte Interphase Formation. *Adv. Energy Mater.* **2017**, *7*, 1700463. [CrossRef]
9. Maça, R.R.; Cíntora Juárez, D.; Castillo Rodríguez, M.; Etacheri, V. Nanointerface-Driven Pseudocapacitance Tuning of TiO_2 Anode for High-Rate, Ultralong-Life and Enhanced Capacity Sodium-Ion Batteries. *Chem. Eng. J.* **2020**, *391*, 123598. [CrossRef]
10. Feng, W.; Maça, R.R.; Etacheri, V. High-Energy-Density Sodium-Ion Hybrid Capacitors Enabled by Interface-Engineered Hierarchical TiO_2 Nanosheet Anodes. *ACS Appl. Mater. Interfaces* **2020**, *12*, 4443–4453. [CrossRef]
11. Sanchez, J.S.; Maça, R.R.; Pendashteh, A.; Etacheri, V.; de la Peña O'Shea, V.A.; Castillo-Rodríguez, M.; Palma, J.; Marcilla, R. Hierarchical Co_3O_4 nanorods anchored on nitrogen doped reduced graphene oxide: A highly efficient bifunctional electrocatalyst for rechargeable Zn–air batteries. *Catal. Sci. Technol.* **2020**, *10*, 1444–1457. [CrossRef]
12. Cao, Y.; Xiao, L.; Sushko, M.L.; Wang, W.; Schwenzer, B.; Xiao, J.; Nie, Z.; Saraf, L.V.; Yang, Z.; Liu, J. Sodium ion insertion in hollow carbon nanowires for battery applications. *Nano Lett.* **2012**, *12*, 3783–3787. [CrossRef] [PubMed]
13. Chen, J.; Zou, G.; Hou, H.; Zhang, Y.; Huang, Z.; Ji, X. Pinecone-like hierarchical anatase TiO_2 bonded with carbon enabling ultrahigh cycling rates for sodium storage. *J. Mater. Chem. A* **2016**, *4*, 12591–12601. [CrossRef]
14. Chao, D.; Zhu, C.; Yang, P.; Xia, X.; Liu, J.; Wang, J.; Fan, X.; Savilov, S.V.; Lin, J.; Fan, H.J.; et al. Array of nanosheets render ultrafast and high-capacity Na-ion storage by tunable pseudocapacitance. *Nat. Commun.* **2016**, *7*, 12122. [CrossRef]
15. Rubio, S.; Maça, R.R.; Ortiz, G.F.; Vicente, C.P.; Lavela, P.; Etacheri, V.; Tirado, J.L. Iron Oxide-Iron Sulfide Hybrid Nanosheets as High-Performance Conversion-Type Anodes for Sodium-Ion Batteries. *Appl. Energy Mater.* **2020**, *3*, 10765–10775. [CrossRef]
16. Yabuuchi, N.; Kubota, K.; Dahbi, M.; Komaba, S. Research development on sodium-ion batteries. *Chem. Rev.* **2014**, *114*, 11636–11682. [CrossRef]
17. Dahbi, M.; Nakano, T.; Yabuuchi, N.; Fujimura, S.; Chihara, K.; Kubota, K.; Son, J.Y.; Cui, Y.T.; Oji, H.; Komaba, S. Effect of Hexafluorophosphate and Fluoroethylene Carbonate on Electrochemical Performance and the Surface Layer of Hard Carbon for Sodium-Ion Batteries. *ChemElectroChem* **2016**, *3*, 1856–1867. [CrossRef]
18. Baggetto, L.; Keum, J.K.; Browning, J.F.; Veith, G.M. Germanium as negative electrode material for sodium-ion batteries. *Electrochem. Commun.* **2013**, *34*, 41–44. [CrossRef]
19. Kalubarme, R.S.; Lee, J.Y.; Park, C.J. Carbon Encapsulated Tin Oxide Nanocomposites: An Efficient Anode for High Performance Sodium-Ion Batteries. *ACS Appl. Mater. Interfaces* **2015**, *7*, 17226–17237. [CrossRef] [PubMed]
20. Qian, J.; Chen, Y.; Wu, L.; Cao, Y.; Ai, X.; Yang, H. High capacity Na-storage and superior cyclability of nanocomposite Sb/C anode for Na-ion batteries. *Chem. Commun.* **2012**, *48*, 7070–7072. [CrossRef]
21. Augustyn, V.; Come, J.; Lowe, M.A.; Kim, J.W.; Taberna, P.-L.; Tolbert, S.H.; Abruña, H.D.; Simon, P.; Dunn, B. High-rate electrochemical energy storage through Li+ intercalation pseudocapacitance. *Nat. Mater.* **2013**, *12*, 518–522. [CrossRef] [PubMed]

22. Longoni, G.; Pena Cabrera, R.L.; Polizzi, S.; D'Arienzo, M.; Mari, C.M.; Cui, Y.; Ruffo, R. Shape-Controlled TiO_2 Nanocrystals for Na-Ion Battery Electrodes: The Role of Different Exposed Crystal Facets on the Electrochemical Properties. *Nano Lett.* **2017**, *17*, 992–1000. [CrossRef] [PubMed]

23. Cargnello, M.; Gordon, T.R.; Murray, C.B. Solution-phase synthesis of titanium dioxide nanoparticles and nanocrystals. *Chem. Rev.* **2014**, *114*, 9319–9345. [CrossRef] [PubMed]

24. Oh, S.; Hwang, J.; Yoon, C.S.; Lu, J.; Amine, K.; Belharouak, I.; Sun, Y. High Electrochemical Performances of Microsphere C-TiO_2 Anode for Sodium-Ion Battery. *ACS Appl. Mater. Interfaces* **2014**, *6*, 11295–11301. [CrossRef] [PubMed]

25. Li, W.; Fukunishi, M.; Morgan, B.J.; Borkiewicz, O.J.; Chapman, K.W.; Pralong, V.V.; Maignan, A.; Lebedev, O.I.; Ma, J.; Groult, H.; et al. A Reversible Phase Transition for Sodium Insertion in Anatase TiO_2. *Chem. Mater.* **2017**, *29*, 1836–1844. [CrossRef]

26. Ge, Y.; Jiang, H.; Zhu, J.; Lu, Y.; Chen, C.; Hu, Y.; Qiu, Y.; Zhang, X. High cyclability of carbon-coated TiO_2 nanoparticles as anode for sodium-ion batteries. *Electrochim. Acta* **2015**, *157*, 142–148. [CrossRef]

27. Chen, J.; Ding, Z.; Wang, C.; Hou, H.; Zhang, Y.; Wang, C.; Zou, G.; Ji, X. Black Anatase Titania with Ultrafast Sodium-Storage Performances Stimulated by Oxygen Vacancies. *ACS Appl. Mater. Interfaces* **2016**, *8*, 9142–9151. [CrossRef] [PubMed]

28. Goriparti, S.; Miele, E.; Prato, M.; Scarpellini, A.; Marras, S.; Monaco, S.; Toma, A.; Messina, G.C.; Alabastri, A.; De Angelis, F.; et al. Direct Synthesis of Carbon-Doped TiO_2-Bronze Nanowires as Anode Materials for High Performance Lithium-Ion Batteries. *ACS Appl. Mater. Interfaces* **2015**, *7*, 25139–25146. [CrossRef]

29. Etacheri, V.; Kuo, Y.; Van Der Ven, A.; Bartlett, B.M. Mesoporous TiO_2-B microflowers composed of (1-1 0) facet-exposed nanosheets for fast reversible lithium-ion storage. *J. Mater. Chem. A* **2013**, *1*, 12028–12032. [CrossRef]

30. Mehraeen, S.; Taşdemir, A.; Gürsel, S.A.; Yürüm, A. Homogeneous growth of TiO_2-based nanotubes on nitrogen-doped reduced graphene oxide and its enhanced performance as a Li-ion battery anode. *Nanotechnology* **2018**, *29*, 255402. [CrossRef]

31. Su, D.; Dou, S.; Wang, G.; Su, D.; Dou, S.; Wang, G. Anatase TiO_2: Better Anode Material than Amorphous and Rutile Phases of TiO_2 for Na-ion Batteries. *Chem. Mater.* **2015**, *27*, 6022–6029. [CrossRef]

32. Xu, Y.; Lotfabad, M.; Wang, H.; Farbod, B.; Xu, Z.; Mitlin, D. Nanocrystalline anatase TiO_2: A new anode material for rechargeable sodium ion batteries. *Chem. Commun.* **2013**, *49*, 8973–8975. [CrossRef]

33. Wang, W.; Liu, Y.; Wu, X.; Wang, J.; Fu, L.; Zhu, Y.; Wu, Y.; Liu, X. Advances of TiO_2 as Negative Electrode Materials for Sodium-Ion Batteries. *Adv. Mater. Technol.* **2018**, *3*, 1800004. [CrossRef]

34. Usui, H.; Domi, Y.; Yoshioka, S.; Kojima, K.; Sakaguchi, H. Electrochemical Lithiation and Sodiation of Nb-Doped Rutile TiO_2. *ACS Sustain. Chem. Eng.* **2016**, *4*, 6695–6702. [CrossRef]

35. Tanaka, Y.; Usui, H.; Domi, Y.; Ohtani, M.; Kobiro, K.; Sakaguchi, H. Mesoporous spherical aggregates consisted of Nb-doped anatase TiO_2 nanoparticles for Li and Na storage materials. *ACS Appl. Energy Mater.* **2019**, *2*, 636–643. [CrossRef]

36. Chen, C.; Wen, Y.; Ji, X.; Yan, M.; Mai, L.; Hu, P.; Shan, B.; Hu, X.; Huang, Y. Na+ intercalation pseudocapacitance in graphene-coupled titanium oxide enabling ultra-fast sodium storage and long-term cycling. *Nat. Commun.* **2015**, *6*, 6929–6936. [CrossRef] [PubMed]

37. Chu, C.; Yang, J.; Zhang, Q.; Wang, N.; Niu, F.; Xu, X.; Yang, J.; Fan, W.; Qian, Y. Biphase-Interface Enhanced Sodium Storage and Accelerated Charge Transfer: Flower-Like Anatase/Bronze TiO_2/C as an Advanced Anode Material for Na-Ion Batteries. *ACS Appl. Mater. Interfaces* **2017**, *9*, 43648–43656. [CrossRef]

38. Peled, E.; Menkin, S. Review-SEI: Past, Present and Future. *J. Electrochem. Soc.* **2017**, *164*, A1703–A1719. [CrossRef]

39. Huang, Y.; Zhao, L.; Li, L.; Xie, M.; Wu, F.; Chen, R. Electrolytes and Electrolyte/Electrode Interfaces in Sodium-Ion Batteries: From Scientific Research to Practical Application. *Adv. Mater.* **2019**, *31*, 1808393. [CrossRef] [PubMed]

40. Ponrouch, A.; Marchante, E.; Courty, M.; Tarascon, J.-M.; Palacín, M.R. In search of an optimized electrolyte for Na-ion batteries. *Energy Environ. Sci.* **2012**, *5*, 8572–8583. [CrossRef]

41. Etacheri, V.; Haik, O.; Goffer, Y.; Roberts, G.A.; Stefan, I.C.; Fasching, R.; Aurbach, D. Effect of Fluoroethylene Carbonate (FEC) on the Performance and Surface Chemistry of Si-Nanowire Li-Ion Battery Anodes. *Langmuir* **2012**, *28*, 965–976. [CrossRef] [PubMed]

42. Song, J.; Xiao, B.; Lin, Y.; Xu, K.; Li, X. Interphases in Sodium-Ion Batteries. *Adv. Energy Mater.* **2018**, *8*, 1703082. [CrossRef]

43. Peled, E.; Patolsky, F.; Golodnitsky, D.; Freedman, K.; Davidi, G.; Schneier, D. Tissue-like Silicon Nanowires-Based Three-Dimensional Anodes for High-Capacity Lithium Ion Batteries. *Nano Lett.* **2015**, *15*, 3907–3916. [CrossRef] [PubMed]

44. Ushirogata, K.; Sodeyama, K.; Tateyama, Y.; Okuno, Y.; Tateyama, Y. Additive Effect on Reductive Decomposition and Binding of Carbonate-Based Solvent toward Solid Electrolyte Interphase Formation in Lithium-Ion Battery. *J. Am. Chem. Soc.* **2013**, *135*, 11967–11974. [CrossRef] [PubMed]

45. Jaumann, T.; Balach, J.; Langklotz, U.; Sauchuk, V.; Fritsch, M.; Michaelis, A.; Teltevskij, V.; Mikhailova, D.; Oswald, S.; Klose, M.; et al. Lifetime vs. rate capability: Understanding the role of FEC and VC in high-energy Li-ion batteries with nano-silicon anodes. *Energy Storage Mater.* **2017**, *6*, 26–35. [CrossRef]

46. Wang, L.; Światowska, J.; Dai, S.; Cao, M.; Zhong, Z.; Shen, Y.; Wang, M. Promises and challenges of alloy-type and conversion-type anode materials for sodium–ion batteries. *Mater. Today Energy* **2019**, *11*, 46–60. [CrossRef]

47. Liu, Q.; Mu, D.; Wu, B.; Wang, L.; Gai, L.; Wu, F. Density Functional Theory Research into the Reduction Mechanism for the Solvent/Additive in a Sodium-Ion Battery. *ChemSusChem* **2017**, *10*, 786–796. [CrossRef]

48. Tao, H.; Zhou, M.; Wang, K.; Cheng, S.; Jiang, K. Glycol Derived Carbon-TiO_2 as Low Cost and High Performance Anode Material for Sodium-Ion Batteries. *Sci. Rep.* **2017**, *7*, 43895–43901. [CrossRef]

49. Song, W.; Jiang, Q.; Xie, X.; Brookfield, A.; McInnes, E.J.L.; Shearing, P.R.; Brett, D.J.L.; Xie, F.; Riley, D.J. Synergistic storage of lithium ions in defective anatase/rutile TiO_2 for high-rate batteries. *Energy Storage Mater.* **2019**, *22*, 441–449. [CrossRef]
50. Yang, P.D.; Zhao, D.Y.; Margolese, D.I.; Chmelka, B.F.; Stucky, G.D. Generalized syntheses of large-pore mesoporous metal oxides with semicrystalline frameworks. *Lett. to Nat.* **1998**, *396*, 152–155. [CrossRef]
51. Yu, X.; Kim, B.; Kim, Y.K. Highly enhanced photoactivity of anatase TiO_2 nanocrystals by controlled hydrogenation-induced surface defects. *ACS Catal.* **2013**, *3*, 2479–2486. [CrossRef]
52. Beuvier, T.; Richard-plouet, M.; Brohan, L. Accurate Methods for Quantifying the Relative Ratio of Anatase and TiO_2(B) Nanoparticles. *J. Phys. Chem. C* **2009**, *113*, 13703–13706. [CrossRef]
53. Wu, L.; Bresser, D.; Buchholz, D.; Giffi, G.; Castro, C.R.; Ochel, A.; Passerini, S.; Giffin, G.A.; Castro, C.R.; Ochel, A.; et al. Unfolding the Mechanism of Sodium Insertion in Anatase TiO_2 Nanoparticles. *Adv. Energy Mater.* **2014**, *5*, 1401142. [CrossRef]
54. Zhang, Y.; Pu, X.; Yang, Y.; Zhu, Y.; Hou, H.; Jing, M.; Yang, X.; Chen, J.; Ji, X. An Electrochemical Investigation of Rutile TiO_2 Microspheres Anchored by Nanoneedle Clusters for its Sodium Storage. *Phys. Chem. Chem. Phys.* **2015**, *17*, 15764–15770. [CrossRef] [PubMed]
55. Randles, J.E.B. Kinetics of Rapid Electrode Reactions. *Discuss. Faraday Soc.* **1947**, *1*, 11–19. [CrossRef]
56. Mogi, R.; Inaba, M.; Jeong, S.-K.; Iriyama, Y.; Abe, T.; Ogumi, Z. Effects of Some Organic Additives on Lithium Deposition in Propylene Carbonate. *J. Electrochem. Soc.* **2002**, *149*, A1578–A1583. [CrossRef]
57. Aurbach, D.; Gamolsky, K.; Markovsky, B.; Gofer, Y.; Schmidt, M.; Heider, U. On the use of vinylene carbonate (VC) as an additive to electrolyte solutions for Li-ion batteries. *Electrochim. Acta* **2002**, *47*, 1423–1439. [CrossRef]
58. Augustyn, V.; Simon, P.; Dunn, B. Pseudocapacitive oxide materials for high-rate electrochemical energy storage. *Energy Environ. Sci.* **2014**, *7*, 1597–1614. [CrossRef]
59. Babu, B.; Shaijumon, M.M. High performance sodium-ion hybrid capacitor based on $Na_2Ti_2O_4(OH)_2$ nanostructures. *J. Power Sources* **2017**, *353*, 85–94. [CrossRef]
60. Lesel, B.K.; Cook, J.B.; Yan, Y.; Lin, T.C.; Tolbert, S.H. Using Nanoscale Domain Size to Control Charge Storage Kinetics in Pseudocapacitive Nanoporous $LiMn_2O_4$ Powders. *ACS Energy Lett.* **2017**, *2*, 2293–2298. [CrossRef]
61. De La Llave, E.; Borgel, V.; Park, K.J.; Hwang, J.Y.; Sun, Y.K.; Hartmann, P.; Chesneau, F.F.; Aurbach, D. Comparison between Na-Ion and Li-Ion Cells: Understanding the Critical Role of the Cathodes Stability and the Anodes Pretreatment on the Cells Behavior. *ACS Appl. Mater. Interfaces* **2016**, *8*, 1867–1875. [CrossRef] [PubMed]
62. Costentin, C.; Porter, T.R.; Save, J. How Do Pseudocapacitors Store Energy? Theoretical Analysis and Experimental Illustration. *ACS Appl. Mater. Interfaces* **2017**, *9*, 8649–8658. [CrossRef] [PubMed]
63. El Ouatani, L.; Dedryvère, R.; Siret, C.; Biensan, P.; Gonbeau, D. Effect of Vinylene Carbonate Additive in Li-Ion Batteries: Comparison of $LiCoO_2$/C, $LiFePO_4$/C, and $LiCoO_2$/$Li_4Ti_5O_{12}$ Systems. *J. Electrochem. Soc.* **2009**, *156*, A468–A477. [CrossRef]
64. Kumar, H.; Detsi, E.; Abraham, D.P.; Shenoy, V.B. Fundamental Mechanisms of Solvent Decomposition Involved in Solid-Electrolyte Interphase Formation in Sodium Ion Batteries. *Chem. Mater.* **2016**, *28*, 8930–8941. [CrossRef]
65. Leggesse, E.G.; Lin, R.T.; Teng, T.F.; Chen, C.L.; Jiang, J.C. Oxidative decomposition of propylene carbonate in lithium ion batteries: A DFT study. *J. Phys. Chem. A* **2013**, *117*, 7959–7969. [CrossRef]
66. Fondard, J.; Irisarri, E.; Courrèges, C.; Palacin, M.R.; Ponrouch, A.; Dedryvère, R. SEI Composition on Hard Carbon in Na-Ion Batteries After Long Cycling: Influence of Salts (NaPF6, NaTFSI) and Additives (FEC, DMCF). *J. Electrochem. Soc.* **2020**, *167*, 070526. [CrossRef]
67. Moshkovich, M.; Gofer, Y.; Aurbach, D. Investigation of the Electrochemical Windows of Aprotic Alkali Metal (Li, Na, K) Salt Solutions. *J. Electrochem. Soc.* **2001**, *148*, E155–E167. [CrossRef]
68. Aurbach, D.; Weissman, I.; Schechter, A.; Cohen, H. X-ray Photoelectron Spectroscopy Studies of Lithium Surfaces Prepared in Several Important Electrolyte Solutions. A Comparison with Previous Studies by Fourier Transform Infrared Spectroscopy. *Langmuir* **1996**, *12*, 3991–4007. [CrossRef]
69. Herstedt, M.; Abraham, D.P.; Kerr, J.B.; Edström, K. X-Ray Photoelectron Spectroscopy of Negative Electrodes from High-Power Lithium-Ion Cells Showing Various Levels of Power Fade. *Electrochim. Acta* **2004**, *49*, 5097–5110. [CrossRef]
70. Muñoz-Márquez, M.A.; Zarrabeitia, M.; Castillo-Martínez, E.; Eguía-Barrio, A.; Rojo, T.; Casas-Cabanas, M. Composition and evolution of the solid-electrolyte interphase in Na2Ti3O7 electrodes for Na-Ion batteries: XPS and Auger parameter analysis. *ACS Appl. Mater. Interfaces* **2015**, *7*, 7801–7808. [CrossRef]
71. Matsuta, S.; Asada, T.; Kitaura, K. Vibrational Assignments of Lithium Alkyl Carbonate and Lithium Alkoxide in the Infrared Spectra. *J. Electrochem. Soc.* **2000**, *147*, 1695–1702. [CrossRef]
72. Tsubouchi, S.; Domi, Y.; Doi, T.; Ochida, M.; Nakagawa, H.; Yamanaka, T.; Abe, T.; Ogumi, Z. Spectroscopic Characterization of Surface Films Formed on Edge Plane Graphite in Ethylene Carbonate-Based Electrolytes Containing Film-Forming Additives. *J. Electrochem. Soc.* **2012**, *159*, A1786–A1790. [CrossRef]

Article

Optimal Siting and Sizing of Battery Energy Storage Systems for Distribution Network of Distribution System Operators

Panyawoot Boonluk [1,2], Apirat Siritaratiwat [1], Pradit Fuangfoo [2] and Sirote Khunkitti [3,*

[1] Department of Electrical Engineering, Faculty of Engineering, Khon Kaen University, Khon Kaen 40002, Thailand; panyawoot@kkumail.com (P.B.); apirat@kku.ac.th (A.S.)

[2] Provincial Electricity Authority, Bangkok 10900, Thailand; pradit.fua@pea.co.th

[3] Department of Electrical Engineering, Faculty of Engineering, Chiang Mai University, Chiang Mai 50200, Thailand

* Correspondence: sirote.khunkitti@cmu.ac.th; Tel.: +66-868589799; Fax: +66-5322-1485

Received: 10 September 2020; Accepted: 12 November 2020; Published: 19 November 2020

Abstract: In this work, optimal siting and sizing of a battery energy storage system (BESS) in a distribution network with renewable energy sources (RESs) of distribution network operators (DNO) are presented to reduce the effect of RES fluctuations for power generation reliability and quality. The optimal siting and sizing of the BESS are found by minimizing the costs caused by the voltage deviations, power losses, and peak demands in the distribution network for improving the performance of the distribution network. The simulation results of the BESS installation were evaluated in the IEEE 33-bus distribution network. Genetic algorithm (GA) and particle swarm optimization (PSO) were adopted to solve this optimization problem, and the results obtained from these two algorithms were compared. After the BESS installation in the distribution network, the voltage deviations, power losses, and peak demands were reduced when compared to those of the case without BESS installation.

Keywords: battery energy storage; renewable energy; distribution network; genetic algorithm; particle swarm optimization

---

1. Introduction

Nowadays, renewable energy sources (RESs) have been widely connected to distribution networks according to the advantage of electricity generation from RESs, which is clean energy, to respond to the high increasing demand in electrical power. However, RESs also consist of a major drawback, which is a fluctuation of power generation, due to the uncertainty of natural sources that cannot be controlled, causing an imbalance between the supply and demand of electrical power. As a result, electrical power flows in a reverse way and power loss occurs in distribution networks [1–4], especially the connection of privately-owned RESs to the distribution network systems of distribution network operators (DNO). Owners of RES companies usually sell electrical energy to the distribution networks based on electricity generation depending on the natural sources at that time. In particular, photovoltaics (PVs) can generate electricity only during the daytime, which is an example of the above-mentioned problem.

The important factors for a distribution network is the reliability of the power system and that the power quality meets the standards. Therefore, energy storage systems (ESSs) have an important role and have been used in distribution networks with the connected RESs to overcome the drawbacks of RES. Additionally, the ESS can balance the electrical power supply and demands [5], improve voltage deviation [6,7], reduce power loss [8–11], reduce peak demand by storing electrical energy during an off-peak time and supplying electric power during peak time [12,13], and use for many objectives including voltage deviation improvement, power loss reduction, and peak demand reduction [14–16].

The optimal location and sizing of an ESS installation can improve the power system's efficiency and reliability [17]. I. Naidji et al. proposed the optimal sizing of an ESS installation that considered the minimum cost at the weakest position by considering the contingency sensitivity index (CSI) [18]. M. Nick et al. introduced the minimum investment cost of an ESS together with the optimal siting and sizing of an ESS to reduce the expenses incurred in the power system by using second-order cone programming (SOCP) [19]. The installation of a battery energy storage system (BESS) cannot only improve the power system efficiency, but also increase the flexibility of dealing with the management (purchase and sell) of electric power for the maximum profit of an electricity supplier [20]. The installation of a BESS, together with the connection of RESs to manage power systems, can support the increasing electricity consumption [21–24]. In addition to the installation of the ESS in the optimal siting and sizing, the appropriate schedule control of an ESS plays an important role in improving the efficiency of power systems and is also able to increase the profits from the sale of electricity for the electricity seller [25–27].

N. Jayasekara et al. proposed an appropriate method to find the optimal siting, sizing, and operation pattern of a BESS [14]. However, the costs—consisting of the battery cost, installation cost and maintenance cost of the BESS—are included in the objective function. Therefore, the obtained siting and sizing of the BESS are not truly appropriate in the aim of improving the efficiency such as minimizing voltage deviation, power loss, and peak demand of the distribution network. Moreover, the comparison of different optimization algorithms has not been investigated to verify the obtained simulation results. Hence, this work aimed to find the optimal siting and sizing of a BESS in distribution networks with the connected RES where the load demand is varied across a day. The objective function was to minimize costs incurred in the distribution networks including the costs of voltage deviation, power loss, and peak demand. Therefore, the truly appropriate optimal siting and sizing of the BESS in distribution networks can be provided. The BESS installation was evaluated in the IEEE 33-bus distribution network. The simulation results were provided by two algorithms comprising of the genetic algorithm (GA) and particle swarm optimization (PSO) and were compared to both verify the simulation results and obtain the appropriate algorithm.

The main contribution of this work includes:

1. The optimal siting and sizing of a BESS in distribution networks with the connected RESs can be provided when the objective function is to minimize the costs incurred in distribution networks that comprise of the costs of voltage deviation, power loss, and peak demand.
2. A performance comparison between the genetic algorithm (GA) and particle swarm optimization (PSO) was undertaken to verify the simulation results and choose the appropriate algorithm.

2. Battery Energy Storage Systems Formulation

ESS technology can store electrical energy in several forms. For instance, electrical energy can be stored in the mechanical form such as pumped hydro, compressed air and flywheel; in the electromagnetic form such as a supercapacitor; in the thermal form such as steam accumulator, molten salt storage, and liquid nitrogen engine; and in the electrochemical form such as flow battery, rechargeable battery, and ultra-battery. The selection of each ESS technology depends on the purpose and physical suitability. In this work, the BESS was chosen to be installed in the distribution network with RESs, and the operation of the BESS was simulated by using the Fourier series.

2.1. Battery Energy Storage Systems (BESSs)

Various types of BESSs such as lead-acid, UltraBattery, NaS, Li-ion, Ni-Cd, and vanadium redox batteries have been widely used for storing electrical energy [28–31]. Li-ion batteries are more popularly used to store electrical energy in many countries such as Germany [32]. Additionally, the price of Li-ion batteries has tended to decrease due to the development of their use in electric vehicles (EVs) [33]. The important characteristics of Li-ion batteries are their high capacity and energy per volume, fast charge and discharge, and low self-discharge rate. From the characteristics of the

Li-ion battery above-mentioned, it requires less installation space, has a rapid response to supply or store the electrical energy, and has low energy loss rate from self-discharging. Therefore, this work applied a Li-ion battery as the BESS.

Several factors affecting the lifespan of a Li-ion battery that should be considered include the temperature, number of duty cycles of the battery, and depth of discharge (DOD). Therefore, for a long life Li-ion battery, good heat dissipation is required where the optimum temperature for the Li-ion battery is around 15–35 °C. Frequent charging and discharging should be avoided, and the a suitable value of DOD is 80% of the total capacity of the battery. Charging or discharging rates should not be too high because the higher charging or discharging rates cause a higher battery temperature, which results in the short lifespan of the battery [34]. In addition, the imbalance charging of each battery cell of the BESS due to the initial unequal state of charge (SOC), which is defined by the battery's present amount of charge divided by its rated charge capacity, may cause some damage to the BESS, a capacity reduction of the BESS, and the deterioration of the BESS. However, this can be avoided by using battery charge equalization systems (BCEs) [35].

2.2. Battery Energy Storage System Simulation

The BESS simulation presented in this work considered the rates of charge or discharge of the BESS at equal intervals within the considered period of 24 h. These 24 h can be equally divided such as 1 h, 30 min, or 15 min, which can obtain the rate of charge or discharge of BESS at m values including 24, 48, and 96, respectively. Therefore, the rates of charge and discharge in the considered period (C_{iT}) can be formulated as the following equation.

$$C_{iT} = \begin{bmatrix} E_B(1) \\ \vdots \\ E_B(m) \end{bmatrix} \tag{1}$$

where $E_B(t)$ is the electrical energy in BESS (MWh) at time $t = 1, 2, 3, \ldots, m$.

To find the values of $E_B(t)$ at each time t, the Fourier series is applied to express the state of energy (SOE) in the total considered intervals. The state of energy of the BESS can be obtained by substituting the Fourier coefficient (C_{iF}) from Equation (2) into Equation (3) to find the values of $E_B(t)$.

$$C_{iF} = \begin{bmatrix} a_1, b_1 \\ \vdots \\ a_n, b_n \end{bmatrix} \tag{2}$$

$$E_B(t) = a_0 + a_1 \cos\left(\frac{2\pi t}{T}\right) + b_1 \sin\left(\frac{2\pi t}{T}\right) + \ldots + a_n \cos\left(\frac{2\pi nt}{T}\right) + b_n \sin\left(\frac{2\pi nt}{T}\right) \tag{3}$$

where a_0, a_n, b_n, n, and T are the constant Fourier coefficient, Fourier cosine coefficients, Fourier sine coefficients, the number of Fourier coefficients (set to 8 [14]), and total period, respectively.

The constant Fourier coefficient (a_0) does not affect the charge or discharge power of BESS due to its constant value. It can be added after calculating the Fourier coefficients (C_{iF}) to ensure that the SOE of BESS is not negative or lower than the minimum depth of discharge (DOD_{min}).

To compute the rate of charge and discharge from the SOE of BESS ($E_B(t)$) at each time t, Equations (4)–(6) are calculated. Suppose $P_B(t)$ is the charge rate or discharge rate at time t. The positive value of $P_B(t)$ means the BESS is charging while a negative value means the BESS is discharging.

$$\Delta E_B = E_B(t) - E_B(t-1) \tag{4}$$

$$P_B(t) = \Delta E_B / (\Delta t \times \eta_c), \quad P_B(t) > 0 \tag{5}$$

$$P_B(t) = (\Delta E_B \times \eta_d) / \Delta t, \quad P_B(t) < 0 \tag{6}$$

where $\eta_c = \eta_d = \sqrt{\eta_{bat}}$, $\eta_{bat} = 0.9$ (the total efficiency of the BESS), P_B, Δt are battery charging efficiency, battery discharging efficiency, battery round trip cycling efficiency, battery power (MW), and sampling time interval, respectively.

The optimal size of the BESS can be found by Equation (7).

$$BatterySize \ (unit.h) = \frac{\left|E_B^{max} - E_B^{min}\right|}{DOD_{max}} \tag{7}$$

where $DOD_{max} = 0.8$ is the maximum depth of discharge value; and E_B^{max} and E_B^{min} are the maximum and minimum battery energy values, respectively. The unit of the energy capacity of the BESS depends on the unit of E_B^{max} and E_B^{min}.

Generally, each round of battery discharge where the energy discharging is equal to the battery capacity is called one operation cycle of a battery. However, to simply calculate the battery cycle, the summation of both charging and discharging energies for all considered time intervals are included, and its average value is instead computed by dividing by 2 (charging and discharging) as in Equation (8) [14]. The lifespan of the BESS can then be evaluated from Equation (9) [14].

$$Cycles = \frac{1}{2} \times \left(\frac{\sum_{t=1}^{T} \left| E_B(t) - E_B(t-1) \right|}{DOD_{max} \times BatterySize} \right) \tag{8}$$

$$Lifespan \ (years) = CycleLife/(Cycles \times D) \tag{9}$$

where *Cycles* is the daily cycles of BESS; *CycleLife* = 3221 is the nominal cycles life of the Li-ion battery [34]; $D = 365$ is the number of operating days of BESS; and *Lifespan* is the lifespan of the BESS (years).

3. Problem Formulation

The installation of the BESS in distribution networks with RESs can improve the efficiency, reliability, and power quality of the distribution networks. The considered objective functions to be minimized include voltage deviation, power loss, and peak demand for providing the optimal siting and sizing of thee BESS under the constraints of distribution networks.

3.1. Objective Functions

The major objective function of this work was to minimize several costs incurred in distribution networks (C_{system}) consisting of voltage regulation cost (C_{VR}), power loss cost (C_{Loss}), and peak demand cost (C_P) in terms of infrastructure development deferrals. Equation (10) represents the objective function and several costs can be found by Equations (11)–(14).

$$f(C_{iF}) = min\left(C_{system}\right) \tag{10}$$

$$C_{system} = C_{VR} + C_{Loss} + C_P \tag{11}$$

$$C_{VR} = \left(\sum_{t=1}^{T} \sum_{i=1}^{N} |V_i - V_{ref}| \right) \times \gamma_{VR} \tag{12}$$

$$C_{Loss} = \left(\sum_{t=1}^{T} \sum_{i=1}^{M} (LineLoss) \right) \times \gamma_{loss} \tag{13}$$

$$C_P = P_{max} \times \Delta t \times \gamma_P \tag{14}$$

where N, V_i, V_{ref}, M, *LineLoss*, P_{max}, γ_{VR}, γ_{loss}, and γ_P are total bus number, voltage magnitude (per unit) at the ith bus, reference voltage which is equal to 1 p.u., total branch number, active power loss in each branch, maximum active power at slack bus over the considered period, rate of voltage regulation cost, rate of power loss cost, and rate of peak demand cost (γ_{VR} = 0.142 \$/p.u. [15], γ_{loss} = 0.284 \$/kWh [14], γ_P = 200 \$/kWh/year [14]), respectively.

3.2. Constraints

(1) Voltage Constraint: Voltage of each bus must be limited within the lower and upper limit throughout the consideration period set at ±5% of the reference voltage as the expressed equation.

$$V_{\text{lower}} \leq V_i^t \leq V_{\text{upper}} \tag{15}$$

where V_{lower} and V_{upper} are the lower bound and upper bound of voltage at ith bus, respectively, and V_i^t is the voltage magnitude at ith bus in time t.

(2) Battery Constraint: The power and capacity of the battery are bounded to ensure that the battery operation does not exceed the boundary limits during charging or discharging, which can be presented as the equation below.

$$P_{B-min} \leq P_{cha}^t, P_{dis}^t \leq P_{B-max} \tag{16}$$

$$E_{B-min} \leq E_B^t \leq E_{B-max} \tag{17}$$

where P_{B-min}, P_{B-max} are the minimum and maximum power of BESS, respectively. P_{cha}^t, P_{dis}^t are the power charge rate and power discharge rate of BESS at time t, respectively. E_{B-min}, E_{B-max} are the minimum and maximum capacity of BESS, respectively.

4. Methodology

The optimal siting and sizing of the BESS in distribution networks with RESs are presented in this work. The simulation of distribution networks was conducted in MATLAB and MATPOWER was also adopted. For the optimization process, genetic algorithm (GA) and particle swarm optimization (PSO) were both applied to find the minimum value of the objective function for finding the optimal siting and sizing of BESS and the performance of these algorithms were compared.

4.1. Test System

The distribution network used to evaluate the optimal siting and sizing of the BESS was the IEEE 33-bus distribution network shown in Figure 1, and the detailed data can be obtained from [14]. The values of load power at each bus can be obtained from the voltage magnitude at each bus (in per unit) without any RES connection according to Equations (18) and (19).

$$P_{L-i} = P_{0i}\left(a_p + b_p|V_i| + c_p|V_i|^2\right) \tag{18}$$

$$Q_{L-i} = Q_{0i}\left(a_q + b_q|V_i| + c_q|V_i|^2\right) \tag{19}$$

where P_{L-i}, Q_{L-i} are the active power and reactive power at the ith bus, respectively. P_{0i}, Q_{0i} are the initial active power and reactive power at the ith bus, respectively, $a_p + b_p + c_p = 1$, $a_q + b_q + c_q = 1$, base voltage (V_{base}) = 12.66 kV, base power (MVA_{base}) = 10 MVA. The two types of RESs, wind turbines (WTs) and photovoltaics (PVs), were connected in the distribution network. Two WTs (1 MW) were each located at buses 18 and 24, three PVs (400 kVA) each were installed at buses 5, 21, and 31, and four PVs (500 kVA) each were installed at buses 8, 12, 28, and 33. In this test system, the parameters were set as $a_p = a_q = 0.4$, $b_p = b_q = 0.3$, and $c_p = c_q = 0.3$. Peak active and reactive powers of the test system were 3.556 MW and 2.191 MVar, respectively. The detailed data of load magnitude and the output generations of WTs and PVs can be found in [14,21].

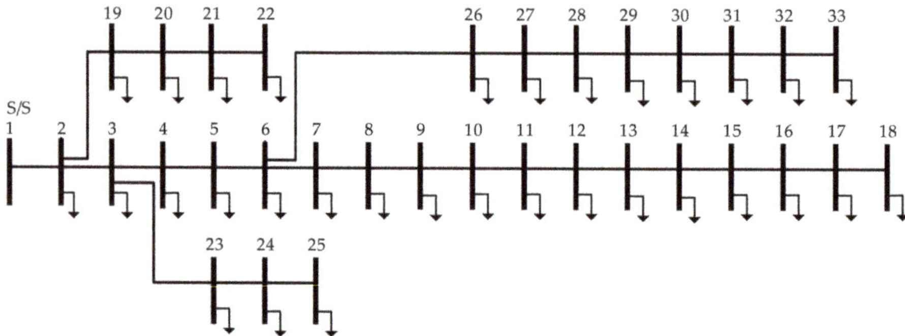

Figure 1. IEEE 33 bus distribution network [14].

4.2. Genetic Algorithm (GA)

Genetic algorithm is inspired and operated by mimicking Charles Darwin's evolution method. First, the initial populations are randomly generated, which are binary numbers. Each population has equal binary numbers to the decision variables (*nVar*) multiplied by the bit number (*nBit*) assigned for each decision variable. The binary numbers for each population are then converted to decimal numbers. Following this, the decimal numbers are compared with the range of each decision variable (*lb*, *ub*) to obtain the real values for each decision variable and then substituting them to find the objective function. After that, three main steps comprising of population selection, cross over, and mutation are considered. In the first step, two populations are randomly selected from the parents' generation, and the cross over is operated between the selected populations in step two where the number of cross over is according to the cross over percentage (*pc*). In the last step, the mutation is proceeded based on the mutation rate (*mu*), and binary numbers are converted to decimal numbers to evaluate the objective function. The best fitness value is updated until the max iteration (*iter_{max}*) is reached [36].

4.3. Particle Swarm Optimization (PSO)

Particle swarm optimization is inspired by the behavior of bird flocking and fish schooling [36]. Initially, the particles are generated with random positions and zero velocity, and the objective function is then evaluated. In the next step, the best positions of each particle, which is called the personal best position ($P_{Pbest}(i)$), are updated and the best position among all particles, which is called the global best position, is updated (P_{Gbest}). The velocity and position of each particle can be updated according to the $P_{Pbest}(i)$ and the P_{Gbest} by the following equations.

$$V_{P(i)} = wP(i) + c_1 r_1 \left(P_{Pbest}(i) - P(i)\right) + c_2 r_2 \left(P_{Gbest} - P(i)\right) \tag{20}$$

$$P(i)_{new} = P(i) + V_{P(i)} \tag{21}$$

where $V_{P(i)}$ is the velocity of the *i*th particle; $P(i)$ is the current position of the *i*th particle; $P_{Pbest}(i)$ is the personal best position of the *i*th particle; P_{Gbest} is the global best position among all particles; $P(i)_{new}$ is the updated position of the *i*th particle, $w = 0.9 - \left(\left((0.9 - 0.4) * it\right) / iter_{max}\right)$; it is the *it*th iteration; $c_1 = c_2 = 2$ and r_1, r_2 are random numbers randomly generated between 0 and 1.

4.4. Application of the Optimal Siting and Sizing of the Battery Energy Storage System (BESS)

The application of the optimal siting and sizing of BESS in this work consisted of the following steps.

1. Define the input data comprising of V_{rated}, R_L, X_L, PV, WT, and MVA_{base} to simulate the considered distribution network in MATPOWER [37,38], and impose candidate buses to find the best location of the BESS installation.

2. Initialize parameters of the algorithm (GA or PSO) and Fourier coefficients, where *lb* and *ub* of the Fourier coefficients are proportionally reduced with the rate $1/n^2$ [39] (*n* is the number of Fourier coefficients).

3. Choose one candidate bus as the location of the BESS installation.

4. Operate the algorithm (as in Section 4.2 for GA or Section 4.3 for PSO) to solve the optimization problem.

5. Conduct power flow where the BESS is installed at one candidate bus.

6. Evaluate the objective function and update the best solution.

7. If the maximum iteration is reached, go to step 8; otherwise, go to step 4.

8. Obtain optimal Fourier coefficients where the BESS is installed at the chosen candidate bus from step 3.

9. If the candidate bus is the last candidate, go to step 10; otherwise, go to step 3.

10. Obtain the optimal siting of the BESS installation, which is the candidate bus providing the minimum value of the objective function.

11. Provide the information of the BESS by substituting the obtained Fourier coefficients in Equations (1)–(9).

The flowchart of the optimal siting and sizing of the BESS method is shown in Figure 2.

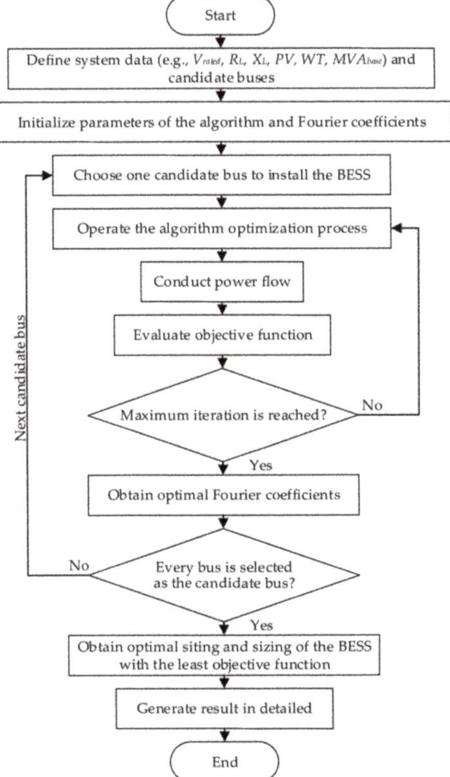

Figure 2. Flowchart of the proposed method.

4.5. System Performance Evaluation

To evaluate the system performance after the BESS installation based on the objective function, voltage deviation, power loss, and peak demand were investigated. The details are as follows.

4.5.1. Voltage Deviation Index (VDI)

To evaluate the voltage profile improvement, the VDI was applied to represent the voltage deviation of each case. The VDI compared the voltage level of each bus (V_{bi} in p.u.) with the reference voltage ($V_{\text{ref}} = 1$ p.u.) and is presented in percent ($\%VDI$). The VDI can be calculated by the expressed equations.

$$\%VDI_i = \max_i^T \left(\frac{|V_{\text{ref}} - V_{bi}|}{V_{\text{ref}}} \right) \times 100 \tag{22}$$

$$\%VDI = \sum_{i=1}^{N} \%VDI_i \tag{23}$$

where $\%VDI_i$ is the VDI of the ith bus, and $\%VDI$ is the total VDI of the distribution network.

4.5.2. Power losses

Power losses including real, reactive, and apparent power losses of the distribution network were compared in the cases with BESS and without BESS. The following equations are computed.

$$P_{\text{L}} = \sum_{i=1}^{T} \sum_{l=1}^{M} P_L^l \tag{24}$$

$$Q_{\text{L}} = \sum_{i=1}^{T} \sum_{l=1}^{M} Q_L^l \tag{25}$$

$$S_{\text{L}} = \sum_{i=1}^{T} \sum_{l=1}^{M} S_L^l \tag{26}$$

where P_{L}, Q_{L}, and S_{L} are the active power loss, reactive power loss, and apparent power loss, respectively. l is the number of branches. T is the total period. M is the total number of branches.

4.5.3. Peak Demand Comparison

In this research, the direction of the active power flow at the slack bus of the distribution network was investigated. If the active power flow is positive, the power flows into the distribution network while the negative active power flow represents the power flow flowing out from the distribution network.

5. Simulation Results

This work aimed to find the optimal siting and sizing of the BESS for a distribution network by minimizing the costs incurred in the distribution network, consisting of voltage regulation costs, costs due to power losses, and peak demand costs within 24 h. The simulation was operated in MATLAB interfaced with MATPOWER 7.0, and the computer specification was an Intel® Core™ i3-6100U 2.30 GHz, RAM 8 GB, Windows 10 Pro 64-bit Operating System. GA and PSO were applied to solve the optimization problems, and their performance was compared. The parameters of the algorithms were as follows: population = 60, maximum iterations = 1000, crossover percentage = 50%, and mutation rate = 0.02. The simulation results are provided in the following subsections.

5.1. Comparison of the Optimal Siting and Sizing of the BESS Installation

The BESS was chosen to be installed at one bus from the 2nd to 33rd bus to find the best location of the BESS installation providing the minimum objective function value. After the BESS installation simulation, three locations providing the minimum objective function values of the GA and PSO are shown in Table 1. It can be seen that the best location of the BESS installation of both GA and PSO was the 6th bus. The power of the BESS of GA was more than that of PSO of about 0.01 MW while the energy capacity of GA was less than that of PSO at around 0.75 MWh. The lifetime of the BESS of each algorithm was equal to 8.8 years.

Table 1. Optimal siting and sizing of the battery energy storage system (BESS) installation.

Case	Location (Bus)	BESS		Lifetime of BESS (Years)
		Power (MW)	Capacity (MWh)	
GA	6	1.99	14.23	8.8
	16	2.09	15.58	8.8
	8	2.11	15.84	8.8
PSO	6	1.98	14.98	8.8
	7	1.98	14.73	8.8
	27	1.85	14.72	8.8

In addition to obtaining the optimal siting and sizing of the BESS, the appropriate operation of the BESS is also important to consider. From Figure 3a, the state of energy (SOE) of the BESS indicating the stored energy in the BESS for 24 h is illustrated. The power of the BESS is shown in Figure 3b, presenting the charging rate or discharging rate of the BESS at each hour. It was noticeable that from 1 a.m. to 8 a.m., the energy of thee BESS gradually decreased because the BESS was in a discharging state at a low rate, and energy was at the minimum capacity from 8 a.m. to 9 a.m. From 9 a.m. to 6 p.m., the BESS changed status from discharging state to charging state, and during this time, the BESS had a maximum charging rate that was equal to 1.99 MW at 2 p.m. for GA and 1.98 MW at 1 p.m. for PSO. The BESS changed status again from 6 p.m. to 12 p.m. where a maximum discharging rate occurred at 10 p.m. for both GA and PSO that was equal to 1.67 MW for GA and 1.73 MW for PSO.

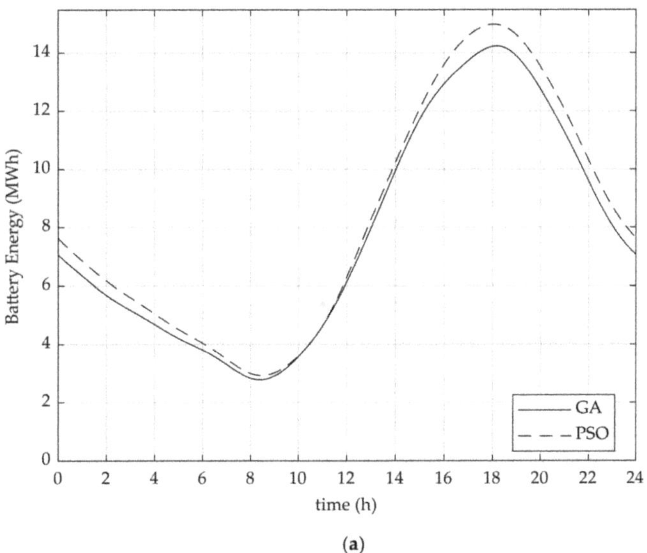

(a)

Figure 3. *Cont.*

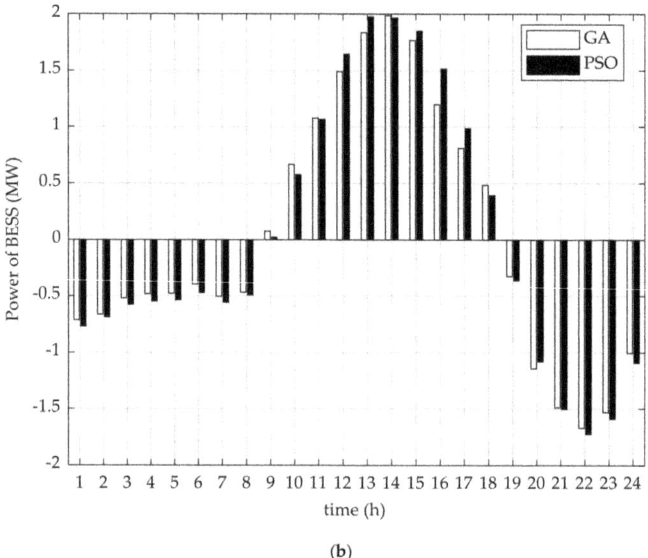

(b)

Figure 3. Battery energy storage system (BESS) operation. (**a**) State of energy (SOE) of BESS; (**b**) Power of BESS.

5.2. Comparison of Voltage Deviation and Power Losses before and after the BESS Installation

After the BESS installation at the 6th bus, it was found that the value of %*VDI* was maximum at the 33rd bus for the base case (without BESS), GA, and PSO at 6.50%, 4.23%, and 4.22%, respectively. The %*VDI* values of the overall distribution network for all three cases are expressed in Table 2. The BESS installation by using PSO gave the minimum %*VDI* value, which means that PSO could better improve the voltage profile in the distribution network than those of the base case and GA. In Figure 4, the voltage profiles at the 33rd bus for 24 h for all three cases are presented to show how the voltage profile was improved when the BESS was installed by comparing the performance of GA and PSO to the base case. It was observed that the voltage profile of the distribution network was below the lower limit for the base case while it was improved to be within the constraint after the installation of the BESS for both GA and PSO. For the power losses shown in Table 2, it was noticed that power losses could be significantly reduced after the BESS installation compared to the base case. It could also be noticed that the BESS installation by GA provided the minimum values of active power and apparent power while the reactive power loss obtained from GA was equal to that of PSO. However, the power losses obtained from GA and PSO after the BESS installation were slightly different.

Table 2. System performance evaluation.

Case	%*VDI*	P_L (MW)	Q_L (MVar)	S_L (MVA)
base	117.56	1.66	1.15	2.02
GA	71.35	1.38	0.98	1.69
PSO	70.40	1.39	0.98	1.70

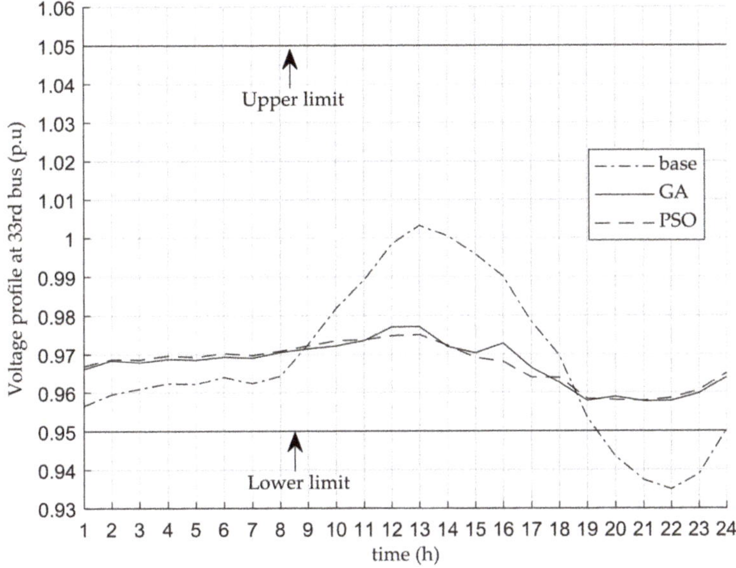

Figure 4. Voltage profile at the 33rd bus.

5.3. Comparison of Peak Demand

In this work, the peak demand was considered in terms of the active power flowing at the slack bus. The higher peak demand represents the higher power flow in the transmission line, so the transmission line of the distribution network needs to be able to support the high power flow in the transmission line. However, if the peak demand can be reduced, the power flow in the transmission line is also decreased. This results in a longer time to improve the transmission line for supporting the higher peak demand in the future. Figure 5 compares the peak demand provided by the base case and presents how the peak demand can be improved after installing a BESS by using GA and PSO. It can be seen in Figure 5 that power flowed in both directions (into and out from the network) for the base case. The maximum active power that was equal to 2.71 MW at 10 p.m. flows into the network, and the maximum active power which was equal to −1.05 MW at 1 p.m. flowed out from the network. After the BESS installation, the active power flow had only one direction (only positive value), which flowed into the distribution network for all 24 h for both GA and PSO. The maximum active power provided by GA was 1.10 MW at 10 a.m. and 1.04 MW at 7 p.m. for PSO, which is depicted in Figure 5. Thus, the BESS installation by using PSO could reduce the peak demand more than that of the GA.

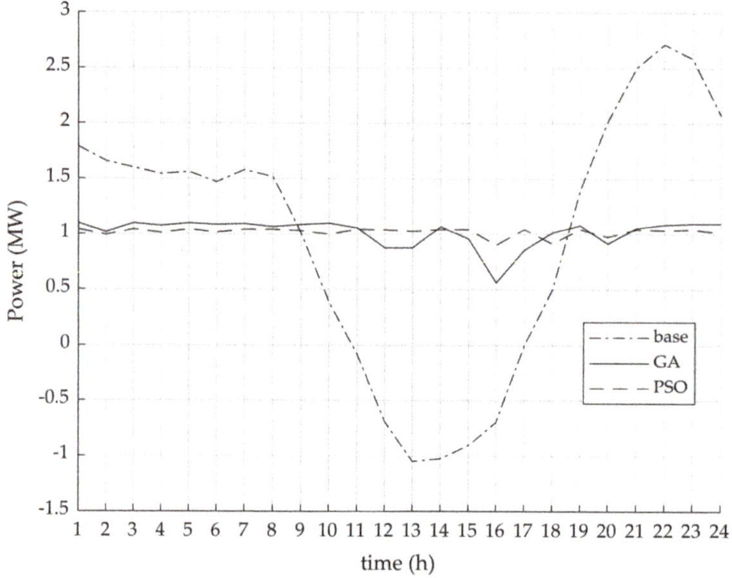

Figure 5. Peak demand profile at the slack bus.

5.4. The Efficiency Comparison between GA and PSO for the Objective Function Optimization

The result comparison of the efficiency for providing the minimum value of the objective function between GA and PSO are shown as the iteration curves in Figure 6. It was observed that during the 0–400 iterations, GA could more quickly find the less objective function value than PSO, but the values were slightly different. After that, the objective value of GA almost remained constant while the objective value of PSO continued decreasing until the 800th iteration before facing a very small decrease until the maximum iteration. At the maximum iteration, PSO could obtain a superior objective value to that of GA, as evident in Table 3. When comparing the number of iterations and the operation time of each algorithm used for the optimization process, it was found that PSO took less time than that of GA, as presented in Table 3. Thus, regarding the overall efficiency for providing the minimum value of the objective function for the considered problem, PSO is more appropriate than GA in terms of both objective value and operation time.

Table 3. The comparison results between a number of iterations and time of use of Genetic Algorithm (GA) and Particle Swarm Optimization (PSO).

Iteration	GA		PSO	
	C_{system} ($)	Time (s)	C_{system} ($)	Time (s)
200	1017.33	1225	1073.11	1219
400	996.57	2465	998.84	2431
600	996.42	3670	979.66	3657
800	996.42	4892	968.13	4864
1000	995.96	6140	967.71	6048

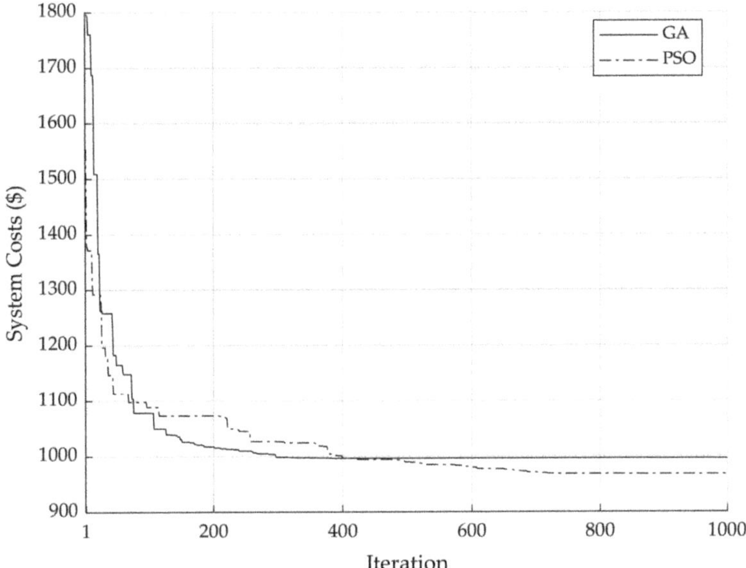

Figure 6. The comparison results between GA and PSO.

After the simulation of the BESS installation in the IEEE 33-bus distribution network with RES by using GA and PSO, the optimal siting of the BESS installation was the 6th bus obtained by both GA and PSO. The size of the BESS was 1.99 MW power and 14.23 MWh capacity provided by GA and 1.98 MW power and 14.98 MWh capacity provided by PSO. It could be observed that the size of the electric power and electric power capacity of the BESS provided by both algorithms had similar values. Thus, the optimal siting and sizing of the BESS for the distribution networks could be chosen by considering the minimum objective value that PSO can find as a better objective value than that of GA. For the BESS operation, it was noticeable that charging and discharging statuses given by GA and PSO for all 24 h were similar, and the electric powers of charging or discharging provided by GA and PSO were slightly different in each time duration.

After the BESS installation by using GA or PSO, it was found that the voltage level in the system could be improved to be in the range of the constraint (±5%), and the voltage deviation was enhanced compared to the base case. Additionally, power losses including active, reactive, and apparent power losses were significantly decreased when compared with the base case. In terms of the peak demand by considering power flow at the slack bus, it could be seen that after the installation of BESS, the power only flowed into the distribution network (one direction) while the power flowed into and out from the network for the base case. This is because the charging and discharging operations of the BESS could balance between the electricity generation and demand. Therefore, the investment in the distribution network expansion to support the increasing electricity demands in the future can be retarded by using the BESS installation.

To compare the efficiency of GA and PSO to solve the problem, the minimum objective function value and operation time were evaluated. From the simulation results, it was found that PSO could provide better objective value than that of GA. PSO also spent less optimization operation time than that of GA to provide the better final solution at the maximum iteration. Therefore, PSO is more appropriate than GA for solving the problem in this work.

6. Conclusions

The optimal siting and sizing of BESS in the distribution network with RESs was presented. The objective function considered to be minimized is the total costs incurred in the distribution network consisting of the costs of voltage deviations, power losses, and peak demands. The BESS installation was operated in an IEEE 33-bus distribution network by using the GA and PSO algorithm for optimizing the objective function, and the results received from both algorithms were compared to verify the accuracy. The results provided by both GA and PSO showed that the BESS installation could improve the efficiency of the distribution network in terms of cost minimization, the reduction of voltage deviations, power losses, and peak demand. It can also well support the RES connection that has a fluctuation in electricity generations. It was also found that PSO is more efficient than GA in terms of the objective function optimization for this problem.

Author Contributions: Conceptualization, P.B., A.S., P.F., and S.K.; Software, P.B.; Validation, P.B., A.S., and S.K.; Formal Analysis, P.B. and S.K.; Investigation, P.B. and S.K.; Resources, P.B.; Data Curation, P.B.; Writing—Original Draft Preparation, P.B. and S.K.; Writing—Review & Editing, P.B., A.S., and S.K.; Visualization, P.B.; Supervision, A.S. and S.K.; Project Administration, A.S. All authors have read and agreed to the published version of the manuscript.

Funding: This research was funded by the Provincial Electricity Authority (PEA) of Thailand.

Acknowledgments: Authors would like to acknowledge Pradit Fuangfoo for his helpful technical discussions.

Conflicts of Interest: The authors declare no conflict of interest.

References

1. Yi, W.; Zhang, Y.; Zhao, Z.; Huang, Y. Multiobjective Robust Scheduling for Smart Distribution Grids: Considering Renewable Energy and Demand Response Uncertainty. *IEEE Access* **2018**, *6*, 45715–45723. [CrossRef]

2. Mortazavi, H.; Mehrjerdi, H.; Saad, M.; Lefebvre, S.; Asber, D.; Lenoir, L. A Monitoring Technique for Reversed Power Flow Detection with High PV Penetration Level. *IEEE Trans. Smart Grid* **2015**, *6*, 2221–2232. [CrossRef]

3. Khani, H.; El-Taweel, N.; Farag, H.E.Z. Real-time optimal management of reverse power flow in integrated power and gas distribution grids under large renewable power penetration. *IET Gener. Transm. Distrib.* **2018**, *12*, 2325–2331. [CrossRef]

4. Méndez Quezada, V.H.; Rivier Abbad, J.; Gómez San Román, T. Assessment of energy distribution losses for increasing penetration of distributed generation. *IEEE Trans. Power Syst.* **2006**, *21*, 533–540. [CrossRef]

5. Laugs, G.A.H.; Benders, R.M.J.; Moll, H.C. Balancing responsibilities: Effects of growth of variable renewable energy, storage, and undue grid interaction. *Energy Policy* **2020**, *139*. [CrossRef]

6. Krata, J.; Saha, T.K. Real-Time Coordinated Voltage Support with Battery Energy Storage in a Distribution Grid Equipped with Medium-Scale PV Generation. *IEEE Trans. Smart Grid* **2019**, *10*, 3486–3497. [CrossRef]

7. Zhang, Y.; Meng, K.; Luo, F.; Yang, H.; Zhu, J.; Dong, Z.Y. Multi-agent-based voltage regulation scheme for high photovoltaic penetrated active distribution networks using battery energy storage systems. *IEEE Access* **2020**, *8*, 7323–7333. [CrossRef]

8. Alzahrani, A.; Alharthi, H.; Khalid, M. Minimization of power losses through optimal battery placement in a distributed network with high penetration of photovoltaics. *Energies* **2019**, *13*, 140. [CrossRef]

9. Wong, L.A.; Ramachandaramurthy, V.K.; Walker, S.L.; Taylor, P.; Sanjari, M.J. Optimal placement and sizing of battery energy storage system for losses reduction using whale optimization algorithm. *J. Energy Storage* **2019**, *26*, 100892. [CrossRef]

10. Da Costa, J.A.; Branco, D.A.C.; Filho, M.C.P.; De Medeiros, M.F.; da Silva, N.F. Optimal sizing of photovoltaic generation in radial distribution systems using Lagrange multipliers. *Energies* **2019**, *12*, 1728. [CrossRef]

11. Nor, N.M.; Ali, A.; Ibrahim, T.; Romlie, M.F. Battery Storage for the Utility-Scale Distributed Photovoltaic Generations. *IEEE Access* **2017**, *6*, 1137–1154. [CrossRef]

12. Zhang, Y.; Campana, P.E.; Lundblad, A.; Yan, J. Comparative study of hydrogen storage and battery storage in grid connected photovoltaic system: Storage sizing and rule-based operation. *Appl. Energy* **2017**, *201*, 397–411. [CrossRef]

13. Sardi, J.; Mithulananthan, N.; Gallagher, M.; Hung, D.Q. Multiple community energy storage planning in distribution networks using a cost-benefit analysis. *Appl. Energy* **2017**, *190*, 453–463. [CrossRef]

14. Jayasekara, N.; Member, S.; Masoum, M.A.S.; Member, S.; Wolfs, P.J.; Member, S. Optimal Operation of Distributed Energy Storage Systems to Improve Distribution Network Load and Generation Hosting Capability. *IEEE Trans. Sustain. Energy* **2016**, *7*, 250–261. [CrossRef]

15. Das, C.K.; Bass, O.; Kothapalli, G.; Mahmoud, T.S.; Habibi, D. Optimal placement of distributed energy storage systems in distribution networks using artificial bee colony algorithm. *Appl. Energy* **2018**, *232*, 212–228. [CrossRef]

16. Das, C.K.; Bass, O.; Mahmoud, T.S.; Kothapalli, G.; Mousavi, N.; Habibi, D.; Masoum, M.A.S. Optimal allocation of distributed energy storage systems to improve performance and power quality of distribution networks. *Appl. Energy* **2019**, *252*, 113468. [CrossRef]

17. Awad, A.S.A.; EL-Fouly, T.H.M.; Salama, M.M.A. Optimal ESS Allocation for Benefit Maximization in Distribution Networks. *IEEE Trans. Smart Grid* **2017**, *8*, 1668–1678. [CrossRef]

18. Naidji, I.; Ben Smida, M.; Khalgui, M.; Bachir, A.; Li, Z.; Wu, N. Efficient allocation strategy of energy storage systems in power grids considering contingencies. *IEEE Access* **2019**, *7*, 186378–186392. [CrossRef]

19. Nick, M.; Cherkaoui, R.; Paolone, M. Optimal allocation of dispersed energy storage systems in active distribution networks for energy balance and grid support. *IEEE Trans. Power Syst.* **2014**, *29*, 2300–2310. [CrossRef]

20. Zheng, Y.; Dong, Z.Y.; Luo, F.J.; Meng, K.; Qiu, J.; Wong, K.P. Optimal allocation of energy storage system for risk mitigation of discos with high renewable penetrations. *IEEE Trans. Power Syst.* **2014**, *29*, 212–220. [CrossRef]

21. Atwa, Y.M.; El-Saadany, E.F. Optimal allocation of ESS in distribution systems with a high penetration of wind energy. *IEEE Trans. Power Syst.* **2010**, *25*, 1815–1822. [CrossRef]

22. Zhang, B.; Dehghanian, P.; Kezunovic, M. Optimal Allocation of PV Generation and Battery Storage for Enhanced Resilience. *IEEE Trans. Smart Grid* **2019**, *10*, 535–545. [CrossRef]

23. Fernández-Blanco, R.; Dvorkin, Y.; Xu, B.; Wang, Y.; Kirschen, D.S. Optimal Energy Storage Siting and Sizing: A WECC Case Study. *IEEE Trans. Sustain. Energy* **2017**, *8*, 733–743. [CrossRef]

24. Khaboot, N.; Srithapon, C.; Siritaratiwat, A.; Khunkitti, P. Increasing benefits in high PV penetration distribution system by using battery enegy storage and capacitor placement based on salp swarm algorithm. *Energies* **2019**, *12*, 4817. [CrossRef]

25. Gong, Q.; Wang, Y.; Fang, J.; Qiao, H.; Liu, D. Optimal configuration of the energy storage system in ADN considering energy storage operation strategy and dynamic characteristic. *IET Gener. Transm. Distrib.* **2020**, *14*, 1005–1011. [CrossRef]

26. Zheng, Y.; Zhao, J.; Song, Y.; Luo, F.; Meng, K.; Qiu, J.; Hill, D.J. Optimal Operation of Battery Energy Storage System Considering Distribution System Uncertainty. *IEEE Trans. Sustain. Energy* **2018**, *9*, 1051–1060. [CrossRef]

27. Luo, L.; Abdulkareem, S.S.; Rezvani, A.; Miveh, M.R.; Samad, S.; Aljojo, N.; Pazhoohesh, M. Optimal scheduling of a renewable based microgrid considering photovoltaic system and battery energy storage under uncertainty. *J. Energy Storage* **2020**, *28*, 101306. [CrossRef]

28. International Renewable Energy Agency (IRENA). *Electricity Storage and Renewables: Costs and Markets to 2030*; IRENA: Abu Dhabi, UAE, 2017; ISBN 978-92-9260-038-9.

29. De Lorenzo, G.; Andaloro, L.; Sergi, F.; Napoli, G.; Ferraro, M.; Antonucci, V. Numerical simulation model for the preliminary design of hybrid electric city bus power train with polymer electrolyte fuel cell. *Int. J. Hydrogen Energy* **2014**, *39*, 12934–12947. [CrossRef]

30. De Luca, D.; Fragiacomo, P.; De Lorenzo, G.; Czarnetski, W.T.; Schneider, W. Strategies for Dimensioning Two-Wheeled Fuel Cell Hybrid Electric Vehicles Using Numerical Analysis Software. *Fuel Cells* **2016**, *16*, 628–639. [CrossRef]

31. Fragiacomo, P.; Astorino, E.; Chippari, G.; De Lorenzo, G.; Czarnetzki, W.T.; Schneider, W. Dynamic modeling of a hybrid electric system based on an anion exchange membrane fuel cell. *Cogent Eng.* **2017**, *4*, 1357891. [CrossRef]

32. Figgener, J.; Stenzel, P.; Kairies, K.P.; Linßen, J.; Haberschusz, D.; Wessels, O.; Angenendt, G.; Robinius, M.; Stolten, D.; Sauer, D.U. The development of stationary battery storage systems in Germany—A market review. *J. Energy Storage* **2020**, *29*, 101153. [CrossRef]
33. Tsiropoulos, I.; Tarvydas, D.; Lebedeva, N. *Li-Ion Batteries for Mobility and Stationary Storage Applications*; Publications Office of the European Union: Luxembourg, 2018; ISBN 978-92-79-97254-6.
34. Lu, L.; Han, X.; Li, J.; Hua, J.; Ouyang, M. A review on the key issues for lithium-ion battery management in electric vehicles. *J. Power Sources* **2013**, *226*, 272–288. [CrossRef]
35. Han, W.; Zou, C.; Zhou, C.; Zhang, L. Estimation of Cell SOC Evolution and System Performance in Module-based Battery Charge Equalization Systems. *IEEE Trans. Smart Grid* **2018**, *10*, 4717–4728. [CrossRef]
36. Boeringer, D.W.; Werner, D.H. Particle swarm optimization versus genetic algorithms for phased array synthesis. *IEEE Trans. Antennas Propag.* **2004**, *52*, 771–779. [CrossRef]
37. Zimmerman, R.D.; Murillo-Sánchez, C.E. *Matpower User's Manual*; Power Systems Engineering Research Center: Tempe, AZ, USA, 2019.
38. Zimmerman, R.D.; Murillo-Sánchez, C.E. *Matpower Optimal Scheduling Tool MOST 1.0.2 User's Manual*; Power Systems Engineering Research Center: Tempe, AZ, USA, 2019.
39. Wolfs, P.; Jayasekera, N.; Subawickrama, S. A Fourier series based approach to the periodic optimisation of finely dispersed battery storage. In Proceedings of the AUPEC 2011, Brisbane, Australia, 25–28 September 2011; pp. 1–6.

Publisher's Note: MDPI stays neutral with regard to jurisdictional claims in published maps and institutional affiliations.

MDPI

St. Alban-Anlage 66

4052 Basel

Switzerland

Tel. +41 61 683 77 34

Fax +41 61 302 89 18

www.mdpi.com

Batteries Editorial Office

E-mail: batteries@mdpi.com

www.mdpi.com/journal/batteries